Bulk Materials Handling Handbook

JACOB FRUCHTBAUM

VAN NOSTRAND REINHOLD COMPANY
New York

Copyright © 1988 by **Van Nostrand Reinhold Company Inc.**
Library of Congress Catalog Card Number: 87-20715
ISBN: 0-442-22684-5

Printed in the United States of America

Van Nostrand Reinhold Company Inc.
115 Fifth Avenue
New York, New York 10003

Van Nostrand Reinhold Company Limited
Molly Millars Lane
Wokingham, Berkshire RG11 2PY, England

Van Nostrand Reinhold
480 La Trobe Street
Melbourne, Victoria 3000, Australia

Macmillan of Canada
Division of Canada Publishing Corporation
164 Commander Boulevard
Agincourt, Ontario M1S 3C7, Canada

16 15 14 13 12 11 10 9 8 7 6 5 4 3 2 1

Library of Congress Cataloging-in-Publication Data
Fruchtbaum, Jacob, 1894–
    Bulk materials handling handbook / Jacob Fruchtbaum.
        p.   cm.
    Includes index.
    ISBN 0-442-22684-5
    1. Bulk solids handling—Handbooks, manuals, etc.   I. Title.
TS180.8.B8F78 1988
    621.8′6—dc19   87-20715
                                                                CIP

Contents

Illustrations

Section 4:   Screw Conveyors

Section 5:   Apron and Mold Conveyors

Section 6:   Flight and Drag Conveyors

Section 7:   Feeders and Vibrating Conveyors

Section 8:   Wire Mesh Conveyors

Section 9:   Drives

Section 10:   Crushers and Screens

Section 11:   Skip Hoists

Section 17:   Package Handling Conveyors

Section 18:   Pneumatic Conveying

Section 19:   Dust Collection

Section 20:   Layout of a Material Handling Plant

Tables

Preface

The handling of bulk materials is a continuously changing science. Since very few schools teach the handling of bulk materials, it is necessary for practicing engineers to develop their own training manuals. This book is an abbreviated version of a manual used for that purpose in our office, and developed over a period of more than 50 years. While some industrial firms follow their own practices, the trend in the past few years has been to adopt the standards of equipment manufacturers' associations and similar organizations. The selection of material and the use of drawings instead of photographs is based on our experience.

It is hoped that users of this text will comment on items of their choosing to make a next edition even more responsive to the needs of the profession. Skilled professionals active in other areas may find the book useful for information and data not usually encountered in their special fields but, in general, the handbook is not for them. It should be of help in training new engineers, and as a reference for engineering students, and those active in the bulk materials handling field.

In 1965, Stanley Snyder joined our staff. A short time later, he was assigned the task of coordinating all of our notes and bringing them up to date. When completely assembled, the result proved too voluminous for practice, and had to be cut down.

During the last four years, various sections of the text were sent, for suggestions and comments, to individuals I respect for their knowledge in their special fields. In addition, some sections were tried out in practice. I felt that great weight should be given to the suggestions of those training younger people, of purchasing agents, plant operators, and maintenance personnel. All the information gathered and suggestions received were considered in preparing the present text. Every effort was made to arrive at an appropriate balance.

The project designs used as illustrations and examples have, with few exceptions, been built and have operated well for many years. To illustrate a desired point, some projects are based on problems presented in manufacturers' catalogs, or slight modifications of actual completed projects. Much of the nomenclature has been brought up to date.

Publication of the material contained herein is not intended as a representation or warranty on the part of the author, publisher, editors, or any other person or firm named herein that it is suitable for any particular use, or free from infringement of any patent or patents.

The text is intended as a guide. When used for any specific project, a competent professional engineer should be retained to verify the assumptions, applicability, calculations, and accuracy of the particular design.

Whenever standards are established by an industry, it is best to follow those standards. This text is intended to replace neither the catalogs supplied by manufacturers nor the texts published by manufacturers' associations. They cover the field in much greater detail than is possible here.

For their help and encouragement, I feel especially indebted to G. Leslie Lardie, former chief engineer of the Electrometallurgical Division of Union Carbide; Arthur Toronski, former chief engineer and president of Speer Carbon Company; and Henry Crosby, former chief engineer of General Mills.

Special thanks are due the Conveyor Equipment Manufacturers Association, and the topflight engineers who work for some of its members.

My gratitude goes to Roger M. Stern, technical and copyeditor of the manuscript, and to Bernice Pettinato, managing editor, from whom I learned much about what happens to a book from the time the author submits the manuscript until bound copies appear. Thanks are also due Darya and Robert E. Emmens, Sr., for their substantial assistance with the proofreading.

Those who have contributed to the text are mentioned therein.

Great thanks also go to Charles N. Brice, Steven Bugay, William N. Carlson, Kenneth McCaskill, Roger Larsh, and Dr. Andrei Reinhorn of our staff for their help.

Thanks are also due

Edward Braun and D. J. Hagen of Allis-Chalmers

William A. Robinson of Babcock & Wilcox Co. (retired)

John A. Klacsman of Clean Sites Inc.

Robert Jorgensen of Buffalo Forge Company and author of *Fan Engineering*

Dr. Robert A. Person of Elkem Company

Clyde E. Ostberg and Harold Swanson of FMC

Roger E. Noel of Goodman Company

R. W. Milk and M. A. Alspaugh of Goodyear Tire and Rubber Company

John J. Keenan of Harnischfeger

H. C. Lautenschlaeger and J. D. DiAntonio of Hewitt-Robins

T. P. Smyre of Jeffrey Manufacturing, Division of Dresser Industries

Florian Schwartzkopf of Kennedy Van Saun

Edward Barnett, formerly of MIT

William Neundorfer of National Air Vibrator Company

John J. Bradley and Lee Doyer of Pennsylvania Crusher Corporation

Don Rearic and Guido Yanniello of Rexnord, Inc.

Arthur Geberin and R. O. Dickey of Screw Conveyor Corporation

Arthur Elmquest of S. K. W.

Glenn Runge, formerly of Speer Carbon Company

James Voelkel of W. S. Tyler Company

Tom Zupo, formerly of Webb Belting

T. H. Anderson of Whiting Corporation.

JACOB FRUCHTBAUM

Introduction

1.1 STANDARDIZATION IN THE INDUSTRY

During the last few years, efforts have been made to standardize the products of the industry. This is a very helpful step for engineers designing the projects and for plant operators who have to store spare parts and service equipment. Every effort has been made to include this standardization in this book.

1.2 COMPUTATION STANDARDS

The methods outlined in the new standards give good results for the given data. They cover nearly every conditon. The design data given in manufacturers' catalogs makes allowances for the components that make up the final design, but the new standards represent the best and latest information made available by all the manufacturers. It is also assumed that all manufacturers will adopt association standards.

Some designs that were made on the basis of previously published data have been included here to assure operators of the safety of the old methods. Manuals published by associations are the most authoritative books in their field; they are well written and present complete data on the subjects covered.

It should be kept in mind that the information given the designer is never carried to the degree of accuracy shown by these computations. This is especially true of capacity. In figuring capacity, it is best to use the lowest weight per cubic foot of the materials, while the highest weight per cubic foot should be used in the selection of motor horsepower and drives. Extra field tests give varying capacities, since materials coming from different sources—and even from the same source—will vary from time to time, especially in the size and percentage of lumps and moisture content. It should be noted further that the formulas and the graphic solutions given in manufacturers' catalogs have produced satisfactory designs when experience and good judgment were used. In unusual applications, conservatism should be used in sizing components.

Some items are not subject to strict analysis. For example, shredded steel tends to coil and to form balls as large as 12 inches in diameter, and the motor horsepower required may be two or three times the computed horsepower. In the long run, experience and good judgment in the use of any data are invaluable.

1.3 PROPERTIES OF MATERIALS

Table 1.1 and the data included in table 1.2 are taken from the CEMA books ''Screw Conveyors'' and ''Belt Conveyors for Bulk Materials.'' Except for the tabulation of F_m, which is used in the design of screw conveyors, and the tabulation of inclination for use in belt conveyor design, the information in these tables is useful for other types of conveyors and, therefore, is included here. The data varies in some books, so the best assumed data has been used here.

Table 1.1 gives the code designation breakdown for column 3 of table 1.2. The first column of table 1.2 lists the materials covered. Column 2 gives the range of density that can be expected in handling that material. The "as conveyed" density is not specifically shown, but often is assumed to be at or near the minimum.

The third column of table 1.2 shows the material code. The code designates the average weight in pounds per cubic foot for the listings; the size letter; the flowability number; the abrasiveness number; and characteristics termed miscellaneous properties or conveyability hazards. Column 4 covers the material factor (F_m) used in the screw conveyor formula to determine the horsepower. The last column gives the recommended maximum inclination angle for use in belt conveyor design.

The information and data in table 1.2 has been compiled by members of CEMA and represents many years of experience in the successful design and application of screw conveyors and belt conveyors for handling the listed materials. The indicated physical characteristics of these materials are not the result of any particular laboratory tests, but were learned from the actual industrial operation of countless screw conveyors.

Table 1.2 includes various grains, seeds, feeds, and so on that are commonly handled in many types of conveyors. The published unit weights, material codes, and material factors (F_m) are for average conditions. For instance, wheat when dry or with a moisture of less than 10% is very free-flowing, and the F_m factor of 0.4 can be used. When higher moistures are prevalent, a material factor of 0.5 or 0.6 is suggested. This phenomena is common to all grains and some other substances. Refer to section 14 for data on some special materials.

Table 1.2 is only a guide. A specific material sample may have properties that vary from those shown in the table. The density ranges may also vary, depending on moisture content as well as its source.

1.4 HANDLING SPECIAL MATERIALS

Suggestions on handling the following materials appear in section 14:

Ashes
Bauxite ore
Carbon chip refuse
Carborundum
Cement
Charcoal
Chrome ore
Coal
Coke
Crushed ice
Crushed stone
Explosive materials
Feed, flour, and grain
Foundry sand
Graphite
Gypsum
Lead and zinc ore
Lime
Logs
Metallurgical coke
Pencil pitch
Perlite and expanded perlite
Petroleum coke
Plastic (fine) materials
Rice
Salt
Sand and gravel
Shale, kaolin (clay)
Soapstone, talc, asbestos, and mica
Soda ash
Steel or cast-iron chips
Sulfur
Wood chips, flour, sawdust
Zirconium

1.5 PROPERTIES OF CHAINS REFERRED TO IN THIS BOOK

Except for some of the tables in section 8, FMC chain nomenclature has been used.

ANSI chains used in connection with wire-cloth conveyors are described in section 8, with the conveyors.

C combination chains (table 3.3) are economical for a variety of conveyors and elevators. Several sizes are available with wear shoes for severe abrasive conditions. SBS bushed chains are widely used for heavy-duty elevators (refer to paragraph 3.8 and table 3.2).

Flight conveyor H chains are listed in table 6.3. Drag conveyor chains, series SD, and WD-480 and WDH-480 chains are listed in table 6.2.

RC chains used for drives are listed in table 9.6.

Series 400 pintle chains are for light-duty service and, for convenience, are listed in table 8.9. Series 700 pintle chains (table 1.3e) are heavier, longer pitch, and are used on long conveyors.

Rivetless chains, nos. 458 and 678, are used on long conveyors, trolley conveyors, and flight and drag conveyors. Their properties are given in paragraph 12.13.2.

Class MR (table 1.3g) and Class 1100 (table 1.3c) chains have cast rollers with offset sidebars and are furnished in malleable iron or Promal. They are used on flat and apron conveyors.

RS and SS (table 1.3f) bushed roller chains are made with either offset or straight sidebars, and are intended for heavy-duty service. Offset sidebar chains ordinarily are used for drives and low-speed applications. Conveyor chains generally are furnished with rollers larger than the sidebars. Some large rollers are provided with flanges. The suffix CR is applied to RS chains having stainless-steel bushings and rollers fitted with polyethylene sleeves. They are used where corrosive conditions exist or where lubrication is impractical.

RO offset sidebar roller steel chains are used for drives on heavy-duty conveyors.

Class 800 ley bushed chains are used on moderate height bucket elevators, handling heavy loads or abrasive material.

Various chain types are listed numerically below with the number of the table giving data for each. Table 1.4 lists chain numbers for types referenced or used in several sections of this book.

Table 1.5 gives chain correction factors for cast and combination chains. Chain correction factors for class SS chains are presented in table 1.6. Chain correction service factors appear in table 1.7.

Table 1.1 Material Classification Code

Major Class	Material Characteristics Included		Code Designation
Density	Bulk density, loose		Actual lb/ft^3
Size	Very fine	no. 200 sieve (0.0029 in.) and under	A_{200}
		no. 100 sieve (0.0059 in.) and under	A_{100}
		no. 40 sieve (0.016 in.) and under	A_{40}
	Fine	no. 6 sieve (0.132 in.) and under	B_6
	Granular	½ in. and under	$C_{1/2}$
		3 in. and under	D_3
		7 in. and under	D_7
	*Lumpy	16 in. and under	D_{16}
		over 16 in. to be specified	
		X = actual maximum size	D_x
	Irregular	stringy, fibrous, cylindrical, slabs, etc.	E
Flowability	Very free-flowing—flow function >10		1
	Free-flowing—flow function >4 but <10		2
	Average flowability—flow function >2 but <4		3
	Sluggish—flow function <2		4
			(Continues)

1.6 REFERENCES

CEMA (Conveyor Equipment Manufacturers Association) publishes standards for light package roller conveyors. These standards cover standard practice, formulas, tables, and charts for the design of units, drives, and horsepower of motors.

Handbooks published by Mathews Conveyors have complete data, which have been used freely in this text with permission.

1.7 ABBREVIATIONS, LETTER SYMBOLS; SI AND OTHER EQUIVALENTS

Table 1.8 lists abbreviations for terms used in the text. Abbreviations for trade associations, professional societies, and manufacturers mentioned in the text appear in table 1.9. Table 1.10 explains letter symbols for units used in this book. SI and other equivalents for units of measure are given in table 1.11.

Table 1.1   (*continued*)

Major Class	Material Characteristics Included	Code Designation
Abrasiveness	Mildly abrasive—Index 1–17	5
	Moderately abrasive—Index 18–67	6
	Extremely abrasive—Index 68–416	7
	Builds up and hardens	F
	Generates static electricity	G
	Decomposes—deteriorates in storage	H
	Flammability	J
	Becomes plastic or tends to soften	K
	Very dusty	L
	Aerates and becomes fluid	M
	Explosiveness	N
Miscellaneous	Stickiness, adhesion	O
Properties	Contaminable, affecting use	P
Or	Degradable, affecting use	Q
Hazards	Gives off harmful or toxic gas or fumes	R
	Highly corrosive	S
	Mildly corrosive	T
	Hygroscopic	U
	Interlocks, mats, or agglomerates	V
	Oils present	W
	Packs under pressure	X
	Very light and fluffy—may be windswept	Y
	Elevated temperature	Z

Source: Courtesy CEMA

*Refer to paragraph 4.27 for lump size limitations.

TABLE 1.2 5

Table 1.2 Material Characteristics

Material	Weight lb/ft³	Material Code	Material Factor (F_m)	Recommended Max Inclination (°)[a]
Aluminum chips, dry	7–15	11E45V	1.2	—
Aluminum chips, oily	7–15	11E45V	0.8	—
Ashes, coal, dry—½ in.	35–45	40C½46TY	3.0	20–25
Ashes, coal, dry—3 in.	35–40	38D₃46T	2.5	—
Ashes, coal, wet—½ in.	45–50	48C½46T	3.0	23–27
Ashes, coal, wet—3 in.	45–50	48D₃46T	4.0	—
Bakelite, fine	30–45	38B₆25	1.4	—
Bark, wood, refuse	10–20	15E45TVY	2.0	27
Bentonite, crude	34–40	37D₃45X	1.2	—
Bentonite—100 mesh	50–60	55A₁₀₀25MXY	0.7	20
Bicarbonate of soda (baking soda)	—	—	0.6	—
Boron	75	75A₁₀₀37	1.0	—
Bran, rice—rye—wheat	16–20	18B₆35NY	0.5	—
Bread crumbs	20–25	23B₆35PQ	0.6	—
Buckwheat	37–42	40B₆25N	0.4	11–13
Calcium carbide, crushed	70–90	80D₃25N	2.0	—
Carbon, activated, dry, fine[b]	—	—	—	—
Carbon black, pelleted[b]	—	—	—	—
Carbon black, powder[b]	—	—	—	—
Carborundum	100	100D₃27	3.0	—
Caustic soda	88	88B₆35RSU	1.8	—
Caustic soda, flakes	47	47C½45RSUX	1.5	—
Cement, clinker	75–95	85D₃36	1.8	18–20
Cement, mortar	133	113B₆35Q	3.0	—
Cement, portland	94	94A₁₀₀26M	1.4	20–23
Charcoal, lumps	18–28	23D₃45Q	1.4	20–25
Chrome ore	125–140	133D₃36	2.5	—
Cinders, blast furnace	57	57D₃36T	1.9	18–20
Cinders, coal	40	40D₃36T	1.8	20
Clay, brick, dry, fines	100–120	110C½36	2.0	20–22
Clay, dry, lumpy	60–75	68D₃35	1.8	18–20
Coal, anthracite, river and culm	55–61	60B₆35TY	1.0	18
Coal, anthracite, sized—½ in.	49–61	55C½25	1.0	16
Coal, bituminous, mined	40–60	50D₃35LNXY	0.9	18
Coal, bituminous, mined, sized	45–50	48D₃35QV	1.0	16
Coal, bituminous, mined, slack	43–50	47C½45T	0.9	22
Coke, loose	23–35	30D₇37	1.2	18
Coke, petroleum, calcined	35–45	40D₇37	1.3	20
Compost	30–50	40D₇45TV	1.0	—
Concrete, pre-mix, dry	85–120	103C½36U	3.0	—
Dolomite, crushed	80–100	90C½36	2.0	22
Flour, wheat	33–40	37A₄₀45LP	0.6	21
Flue dust, boiler H dry	30–45	38A₄₀36LM	2.0	—
Flyash	30–45	38A₄₀36M	2.0	—
Graphite flake	40	40B₆25LP	0.5	—
Graphite flour	28	28A₁₀₀35LMP	0.5	—
Graphite ore	65–75	70Dₓ35L	1.0	—
Gravel, bank run	90–100	—	—	20
Gravel, dry, sharp	90–100	95D27	—	15–17
Gypsum, calcined	55–60	58B₆35U	1.6	—
Gypsum, calcined, powdered	60–80	70A₁₀₀35U	2.0	—
Gypsum, raw—1 in.	70–80	75D₃25	2.0	—
Ice, crushed	35–45	40D₃35O	0.4	—
Ice, flaked	40–45	43C½35O	0.6	—

Table 1.2 (*continued*)

Material	Weight lb/ft^3	Material Code	Material Factor (F_m)	Recommended Max. Inclination (°)[a]
Iron ore concentrate	120–180	150A$_{40}$37	2.2	18–20[c]
Lime, ground, unslaked	60–65	63B$_6$35U	0.6	23
Lime, hydrated	40	40B$_6$35LM	0.8	20[c]
Lime, hydrated, pulverized	32–40	36A$_{40}$35LM	0.6	22
Lime, pebble	53–56	C$_{1/2}$25HU	2.0	17
Limestone, agricultural	68	68B$_6$35	2.0	20
Limestone, crushed	85–90	88D$_x$36	2.0	—
Limestone, dust	55–95	75A$_{40}$46MY	1.6–2.0	—
Oats	26	26C$_{1/2}$25MN	0.4	10
Oat flour	35	35A$_{100}$35	0.5	—
Paper pulp (4% or less)	62	62E45	1.5	—
Paper pulp (6–15%)	60–62	61E45	1.5	—
Perlite, expanded	8–12	10C$_{1/2}$36	0.6	—
Phosphate acid fertilizer	60	60B$_6$25T	1.4	13
Salt, dry, coarse	45–60	53C$_{1/2}$36TU	1.0	18–22
Salt, dry, fine	70–80	75B$_6$36TU	1.7	11
Sand, dry bank (damp)	110–130	120B$_6$47	2.8	20–22
Sand, dry bank (dry)	90–110	100B$_6$37	1.7	16–18
Sand, dry silica	90–100	95B$_6$27	2.0	10–15
Sand, foundry (shake out)	90–100	95D$_3$37Z	2.6	22
Sawdust, dry	10–13	12B$_6$45UX	0.7	22
Shale, crushed	85–90	88C$_{1/2}$36	2.0	22
Silica, flour	80	80A$_{40}$46	1.5	—
Slag, blast furnace, crushed	130–180	155D$_3$37Y	2.4	10
Slag, furnace, granular, dry	60–65	63C$_{1/2}$37	2.2	13–16
Slate, crushed—½ in.	80–90	85C$_{1/2}$36	2.0	15
Slate, ground—⅛ in.	82–85	84B$_6$36	1.6	15
Sludge, sewage, dried	40–50	45E47TW	0.8	—
Sludge, sewage, dry, ground	45–55	50B46S	0.8	—
Soapstone, talc, fine	40–50	45A$_{200}$45XY	2.0	—
Soda ash, heavy	55–65	60B$_6$36	1.0	19
Soda ash, light	20–35	28A$_{40}$36Y	0.8	22
Sulfur, crushed—½ in.	50–60	55C$_{1/2}$35N	0.8	—
Sulfur, lumpy—3 in.	80–85	83D$_3$35N	0.8	—
Sulfur, powdered	50–60	55A$_{40}$35MN	0.6	—
Talcum—½ in.	80–90	85C$_{1/2}$36	0.9	—
Talcum powder	50–60	55A$_{200}$36M	0.8	—
Tricalcium phosphate	40–50	45A$_{40}$45	1.6	—
Vermiculite, expanded	16	16C$_{1/2}$35Y	0.5	—
Vermiculite ore	80	80D$_3$36	1.0	20
Wheat	45–48	47C$_{1/2}$25N	0.4	12
Wheat, cracked	40–45	43B$_6$25N	0.4	—
Wheat germ	18–28	23B$_6$25	0.4	—
Wood chips, screened	10–30	20D$_3$45VY	0.6	27
Wood flour	16–36	26B$_6$35N	0.4	—
Wood shavings	8–16	12E45VY	1.5	—
Zinc, concentrate residue	75–80	78B$_6$37	1.0	—
Zinc oxide, heavy	30–35	33A$_{100}$45X	1.0	—
Zinc oxide, light	10–15	13A$_{100}$45XY	1.0	—

Source: Courtesy CEMA

[a] For screw conveyors (1980).
[b] Consult manufacturer.
[c] Consult engineer.

Table 1.3 Chain Details

Chain Number	Pitch (in.)	Average Ultimate Strength (lb)	Allowable Chain Pull (lb)
a: Roller chain; straight sidebar; large flanged roller; rollers of malleable, white or Flint-Run iron			
SS-658	6.000	26,000	4,650
SS-922	9.000	41,000	7,200
SS-933	9.000	56,000	9,200
SS-940	9.000	56,000	9,200
SS-1222	12.000	41,000	7,200
SS-1233	12.000	56,000	9,200
SS-1240	12.000	56,000	9,200
SS-1244	12.000	74,000	12,700
SS-1822	18.000	41,000	7,200
SS-1833	18.000	56,000	9,200
b: Roller chain; straight sidebar; small roller			
RS-953	6.000	38,000	5,600
RS-944 plus	6.000	60,000	5,900
SS-928	9.000	41,000	7,200
SS-942	9.000	56,000	9,200
SS-1242	12.000	56,000	9,200
c: 1100 class cast roller chain			
1120	4.000	6,500	1,080
1124	4.000	12,000	1,870
1113	4.040	20,000	3,220
1130	6.000	28,000	3,750
d: 800 ley bushed pintle chain[a]			
823	4.000	19,000	3,170
825	4.000	30,000	5,000
830	6.000	30,000	5,000
844	6.000	40,000	6,670
e: 700 pintle chain[b]			
720	6.000	27,500	3,220
730	6.000	40,000	3,750
f: RS and SS roller chain; straight sidebar; large roller			
RS-3013	3.000	13,000	2,100
RS-4019	4.000	19,000	2,450
RS-60	4.040	33,000	4,950
RS-6238	6.000	38,000	5,600
SS-927	9.000	41,000	7,200
RS-1114	6.000	28,000	3,800
RS-4013	4.000	13,000	2,150
SS-1227	12.000	41,000	7,200
SS-1827	18.000	41,000	7,200

Table 1.3 (*continued*)

Chain Number	Pitch (in.)	Average Ultimate Strength (lb)	Allowable Chain Pull (lb)	

g: Cast rollers; cast offset sidebar with integral barrel; fabricated steel pins or rivets; abutting barrel halves

MR-1½	2.970	19,000	2,820

h: Large, plastic-lined rollers; bushings of stainless steel

RS-1736 CR	6.000	28,000	3,150

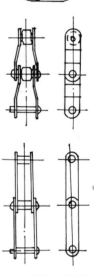

j: Roller bushed chain; offset sidebar; small roller; steel chain

RO-882	2.609	26,000	2,500
RO-40	3.075	28,000	4,650

k: Bushed chain; straight sidebar; small roller

SS-1730	6.000	45,000	6,900

[a] For malleable iron chain. Promal iron chains approximately 25% stronger.
[b] For pearlitic malleable iron (comparable to Promal iron).

Table 1.4 Chain Numbers Used Throughout the Book

Chain Type	Refer to Table:	Chain Type	Refer to Table:
MR-1½	1.3g	H-78	17.4
SM-12	17.4	H-79	8.9
SD-19	6.2	RC-80	9.5
SD-21	6.2	RC-100	9.5
SD-27	6.2	SBS-102½	3.2
SD-28	6.2	C-102B	3.3
SD-29	6.2	SBS-102B	3.2
		C-102½	3.3
RC-35	8.7		
RC-40	9.5	H-104	6.3
RO-40	1.3j	C-110	3.3
RC-50	9.5	SBS-110	3.2
RC-60	9.5	C-111	3.3
RS-60	1.3f	SBS-111	3.2

TABLE 1.4   9

Table 1.4   (*continued*)

Chain Type	Refer to Table:	Chain Type	Refer to Table:
H-112	6.3	SS-933	1.3a
H-116	6.3	SS-940	1.3a
		SS-942	1.3b
		RS-944 plus	1.3b
RC-120	9.5	RS-953	1.3b
C-131	17.4	1113	1.3c
C-132	3.3	RS-1114	1.3f
RC-140	9.5	1120	1.3c
SBS-150 plus	3.2	1124	1.3c
RC-160	9.5	1130	1.3c
C-188	3.3		
SBS-188	3.2		
348 mod	17.4	SS-1222	1.3a
SR-420	17.4	SS-1227	1.3f
		SS-1233	1.3a
		SS-1240	1.3a
442	8.9	SS-1242	1.3b
445	8.9	SS-1244	1.3a
452	8.9	SS-1730	1.3k
455	8.9	RS-1736 CR	1.3h
458	(see para 12.13.2)	SS-1822	1.3a
X458	17.4	SS-1827	1.3f
462	8.9	SS-1833	1.3a
468	17.4		
477	8.9		
H-480	6.3	C-2040	8.8
WD-480	6.2	C-2042	8.8
WDH-480	6.2	C-2050	8.8
488	8.9	C-2052	8.8
SR-620	17.4	C-2060	8.8
SR-625	17.4	C-2062	8.8
SS-658	1.3a	C-2080	8.8
678	(see para 12.13.2)	C-2082	8.8
720	1.3e	C-2100	8.8
730	1.3e	C-2102	8.8
		C-2120	8.8
		C-2122	8.8
823	1.3d		
825	1.3d		
830	1.3d	RS-3013	1.3f
844	1.3d	RS-4013	1.3f
SBS-856	3.2	RS-4019	1.3f
RO-882	1.3j	4103	8.9
SS-922	1.3a	RS-4113	8.6
SS-928	1.3b	4124	8.9
SS-927	1.3f	RS-6238	1.3f

Table 1.5 Chain Correction Factors for Cast and Combination Chains

Number of Teeth in Driving Sprocket	Correction Factor												
	Chain speed (fpm)												
	10	25	50	75	100	125	150	175	200	225	250	275	300
6	1.05	1.25	1.57	1.92	2.28	2.75	3.31	4.08	5.03	—	—	—	—
7	0.971	1.10	1.29	1.46	1.64	1.84	2.07	2.34	2.62	2.98	3.39	3.92	4.52
8	0.935	1.04	1.19	1.32	1.44	1.57	1.71	1.86	2.02	2.20	2.40	2.62	2.85
9	0.909	0.990	1.12	1.23	1.34	1.44	1.55	1.66	1.77	1.89	2.01	2.15	2.29
10	0.885	0.962	1.07	1.16	1.25	1.33	1.41	1.49	1.57	1.66	1.75	1.84	1.92
11	0.870	0.935	1.02	1.10	1.18	1.25	1.32	1.39	1.46	1.53	1.60	1.68	1.74
12	0.847	0.901	0.990	1.06	1.13	1.20	1.26	1.32	1.38	1.45	1.51	1.56	1.62
14	0.840	0.885	0.952	1.01	1.06	1.12	1.17	1.22	1.27	1.32	1.37	1.42	1.46
16	0.830	0.870	0.926	0.971	1.02	1.07	1.11	1.15	1.20	1.24	1.28	1.33	1.37
18	0.824	0.862	0.909	0.952	1.00	1.04	1.08	1.12	1.15	1.19	1.23	1.27	1.30
20	0.820	0.855	0.901	0.943	0.980	1.02	1.05	1.09	1.12	1.16	1.19	1.23	1.26
24	0.813	0.840	0.877	0.909	0.943	0.971	1.00	1.03	1.06	1.09	1.12	1.16	1.19

Source: Courtesy Link-Belt

Table 1.6 Chain Correction Factors for SS Class Chains

Number of Teeth in Driving Sprocket	Correction Factor												
	Chain speed (fpm)												
	10	25	50	75	100	125	150	175	200	225	250	275	300
6	0.917	1.09	1.37	1.68	2.00	2.40	2.91	3.57	4.41	—	—	—	—
7	0.855	0.971	1.13	1.27	1.44	1.61	1.81	2.04	2.29	2.60	2.96	3.42	3.95
8	0.813	0.909	1.04	1.16	1.26	1.37	1.49	1.63	1.76	1.93	2.10	2.29	2.48
9	0.794	0.870	0.980	1.07	1.17	1.26	1.36	1.45	1.55	1.65	1.76	1.88	2.00
10	0.775	0.840	0.943	1.02	1.09	1.16	1.24	1.31	1.37	1.45	1.53	1.61	1.68
11	0.758	0.820	0.901	0.971	1.03	1.09	1.15	1.22	1.28	1.34	1.40	1.46	1.52
12	0.741	0.787	0.862	0.926	0.990	1.05	1.10	1.16	1.21	1.26	1.32	1.37	1.42
14	0.735	0.769	0.833	0.885	0.935	0.980	1.02	1.07	1.11	1.15	1.19	1.24	1.28
16	0.725	0.763	0.813	0.855	0.893	0.935	0.971	1.01	1.05	1.08	1.12	1.16	1.19
18	0.719	0.752	0.800	0.833	0.877	0.909	0.943	0.980	1.01	1.04	1.08	1.11	1.14
20	0.717	0.746	0.787	0.826	0.855	0.893	0.917	0.952	0.980	1.01	1.04	1.07	1.10
24	0.714	0.735	0.769	0.800	0.820	0.847	0.877	0.901	0.935	0.962	0.980	1.01	1.04

Source: Courtesy Link-Belt

TABLE 1.8   11

Table 1.7   Service Factors

Operating Conditions	Service Factor	Operating Conditions	Service Factor
Load characteristics		*Atmospheric conditions*	
Uniform or steady load	1.0	Relatively clean and moderate temperatures	1.0
Moderate shock load	1.2[a]	Moderately dusty and moderate temperatures	1.2
Heavy shock load	1.5	Exposed to weather, very dusty, abrasive, mildly corrosive, and reasonably high temperatures	1.4
Frequency of shock		*Operation*	
Infrequent shock	1.0	8–10 hr per day	1.0
Frequent shock	1.2	10–24 hr per day	1.2

Source: Courtesy Link-Belt

[a] Link-Belt uses 1.2 but 1.1 is used in this text.

Table 1.8   Abbreviations for Terms Used in This Book

addl	additional	intmt	intermittent
aprx	approximately	ks	keyseat
aux	auxiliary	l	long
bf	between faces	ld	load
bm	bending moment	lg	length
brg	bearing	lin	linear
c to c	center to center	lub	lubricate
coef	coefficient	MP	multiple-ply (belt)
conv	conveyor	o to o	outside to outside
cr	corrosion resistant	OD	outside diameter
crcmf	circumference	p	pitch
crs	cold-rolled steel	para	paragraph
crt	cathode-ray tube	pd	pitch diameter
ctr	center	ph	phase
d	deep	piw	pounds per inch width (belt)
deg	degree	pnl	panel
dia	diameter	press	pressure
dim	dimension	RC	roller chain
dist	distance	rf	rolling friction
distr	distribute	recm	recommend
dp	depth	rmvbl	removable
dpt	depart	rom	run-of-mine
dptr	departure	rwy	railway
eff	effective	set	setting
ent	entry	sp	static pressure
eqpt	equipment	spa	space
equiv	equivalent	ss	setscrew
fig	figure	std	standard
fmr	former	stl	steel
frict	friction	suct	suction
ga	gauge	svce	service
h	high	t	tooth
ht	height	T	tension (lb)
ID	inside diameter	tbm	total bending moment
intmd	intermediate	tefc	totally enclosed, fan cooled

Table 1.8   (*continued*)

tm	torsional moment	w	wide
ttm	total torsional moment	W&M	Washburn & Moen (gauge)
TU	takeup	wd	width
typ	typical	wg	water gauge
V	velocity	wp	working point
VP	velocity pressure	wt	weight
vs	vibrating screen		

Note: Unless otherwise shown, the same abbreviation is used for all tenses, all endings, the singular and plural, the possessive case, and the noun and modifying forms of a term.

Table 1.9   Abbreviations for Trade Associations, Professional Societies, and Manufacturers Mentioned in Text

A-C	Allis-Chalmers
ACA	American Chain Association
ACECO	American Crane and Equipment Corp
ACGIH	American Conference of Governmental Industrial Hygienists
AFBMA	Antifriction Bearing Manufacturers Association
AGMA	American Gear Manufacturers Association
AISC	American Institute of Steel Construction
AISE	Association of Iron and Steel Engineers
AMCA	Air Movement and Control Association
ANSI	American National Standards Institute
ARA	American Railway Association
ASHRAE	American Society of Heating, Refrigeration, and Air-Conditioning Engineers
ASME	American Society of Mechanical Engineers
ASTM	American Society for Testing and Materials
Beth	Bethlehem Steel
CEMA	Conveyor Equipment Manufacturers Association
CMAA	Crane Manufacturers Association of America
COIV	Committee on Industrial Ventilation (p/o American Conference of Governmental Industrial Hygienists)
CVC	Cleveland Vibrator Company
FMC	Food Machinery and Chemical Corp
GTR	Goodyear Tire and Rubber Company
KVS	Kennedy Van Saun Corp (subsidiary of McNally Pittsburgh)
L-B	Link-Belt (div of FMC Corp)
MMA	Monorail Manufacturers Association
NEMA	National Electrical Manufacturers Association
OSHA	Occupational Safety and Health Administration (US Govt)
P&H	Harnischfeger Corp
RMA	Rubber Manufacturers Association
SAE	Society of Automotive Engineers
USX	United States Steel

TABLE 1.11   13

Table 1.10   Letter Symbols for Units

A	ampere		K	Kelvin (SI temperature unit)
Btu	British thermal unit		kg	kilogram
cfm	cubic feet per minute		kPa	kilopascal
cm	centimeter		L	liter
cm^2	square centimeter		lb	pound
cm^3	cubic centimeter		mi	mile
°C	degree Celsius		mi^2	square mile
°F	degree Farenheit		μm	micrometer
fpm	feet per minute		mm	millimeter
ft	foot		NEC	National Electrical Code
ft^2	square foot		psig	pound per square inch gauge
ft^3	cubic foot		rad	radian
ft-lb	foot-pound		rpm	revolutions per minute
gal	gallon		t	ton (metric)
hp	horsepower		thm	therm
Hz	hertz		tph	tons per hour
in.	inch		V	volt
in^2	square inch		VP	vapor pressure (in.)
in^3	cubic inch		W	watt
inHg	inches of mercury		yd	yard
in-lb	inch-pound		yd^2	square yard

Notes: The same symbol form applies for both singular and plural use.
Letter symbols for quantities are defined at appropriate points in the text.

Table 1.11   SI and Other Equivalents

Temperature

$$°F = 9/5 \times °C + 32$$
$$°C = 5/9 \times (°F - 32)$$
$$K = °C + 273.5$$

Pressure

1 atmosphere	= 29.92 inHg
	= 76 cm Hg
	= 33.9 ft H_2O
1 atmosphere	= 1.033227 kg/cm^2
	= 14.69 lb/in^2
1 kg/cm^2	= 14.223 lb/in^2
1 lb/in^2	= 0.07031 kg/cm^2
	= 6.895 kPa
water pressure	= head (ft) × 0.434 lb/in^2
density of air at	
70°F and 29.92 inHg	= 0.075 lb/ft^3

Acceleration

1 m/s^2	= 3.28084 ft/s^2
1 ft/s^2	= 0.3048 in./s^2

Length and Area

1 m	= 3.281 ft
	= 39.37 in.

Length and Area (continued)

1 yd	= 0.914 m
1 ft	= 0.3048 m
1 mi	= 1.609 km
1 km	= 0.6214 mi
1 in.	= 2.54 cm
1 micron (deprecated)	= 1 μm
	= 0.001 mm
1 angstrom (Å)	= 0.0001 μm
1 acre	= 43,560 ft^2
	= 4046 m^2
1 m^2	= 1.196 yd^2
1 yd^2	= 0.836 m^2
1 mi^2	= 2.590 km^2

Volume

1 ft^3	= 0.0283 m^3
	= 7.48 gal
1 m^3	= 35.31 ft^3
	= 1.308 yd^3
1 in^3	= 16.387 cm^3
1 gal	= 3.785 L
	= 231 in^3
1 L	= 1.057 qt
	= 0.264 gal
1 qt	= 0.946 L

(Continues)

Table 1.11 *(continued)*

Weight, Mass, Density		Miscellaneous Units	
1 kg	= 2.205 lb	1 Btu	= 1055 joules
1 kg/m^3	= 0.06243 lb/ft^3	1 hp	= 746 W
1 ton (short)	= 0.907 t (metric)		= 550 ft-lb/s
1 ton (long)	= 1.0161 t (metric)	1 rad	= 57.3°
1 t (metric)	= 1000 kg	1 thm	= 100,000 Btu
	= 1.103 tons (short)	V of sound in air	
1 lb/ft^3	= 16.02 kg/m^3	at sea level	= 1100 ft/s
1 lb	= 0.4536 kg	g	= 32.2 ft/sec^2
			= 980.6 cm/sec^2
		0.018 Btu	= heat required to raise 1 ft^3 air 1°F

Belt Conveyors

2.1 GENERAL

A belt conveyor is used for transporting materials, generally horizontally, from one location to another. It is made up of an endless belt on top of idlers (rollers). The idlers are arranged in a trough on the carrying side and flat on the bottom, or return side. The driving unit moves one or more pulleys which, in turn, move the belt. The whole conveyor is supported on a steel (seldom wood) structure, which supports and keeps the idlers in line, as well as supports the pulleys and the drive. On the troughed or conveyor side, a tripper is located wherever it is necessary to discharge material at an intermediate location (see figure 2.1).

Belt conveyors are capable of handling a wide variety of bulk materials. They are suitable for handling certain corrosive materials that would quickly attack parts of an all-metal conveyor. The range of sizes of material that may be handled on belt conveyors is limited only by the width of the belt, and the materials may vary from very fine chemicals to lump ore, stone, coal, or pulpwood logs. The moisture content of the material can vary from wet to dry and dusty, as well as those that are sticky, or have a tendency to pack due to moisture. Friable materials may be handled on belt conveyors with minimum degradation. Belt conveyors generally are so-called self-cleaning units when handling fairly dry materials; if material is slightly damp or very fine, a very light film of material may be found on the return run if the chemical makeup of the material has a slight affinity for the rubber cover. Belt conveyors are very durable when handling such materials as crushed stone, sand, or gravel.

Flat belts are seldom used for carrying belt conveyors. For their use in package conveyors, refer to section 17; for their use in elevators, refer to section 3.

2.2 INFORMATION REQUIRED FOR DESIGNING OR ESTIMATING BELT CONVEYORS

The objective is to design a belt conveyor that will deliver maximum performance at minimum cost per ton of material handled. To do so, it is necessary to know the answers to the following questions:

1. What is the horizontal distance over which the material is to be conveyed?
2. What is the vertical height that the material is to be lifted or lowered?

The answers to questions 1 and 2 will determine whether or not a belt conveyor will do the work at minimum cost, as opposed to, say, elevators and screw conveyors.

3. What kind of material is to be handled and what is its weight per cubic foot?
4. What is the average required capacity in tons per hour? And the maximum in tons per hour?
5. How will the flow of material to the belt be controlled?
6. What are the dimensions of the largest lumps?
7. What percentage of the total volume to be handled will consist of the largest lump size?

The answers to questions 3 through 7 will determine the

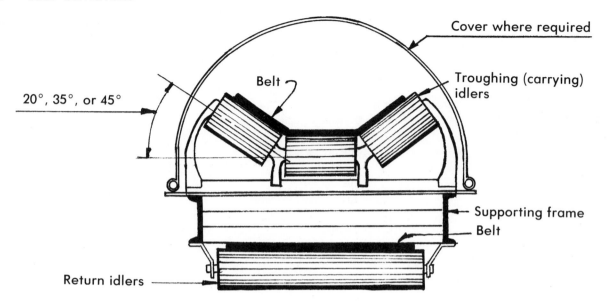

Figure 2.1 Cross-section of belt conveyors.

speed of the conveyor in feet per minute, width of belt to be used, power required to drive the conveyor, type of drive to be used, number of plies in belt, diameter of pulleys and shafts, type and spacing of troughing idlers.

8. Is the material hot, cold, wet, dry, sticky, oily, abrasive, or corrosive? To what degree?
9. How many loading points are there and where are they located?
10. How is the material to be discharged from conveyor over the head pulley, or through a tripper?
11. How many discharge points are there and where are they located?

Answers to questions 3 through 11 will determine the quality and thickness of the rubber cover on the belt.

12. Where would be the most convenient location for the drive?
13. Would an electric motor or some other type of engine be used to drive the conveyor?
14. If an electric motor, what are the current and voltage characteristics? What is its output speed, size and keyseat of output shaft?
15. Is the material to be weighed in transit on the belt? If so, about what location?
16. Is any tramp iron to be removed from the material as it passes over the discharge pulley?
17. How many hours (daily) will conveyor operate?
18. What are the climatic conditions? (Ambient or surrounding temperatures; maximum and minimum)
19. Is conveyor intended to be reversible?
20. Will conveyor be enclosed, or will it be exposed to the weather?

After the engineer has the answers to the preceding questions, he can determine whether a belt conveyor can

do the complete work intended, or can partially handle the material for a certain distance and then discharge it to another mechanical device, such as a bucket elevator, to deliver the material where intended. In some cases, the conveyor layout would require an angle or rise greater than the maximum safe inclination for the material handled. If so, a combination of a belt conveyor and, possibly, a bucket elevator, can accomplish the desired result. Many manufacturers of belt conveyor parts, including the rubber belt companies, publish complete engineering data on safe inclinations of troughed belt conveyors for handling various bulk materials. Typical belt conveyor designs are found on the following pages, but there are many variations of these typical designs. The actual determining factor as to the design of various parts must come either from the formulae set down by CEMA, from a belt conveyor manufacturer's book, or from the charts developed for quick figuring. These charts are generally accurate and reliable, and can be used safely on industrial belt conveyor installations, if allowance is made for special conditions.

2.3 ANGLE OF INCLINE

Belt conveyors can be used at angles up to 15°, although in some cases, they have been inclined 25° or even 30° from the horizontal.

The angle of inclination at which a belt conveyor will convey a bulk material depends upon the material particle size, shape of lumps, moisture content, angle of repose, and flowability. Design factors will affect the behavior of materials on an inclined belt, including belt speed, whether material is ascending or descending, how fully the belt is loaded and whether it is loaded continuously, uniformly, and centrally. When the incline is too steep, some part of the bed of material may slide, flow, or roll back, resulting in spillage. For example, in crushed stone plants, as the rock goes through the crusher, most of the crushed pieces, large or small, fracture to sharp edges. These edges tend to cling together on an inclined belt and, once the load is discharged onto the belt, it usually stays where it drops until it reaches the discharge point. Refer to table 1.2 for recommended maximum inclination for the material being handled.

Loading should occur at relatively low angles, so that the material will settle and rest prior to the incline, and maximum inclines can be obtained only if the material is at rest on the belt prior to steep slopes.

Belts can be made with a serrated or diamond pattern rubber cover that will help increase the safe incline angle.

2.4 BELT CONVEYOR ELEMENTS

Belt conveyors consist of ten essential elements.

1. The belt that forms the moving and supporting surface on which the conveyed material rides.
2. The idlers that form the supports for the troughed carrying run of the belt and the flat return run.
3. The pulleys that support and direct the belt and control its tension.
4. The drive that imparts power through one or more pulleys to move the belt and its load.
5. The structure that supports and maintains alignment of the idlers, pulleys, and drive.
6. Belt tripper, if necessary.
7. Belt cleaners.
8. Belt-training device.
9. Band brakes or holdbacks when conveyor is inclined, so as to hold a loaded belt in the event of a power interruption, to prevent the belt from running backward.
10. Proper type of takeup, to keep belt in tension (gravity takeups are used on many installations and when space permits).

2.4.1 Belt Characteristics

The type of belt is determined by the material handled, the longitudinal tensile stresses, spacing of idlers, weight of material, angular degree of idlers, and transverse flexibility of belt.

Transversely, the belt must be flexible enough to trough with the idlers and strong enough to span across the space between the idlers. It must also be flexible enough, without the load, to rest on the middle idler.

The belt carcass, which takes the tension and supports the load, consists essentially of cotton, rayon, nylon, polyester, or a combination of these, covered with rubber and reinforced at times with wire or fiberglass.

The rubber covers are made to resist abrasion and cutting. Special covers are made to resist oil, heat, combinations of heat and oil (up to 400°F), fire, and various chemicals. They can be made static-conductive. Given the proper information, the rubber or equipment manufacturer can recommend the best type for the particular purpose. For use with sharp, abrasive material, and for resistance to ozone and acids, the coating should be RMA (Rubber Manufacturers Association) Grade 1. RMA Grade 2 is the best buy for the majority of abrasion-resistant installations. Both grades have a skim coat between plies to provide for maximum flexure. A neoprene cover is recommended for petroleum oils; buna N for petroleum, vegetable, and animal oils. For high temperatures, the stress on the carcass fabric should be reduced to 75%.

For belts over 8 ft wide, and/or travelling over 750 fpm, it is best at the present time to involve the belting and the conveyor manufacturers in the design. Serious problems can result with the belting and idlers, especially at transfer points and in handling of lumpy, abrasive materials. The carrying side of belt conveyors, except for picking belts or package-carrying conveyors, is troughed 20°, 35°, or 45°.

2.4.2 Rating of Multiple-Ply Belts

Multiple-ply belts are now rated on the basis of tensile strength (in pounds-per-inch-width per ply) when the splices are vulcanized. With normal mechanical-fastener splices, the rating is reduced to approximately three-quarters of the vulcanized values. Some belt manufacturers can get adequate fastener holding due to their weave, and no reduction in the rating is required. Table 2.1 refers only to allowable belt tension.

2.4.3 Reduced-Ply Belting

Reduced-ply belts have carcasses of high-strength fibers, with either fewer plies or special weaves. No standards are available, but reduced-ply belting tension ratings can be made for all values of multiple-ply belting.

2.4.4 Steel Cable Belting

Steel cable belting consists of a single layer of steel cables embedded in rubber. Their use is limited to high tension requirements.

Table 2.1   Tension Rating of Multiple-Ply Belts

Fabric Identification	Tension Ratings, lb per inch per ply[a]	
	Normal Mechanical-Fastener Splice	Normal Vulcanized Splice
Multiple-ply 35	27	35
Multiple-ply 43	33	43
Multiple-ply 50	40	50
Multiple-ply 60	45	60
Multiple-ply 70	55	70
Multiple-ply 90	—	90
Multiple-ply 120	—	120
Multiple-ply 155	—	155
Multiple-ply 195	—	195
Multiple-ply 240	—	240

Source: Courtesy CEMA

[a]Allowable tension = rating per ply × no. of plies × width of belt in inches

2.4.5   Troughing Empty Belts and Load Supports

Table 2.2 gives the maximum number of plies for various idler angles so that the empty belt will trough properly.

Table 2.3 gives the minimum widths for empty troughing of reduced-ply belts (paragraph 2.4.3).

2.4.6   Lagging of Pulley

When the pull on the head or drive pulley indicates that the pulley should be lagged, or covered, to prevent slippage, it is important to select a cover that will not tear off the pulley. Practically all pulley lagging is now vulcanized.

One way to lag a pulley is to stretch a piece of rubber around the pulley and rivet it tightly to the pulley face, as shown on figure 2.2. Lagging should be stretched around the pulley by a hand winch to avoid pulling loose.

Instead of bolting, a vulcanized rubber coating (usually a rubber cover reinforced with several plies of fabric construction) can be tack welded or cemented to the

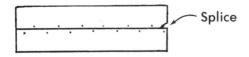

Figure 2.2   Lagging of a pulley.

pulley, or slid on (a rubber molded lagging) and vulcanized on steel backing plates.

Rubber on drive pulleys should have a durometer hardness rating of 55 to 65 Shore A scale. On snub or belt pulleys, a lower durometer may be used.

With a heavy pull on the head pulley, the herringbone-type cut rubber lagging made by most belt manufacturers can be used. The grooves are molded into the rubber and a pulley manufacturer can attach this lagging to a pulley with a special machine tool in the shop and apply the lagging with a special adhesive cement. This effect gives an excellent grip to the belt in all kinds of weather. It is on the expensive side, but does the work and, in many cases, depending on the size of the pulley, the lagging will cost approximately the same as the pulley.

Lagging should be grooved when handling wet material.

2.4.7   Belt Splices

All splices should preferably be vulcanized, for both economy and quality of joint. If mechanical fasteners are used, the ends must be cut square. Notching the edge will help. The fasteners come in several types, such as bolted plates, hinged plates, and riveted hinged plates. Some fasteners are made with hook-type cables. The fasteners should be of a metal suitable for local conditions (see figures 2.3 and 2.4).

The mechanically fastened splices can be installed fast, and the cost is a fraction of that for the vulcanized joint.

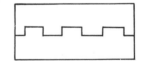

Figure 2.3   Vulcanized joint.

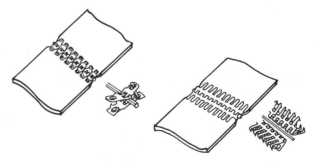

Figure 2.4   Mechanical joints. *(Courtesy Goodyear Tire & Rubber Co.)*

Table 2.2 Maximum Plies for Troughing of Empty MP (Multiple-Ply) Type Belts

Belt Width (in.)	Idler Angle	MP 35 MP 43	MP 50 MP 60 MP 70	MP 90 MP 120	MP 155	MP 195 MP 240
18	20°	5	—	—	—	—
	35°	4	—	—	—	
	45°	—	—	—	—	—
24	20°	6	5	4	4	—
	35°	5	4	4	—	—
	45°	4	—	—	—	—
30	20°	7	6	5	5	4
	35°	6	5	5	4	4
	45°	5	4	4	—	—
36	20°	8	7	6	6	5
	35°	7	6	6	5	5
	45°	6	5	5	4	4
42	20°	8	8	7	6	6
	35°	8	7	7	6	5
	45°	7	6	6	5	5
48	20°	8	8	8	7	6
	35°	8	8	7	6	6
	45°	8	7	6	5	5
54	20°	8	8	8	8	7
	35°	8	8	8	7	6
	45°	8	8	7	6	6
60	20°	8	8	8	8	8
	35°	8	8	8	8	7
	45°	8	8	8	7	6
72–84–96	20°	8	8	8	8	8
	35°	8	8	8	8	8
	45°	8	8	8	8	7

Source: Courtesy CEMA

Note: The maximum number of plies is limited to eight by practical manufacturing and belt-handling considerations.

2.5 SELECTION OF BELT; SIZE AND SPACING OF IDLERS

Tables 2.4 and 2.5 are CEMA recommendations, and give the minimum plies for load support, based on multiple-ply strength of belts.

Table 2.6 gives Goodyear Tire and Rubber Company minimum-ply recommendations for various fabric groups. These tables should be checked during belt selection.

Table 2.7 should be used for checking the capacity of the belt conveyor. The surcharge capacity is covered in table 2.8.

Table 2.3 Troughability of Reduced-Ply Type Belts

Minimum Widths for Empty Troughing

Rated Belt Tension	20° Idlers	35° Idlers	45° Idlers
To 150 PIW	14	18	18
To 200 PIW	16	24	24
To 250 PIW	24	24	30
To 300 PIW	30	30	30
To 500 PIW	36	36	36
To 700 PIW	42	42	48

Source: Courtesy CEMA
Notes: PIW = lb/in. width.
The above table is *only* a guide. Because of the wide variety of belt constructions offered by manufacturers, it is important that they be consulted for more accurate values.

Table 2.4 Minimum Plies for Load Support, 20° Idlers, MP (Multiple-Ply) Belts

Belt Width (in.)	25–49 pcf Material MP fabric ID 43	50	70	120	155	195	240	50–74 pcf Material MP fabric ID 43	50	70	120	155	195	240	75–99 pcf Material MP fabric ID 43	50	70	120	155	195	240	100–150 pcf Material MP fabric ID 43	50	70	120	155	195	240
18	3	3	3					3	3	3	3	3			4	3	3	3	3			4	4	3	3	3		
24	3	3	3					4	3	3	3	3			5	4	3	3	3			5	4	4	4	4		
30	4	3	3	3	3	3	3	4	4	3	3	3	3	3	5	4	4	4	4	4	4	6	5	4	4	4	4	4
36	4	4	3	3	3	3	3	5	4	4	4	4	4	4	6	5	4	4	4	4	4	6	6	5	5	5	5	5
42	4	4	4	4	4	3	3	5	5	4	4	4	4	4	6	5	5	5	5	5	4	7	6	6	6	6	5	5
48	5	4	4	4	4	4	3	6	5	5	5	5	4	4	7	6	6	6	5	5	5	7	7	6	6	6	6	5
54	5	5	4	4	4	4	4	6	6	5	5	5	5	4	7	7	6	6	6	5	5	8	8	7	7	7	6	6
60	6	5	5	5	5	4	4	7	6	6	6	5	5	4	8	7	7	7	6	6	5			8	7	7	7	6
72	6	6	5	5	5	5	4	8	7	6	6	6	6	5		8	7	7	7	7	6				8	7	7	7
84	7	7	6	6	6	6	5		8	7	7	6	6	6			8	8	7	7	6					8	8	7
96	8	8	7	7	7	7	6			8	8	7	7	6				8	8	7								8

Source: Courtesy CEMA   *Note:* pcf = lb/ft³

Table 2.5 Minimum Plies for Load Support, 35–45° Idlers, MP (Multiple-Ply) Belts

Belt Width (in.)	25–49 pcf Material MP fabric ID 43	50	70	120	155	195	240	50–74 pcf Material MP fabric ID 43	50	70	120	155	195	240	75–99 pcf Material MP fabric ID 43	50	70	120	155	195	240	100–150 pcf Material MP fabric ID 43	50	70	120	155	195	240
24	4	3	3	3				4	4	4	4	4			5	4	4	4	4	4		5	5	4	4	4		
30	4	4	3	3				5	4	4	4	4	4	4	5	5	4	4	4	4	4	6	5	5	5	5	5	4
36	4	4	4	4	4	3	3	5	5	5	5	4	4	4	6	5	5	5	5	5	4	6	6	6	6	6	6	5
42	5	4	4	4	4	4	3	6	5	5	5	5	5	4	6	6	6	6	6	5	5	7	7	6	6	6	6	5
48	5	4	4	4	4	4	4	6	6	6	6	5	5	5	7	6	6	6	6	6	5	8	7	7	7	7	7	6
54	6	5	4	4	4	4	4	7	6	6	6	6	5	5	8	7	7	7	7	6	6	8	8	7	7	7	7	6
60	6	5	5	5	5	4	4	7	7	6	6	6	6	5	8	7	7	7	7	7	6			8	8	7	7	7
72	6	6	5	5	5	5	5	8	7	7	7	7	6	6		8	7	7	7	7	6						8	7
84	7	7	6	6	6	6	6		8	8	8	8	7	7			8	8	8	8	7							8
96	8	8	7	7	7	7	7						8	8							8							

Source: Courtesy CEMA   *Note:* pcf = lb/ft³

Table 2.6 Goodyear Tire & Rubber Company's Minimum-Ply Recommendations

Width (in.)	Idlers (see below)	To 50 pcf: coke, carbon, grain, wood chips, peat, soda ash			50 to 75 pcf: borax, dry lumpy clay, coal, lime, sulphur, sugar, alumina						75 to 100 pcf: slag, bauxite, slate, Portland cement, limestone, sand-gravel, gypsum						More than 100 pcf: quartz, most ores, rock, wet sand talc, sinter, Taconite					
		2	3	4	2	3	4	5	6	7	2	3	4	5	6	7	2	3	4	5	6	7
18–20	A	2	2	2	3	2	2	2			3	3	3	3			3	3	3	3		
	B	3	3	2	3	3	3	3			4	3	3	3			4	4	3	3		
24	A	3	3	2	3	3	3	3			4	3	3	3			4	4	4	3		
	B	3	3	3	4	3	3	3			4	4	3	3			5	4	4	4		
	C	4	3	3	4	4	4	3			5	4	4	3			5	4	4	4		
	D	4	4	4	5	4	4	4			5	5	5	4			6	5	5	4		
30	A	3	3	3	4	3	3	3	3	3	4	3	3	3	3	3	5	4	4	4	3	3
	B	4	3	3	4	4	3	3	3	3	5	5	4	4	4	4	5	5	4	4	4	4
	C	4	4	3	5	4	4	4	4	4	5	5	4	4	4	4	6	5	4	4	4	4
	D	5	4	4	6	5	4	4	4	4	6	6	5	4	4	4	6	6	5	5	4	4
36	A	4	3	3	4	4	3	3	3	3	5	4	4	4	4	4	5	5	4	4	4	4
	B	4	4	3	5	4	4	4	4	4	5	5	4	4	4	4	6	5	5	5	5	5
	C	4	4	4	5	4	4	4	4	4	6	5	5	5	4	4	6	6	6	5	5	5
	D	5	5	4	6	5	5	4	4	4	6	6	5	5	5	4	7	6	6	6	5	5
42	A	4	4	4	4	4	4	4	4	4	5	5	4	4	4	4	6	5	5	5	4	4
	B	4	4	4	5	5	4	4	4	4	6	5	5	5	5	4	6	6	6	6	5	5
	C	5	4	4	5	5	4	4	4	4	6	6	5	5	5	5	7	6	6	6	6	5
	D	5	5	5	6	6	5	5	4	4	7	6	6	6	5	5	7	7	6	6	6	5
48	A	4	4	4	5	4	4	4	4	4	6	5	5	4	4	4	6	6	6	5	5	4
	B	5	4	4	5	5	5	5	4	4	6	6	6	5	5	5	7	6	6	6	6	5
	C	5	5	4	5	5	5	5	5	4	7	6	6	6	5	5	8	7	7	6	6	6
	D	6	5	5	6	6	6	5	5	5	7	6	6	6	6	5	8	7	7	6	6	6
54	A	5	5	4	5	5	4	4	4	4	6	6	5	5	5	4	7	6	6	6	5	5
	B	5	5	5	6	5	5	5	5	4	7	6	6	6	5	5	8	7	7	7	6	6
	C	5	5	5	7	6	6	5	5	5	8	7	7	6	6	5	8	8	7	7	7	6
	D	6	6	5	7	6	6	6	5	5	8	7	7	7	6	6	8	8	7	7	7	6
60–66	A	5	5	5	6	5	5	5	4	4	7	6	6	6	5	5	7	7	7	7	6	5
	B	6	5	5	7	6	6	5	5	4	8	7	7	6	6	5	8	8	7	7	7	6
	C	6	5	5	7	7	6	6	5	5	8	8	7	7	6	6		8	7	7	7	7
	D	6	6	6	7	7	6	6	6	5	8	8	7	7	7	6		8	7	7	7	7

(Continues)

Table 2.6 (*Continued*)

Width (in.)	Idlers (see below)	To 50 pcf coke, carbon, grain, wood chips, peat, soda ash			50 to 75 pcf borax, dry lumpy clay, coal, lime, sulphur, sugar, alumina						75 to 100 pcf slag, bauxite, slate, Portland cement, limestone, sand-gravel, gypsum						More than 100 pcf quartz, most ores, rock, wet sand talc, sinter, Taconite					
		2	3	4	2	3	4	5	6	7	2	3	4	5	6	7	2	3	4	5	6	7
72–78	A	5	5	5	7	6	5	5	5	4	8	7	7	6	6	5	8	8	7	7	6	6
	B	6	5	5	8	7	6	6	6	5		8	7	7	7	6		8	8	7	7	7
	C	6	6	5	8	7	7	6	6	6		8	8	7	7	6			8	8	8	7
	D	7	6	6	8	7	7	7	6	6		8	8	7	7	6			8	8	8	7
84–96	A		6	6		7	6	6	6	5		7	7	6	6	5		8	7	7	6	6
	B		7	6		7	7	6	6	6		8	7	7	7	6		8	8	7	7	7
	C,D		7	7		8	7	7	7	6		8	8	7	7	6			8	8	8	7

Idlers

A: 20° picking idlers less 1 ply on 5 plies or more. Do not reduce 3- and 4-ply belts. Capacities must not exceed those shown in table 2.7, with maximum surcharge angle of 30°

A: 20° under 2/3 loaded*
B: 20° over 2/3 loaded*
   30–35° under 2/3 loaded*
C: 30–35° over 2/3 loaded*
   45° under 2/3 loaded*
D: 45° over 2/3 loaded*

Fabric groups

HD = heavy duty; N = nylon; F = fabric; H = heavy; XH = extra heavy; RN = rayon
2: 32 HDNF
3: 36 HDNF, H and XH rayon, HDRN-10, 20
4: 42 and 48 HDNF
5: HDRN-40
6: HDRN-50
7: HDRN-70

Important: The percent loading must be calculated against the peak loading rate unless a reduced, uniform controlled rate can be guaranteed. For example, assume a belt that has a volumetric capacity of 500 tph is delivering an average of 250 tph. This belt is not 50% loaded unless it is evenly loaded at all times. If it is sporadically loaded, then its peak load is over 50% and could be as high as 100%. When the peak load cannot be established, 100% must be assumed.

Note: If uniformly loaded under 45%, subtract 1 ply from 5 ply or heavier belts. For systems using idler combinations of 45° load point/35° carrying, or 45° load point/20° carrying, use the smaller angle for the above table.

Table 2.9 gives spacing and types of idlers. Checking the tension in the belt can be done only when the maximum tension in the belt is computed.

Reduced-ply belts can be used with the CEMA tables when the tension in the belts is above the values allowed.

As to types of idlers, "Belt Conveyors," by CEMA, classifies idlers as follows: A and B idlers are for light duty, and are made in 4- and 5-in. diameter sizes. C and D idlers are for medium duty. The C idlers are made in 4-, 5-, and 6-in. diameter sizes. The D idlers are made in 5- and 6-in. diameter sizes. E idlers are for heavy duty and are made in 6- and 7-in. diameter sizes.

Table 2.9 gives the recommended spacing of idlers. The type of idler selected is based on maximum loading conditions (refer to paragraph 2.30). Where large or heavy lumps are involved, it is best to use closer spacing. Closer spacing (about 12 in. center to center) should be used at dump hoppers.

2.6 DRIVE (see section 9)

Standard drives may be

1. roller-chain drive, speed reducer, coupling, and standard motor
2. direct-connected motorized reducer and coupling to head or drive shaft of belt conveyor
3. speed reducer mounted on drive shaft of belt conveyor with V-belt drive to motor

Arrangements 1 and 3 may be desirable because they make possible the increase or decrease in the speed of the conveyor.

The location of the drive should take into account the

Table 2.7 Capacity of Belt Conveyors

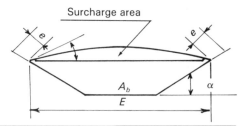

Trough Angle		$\alpha = 20°$		$\alpha = 35°$		$\alpha = 45°$	
Width of Belt (in.)	Edge e (in.)	E (in.)	A_b (ft^2)	E (in.)	A_b (ft^2)	E (in.)	A_b (ft^2)
18	1.89	13.875	0.089	12.914	0.143		
24	2.22	18.931	0.174	17.469	0.278	16.504	0.328
30	2.55	24.073	0.287	22.420	0.459	20.884	0.540
36	2.88	29.215	0.428	27.167	0.684	25.264	0.803
42	3.21	34.361	0.597	31.926	0.952	29.662	1.117
48	3.54	39.504	0.794	36.673	1.265	34.042	1.484
54	3.87	44.646	1.02	41.421	1.623	38.422	1.903
60	4.20	49.792	1.27	46.179	2.024	42.821	2.372
72	4.68	60.077	1.86	55.674	2.962	51.581	3.468
84	5.52	70.248	2.62	64.829	4.156	59.792	4.846

Notes: The capacity (A_b, with troughing idlers) will vary slightly with the type of idlers used. The thickness of the belt is disregarded and no allowance is made for the sag of the belt.

Edge distance, e, = 0.055 width of belt + 0.9 in. For a different edge distance, correct the value of E.

Area of surcharge = $F \times E^2$

Total area, A_T, = area in table + FE^2 (ft^2)

Total capacity in tons =

$$\frac{A_T w v}{2000}$$

where w = weight in lb/ft^3 and v = velocity in fpm.

Table 2.8 Surcharge Capacity

Angle of Surcharge β	Factor F
5°	0.01454
10°	0.02918
15°	0.04402
20°	0.05916
25°	0.07466
30°	0.0906

ease of servicing it. Manufacturers' catalogs are generally based on operating 8–10 hours per day. If the conveyor operates more than one shift, the size of the various parts should be increased. Refer to table 9.1 for correction factors for longer operating hours.

2.7 GREASING OF IDLERS

Most idler rolls have through shafts, providing large grease reservoirs, which are completely filled at the fac-

tory. Both ball bearings and roller bearings are protected by cartridge seals to keep grease in and dirt out over a long period of time, whether outdoors or indoors. Any well designed idler roll will do its work for a period of 6 months or more, because each roll will hold between 1¼ and 1½ lb of grease. Some tests have shown that properly designed and correctly made idlers will not need additional grease for several years, even when handling coke or sintered iron-ore mix. Idlers should not be overgreased. Opinions vary regarding the required frequency of greasing the idlers. It is wise to follow manufacturers' recommendations.

2.8 BELT TRIPPERS

Many belt conveyors are equipped with trippers that discharge material at different locations, such as bins, silos, and outdoor storage.

Operators of some small-capacity installations which operate only a few hours a day will find it economical to use a hand-propelled tripper. A hand-propelled trip-

Table 2.9 Spacing and Types of Idlers

Width of Belt (in.)		Weight of Material handled (lb/ft³)[a]					Return Idlers
		50	75	100	150	200	
18	Spacing[b]	5.0 ft	5.0 ft	5.0 ft	4.5 ft	4.5 ft	10.0 ft
	Type	A	A	A	A	B	
	Weight	12	12	14	17	17	
24	Spacing	4.5 ft	4.5 ft	4.0 ft	4.0 ft	4.0 ft	10.0 ft
	Type	A	A	B	B	C	
	Weight	16	16	19	23	23	
30	Spacing	4.5 ft	4.5 ft	4.0 ft	4.0 ft	4.0 ft	10.0 ft
	Type	A	B	C	C	C	
	Weight	20	21	24	29	29	
36	Spacing	4.5 ft	4.0 ft	4.0 ft	3.5 ft	3.5 ft	10.0 ft
	Type	B	C	C	C	D	
	Weight	28	32	35	41	48	
42	Spacing	4.5 ft	4.0 ft	3.5 ft	3.0 ft	3.0 ft	10.0 ft
	Type	C	C	C	C	D	
	Weight	34	36	42	49	59	
48	Spacing	4.0 ft	4.0 ft	3.5 ft	3.0 ft	3.0 ft	10.0 ft
	Type	C	C	D	D	E	
	Weight	41	48	51	69	77	
54	Spacing	4.0 ft	3.5 ft	3.5 ft	3.0 ft	3.0 ft	10.0 ft
	Type	C	D	D	E	E	
	Weight	48	51	58	78	89	
60	Spacing	4.0 ft	3.5 ft	3.0 ft	3.0 ft	3.0 ft	10.0 ft
	Type	D	D	E	E	E	
	Weight	60	67	70	87	99	
72	Spacing	3.5 ft	3.5 ft	3.0 ft	2.5 ft	2.5 ft	8.0 ft
	Type	E	E	E	E	E	
	Weight	74	105	83	113	130	
84	Spacing	3.5 ft	3.0 ft	2.5 ft	2.5 ft	2.0 ft	8.0 ft
	Type	E	E	E	E	E	
	Weight	102	115	127	149	165	

Notes: Values are adjusted to CEMA data. Only sizes generally involved in bulk material handling are given. For material weighing 150 lb/ft³ or over, apron conveyors should be considered.

[a]The weight of the material carried is not included in the weights given.

[b]Where large or heavy lumps are involved, closer spacing is best. The spacing may be limited by the load rating of the idler (refer to paragraph 2.30).

per is arranged so that a man walking alongside the trip-per can manually operate a handwheel or handcrank on a small shaft which, in turn, operates a roller-chain drive connected to one of the axle shafts. This type of tripper is not easily moved, due to its weight, and should be used only when the discharge from a belt goes to two or three openings, and then not too far apart. It can operate in either direction.

A belt-propelled tripper is a standard in industrial work. It is easily operated and is used where travel distance and direction can be manually controlled. Power to move the tripper is obtained from the conveyor belt and is transmitted from a pulley shaft to a countershaft through manually engaged friction wheels, and from the countershaft by a chain drive to the tripper wheels. Normally, these trippers are provided with a hand-operated, quick-acting rail clamp for holding the tripper in a fixed position. An operator's platform and a safety guard over the friction-wheel drive also are provided. Tripper travel is about one-tenth of the speed of the conveyor belt and is reversible.

A motor-propelled tripper should be used where continuous and uniform distribution of material along the conveyor is required, or where the tripper is to be moved or reversed frequently. Propelled by an independent electric motor, it can be automatically reversed at each end of its travel by limit switches carried on the tripper, and actuated by stops placed where desired along the runway. The tripper can be moved while the conveyor belt is moving or at rest. The operator's platform can be placed on either side. Normal speed of this unit is 30 fpm.

Two-way discharge chutes are provided on all trippers with a manually operated flopgate. On rare occasions, a three-way discharge chute has been furnished, resulting in a discharge to either side of the belt and over the discharge pulley back onto the belt again. Such a condition exists in the handling of grain into silos.

Several methods of bringing power to the tripper have been used. One method uses overhead trolley wires, which must be protected for a person riding on the tripper platform. A more accepted method that is considered safe, and is now in common use, is to use a cable reel carried on the tripper (see figure 2.5) so the cable wire can be plugged into a power outlet about every 50 ft or so. This would appear to require constantly jumping up and down from the tripper to plug in along the run, but it does not seem to work out that way. The safety features of the reel overcome any small objections.

If a tripper is used on the belt, the pulleys should be inspected regularly to see that the belt does not scrape against the frame and that the belt runs true over the pulleys on the tripper.

2.9 PLOW VERSUS TRIPPER

There are several factors entering into the use of a plow. The belt has to be flat where a plow is used. This is not always possible when using a trough belt conveyor. It takes considerable space to flatten a belt from leaving trough idlers, and after flattening to return to trough idlers again.

Flat carrying idlers are used to keep the belt flat at the discharge point. In foundry work, where a belt conveyor distributes prepared sand to a series of closely spaced storage bins, a flat belt is used for the entire length. These bins are located over molding machines, and the molder operator controls the operation of the plow over his bin. This stationary, steel V-shaped plow can be manually operated or air-operated from the floor level. When the plow is dropped into position, it will just about touch the belt, plowing off the sand intended for the particular bin. When the bin is filled, as indicated by a red light in the electrical system, the plow is raised automatically, and sand travels on to the next bin.

In a foundry, if the power fails when the plow is down on the belt and the bin is full, the molder below may get a little shower of sand which indicates to him that something is wrong, and he can push a button, stopping the overhead belt. To prevent any jam, if the electrical system fails when all the plows are raised and the belt is operating with a full load of sand on it, an overflow chute at the end of the belt is provided, diverting any surplus sand to a container or barrel on the floor, or sometimes, this overflow chute will extend to a shake-out conveyor under the floor, allowing the sand to be reconditioned and to get back into the prepared sand conveyor. The standard design of a belt tripper is very seldom used in old foundries because of limited space in buildings already built. Some newly designed foundries have ample space for a tripper.

An inexpensive arrangement of a movable plow that is still using troughing idlers is shown on figure 2.6. Note that the complete plow assembly is supported by a troughing idler and is moved along the belt by cables attached to motor-operated winding drums at both ends of the belt. As the belt leaves the troughing idlers, it flattens, so the V-shaped plow can discharge to the bin. This plow is manually operated. The system works well and is used where a unit of this kind does not have to be moved too often. It is much less expensive than a tripper of any kind.

2.10 MAGNETIC PULLEY

Magnetic pulleys are inserted to remove foreign steel or tramp iron in the material being conveyed on the belt. They usually operate on 220 V dc.

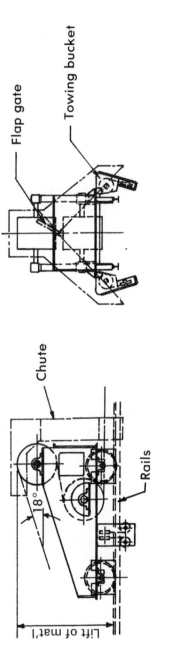

CABLE PROPELLED TRIPPER

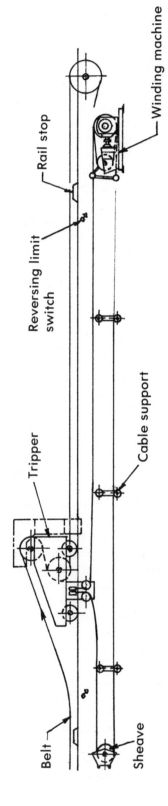

Figure 2.5 Arrangement of a cable-propelled belt tripper.

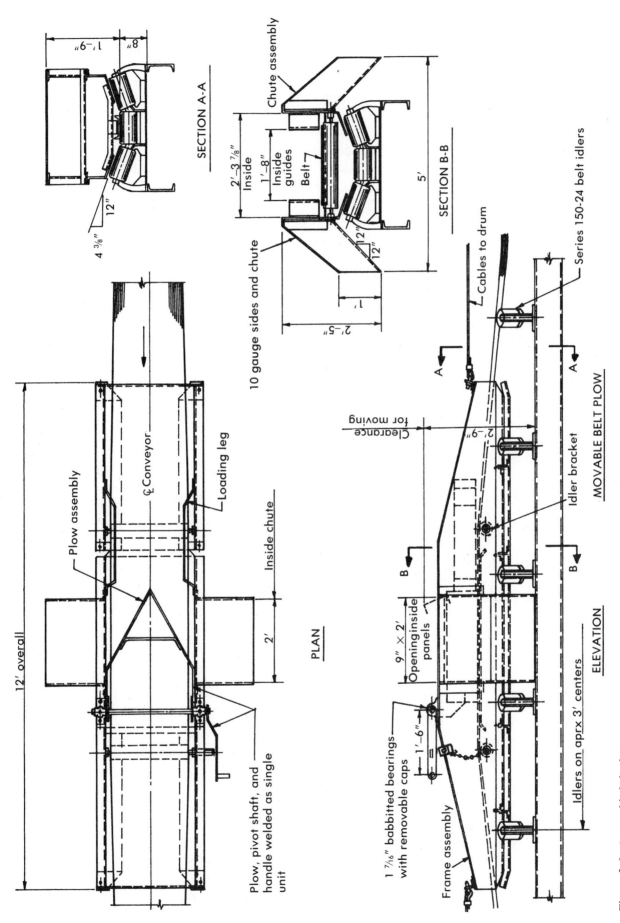

SECTION A-A

1'-9"   8"

12"

4 3/8"

SECTION B-B

Chute assembly

2'-3 7/8" Inside

1'-8" Inside

Inside guides

Belt

12"

12"

10 gauge sides and chute

1'

2'-5"

PLAN

Plow assembly

℄ Conveyor

Loading leg

Inside chute

2'

12' overall

Plow, pivot shaft, and handle welded as single unit

ELEVATION

MOVABLE BELT PLOW

Cables to drum

Series 150-24 belt idlers

Clearance for moving

2'-9"

Idler bracket

B

B

A

A

9" × 2' Opening inside panels

1'-6"

1 7/16" babbitted bearings with removable caps

Frame assembly

Idlers on aprx 3' centers

Figure 2.6 A movable belt plow.

27

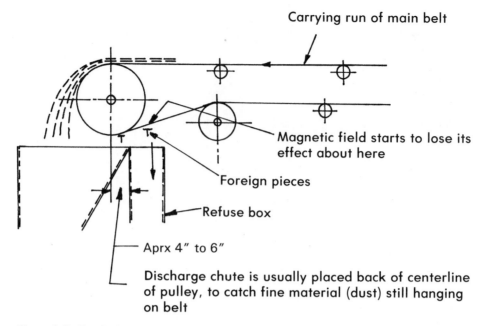

Carrying run of main belt

Magnetic field starts to lose its effect about here

Foreign pieces

Refuse box

Aprx 4" to 6"

Discharge chute is usually placed back of centerline of pulley, to catch fine material (dust) still hanging on belt

Figure 2.7 Standard-type arrangement of a magnetic pulley.

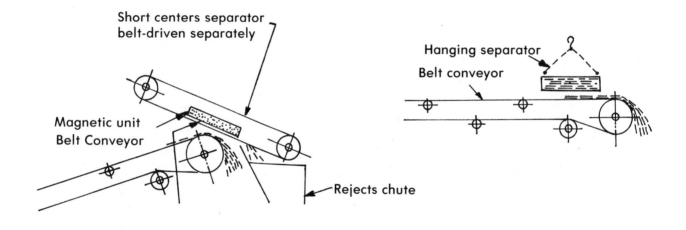

Short centers separator belt-driven separately

Magnetic unit Belt Conveyor

Rejects chute

Hanging separator

Belt conveyor

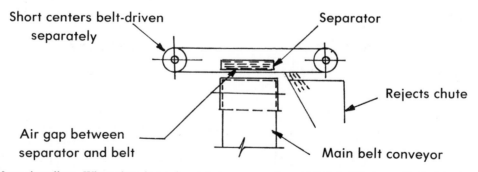

Short centers belt-driven separately

Separator

Rejects chute

Air gap between separator and belt

Main belt conveyor

Figure 2.8 Magnetic pulleys. Where there is an air gap between separator and belt (which is usually 3-ply), there is very little loss of the magnetic field of efficiency. (Note: Before inspecting any separator that has wires connected to it, be sure to turn off the power.)

Normally, the depth of the load on the belt should not exceed 6 in., and should be even less, even if large lumps are involved. General arrangements are shown on figures 2.7 and 2.8.

These pulleys are manufactured in two types: in one, the wires come to the pulley through a hollow shaft upon which the pulley is mounted; in the other, there are permanent-magnet type pulleys that do not require any wires. They should not be covered.

A permanent-magnet type has come into use and gained in popularity. It requires no rectifier, is less costly, easier to maintain, and requires no expense for power. Permanent-magnet pulleys are available with either a radial field, an axial field, or a criss-cross field. The criss-cross circuit design eliminates any magnetic dead spots, which could be present in either axial or radial design.

On a long belt conveyor that requires the head pulley to be rubber-covered for maximum pulling of the belt, a magnetic pulley is of little benefit. In this case, a suspended separation magnet should be placed above the discharge pulley, as shown on figure 2.8.

It should be kept in mind that any steel skirt boards or chutes designed to fit closely around a magnetic pulley can, in time, become magnetized and render the whole arrangement useless. To overcome this, plenty of clearance should be provided around the pulley, or the steel should be covered with a nonmagnetic material.

2.11 SKIRTBOARDS

Skirtboards should be arranged and set so that the space between the belt and the board or skirt liner increases in the direction of belt travel (see figure 2.9), thus permitting any material to work free and not gouge the lining or push the belt to one side of the idlers.

The lining will prevent material from creeping to the edge where the load on the belt is deep. The skirt rubber should be on the firm side and easily maintain its shape. If any foreign material such as ore or gravel should force its way under the rubber edge, the rubber will usually give way, but springs back immediately.

Because of the close spacing of idlers at feeding locations, the belt will not sag under the load. The skirtboard with the rubber lining will reduce the material leakage. The removable rubber strip can be replaced when worn.

2.12 TRAINING OF BELT ON CONVEYOR

Belts should always run true and be centered on the carrying and return idlers, when loaded as well as when empty. If one part of the belt runs true, and another part

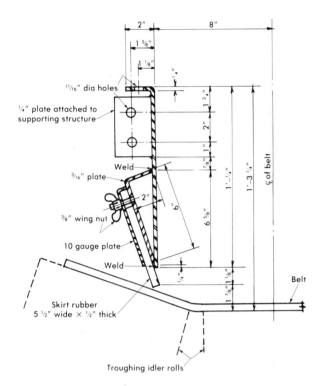

Figure 2.9 Skirtboard details for a 24-in. belt.

is out of line, it shows that the belt is crooked. If it runs out of line at the splice, it is likely that the splice or the joint is not square. If the belt runs steadily out at any one point in the conveyor, and it keeps running out at the same idlers, it is the idlers that are out of line, usually the idler just behind the point where the belt starts running out of line. Belts should be trained while empty, and if the belt runs out of line when loaded, it is probably because of uneven loading, and the loading chute should be corrected. Training troughing idlers is recommended to keep the belt in line on runs, say, over 75–100 ft.

Belts cannot be made to run true by adjustments of takeup screws or head pulley. Both head and tail pulleys should be lined up correctly and kept so. All training should be done by the idlers. The center roll of the troughing idler must always be in contact with the belt, because it is the one that really steers or trains the belt, unless the belt is over-plied or too stiff to assume the shape of the idlers when empty.

Idlers should be kept square with the belt, and when training the belt by lining up idlers, always work with the direction of belt travel. Stuck or sluggish idler rolls tend to deflect the belt. Side guide or self-aligning idlers should never be used while training a belt. Guide idlers bearing against the edge of the belt are frequently the cause of excessive wear and failure. They loosen up the edges, permitting the entrance of moisture and dirt between the fabric plies. Poor foundations or conveyor supports are sometimes the cause of bad alignment.

It is possible to train a belt by slightly shifting some of the conveyor idlers. Many of the troughing idlers have slotted holes in the mounting brackets, so that when training a conveyor belt, it is possible to loosen the mounting bolt slightly and then shift one side of the idler forward or backward to line up the belt properly.

Training a belt by increasing the tension not only strains the driving machinery, but puts an unnecessary stress on the belt, weakens the splice, takes out the troughing of the belt, and eliminates the effective training action of the center idler roll. It is recommended sometimes on a temporary basis to accelerate belt break-in and, therefore, help train.

On conveyors over 100 ft long, it is good practice to install belt-training swivel idlers on the carrying side to ensure a true-running belt. If, for some unforeseen reason, the belt tends to run to one side of a number of idlers, the training idler will force the belt back in a straight line. These training idlers are placed 10–20 ft from head and tail pulleys, and then 50–75 ft apart on the carrying side. If trouble occurs on the return side, swivel training idlers can be placed on this run and spaced as on the carrying run. Some types cannot be used on reversible belts.

2.12.1   Belt-Training Idlers

Mr. Roger E. Noel of Goodman Conveyor Company has prepared a training manual entitled "Conveyor Belt Training." The remainder of this paragraph, as well as paragraph 2.12.2, are reproduced here with his permission.

If the conveyor has been installed perfectly straight and level, and all pulleys and idlers are square with the conveyor, and the belt splices are correct and square, and there are no defects in the conveyor belting, then the belt should require no additional training.

All of the above conditions are rarely present, however, and some training of the belt is usually required.

2.12.2   Training of the Belt Should be Done by One Person

Train the return strand first, starting at the head or discharge end. (Temporary training may be necessary near the tail section if the belt is running into the structure.) To correct misalignment, go back several return rolls and slide the return roll hanger bracket slightly (fractions of an inch).

If the return belt is running to the right side, looking toward the tail, then move either the right side hanger toward the tail, or move the left side hanger toward the discharge. Do not attempt to correct by moving only one roll. Correction should be made by adjusting several rolls slightly. Whenever the rolls are not square with the belt, there is sliding action between the belt and roll shell. This sliding action will cause premature wearing of the roll shell and belt covers. This rate of wear will increase as the angle is increased. Do not create a training

problem at one point by overtraining at another point. To prevent this overtraining, the belt should be allowed to make a complete revolution after each adjustment is made. (Example: a 3000-ft conveyor operating at 500 fpm requires 12 minutes for one revolution.)

After the belt has been trained the full length on the return side, repeat the process on the carrying side, starting at the tail section.

The complete run-in process includes aligning the belt both unloaded and loaded. When adjusting the position of the troughing idlers and return idlers, be sure to loosen the idler mounting bolts before moving, and retighten them immediately after moving. Be very safety-minded: avoid contact with the moving belt.

Initial loading of the belt should be in gradual steps, with the belt alignment checked during each step, until the ultimate loading is reached.

When making corrections to the belt alignment in the various loaded conditions, recheck the idlers previously adjusted out-of-square to see if squaring these idlers will improve alignment under load.

Note: Initial startup and movement of the belt will indicate where corrections are required. The first corrections must be at points where the belt is in danger of being damaged. (This is usually behind drive and at tail end.) Once the belt is clear of all danger points, a sequence of training can be followed.

2.12.3   CEMA Data

In "Belt Conveyors for Bulk Materials," CEMA states

The usual training idler has the carrying roll frame mounted on a central pivot approximately perpendicular to the conveyor belt. Means are provided to cause the carrying rolls to become skewed with respect to the centerline of the conveyor. As the belt traverses the skewed rolls, they urge the displaced belt to return to the conveyor centerline and, in doing so, the rolls are urged to return to proper alignment.

Fixed guide rolls placed perpendicular to the edge of the conveyor belt are not recommended, because continuous contact with the conveyor belt edge accelerates belt edge wear, appreciably reducing belt life [see figure 2.10].

2.13   WEIGHING MATERIAL IN MOTION

All scales should be specified to meet the tolerances, tests, and loading as required by the latest applicable regulations of the "National Conference on Weights and Measures" for the particular type of scale furnished.

The *U.S. Industrial Directory* lists several pages of various scales including automatic, bag filling and batching, electronic, conveyor, and other specialty scales. The selection is limitless. Unless you have data from or experience with some special make, the local telephone directory will provide names for many desirable types.

The Merrick Corporation makes conveyor scales for weighing material in motion; these are available in $\pm 1/8\%$ and $\pm 1/2\%$ accuracies. They also make a line of

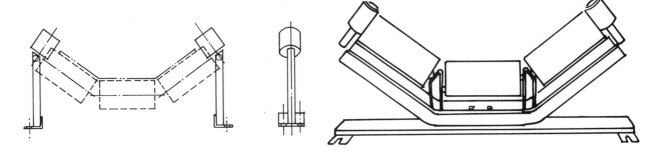

Figure 2.10 Troughed idlers. *(Courtesy Goodman Conveyor Company)*

gravimetric belt-weigh feeders for automatic batching and proportioning of material.

Fairbanks Weighing Division of Colt Industries, Ramsey Engineering Co., and A.B.C. Scale are among others making similar equipment.

A Tecnetic Industries' scale can be introduced into a belt conveyor system to read instantaneously tons per hour travelling on the belt. Power can be a 12-V battery or 115/230-V, 60-Hz, single-phase. Ambient temperature is −20°F to +120°F. It is made in widths of standard belts up to 72 in.

Howe-Richardson Scale Company uses electronic controls for automatic batching and proportioning by weight of material fed into a system. The flow has to be evenly controlled. It is capable of flexible requirements and controls.

2.13.1 Location of Scale

The scale mechanism (figure 2.11) should be placed at the tail end of the belt conveyor, where the belt tension is a minimum, but not closer than the distance from the chute or spout feeding the belt, governed by the formula

$$\frac{3 \times W \times V}{100}$$

where W is the width of the belt and V is the belt speed in fpm. The Merrick Corporation issues an Application Data Sheet, no. ADS 102, for their Weightometer® installation. Instead of the formula given above, they state

The material must settle on the conveyor belt before it is weighed, in order to achieve accurate weight measurement. The distance required from the end of the infeed skirtboards

to the idler adjacent to the Weightometer weighbridge is as follows:

Belt speed (fpm)	Q = Distance required (ft)
100–300	3–4
300–400	6–8
450–over	9–12

If the conveyor is short (aprx 50 ft or less) and/or the intended scale location is near the head pulley, there must be a minimum of two idlers for flat or 20° troughed, or three idlers for 35° (one-35°, two-20°) troughed idlers between the Weightometer weighbridge and the head pulley. Gradual transition is preferred to prevent a lifting effect of the belt.

If the conveyor is longer than 75 ft, a gravity takeup must be provided to maintain a uniform belt tension. The takeup must not introduce excessive belt tension and should be designed to allow the addition or removal of counterweights.

The conveyor structure (particularly in the area of the scale) must be rigid enough to resist deflection under maximum load. Rope conveyors are not acceptable, unless a rigid section is provided for the Weightometer.

Merrick further recommends that:

1. Wherever there is a change in either vertical or horizontal alignment, the unit should be placed 40 ft from the point of tangency for concave curves, and
2. 16 ft for 20° idlers, and 20 ft for 35° idlers on convex curves.
3. If the location is subject to winds over 5 mph, wind breaks should be provided for a distance of 40 ft on either side of the unit.
4. If the conveyor is inclined, the inclination must be fixed at one specific angle. Portable conveyors must be provided with some means of readjustment to obtain the design inclination.
5. The conveyor inclination must not exceed the material angle of slide, to prevent reweighing of the material.
6. Always use rigid 3-roll inline-type idlers on the scale weighbridge and 3 or 4 idlers either side.
7. If training idlers are used, they must not be located within 40 ft either side of the scale.
8. If the feed to the conveyor is not uniform, a series of diagonally placed plows should be used to eliminate hills and valleys in the material.
9. Avoid trying to weigh accurately in both directions on reversible conveyors. The use of two scales is a better solution.

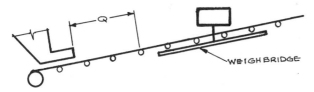

Figure 2.11 Location of the scale.

10. Normally, shimming the idlers adjacent to the weigh-bridge is necessary to permit installation [as shown on figure 2.12a].
11. Shimming can be eliminated by depressing the stringer section under the weighbridge [figure 2.12b].
12. Shimming can also be eliminated by notching the stringer section [figure 2.12c].

2.13.2 Weigh Larries
(figure 2.13)

Wherever accurately weighed material is required, a weigh larry is used to measure and deliver previously determined quantities of material.

The larry will permit close spacing of discharge gates and, in handling coal, will distribute it uniformly in the stoker hopper, preventing segregation of lumps and fines.

Larries are built with capacities of 1000–10,000 lb and may be hand-propelled, floor-operated, or motor-propelled from the floor, platform, or cage.

The hopper is built of steel plate, with a discharge chute fitted with a pivoted gate. The frame and beam box are rigidly built. The wheels are cast-iron, flanged, chill rims, with roller bearings.

The scale is beam type, equipped with a ticket re-corder that punches or prints ticket with accurate weight readings.

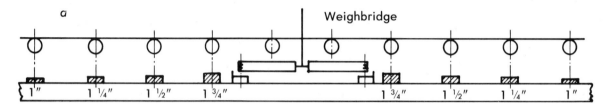

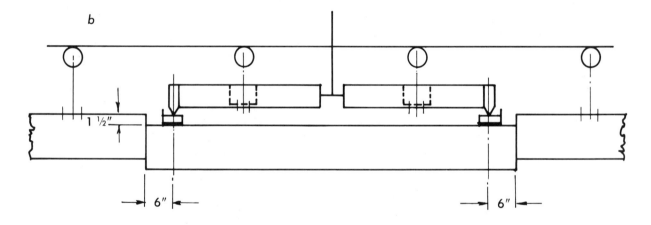

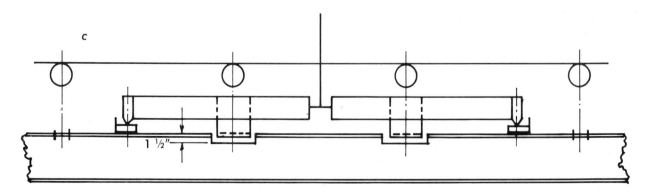

Figure 2.12 Shimming of idlers.

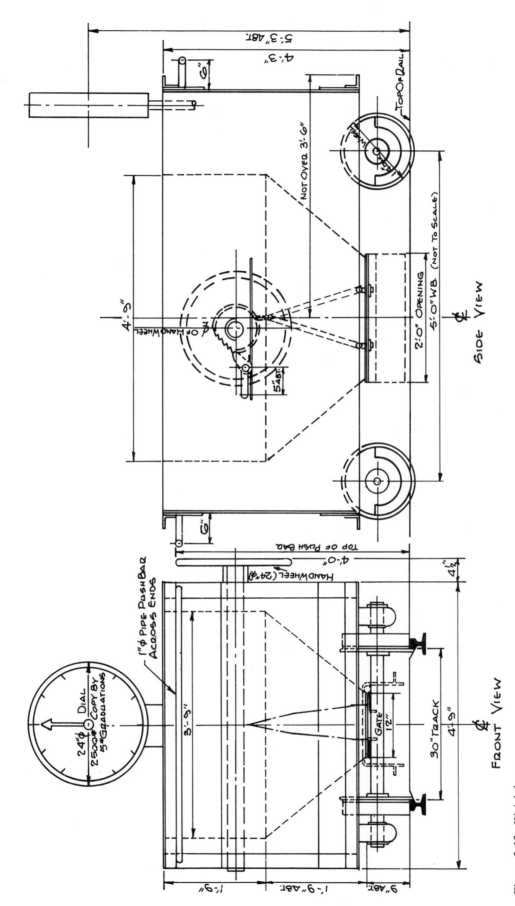

Figure 2.13 Weigh larry.

The hand-operated larries are propelled by means of a coil chain running over a pocket wheel, keyed to one of the truck axles. The 2000-lb larry has a countershaft with cut spur gearing to aid propulsion.

Motor-driven larries are provided with pocket wheel and chain for emergency hand propulsion.

The floor- and platform-operated, motor-driven larries operate at a speed of 125 fpm, and are driven by high-torque squirrel-cage motors with pushbutton control.

The cage-operated, motor-driven larries have a speed of 3000 fpm, and are driven by slip-ring motors with spring handle controller to fill hoppers at slow speed.

Tracks for hand-propelled larries consist of steel channels stiffened with angles. The motor-propelled larries run on Tee rails supported on wide flange beams. All larries are provided with overflow protection and safety devices to prevent hopper from falling in case of failure of scale rods, and with overrun protection.

Interlocking gate mechanisms can be provided where the larry receives material from a number of gates. The interlocking mechanism should be operated from the larry to eliminate individual operating chains and levers on each gate, and thus to prevent the gates from being opened unless the larry is directly beneath the gate.

2.14 SHUTTLE BELT CONVEYOR

This type of conveyor is mounted on a structural frame supported by single-flange rollers rolling on a Tee rail track. Normally, the belt is fed by a chute or conveyor.

While the shuttle belt does not come under the direct influence of a tripper, many times it can replace a tripper and simplify the design. The carrying belt is reversible, and also the supporting structural frame.

These belts can be made any width. Up to 25 ft long, they can be moved by hand. Commercially, they do not go over 75 ft in length, but can be designed longer, if necessary. When over 25 ft long, a separate small motor drive is arranged to connect to one axle shaft which, in turn, allows the shuttle belt to travel on the Tee rail track between 30 fpm and 50 fpm (figures 2.14 and 2.15). A shuttle belt can be arranged on a revolving frame and track.

Figure 2.16 shows a belt conveyor delivering ore to a storage pile by a combination wing-type movable tripper and shuttle belt. This arrangement allows for considerable additional storage of material.

2.15 PINION–SWIVEL ARRANGEMENT FOR FOOT OF BELT CONVEYOR

Figure 2.17 shows a 30-in. belt conveyor pinion arrangement that can be used for an inclined or horizontal belt conveyor that was intended to swivel in the path of a circle. The arrangement can be applied to any width belt.

2.16 TROUGHING AND RETURN IDLERS

The spacing of the troughed belt idlers for handling various materials can be obtained from manufacturers' catalogs (refer also to table 2-9). These values are the results of combined efforts by the belt and equipment manufacturers. Return idlers usually are spaced about 10 ft apart. Swivel belt-training idlers are spaced beginning 10 ft from the head and 10 ft from the tail pulleys, and then about 100 ft apart. Swivel flat-roll return idlers are necessary on belts over 100 ft on centers, to keep the return belt from running off the return idlers. These can be spaced about the same as the swivel troughing idlers. Some self-aligning effect of the carrying run can be obtained by installing the carrying idlers with a forward tilt, in the direction of belt travel of not more than 2°. This has, however, resulted in increased wear on the rubber cover of the belt and the belt edges, as well as the troughing rolls. If the belt is reversible, the tilting idler method cannot be used.

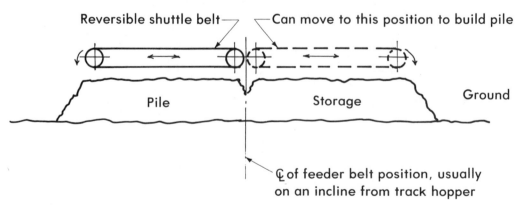

Figure 2.14. Shuttle belt conveyor diagram

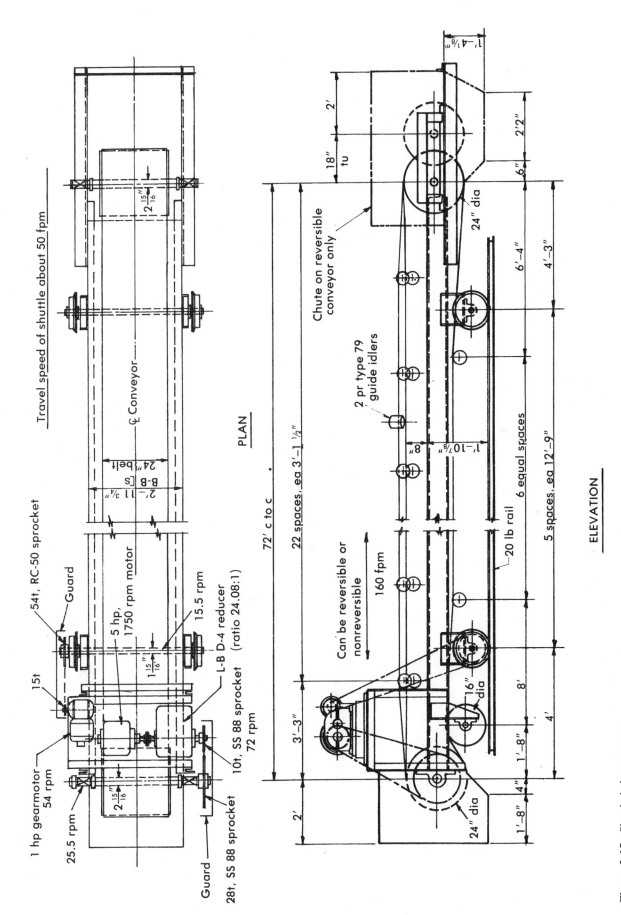

Figure 2.15 Shuttle belt conveyor.

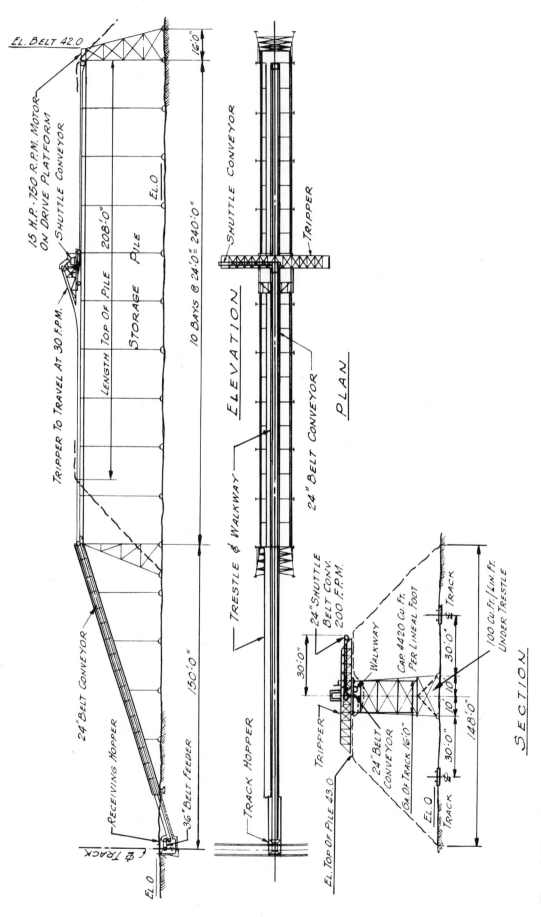

Figure 2.16 Outdoor ore storage system using a traveling wing tripper and shuttle belt.

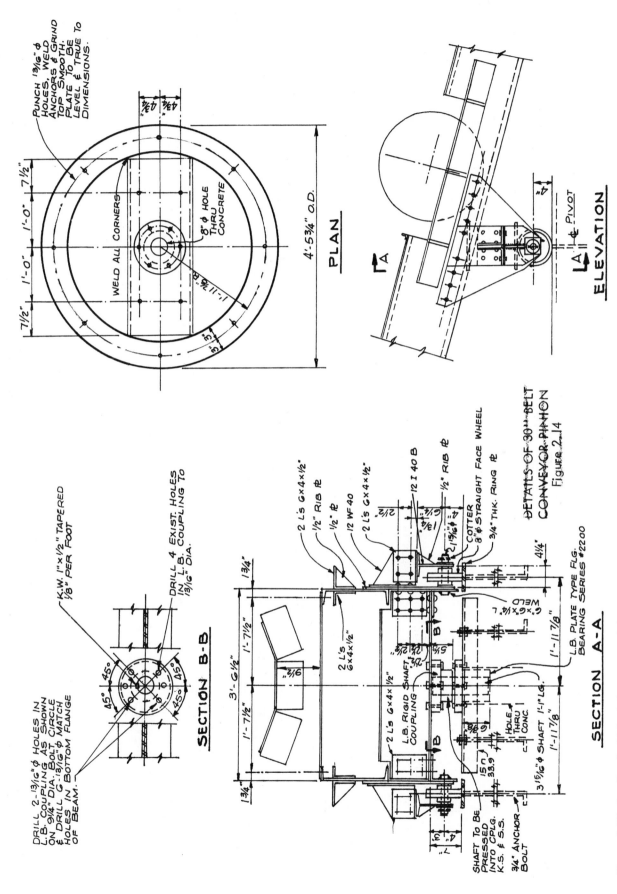

Figure 2.17 Details of a 30-in. belt-conveyor pinion.

PLAN

ELEVATION

Ҁ PIVOT

SECTION B-B

SECTION A-A

DETAILS OF 30" BELT
CONVEYOR PINION
Figure 2.14

PUNCH 13/16" φ HOLES, WELD ANCHORS & GRIND TOP SMOOTH. PLATE TO BE LEVEL & TRUE TO DIMENSIONS.

WELD ALL CORNERS

8" φ HOLE THRU CONCRETE

4'-5¾" O.D.

DRILL 2-13/16" φ HOLES IN L.B. COUPLING AS SHOWN ON 9¼" DIA. BOLT CIRCLE & DRILL 6-13/16" MATCH HOLES IN BOTTOM FLANGE OF BEAM.

K.W. 1"×½" TAPERED ⅛" PER FOOT

DRILL 4 EXIST. HOLES IN L.B. COUPLING TO 13/16" DIA.

2 L's 6×4×½"
½" RIB ℟
½" ℟
12 WF 40
2 L's 6×4×½"
12 I 40 B
½" RIB ℟
COTTER
8" φ STRAIGHT FACE WHEEL
¾" THK. RING ℟
6"×6"×⅜" L
WELD
L.B. PLATE TYPE FLG. BEARING SERIES #2200

2 L's 6×4×½"
L.B. RIGID SHAFT COUPLING
HOLE THRU CONC.
3¹⁵/₁₆" φ SHAFT 1'-1" LG.
15 ∏ J
33.9
SHAFT TO BE PRESSED INTO CPLG. K.S. & S.S.
¾" ANCHOR BOLT

Usage has shown that the use of three-roll idlers results in less maintenance. It is practically standard. Some old installations may have five-roll idlers. In some installations, flat troughing idlers with flared ends have been used. They worked well on runs up to 350 feet. The belt, however, showed extra wear, because the flared ends turned at a greater speed than the smaller center section.

Spacing of idlers is important as it influences greatly the life of the belt. Any of the manufacturers' catalogs show the proper type and spacing of idlers. The data usually is based on the weight of the material per cubic foot, size of lumps, and belt width. Most manufacturers can furnish about the same type of belt idler, some varying in details, such as the construction of the steel or cast iron shell (roller), supports, and method of lubrication of antifriction roller or ball bearings.

Idler rolls are made of rolled steel, seamless pipe, cast iron, and rubber-covered to absorb shock of load onto a belt under a hopper or bin.

Training swivel troughing and return idlers are used on long belts to keep belt centered on idlers. Any of the manufacturers' catalogs will give data as to where to locate these units.

2.17 SUSPENDED IDLERS

2.17.1 Catenary Type

The catenary type idlers require little maintenance and are easily replaceable. See figure 2.18 for a project furnished by Hewitt-Robins.

The idlers are made in belt widths of 18–48 in. In general, they are flexible, the main support being a steel wire cable, urethane- or neoprene-covered. The ends of the cable fit into tapered roller or ball bearings for 18- to 36-in. idlers or tapered or double-row ball bearings for over 36 in. Bearings are sealed and lubricated for life at the factory and, under normal conditions, a troughing or return idler may last 5–8 yr, depending on installation, whether outdoors and not covered, or indoors operating under dusty and abrasive conditions.

2.17.2 Garland Type

The suspended garland type idlers are often used on return runs and for troughing along the conveyor. Training idlers are not required. Generally, there are three

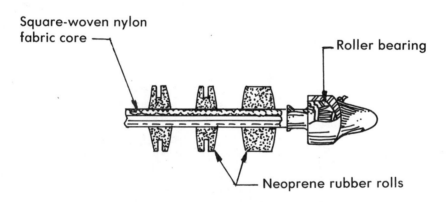

SECTION SHOWING ROLL CONSTRUCTION

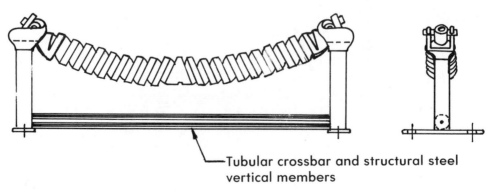

Figure 2.18 Catenary-type suspended idler.

rolls per idler, but five-roll idlers are often used at loading points. These idlers are easily replaceable. The (garland) idlers are available in a three-roll assembly for 35° angle, and a five-roll assembly for 55° angle. Some manufacturers make a two-roll return idler assembly for 10° and 15° slopes. The three-roll, and especially the five-roll, idlers are used for conveying heavy ores and at loading and transfer points.

Suspended (garland) idlers are made in belt widths of 18–96 in., with either 6- or 7-in. rolls.

Many engineers use the terms catenary and garland interchangeably. Both are suspended types but, strictly speaking, the garland consists of two, three, or five rolls.

2.18 BELT CLEANERS

Belt cleaners are necessary where material conveyed is, at times, wet or damp, tacky or sticky. They are not always completely satisfactory, but they do help. Many fine particles, sometimes abrasive enough to cause considerable wear on the bed pulley of a vertical gravity takeup, by building up a crust on the return idler, will eventually harm the rubber cover of the belt. The cleaners prevent return of material on the return run. Good maintenance is very important for efficient operation, and for the life of the belt.

A high-speed bristle brush, with a supporting shaft rotating at approximately 450 rpm, resulting in a longitudinal surface speed of 1200 fpm, gives a satisfactory performance if the material is of a dry nature and does not stick to rubber. With slightly moist or damp material, the brush will fill up with fines and become ineffective. The brush arrangement is usually installed directly under the head discharge pulley, and is operated by a chain drive from the head shaft.

A beater-type brush, operating at 700 fpm, often will do a good job of dislodging caked or frozen material from a belt. It consists of three steel discs with reinforced rubber strips. It has been successful on outdoor installations in stone quarries, where the stone has some overburden with it.

Another arrangement is based on the operation of a fast-travelling scraper flight under the return run of the belt. These flights are equipped with a double strand of chain travelling about three times the speed of the return surface. They are equipped with a rubber scraper, as shown on figure 2.19. This unit can be driven from the head shaft of the main belt, or separately.

This unit will keep the return run of the main belt fairly clean of damp or wet materials. The material drops into a collecting hopper below, and the whole arrangement can be enclosed. This unit will take up some space, and is more expensive than some of the others.

A simple way to clean a belt is by the use of a counterweighted, rubber-lined, stationary unit, without any moving parts. The rubber scraper blade can hug the rubber cover of the belt and is placed directly below the head pulley. In extreme cases, another blade is placed about 12 in. away. What fine particles the first blade misses, the second blade usually gets and, in most cases, the return run of the belt is clear of material. This arrangement does a good job on moist, damp, or wet ma-

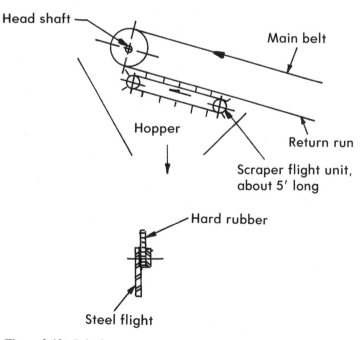

Figure 2.19 Belt cleaner.

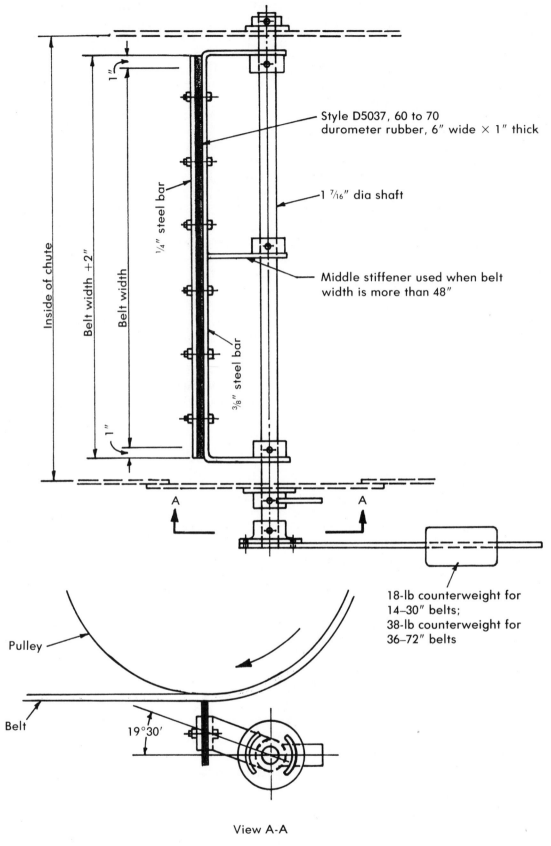

Figure 2.20 Arrangement of a belt scraper.

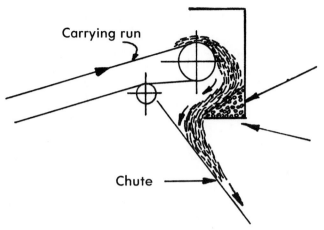

Carrying run

Chute

Stone pocket forms part of steel supporting framework

First, this shelf fills up with material until it reaches its angle of repose; then acts as a cushion to soften the blow of discharge; finally, material falls with less force to a chute

Figure 2.21 Transfer from belt to belt.

terials. These scrapers are inexpensive compared with the bristle and beater arrangement and scraper conveyor unit. It is shown on figure 2.20.

2.19 TRANSFER OF MATERIAL FROM BELT TO BELT

The conventional transfer method is to discharge material by steel chute, sometimes lined with removable steel or hard rubber liners. A fast-moving short-centers belt feeder works well, and gives uniform flow. The feeder should be of the same width as the belt conveyor and run parallel with it but at a speed at least 50% greater. With heavy abrasive lumps, this belt will need frequent replacements, but it will be less costly than replacing chutes, and it will save in height at the transfer point.

Many belt conveyors of high capacity, particularly in quarry operations, operate at 400–600 fpm, handling rock with large pieces discharged directly into a stone pocket before the rock falls to a chute, as shown on figure 2.21.

Belt conveyors have been built that are over 2 miles long. Steel-cable reinforced belts have made these runs possible. Specially designed driving machinery and special belting is required for these long runs.

2.20 TAKEUPS AND BACKSTOPS

2.20.1 Takeups

For normal capacity runs up to, say, 150-ft centers, the hand-adjusted protected screw takeup is generally used. The adjusting screw is protected under an inverted steel angle. For conveyors of short centers and small capac-

ity, light steel frame hand-adjusted, babbitted bearing takeups can be used. They can be furnished with babbitt or antifriction roller bearings.

For lengths of over 100 ft, a floating gravity takeup can be used, if space permits its installation. This type of takeup is located, if possible, near the head pulley and is designed for vertical operation. It automatically maintains constant and controlled belt tension. The takeup is generally placed at a point between the head and tail pulley (refer to table 2.10).

On very-large-capacity high-speed conveyors of long centers, and located near the ground, with belts 48, 54, or 60 in. wide, horizontal roller carriages are provided. These consist of a steel frame, roller-bearing equipped, very ruggedly made, usually with 4 ft 8½ in. gauge supporting rollers.

Details of various takeups are shown on figure 2.22.

Table 2.10 Recommended Takeup Travel in Percent of Center Distance

	100% Rated Tension	*75% or Less Rated Tension*
Fastened Splices		
Screw takeup	1½%	1%
Auto takeup	2%	1½%
Vulcanized Splices		
Screw takeup	4%	3%
Auto takeup	2½% + 2 ft	

Source: Courtesy The Goodyear Tire & Rubber Company
Note: Only short endless feeder belts and the like should be vulcanized on conveyors with screw takeups.

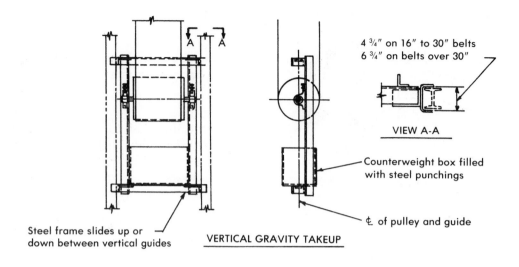

4 ³⁄₄" on 16" to 30" belts
6 ³⁄₄" on belts over 30"

VIEW A-A

Counterweight box filled
with steel punchings

℄ of pulley and guide

Steel frame slides up or
down between vertical guides

VERTICAL GRAVITY TAKEUP

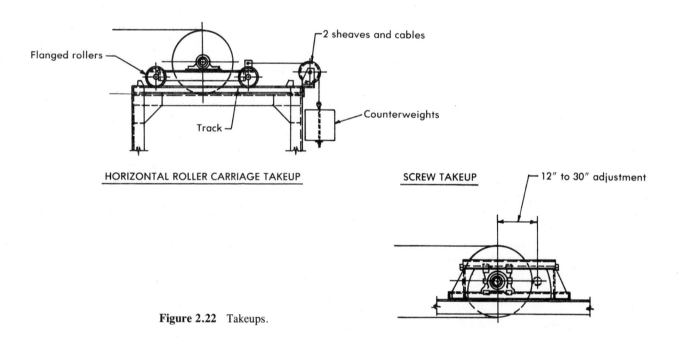

Flanged rollers

2 sheaves and cables

Counterweights

Track

HORIZONTAL ROLLER CARRIAGE TAKEUP

SCREW TAKEUP

12" to 30" adjustment

Figure 2.22 Takeups.

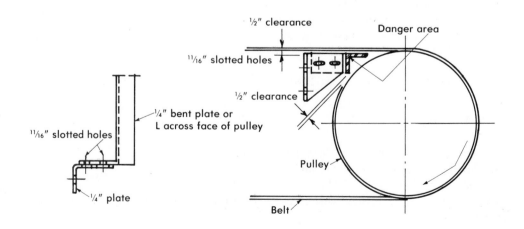

½" clearance

Danger area

¹¹⁄₁₆" slotted holes

½" clearance

¹⁄₄" bent plate or
L across face of pulley

¹¹⁄₁₆" slotted holes

Pulley

¹⁄₄" plate

Belt

Figure 2.23 Nip guard.

2.20.2 Backstops

Brakes are recommended for inclined belt conveyors. Band brakes have generally been replaced by Marland, Formsprag, Stephens-Adamson, and other one-way clutches, tied into the head frame steel. These units are made up of a series of sprags or rollers, free-running on a shaft in one direction and selflocking on reversal. For a full treatment of brakes and backstops, refer to "Belt Conveyors for Bulk Materials" by CEMA.

2.21 COVER

In handling material that is likely to put dust in the air, it is wise to put a cover over the belt, as shown on figure 2.1. The cover should be hinged on one side and latched on the other, with the latch on the operator's side. The arrangement can be applied to any width belt.

2.22 SAFETY PROTECTION AT PULLEYS

The scheme shown on figure 2.23 is to help prevent a maintenance person from getting a hand or arm where a belt starts curving around a pulley. Many people, careless or otherwise, have been caught in such a position; some have had fingers crushed or an arm flattened or taken off. The insurance laws now require protection at this location, and this simple arrangement, called Nip Guards, has proved very effective. It can be installed in front of any pulley. In the case of a magnetic pulley, the material used should be copper or some other nonmagnetic material.

2.23 TENSION FOR VARIOUS LAYOUTS

In addition to calculation of the effective belt tension, T_e, a designer must consider the belt tension values that occur at other points of the conveyor's belt path.

Figures 2.24 through 2.32 illustrate various possible belt conveyor layouts and profiles, and the appropriate tension formulas. They are reproduced from CEMA's publication, "Belt Conveyors for Bulk Materials," with permission. Some of these examples are more commonly applied than others; the order of presentation is not intended to imply any preference of design. Many of these diagrams illustrate the takeup (TU) in two locations only to show alternatives. Two automatic takeups cannot function properly on the same conveyor.

Nondriving pulley frictions have been omitted. There is no change in belt tension at takeups or at snub pulleys. Lowering regenerative loads are indicated by minus ($-$) signs. All the nomenclature is defined in paragraph 2.25.

$$T_e = T_1 - T_2$$
$$T_2 = C_w \times T_e \quad \text{or} \quad T_2 = T_t + T_b - T_{yr}$$
Use the larger value of T_2

$$T_t = T_0 \quad \text{or}$$
$$T_t = T_2 - T_b + T_{yr}$$
Use the larger value of T_t
$$T_t = T_{min} \quad T_1 = T_{max}$$

$$T_{cx} = T_t + T_{wcx} + T_{fcx}$$
$$T_{rx} = T_t + T_{wrx} - T_{frx}$$

Takeup on return run or at tail pulley

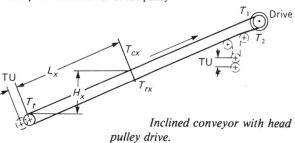

Inclined conveyor with head pulley drive.

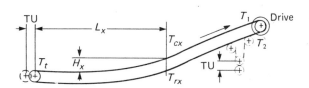

Horizontal belt conveyor with concave vertical curve, and head pulley drive.

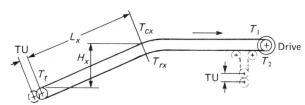

Horizontal belt conveyor with convex vertical curve , and head pulley drive.

Figure 2.24 Head pulley drive—horizontal or elevating. *(Courtesy CEMA)*

$T_e = T_1 - T_2$
$T_2 = C_w \times T_e$ or $T_2 = T_0 - T_e$
Use the larger value of T_2

$T_1 = T_e + T_2$
$T_t = T_2 + T_b + T_{yr}$
$T_{max} = T_t$ or T_1
$T_{min} = T_t$ or T_1
$T_{cx} = T_t - T_{wcx} + T_{fcx}$
$T_{rx} = T_t - T_{wrx} - T_{frx}$

$T_e = T_1 - T_2$
$T_2 = C_w \times T_e$ or $T_2 = T_0$
Use the larger value of T_2

$T_t = T_{max} = T_1 + T_b + T_{yr}$
$T_{min} = T_2$
$T_{cx} = T_t - T_{wcx} + T_{fcx}$
$T_{rx} = T_t - T_{wrx} - T_{frx}$

Takeup on return run not recommended to
avoid driving through the takeup

Takeup on return run or at tail pulley

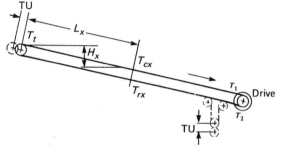

*Declined belt conveyor with head
pulley drive. Lowering without
regenerative load.*

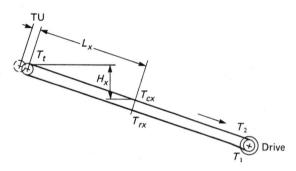

*Declined belt conveyor with head
pulley drive. Lowering with regenerative
load.*

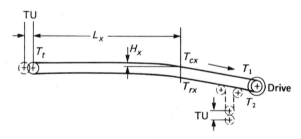

*Conveyor with convex vertical
curve, head pulley drive. Lowering
without regenerative load.*

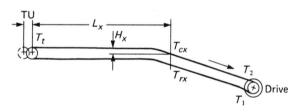

*Conveyor with convex vertical
curve, head pulley drive. Lowering with
regenerative load.*

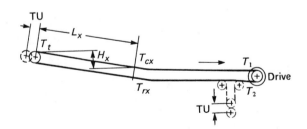

*Conveyor with concave vertical
curve, head pulley drive. Lowering
without regenerative load.*

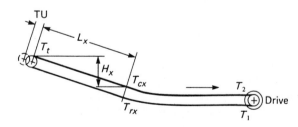

*Conveyor with concave vertical
curve, head pulley drive. Lowering with
regenerative load.*

Figure 2.25 Head pulley drive—lowering without regeneration. *(Courtesy CEMA)*

Figure 2.26 Head pulley drive—lowering with regeneration. *(Courtesy CEMA)*

$T_e = T_1 - T_2$

$T_2 = C_w \times T_e$ or $T_2 = T_0$

Use the larger value of T_2

$T_t = T_2$

$T_{min} = T_2$

$T_{hp} = T_1 - T_{yr} + T_b$

$T_{max} = T_1$

$T_{cx} = T_2 - T_{wcx} + T_{fcx}$

$T_{rx} = T_1 - T_{wrx} - T_{frx}$

$T_e = T_1 - T_2$

$T_2 = C_w \times T_e$ or $T_2 = T_{hp} + T_b + T_{yr} - T_e$

Use the larger value of T_2

$T_{hp} = T_0$ or $T_{hp} = T_1 - T_b - T_{yr}$

Use the larger value of T_{hp}

$T_{hp} = T_{min}$

$T_1 = T_e + T_2 = T_{max}$

$T_{cx} = T_2 - T_{wcx} + T_{fcx}$

$T_{rx} = T_1 - T_{wrx} - T_{frx}$

Takeup on return run not recommended. For arrangement which is preferred to above, drive is on return run, Figure 6.15A, page 114.

Takeup on return run not recommended to avoid driving through the takeup

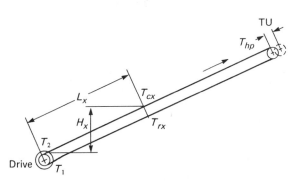

Inclined conveyor with tail pulley drive.

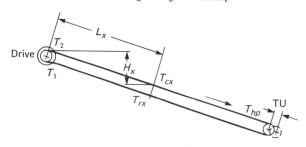

Declined belt conveyor with tail pulley drive. Lowering without regenerative load.

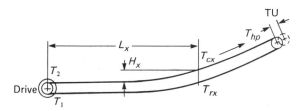

Horizontal belt conveyor with concave vertical curve, and tail pulley drive.

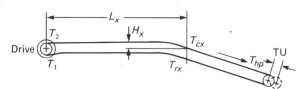

Conveyor with convex vertical curve, tail pulley drive. Lowering without regenerative load.

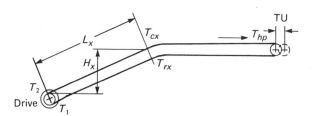

Horizontal belt conveyor with convex vertical curve, and tail pulley drive.

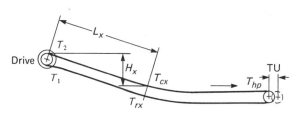

Conveyor with concave vertical curve, tail pulley drive. Lowering without regenerative load.

Figure 2.27 Tail pulley drive—horizontal or elevating. *(Courtesy CEMA)*

Figure 2.28 Tail pulley drive—lowering without regeneration. *(Courtesy CEMA)*

$T_e = T_1 - T_2$

$T_2 = C_w \times T_e$ or $T_2 = T_0 + T_b + T_{yr}$

Use the larger value of T_2

$T_{hp} = T_2 - T_b - T_{yr}$ or $T_{hp} = T_0$

$T_{hp} = T_{min}$

$T_1 = T_{max} = T_e + T_2$

$T_{cx} = T_1 - T_{wcx} + T_{fcx}$

$T_{rx} = T_2 - T_{wrx} - T_{frx}$

$T_e = T_1 - T_2$

$T_2 = C_w \times T_e$ or

$T_2 = T_0 - 0.015\, W_b L_s + W_b H_d$

here

   H_d = lift to the drive pulley

Use the larger value of T_2

$T_t = T_{min}$ and $T_t = T_0$

$T_t = T_2 + 0.015\, W_b L_s - W_b H_d$

Use the larger value of T_t

$T_{hp} = T_t + T_{fcx} + T_{wcx}$ where $L_x = L$

$T_{hp} = T_{max}$

$T_{cx} = T_t + T_{fcx} + T_{wcx}$

$T_{rx} = T_t - T_{frx} + T_{wrx}$

Takeups on return run or at tail pulley.

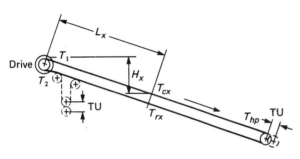

Declined belt conveyor with tail pulley drive. Lowering with regenerative load. "Calculate belt tension required at takeup during acceleration and make takeup adequate to prevent lift-up. See page 126."

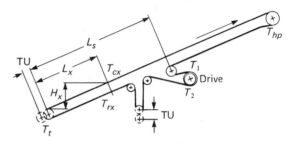

Inclined conveyor with drive on return run.

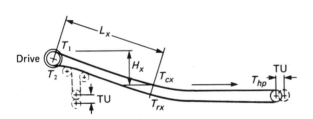

Conveyor with concave vertical curve, tail pulley drive. Lowering with regenerative load.

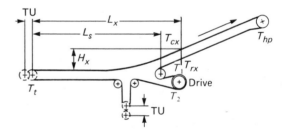

Horizontal belt conveyor with concave vertical curve, and drive on return run.

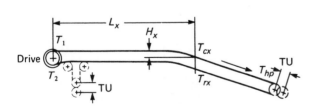

Conveyor with convex vertical curve, tail pulley drive. Lowering with regenerative load.

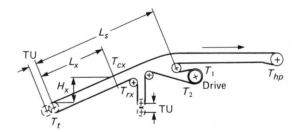

Horizontal belt conveyor with convex vertical curve, and drive on return run.

Figure 2.29   Tail pulley drive—lowering with regeneration. *(Courtesy CEMA)*

Figure 2.30   Drive on return run—horizontal or elevating. *(Courtesy CEMA)*

$T_e = T_1 - T_2$

$T_2 = C_w \times T_e$ or $T_2 = T_0 - T_e$

Use the larger value of T_2

$T_1 = T_e + T_2$

$T_t = T_2 + 0.015 W_b L_s - W_b H_d$

here

H_d = lift to the drive pulley, or $T_t = T_0$

Use the larger value of T_t

$T_{max} = T_t$ or T_1

$T_{min} = T_t$ or T_1

$T_{hp} = T_t + T_{fcx} - T_{wcx}$

here

$L_x = L$

$T_{cx} = T_t + T_{fcx} - T_{wcx}$

$T_{rx} = T_t - T_{frx} - T_{wrx}$

Takeup on return run or at tail pulley

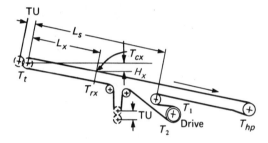

Declined conveyor, with drive on return run. Lowering without regenerative load.

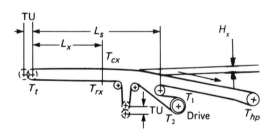

Conveyor with convex vertical curve, drive on return run. Lowering without regenerative load.

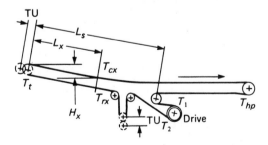

Conveyor with concave vertical curve, drive on return run. Lowering without regenerative load.

Figure 2.31 Drive on return run—lowering without regeneration. *(Courtesy CEMA)*

$T_e = T_1 - T_2$

$T_2 = C_w \times T_e$ or $T_2 = T_0$

Use the larger value of T_2

$T_1 = T_e + T_2$

$T_t = T_{max} = T_1 + 0.015 W_b L_s + W_b H_d$

here

H_d = lift to the drive pulley

$T_{hp} = T_0$

$T_{hp} = T_2 - 0.015 W_b (L - L_s) - W_b (H - H_d)$

Use the larger value of T_{hp}

$T_{cx} = T_t - T_{wcx} + T_{fcx}$

$T_{rx} = T_t - T_{wrx} - T_{frx}$

Takeup on return run or at head pulley

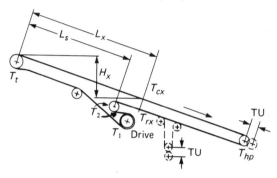

Declined conveyor with drive on return run. Lowering with regenerative load.

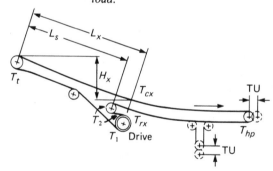

Conveyor with concave vertical curve, drive on return run. Lowering with regenerative load.

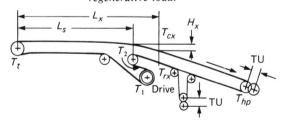

Conveyor with convex vertical curve, drive on return run. Lowering with regenerative load.

Figure 2.32 Drive on return run—lowering with regeneration. *(Courtesy CEMA)*

2.24 RECOMMENDED BELT SPEEDS AND BELT WIDTHS

The values given in table 2.11 have been used in many designs and have been found to be satisfactory, if good judgment or previous experience is followed. Recently, speeds up to 1200 fpm have been reached, and even higher speeds are in the offing at the time of this writing. Belt widths up to 10 ft are being discussed.

Because of wear and training of the belt, bearing capacity of the rollers, values of friction of greater speed, and unloading from the belt to other belts or hoppers, it is wise to consult the belt and conveyor manufacturers before exceeding the limits in the text.

It is important to evaluate the cost of a shutdown of a high-capacity conveyor, especially if carrying a variable load. The use of two conveyors, each operating at a lower capacity, should be considered.

Tables 2.11 and 2.12 cover the same field. Note that both make an allowance in some items for an increase in speed with an increase in belt width. Table 2.11 gives a wider choice of belt width and a more detailed relation between belt width and speed. Table 2.12 represents a later development and agreement in thought among manufacturers, and it omits the use of the narrow belt widths. The tables actually do not vary much.

2.25 DESIGN OF A BELT CONVEYOR

It is necessary to estimate the weight of the belt and idlers before proceeding with the design of the conveyor. Tables 2.13 and 2.14 give close approximations.

The horsepower can be determined (1) by figuring belt tension first, (2) by graphic methods, or (3) by using the formulas for horsepower.

Horsepower and belt tension are related by the formula:

$$hp = \frac{\text{belt tension} \times \text{speed (in ft/min) ``V''}}{33,000}$$

Table 2.11 Maximum Recommended Belt Speed as Determined by Material Handled and Belt Width

Material			Recommended Belt Speed (fpm)[a]												
			Belt Width (in.)												
Characteristics		Material Example	14[c]	16[c]	18	20	24	30	36	42	48	54	60	72	84
Maximum size lumps, sized or unsized[b]	Mildly abrasive	Coal, earth	300	300	400	400	450	500	550	600	600	650	650	650	650
	Very abrasive, not sharp	Bank gravel	300	300	400	400	450	500	550	550	600	600	600	600	600
	Very abrasive, sharp and jagged	Stone, ore	250	250	300	350	400	450	500	500	550	550	550	550	550
Half max lumps, sized or unsized	Mildly abrasive	Coal, earth	300	300	400	400	500	600	650	700	700	700	700	700	700
	Very abrasive	Slag, coke, ore, stone, cullet	300	300	400	400	500	600	650	650	650	650	650	650	650
Flakes		Wood chips, bark, pulp	400	450	450	500	600	700	800	800	800	800	800	800	800
Granular, ⅛″ to ½″ lumps		Grain, coal, cottonseed, sand	400	450	450	500	600	700	800	800	800	800	800	800	800
Fines	Light, fluffy, dry, dusty	Soda ash, pulverized coal	220–250[d]												
	Heavy	Cement, flue dust	250–300[d]												
Fragile, where degradation is harmful		Coke, coal	200–250[d]												
		Soap chips	150–200[d]												

Source: Courtesy FMC

[a]Maximum for belts traveling horizontally on ball- or roller-bearing idlers. For picking belts, speed is usually 50–100 fpm. Belts with discharge plows should not travel faster than 200 fpm. For trippers, such as the type shown in Catalog 1050, recommended speed is 300–400 fpm. Trippers for higher speed applications can be furnished. A speed of at least 300 fpm should be maintained for proper discharge when using 35° and 45° idlers, and also for materials tending to cling to belt.

[b]*Unsized* means a uniform mixture of material in which not more than 10% is lumps ranging from maximum size to ½ maximum size, at least 15% is fines or lumps smaller than ¹⁄₁₀ maximum, and remaining 75% is lumps of any size smaller than ½ maximum. *Sized* means a uniform mixture in which not more than 20% is lumps ranging from maximum size to ½ maximum size, and remaining 80% is lumps no larger than ½ maximum size and no smaller than ¹⁄₁₀ maximum size.

[c]Minimum speed for 35° and 45° idlers is 300 fpm.

[d]Available in 20° only.

Table 2.12 Recommended Maximum Belt Speeds and Widths Based on Material Handled

Material	Belt Speeds (fpm)	Belt Width (in.)
Grain or other free-flowing, nonabrasive material	500	18
	700	24–30
	800	36–42
	1000	48–96
Coal, damp clay, soft ores, overburden and earth, fine-crushed stone	400	18
	600	24–36
	800	42–60
	1000	72–96
Heavy, hard, sharp-edged ore, coarse-crushed stone	350	18
	500	24–36
	600	Over 36
Foundry sand, prepared or damp; shakeout sand with small cores, with or without small castings (not hot enough to harm belting)	350	Any width
Prepared foundry sand and similar damp (or dry abrasive) materials discharged from belt by rubber-edged plows	200	Any width
Nonabrasive materials discharged from belt by means of plows	200, except for wood pulp, where 300 to 400 is preferable	Any width
Feeder belts, flat or troughed, for feeding fine, nonabrasive, or mildly abrasive materials from hoppers and bins	50–100	Any width

Source: Courtesy CEMA

Table 2.13 Estimated Average Belt Weight, Multiple- and Reduced-Ply Belts (lb/ft)

Belt Width (b) (in.)	Material Carried (lb/ft³)		
	30–74	75–129	130–200
18	3.5	4	4.5
24	4.5	5.5	6
30	6	7	8
36	9	10	12
42	11	12	14
48	14	15	17
54	16	17	19
60	18	20	22
72	21	24	26
84	25	30	33
96	30	35	38

Source: Courtesy CEMA
Notes: For steel-cable belts, increase above value by 50%.
Actual belt weights vary with different constructions, manufacturers, cover gauges, etc. Use above values for estimating. Obtain actual values from belt manufacturers whenever possible.

Table 2.14 Weight per Linear Foot of Belt and Revolving Idler Parts (lb)

Belt Width, b (in.)	Material Weight (lb/ft³)			
Inches	50	100	150	200
18	12	14	17	17
24	16	19	23	23
30	20	24	29	29
36	28	35	41	48
42	34	42	49	59
48	41	51	69	77
54	48	58	78	89
60	60	70	87	99
72	74	83	113	130
84	102	127	149	165
96	117	143	181	181

Source: Courtesy CEMA
Note: This table of weights is representative of average weights of revolving idler parts. Where actual weights are known, they should be used in the graphical solution.

2.25.1 Data (see figure 2.33)

Height of lift in feet: $H = 46$ ft center to center pulleys
Length of conveyor: $L = 400$ ft center to center pulleys
Material to be handled: crushed limestone
Weight of material: $W = 90$ lb/ft^3
Capacity: 300 tph $= 600,000/90 = 6667$ ft^3/hr
Lumps: 2 in. maximum
Maximum surcharge angle: 30° (by experiment)
Operation: 8 hr/day
Conveyor: housed
Temperature: above freezing
Degradation: not important
Tan angle: $46/400 = 0.115$
Angle: 6° 35′ ($< 18°$ which is maximum recommended for stone)
Recommended maximum speed (table 2.12): 500 fpm
Recommended belt width (table 2.12): 24–36 in.

A 24-in. belt was chosen and 20° idlers used to permit thicker belts for abrasion resistance. Based on idler spacing in table 2.9, letters A to E, refer to the type of idler. Where large or heavy lumps are involved, it is best to use closer spacing. Values have been adjusted according to CEMA data. The weight of the material carried is not included in the weights given in table 2.9. Refer to paragraph 2.30 for idler selection data.

Capacity (tables 2.7 and 2.8):

$$0.174 + \left[0.0906 \left(\frac{18.931}{12} \right)^2 \right]$$

$$= 0.174 + 0.225$$

$$= 0.399 \text{ ft}^3/\text{ft}$$

Required speed:

$$\frac{6667}{0.399 \times 60} = 278 \text{ fpm (say, 280 fpm)}$$

2.25.2 Belt Tension Method (Using the CEMA Formulas for Belt Tension)

a. T_x = carrying idler friction in lb
    $= L \times K_x K_t$

where L = length in feet and K_t is the ambient temperature correction factor (see below). A_1 values for figuring K_x are listed below. For return idlers, $K_x = 0.015$. Values of A_1 for carrying idlers are:

7-in. dia idler rolls, CEMA E7 = 2.4
6-in. dia idler rolls, CEMA E6 = 2.8
6-in. dia idler rolls, CEMA C6, D6 = 1.5
5-in. dia idler rolls, CEMA A5, B5, C5, D5 = 1.8
4-in. dia idler rolls, CEMA A4, B4, C4 = 2.3

For regenerative declined conveyors, $A_1 = 0.0$. For 5-in. idlers, type B5 (table 2.9), $A_1 = 1.8$ (see above).

S_i = spacing of idlers in ft
  = 4.0 ft (table 2.9)
W_b = weight of belt in lb/ft of belt length
  = 5.5 lb for 24-in. belt (table 2.13)
W_m = weight of material in lb/ft of belt length
  $= \dfrac{300 \text{ tons} \times 2000}{60 \text{ min} \times 280 \text{ fpm}} = 35.7 \text{ lb/ft}$
K_x = lb tension/ft of belt
  $= 0.00068 (W_b + W_m) + (A_1/S_i)$
  $= 0.00068 (5.5 + 35.7) + (1.8/4.0)$
  $= 0.478$

b. K_t (the ambient temperature correction factor) can be assumed to be 1.0 when operating in an ambient temperature of 20°F or above. For operating temperatures below that, it is best to consult the equipment manufacturer. Thus:

$$T_x = 400 \times 0.478 \times 1.0 = 191 \text{ lb}$$

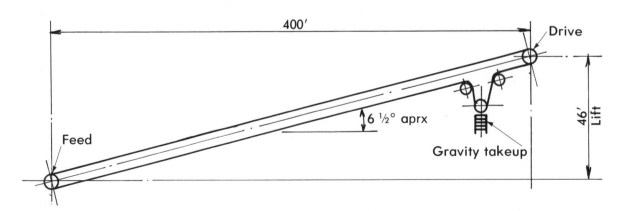

Figure 2.33 Design diagram of a belt conveyor. (The actual design dimensions were changed to agree with those of an illustration in Link-Belt Catalog 1000.)

2.25.3 Belt Flexure

a. Idlers: value of K_y. Given a belt angle of 6°35′, a 400-ft length, and $W_b + W_m = 41.2$, $K_y = 0.029$ (table 2.15). According to the footnote to table 2.15, the tabular spacing is 4.5 ft. The corrected K_y for 4.0-ft spacing is

$$\frac{0.0259 + 0.0278}{2}$$

$$= 0.02685, \text{ say } 0.027 \text{ (table 2.16)}$$

T_{yb}, the tension in lb resulting from the resistance of the belt to flexure as it rides over both the carrying and return idlers,

$$= T_{yc} + T_{yr}$$
$$= (L \times K_y \times W_b \times K_t)$$
$$\quad + (L \times 0.015 \times W_b \times K_t)$$
$$= (L \times W_b \times K_t)(K_y + 0.15)$$
$$= (400 \times 5.5 \times 1.0)(0.027 + 0.015)$$
$$= 92 \text{ lb} \qquad \textbf{Formula (2)}$$

where

L = 400 ft (length)
K_t = 1.0 (paragraph 2.25.2)
K_y = 0.027 (see above)
W_b = 5.5 lb (paragraph 2.25.2)
W_m = 35.7 lb

T_{ym}, the tension resulting from the resistance of the material to flexure as it rides with the belt over the carrying idlers,

$$= L \times K_y \times W_m$$
$$= 400 \times 0.027 \times 35.7$$
$$= 386 \text{ lb} \qquad \textbf{Formula (3)}$$

Refer to the note at table 2.15.

b. Material lift: T_m

$$= W_m \times (\pm H)$$
$$= 35.7 \times 46$$
$$= 1642 \text{ lb} \qquad \textbf{Formula (4)}$$

Table 2.15 Factor K_y Values

Conveyor Length (ft)	$W_b + W_m$ (lb/ft)	Percent Slope						
		0	3	6	9	12	24	33
		Approximate Degrees						
		0	2	3.5	5	7	14	18
250	20	0.035	0.035	0.034	0.031	0.031	0.031	0.031
	50	0.035	0.034	0.033	0.032	0.031	0.028	0.027
	75	0.035	0.034	0.032	0.032	0.030	0.027	0.025
	100	0.035	0.033	0.032	0.031	0.030	0.026	0.023
	150	0.035	0.035	0.034	0.033	0.031	0.025	0.021
	200	0.035	0.035	0.035	0.035	0.032	0.024	0.018
	250	0.035	0.035	0.035	0.035	0.033	0.021	0.018
	300	0.035	0.035	0.035	0.035	0.032	0.019	0.018
400	20	0.035	0.034	0.032	0.030	0.030	0.030	0.030
	50	0.035	0.033	0.031	0.029	0.029	0.026	0.025
	75	0.034	0.033	0.030	0.029	0.028	0.024	0.021
	100	0.034	0.032	0.030	0.028	0.028	0.022	0.019
	150	0.035	0.034	0.031	0.028	0.027	0.019	0.016
	200	0.035	0.035	0.033	0.030	0.027	0.016	0.014
	250	0.035	0.035	0.034	0.030	0.026	0.017	0.016
	300	0.035	0.035	0.034	0.029	0.024	0.018	0.018

Table 2.15 (*Continued*)

Conveyor Length (ft)	$W_b + W_m$ (lb/ft)	Percent Slope						
		0	3	6	9	12	24	33
		Approximate Degrees						
		0	2	3.5	5	7	14	18
500	20	0.035	0.033	0.031	0.030	0.030	0.030	0.030
	50	0.034	0.032	0.030	0.028	0.028	0.024	0.023
	75	0.033	0.032	0.029	0.027	0.027	0.021	0.019
	100	0.033	0.031	0.029	0.028	0.026	0.019	0.016
	150	0.035	0.033	0.030	0.027	0.024	0.016	0.016
	200	0.035	0.035	0.030	0.027	0.023	0.016	0.016
	250	0.035	0.035	0.030	0.025	0.021	0.016	0.015
	300	0.035	0.035	0.029	0.024	0.019	0.018	0.018
600	20	0.035	0.032	0.030	0.029	0.029	0.029	0.029
	50	0.033	0.030	0.029	0.027	0.026	0.023	0.021
	75	0.032	0.030	0.028	0.026	0.024	0.020	0.016
	100	0.032	0.030	0.027	0.025	0.022	0.016	0.016
	150	0.035	0.031	0.026	0.024	0.019	0.016	0.016
	200	0.035	0.031	0.026	0.021	0.017	0.016	0.016
	250	0.035	0.031	0.024	0.020	0.017	0.016	0.016
	300	0.035	0.031	0.023	0.018	0.018	0.018	0.018
800	20	0.035	0.031	0.030	0.029	0.029	0.029	0.029
	50	0.032	0.029	0.028	0.026	0.025	0.021	0.018
	75	0.031	0.029	0.026	0.024	0.022	0.016	0.016
	100	0.031	0.028	0.025	0.022	0.020	0.016	0.016
	150	0.034	0.028	0.023	0.019	0.017	0.016	0.016
	200	0.035	0.027	0.021	0.016	0.016	0.016	0.016
	250	0.035	0.026	0.020	0.017	0.016	0.016	0.016
	300	0.035	0.025	0.018	0.018	0.018	0.018	0.018

Source: Courtesy CEMA

Notes: Tables 2.15 and 2.16 apply to conveyors 800-ft long and less, having a single slope, less than 3% sag of belt between idlers, and average belt tension under 16,000 lb. CEMA's "Belt Conveyors for Bulk Materials" (Second Edition, 1979) has data for requirements beyond those covered here.

Idler spacing: The above values of K_y are based on the following idler spacing (for other spacing, see table 2.16):

$(W_b + W_m)$, lb/ft:	less than 49	50–99	100–149	above 150
S_i, ft:	4.5	4.0	3.5	3.0

Table 2.16 Corrected Factor K_y Values When Other Than Tabular Carrying Idler Spacings Are Used

$W_b + W_m$ (lb/ft)	S_i (ft)	Reference Values of K_y for Interpolation									
		0.016	0.018	0.02	0.022	0.024	0.026	0.028	0.030	0.032	0.034
	3.0	0.016	0.016	0.016	0.0168	0.0183	0.0197	0.0212	0.0227	0.0242	0.0257
	3.5	0.016	0.0160	0.0169	0.0189	0.0207	0.0224	0.0241	0.0257	0.0274	0.0291
50	**4.0**	**0.016**	**0.0165**	**0.0182**	**0.0204**	**0.0223**	**0.0241**	**0.0259**	**0.0278**	**0.0297**	**0.0316**
	4.5	0.016	0.018	0.02	0.022	0.024	0.026	0.028	0.030	0.032	0.034
	5.0	0.0174	0.0195	0.0213	0.0236	0.0254	0.0273	0.0291	0.031	0.0329	0.0348
	3.0	0.016	0.0165	0.0185	0.0205	0.0222	0.024	0.0262	0.0281	0.030	0.0321
	3.5	**0.016**	**0.018**	**0.02**	**0.022**	**0.024**	**0.026**	**0.028**	**0.030**	**0.032**	**0.034**
100	4.0	0.0175	0.0193	0.0214	0.0235	0.0253	0.0272	0.0297	0.0316	0.0335	0.035
	4.5	0.0184	0.021	0.0230	0.0253	0.027	0.029	0.0315	0.0335	0.035	0.035
	5.0	0.0203	0.0225	0.0249	0.027	0.0286	0.0306	0.033	0.035	0.035	0.035
	3.0	**0.016**	**0.0164**	**0.0186**	**0.0205**	**0.0228**	**0.0246**	**0.0267**	**0.0285**	**0.0307**	**0.0329**
	3.5	0.016	0.018	0.02	0.022	0.024	0.026	0.028	0.030	0.032	0.034
150	4.0	0.0175	0.0197	0.0213	0.0234	0.0253	0.0277	0.0295	0.0312	0.033	0.035
	4.5	0.0188	0.0213	0.0232	0.0253	0.0273	0.0295	0.0314	0.033	0.0346	0.035
	5.0	0.0201	0.0228	0.0250	0.0271	0.0296	0.0316	0.0334	0.035	0.035	0.035
	3.0	**0.016**	**0.018**	**0.02**	**0.022**	**0.024**	**0.026**	**0.028**	**0.030**	**0.032**	**0.034**
	3.5	0.0172	0.0195	0.0215	0.0235	0.0255	0.0271	0.0289	0.031	0.0333	0.0345
200	4.0	0.0187	0.0213	0.0235	0.0252	0.0267	0.0283	0.0303	0.0325	0.0347	0.035
	4.5	0.0209	0.023	0.0253	0.0274	0.0289	0.0305	0.0323	0.0345	0.035	0.035
	5.0	0.0225	0.0248	0.0272	0.0293	0.0311	0.0328	0.0348	0.035	0.035	0.035
	3.0	**0.016**	**0.018**	**0.02**	**0.022**	**0.024**	**0.026**	**0.028**	**0.030**	**0.032**	**0.034**
	3.5	0.0177	0.0199	0.0216	0.0235	0.0256	0.0278	0.0295	0.031	0.0327	0.0349
250	4.0	0.0192	0.0216	0.0236	0.0256	0.0274	0.0291	0.0305	0.0322	0.0339	0.035
	4.5	0.021	0.0234	0.0253	0.0276	0.0298	0.0317	0.0331	0.0347	0.035	0.035
	5.0	0.0227	0.0252	0.0274	0.0298	0.0319	0.0338	0.035	0.035	0.035	0.035

Source: Courtesy CEMA
Notes: For tension in belt exceeding 16,000 lb, use $K_y = 0.016$.
See also footnote to table 2.15 on idler spacing.

c. Pulley resistance: T_p

T_p, the total of the belt tensions required to rotate each of the pulleys on the conveyor (table 2.17),

$$= 150 \text{ lb for figuring belt tension and}$$

$$200 + 150 + 100 + 100 + 150$$

$$= 700 \text{ lb for figuring power}$$

Head + Tail + Snub + Snub + Takeup

Formula (5)

d. Tension required to accelerate material: T_{am}

$$= 2.8755 \times 10^{-4} \times Q \times \frac{V - V_o}{t}$$

$$= 2.8755 \times 10^{-4} \times 300 \times 280$$

$$= 24 \text{ lb} \qquad \textbf{Formula (6)}$$

where $Q = 300$ tons per hour, $t =$ time, $V = 280$ fpm, and $V_o = 0$, since there is no initial velocity in the direction of flow.

Table 2.17 Belt Tension to Rotate Pulleys

Location of Pulleys	Degrees Wrap of Belt	Tension at Belt Line[a] (lb per pulley)
Tight side	150–240	200
Slack side	150–240	150
All other pulleys	less than 150	100

Source: Courtesy CEMA

[a]Double these values for pulley shafts not operating in antifriction bearings.

e. Skirtboard friction: T_{sb}

$$= L_b(C_s h_s^2)$$

$$= 3(0.128 \times 10^2)$$

$$= 38$$

where

L_b = skirtboard length in ft (from drawing) = 3 ft
C_s = 0.1280 (table 2.18)
h_s = depth of material touching skirtboard
= 10 in.

If rubber edging were used, 3 lb per linear foot of edging should be added to overcome additional friction. In this case it would be $2L_b \times 3 = 6L_b = 6 \times 3 = 18$ lb. In the particular design illustrated, only a skirtboard without rubber edging was used.

f. Plow: single plow acting

T_{pl} = 5.0 lb for full-Vee or single-slant plow, removing all material from the belt
= 3.0 lb for partial-Vee or single-slant plow removing half the material from the belt

Since no plows are involved, nothing is added. If there were a plow, $T_{pl} = 5.0 \times 24 = 120$ lb or $3.0 \times 24 = 72$ lb.

g. Belt-cleaning devices: T_{bc}

Figure 2–3 lb per inch width of scraper blade. In the particular design illustrated, no allowance was made for belt-cleaning devices.

h. Horsepower
Total T_e

$$= 191 + 92 + 386 + 1642 + 700 + 24 + 38$$

$$= 3073 \text{ lb for hp}$$

$$\text{hp} = \frac{3073 \times 280}{33,000} = 26.07$$

Table 2.18 Skirtboard Friction Factor, C_s

Material	Factor C_s
Alumina, pulv., dry	0.1210
Ashes, coal, dry	0.0571
Bauxite, ground	0.1881
Beans, Navy, dry	0.0798
Borax	0.0734
Bran, granular	0.0238
Cement, Portland, dry	0.2120
Cement clinker	0.1228
Clay, ceramic, dry fines	0.0924
Coal, anthracite, sized	0.0538
Coal, bituminous, mined	0.0754
Coke, ground fine	0.0452
Coke, lumps and fines	0.0186
Copra, lumpy	0.0203
Cullet	0.0836
Flour, wheat	0.0265
Grains, wheat, corn or rye	0.0433
Gravel, bank run	0.1145
Gypsum, ½″ screenings	0.0900
Iron Ore, 200 lb per cu ft	0.2760
Lime, burned, ⅛″	0.1166
Lime, hydrated	0.0490
Limestone, pulv., dry	0.1280
Magnesium chloride, dry	0.0276
Oats	0.0219
Phosphate rock, dry, broken	0.1086
Salt, common, dry fine	0.0814
Sand, dry, bank	0.1378
Sawdust, dry	0.0086
Soda ash, heavy	0.0705
Starch, small lumps	0.0623
Sugar, granulated, dry	0.0349
Wood chips, hogged fuel	0.0095

Source: Courtesy CEMA

Note: To this skirtboard friction must be added 3 lb for every linear ft of each skirtboard, to overcome friction of the rubber skirtboard edging, when used, with the belt. Then,

$$T_{sb} = T + 2L_b \times 3$$

$$= C_s L_b h_s^2 + 2L_b \times 3$$

$$= L_b(C_s h_s^2 + 6)$$

Maximum belt tension for belt design

$$= 191 + 92 + 386 + 1642 + 150 + 24 + 38$$

$$= 2523 \text{ lb}$$

Table 2.19 Recommended Belt Sag Percentages for Various Full Load Conditions

Idler Troughing Angle	Material		
	All Fines	One-half the Maximum Lump Size	Maximum Lump Size
20°	3%	3%	3%
35°	3%	2%	2%
45°	3%	2%	1½%

Source: Courtesy CEMA

Note: Reduced-load cross-sections will permit an increase in the sag percentage, resulting in a decreased minimum tension. Such a choice may lead to a more economical belt selection, as the maximum tension will be reduced accordingly.

i. Sag: y

Sag (y), the vertical drop in feet between idlers,

$$= S_i^2 \times \frac{W_b + W_m}{8T_o}$$

$$= 4^2 \times \frac{41.2}{8(2562)}$$

$$= 0.032$$

where $W_b + W_m = 5.5 + 35.7 = 41.2$ lb; $T_o = 2562$ lb; and, since carrying idlers are spaced 4 ft on centers (refer to table 2.9), $S_i = 4.0$ ft. Then,

$$\frac{0.032}{4.0} = 0.008 < 1\% \text{ sag}$$

(refer to table 2.19).

For 3% assumed sag of the belt, $T_o = 4.2 \times S_i (W_b + W_m)$. For 2% sag, $T_o = 6.25 \times S_i (W_b + W_m)$, and for 1.5% sag, $T_o = 8.4 \times S_i (W_b + W_m)$.

2.25.4 Slack Slide Tension (T_2)

The minimum tension (T_2) required to drive the belt (tail pulley drive, figure 2.27) without slippage is

$$T_2 = C_w \times T_e \qquad \textbf{Formula (8)}$$

or $T_2 = T_o$. (C_w is the wrap factor. See table 2.20.)

For elevating horizontal runs with a head pulley drive,

$$T_2 = T_o \pm T_b \pm T_{yr} \qquad \textbf{Formula (9)}$$

Use the larger values for T_2.

For example, consider a problem where $C_w = 0.84$

Table 2.20 Wrap Factor, C_w

Type of Pulley Drive	θ Wrap	Bare Pulley	Lagged Pulley
Automatic Takeup			
Single, no snub	180°	0.84	0.50
Single, with snub	200°	0.72	0.42
	210°	0.66	0.38
	220°	0.62	0.35
	240°	0.54	0.30
Dual	380°	0.23	0.11
	420°	0.18	0.08
Manual Takeup			
Single no snub	180°	1.2	0.8
Single with snub	200°	1.0	0.7
	210°	1.0	0.7
	220°	0.9	0.6
	240°	0.8	0.6
Dual	380°	0.5	0.3
	420°	—	—

Source: Courtesy CEMA

Notes: For wet belts and smooth lagging use bare-pulley factor; for wet belts and grooved lagging, use lagged-pulley factor.

If wrap is unknown, assume the following: single, no snub = 180°; single, with snub = 210°; dual = 380°.

for automatic takeup, with a bare pulley and 180° wrap (table 2.20):

$T_b = HW_b = 46$ ft $\times 5.5$ lb $= 253$ lb
(where H = net change in elevation in feet)
T_{yr} (the return belt friction)
$\quad = 0.015 \times L \times W_b \times K_t$
$\quad = 0.015 \times 400 \times 5.5 \times 1.0$
$\quad = 33$ lb
(for K_t, see paragraph 2.25.2)
T_o (based on 2% sag)
$\quad = 6.25 \times S_i \times (W_b + W_m)$
$\quad = 6.25 \times 4 \times (5.5 + 35.7)$
$\quad = 1030$ lb

Then,

$$T_2 = 0.84 \times 2523 = 2119 \text{ lb}$$

or

$$= 1030 + 253 - 33 = 1250 \text{ lb}$$

Use the larger value, 2119 lb.

$$T_{max} = T_e + T_2$$
$$= 2523 + 2119$$
$$= 4642 \text{ lb}$$

or

$$= 4642 \text{ lb}/24 \text{ in.}$$
$$= 193 \text{ piw}$$

$$193 \text{ piw}/5 \text{ ply} = 39 \text{ lb/ply}$$

Therefore, use MP-50, 5-ply belt.

Since the takeup is near the head drive, the T_2 tension of 2119 lb can be reduced by the weight of about 3 ft of belt or left at 2119 lb. If the takeup were near the tail pulley, the takeup tension would be

$$T_2 + T_{yr} - T_b = 2119 + 33 - 253 = 1899 \text{ lb}$$

This sample problem is based on an existing conveyor. In most cases, it is preferable to lag all drive pulleys, which usually results in less total tension. In the example, if lagged pulleys had been used, $C_w = 0.5$ and

$$T_2 = 0.5 \times 2523$$
$$= 1262 \text{ lb}$$
$$T_{max} = 2523 + 1262$$
$$= 3785 \text{ lb}$$

or

$$= 3785 \text{ lb}/24 \text{ in.}$$
$$= 158 \text{ piw}$$

Table 2.21 Wrap Limits

Type of Pulley Drive	Wrap Limits[a]
Single, no snub	180°
Single, with snub	180–240°
Dual	360–480°

Source: Courtesy CEMA
[a]Applies to either bare or lagged pulleys.

This could, in some cases, result in a more economical belt selection, in addition to decreasing loads on the pulleys, shafts, and pillow blocks. Refer to table 2.21 for wrap limits.

2.25.5 Pulleys

Head Pulley. For 90 lb/ft^3 material, with a 24-in. belt, table 2.4 calls for a 4-ply minimum MP-50 belt, but 213 lb/50 in. (paragraph 2.25.4) requires a 5-ply belt. Table 2.22 shows that, for a 5-ply MP-50 belt with over 80% tension, a 30-in. pulley is required.

$$\frac{\text{actual tension}}{\text{allowable tension}} = \frac{213}{5 \times 50 \text{ lb}} = 85\%$$

Table 2.23 shows that 24-in. pulleys are required. For crushed stone, a $\frac{1}{8}$-in. top rubber cover and $\frac{1}{32}$-in. bottom rubber cover appear satisfactory. If the design were done now, a 30-in. head pulley would have been used.

If the percentage of actual tension to allowable tension is less than or equal to 80%, use the size given in

Table 2.22 Minimum Pulley Diameters for Multiple-Ply Type Belts (in.)

Number of Plies	MP 35%			MP 43, 50			MP60, 70, 90, 120			MP 155			MP 195, 240		
	% Tension			% Tension			% Tension			% Tension			% Tension		
	80 to 100	60 to 80	40 to 60	80 to 100	60 to 80	40 to 60	80 to 100	60 to 80	40 to 60	80 to 100	60 to 80	40 to 60	80 to 100	60 to 80	40 to 60
3	18	14	12	20	18	14	24	20	16	30	24	20	36	30	24
4	20	18	16	24	20	18	30	24	20	36	30	24	42	36	30
5	24	20	18	30	24	20	36	30	24	42	36	30	48	42	36
6	30	24	20	36	30	24	42	36	30	48	42	36	54	48	42
7	36	30	24	42	36	30	48	42	36	54	48	42	60	54	48
8	42	36	30	48	42	36	54	48	42	60	54	48	66	60	54

Source: Courtesy CEMA
Note: Regarding tables 2.22, 2.24, and 2.25, reverse bending encountered on dual pulley drives or on a high-tension reverse bend before a single-drive pulley on the return run, requires pulley diameters 6 in. larger than listed.

Table 2.23 Minimum Pulley Diameters (in.)

Belt Width	*Weight of Material (lb/ft³)*			
	50	*75*	*100*	*150*
18	18	18	24	30
24	18	18	24	30
30	18	18	24	30
36	24	24	30	36
42	24	24	36	42
48	24	36	42	42
56	24	36	42	42
60	36	36	42	42
72	—	42	42	48
84	—	—	48	54

Source: Courtesy CEMA

table 2.23. For any other percentage, x, the size of the pulley can be reduced to $x/80$ of the value given in the table.

Some engineers use the next larger size to five times the number of plies in inches for the tight side and four times for the light side.

Tail Pulley. Table 2.22 shows that, in the column for 40–60%, for a 5-ply belt, a 20-in. pulley is required.

If slack (return) side tension is 2442 lb,

$$\frac{2442}{6000} = 40.7\% \text{ allowable tension on the belt}$$

$$= 50 \times 24 \times 5 \text{ ply} = 6000$$

Since the belt wrap is over 6 in., a 20-in. pulley is the minimum required. In this case, however, a 24-in. pulley was used, the same as at the head shaft. A 20-in.

Table 2.24 Minimum Pulley Diameters for Reduced-Ply Type Belts (in.)

Maximum Belt Tension	80–100% Tension	60–80% Tension	40–60% Tension
To 100 PIW	14	12	12
To 150 PIW	16	14	12
To 200 PIW	18	16	14
To 300 PIW	24	20	18
To 400 PIW	30	24	20
To 500 PIW	36	30	24
To 700 PIW	42	36	30

Source: Courtesy CEMA
Notes: PIW = pounds per inch width of belt.
See also footnote to table 2.22.

Table 2.25 Minimum Pulley Diameters for Steel-Cable Belts (in.)

Rated Tension	80–100% Tension	60–80% Tension	40–60% Tension
To 1,000 PIW	30	30	24
To 1,800 PIW	42	36	30
To 2,400 PIW	48	36	30
To 2,800 PIW	54	42	36
To 3,500 PIW	54	48	36

Source: Courtesy CEMA
Notes: PIW = pounds per inch width of belt.
See also footnote to table 2.22.

diameter pulley was used in the takeup, with two 14-in. idler snub pulleys. Refer to table 2.24 for minimum pulley diameters for reduced-ply belts, and to table 2.25 for steel cable belts.

Table 2.6 shows that, for a 24-in. belt, idler "B" (20°, over two-thirds loaded), 75–100 lb material, a 4-ply fabric (42 oz., heavy duty, nylon fabric) should be figured.

If all the requirements of CEMA are followed, it is rare to check further. If experience is lacking, it is best to follow and check table 2.2, which calls for 5 plies maximum, and table 2.3, which calls for a minimum 14-in. belt width for reduced-ply belting (GTR selection).

2.25.6 Graphic Solution

Belt: 24 in.
Capacity: 300 tph
Weight of material: $W = 90 \text{ lb/ft}^3$
Weight per linear foot of belt + revolving idler parts: 19 lb (table 2.14)
From figure 2.34, for $L = 400$ ft and $W = 19$ lb,

$$hp = \frac{0.7 \times 300}{100}$$

$$= 2.1$$

From figure 2.35, for 300 tons read $0.3 \times 46 = 13.8$ hp.
From figure 2.36,

$$1.5 \times \frac{300}{100} = 4.5 \text{ hp}$$

The total hp is $2.1 + 13.8 + 4.5 = 20.4$, against a computed value of 20.24 hp.

The equations for the curves on figures 2.34, 2.35, and 2.36 are given below, along with numerical solutions to the problem.

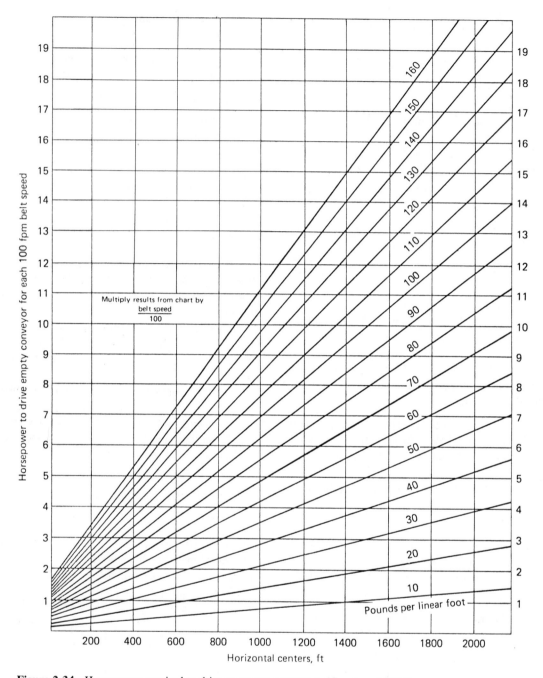

Figure 2.34 Horsepower required to drive an empty conveyor. *(Courtesy CEMA)*

a. Hp to Drive Empty Conveyor
    (see figure 2.34)

$$= 0.05w + \frac{0.02 \times L \times w \times V}{33,000}$$

$$= 0.05(19) + \frac{0.02 \times 400 \times 19 \times 280}{33,000}$$

$$= 0.95 + 1.29$$

$$= 2.24$$

where

w = weight per ft of belt and revolving parts: 19 lb (table 2.14)

0.02 = friction factor for empty conveyor (refer to table 2.26)

L = length of conveyor in ft: 400

V = velocity in fpm: 280

b. Hp to Elevate 1 tph to 1 ft
    (see figure 2.35)

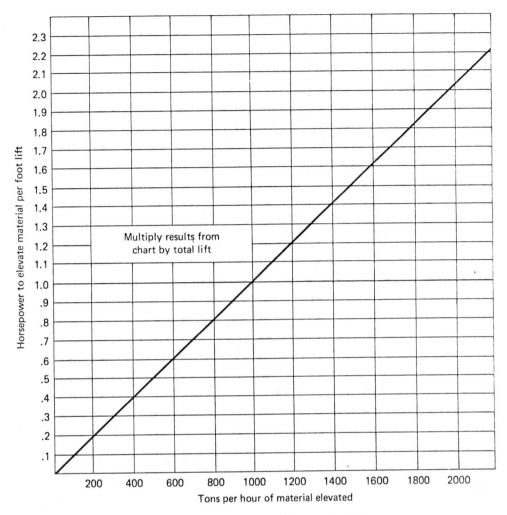

Figure 2.35 Horsepower required to elevate material. *(Courtesy CEMA)*

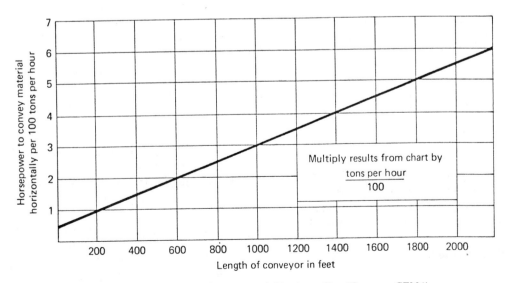

Figure 2.36 Horsepower required to convey material horizontally. *(Courtesy CEMA)*

Table 2.26 Friction Factors

Part in which friction originates	Friction factor, C
Revolving parts of ball or roller bearing belt idlers	0.02
Empty belt on ball or roller bearing belt idlers	0.02[a]
Material only on belt on antifriction belt idlers	0.25[a]
Tail, bend and snub pulleys and shafts, ball and roller bearings	0.01
Tail, bend and snub pulleys and shafts, sleeve bearings	0.02
Takeup pull or counterweight, ball and roller bearings	0.01
Takeup pull or counterweight, sleeve bearings	0.02
Drive pulley assembly, ball and roller bearings	0.01
Drive pulley assembly, sleeve bearings	0.02
Fabric belt, including material carried, sliding on polished steel	0.25–0.30
Friction-surfaced belt and material carried, sliding on polished steel	0.30–0.35
Rubber-surfaced belt and material carried, sliding on polished steel	0.45–0.55
Fabric-surfaced belt and material carried, sliding on polished wood	0.30–0.35
Friction-surfaced belt and material carried, sliding on polished wood	0.35–0.40
Rubber-surfaced belt and material carried, sliding on polished wood	0.40–0.50

Source: Courtesy FMC

Note: This table shows recommended friction factors based on reasonable care, maintenance, lubrication, and alignment, and for operation in temperatures above −20°F.

[a]The difference between the friction factors of the empty belt and the material on the belt is caused by the resistance of the material to the slight but repetitive disturbance to which it is subjected as it passes over the belt idlers.

$$= \frac{2000}{60} \times \frac{1}{33,000} = 0.00101$$

say, 0.001 per ton per hour.

Thus, for 300 tph, for a 46-ft lift, hp is

$$0.001 \times 300 \times 46 = 13.8$$

c. Hp to Convey Material Horizontally
   (see figure 2.36)

$$= 0.004T + 0.000025TL$$

$$= (0.004 \times 300) + (0.000025 \times 300 \times 400)$$

$$= 1.2 + 3.00$$

$$= 4.20$$

where T = tons per hour, L = length in feet, and 0.025 is the friction factor for material on the belt (refer to table 2.26).

Total hp at the drive shaft = 2.24 + 13.8 + 4.20 = 20.24. The equations given above lend themselves to plotting curves. Figure 2.34 shows curves for hp required to drive an empty conveyor (a); figure 2.35, for hp required to elevate the material (b); and figure 2.36 for hp required to convey the material horizontally (c).

2.25.7 Jeffrey Formula

The various manufacturers have their own formulas for figuring horsepower. The original computations were based on the Jeffrey (Dresser Industries) formula, and are repeated here.

To drive an empty conveyor horizontally, hp is figured by

$$\frac{F(1.07L + 50)(0.03QS)}{1000}$$

where F = the coefficient of friction, 0.03; S = 280 fpm; and $Q = 2$

$$2B + \frac{W_c}{S_c} + \frac{W_r}{S_r}$$

$$= 2(5.5) + \frac{24.6}{4.0} + \frac{21.2}{10.0}$$

$$= 19.27$$

where B = weight of the belt in lb/ft, 5.5 lb; W_c = weight of the rotating parts of the carrying idler; W_r = weight of the rotating parts for the return idler; and S_c and S_r = spacing of the idlers. Therefore,

$$hp = \frac{0.03(1.07 \times 400 + 50)(0.03 \times 19.27 \times 280)}{1000}$$

$$= 2.32$$

to drive an empty conveyor horizontally.

To move the material horizontally, figure

$$hp = \frac{F(1.07L + 50)C}{1000}$$

$$= \frac{0.03(1.07 \times 400 + 50)300}{1000}$$

$$= \frac{0.03 \times 478 \times 300}{1000}$$

$$= 4.30$$

where C = tons per hour, 300, and F is assumed to be 0.03, as before (refer to paragraph 2.26.2). F can be 0.025 for well maintained units under ideal conditions.

To raise the material 46 ft, figure

$$hp = \frac{CH}{1000}$$

$$= \frac{300 \times 46}{1000}$$

$$= 13.8$$

This last item is often neglected.

The total hp = 2.32 + 4.30 + 13.8 = 20.42. No allowance is made for trippers.

2.25.8 Comparison of Methods

Omitting the value of the skirtboard from the CEMA formulas gives the following values for the required horsepower at the drive shaft:

a. CEMA: 26.4 hp
b. Graphic: 20.24 hp
c. Jeffrey: 20.42 hp

Methods b and c can be used for preliminary design. Experience is required to determine the factor to be added to the results obtained. In view of the standardization of the industry, it is best to follow the formulas in the CEMA handbook.

2.25.9 Friction Factors

In estimating a belt conveyor for industrial work, it is common practice to figure a friction factor of 0.03. This usually takes care of all conditions in areas subject to summer and winter conditions. Table 2.26 is a reliable guide.

2.25.10 Conveyor Bill of Material

Inclined belt conveyor, 24″ wide × 402′7″, 30 hp at 280 fpm (45 rpm).

1 head pulley, 24″ diameter × 26″ cfws, with ⅜″ thick vulcanized lagging, bored 4⁷⁄₁₆″ diameter kw and ss
1 head pulley shaft, 4⁷⁄₁₆″ diameter × 5′2″ long, with 3 keyways
2 head pulley shaft bearings, 4⁷⁄₁₆″ diameter spherical roller-bearing pillow blocks
1 tail pulley, 24″ diameter × 26″ cf self-cleaning, bored 2¹⁵⁄₁₆″ diameter
1 tail pulley shaft, 2¹⁵⁄₁₆″ diameter × 3′8″ long, with 2 keyways
2 tail pulley shaft bearings, 2¹⁵⁄₁₆″ diameter spherical roller-bearing pillow blocks
85 troughing idlers 20° (5″ steel rollers with roller bearings)
3 troughing idlers 20° (5″ rubber rollers with roller bearings)
2 troughing idlers 20° training, 5″ diameter rollers
40 return idlers 5″ diameter rollers
825 ft conveyor belt 24″ wide, ³⁄₁₆″ × ¹⁄₁₆″ covers, 5 ply, MP-50
1 gravity takeup assembly, consisting of:
    1 takeup pulley, 20″ diameter × 26″ cfws, bored 2-¹⁵⁄₁₆″ diameter
    1 takeup pulley shaft, 2¹⁵⁄₁₆″ diameter × 3′8″ long, with 2 keyways
    2 takeup pulley shaft bearings, 2¹⁵⁄₁₆″ diameter spherical roller-bearing pillow blocks
    2 idler pulleys, 14″ diameter × 26″ cfws, bored 2³⁄₁₆″ diameter
    2 idler pulley shaft, 2³⁄₁₆″ diameter × 3′8″ long, with 2 keyways
    4 idler pulley shaft bearing, 2³⁄₁₆″ diameter spherical roller-bearing pillow blocks
    1 fabricated takeup frame with guideposts
    1 lot, fabricated conveyor frames with supports

2.25.11 Conveyor Drive Bill of Materials (refer to section 9)

1 electric motor, 30 hp, 1750 rpm, totally enclosed, 230/460 V, 3 ph, 60 Hz
1 shaft-mounted reducer (24:1 ratio) with motor mount (Class II service)
1 integral backstop for above reducer
1 V-belt drive (1.62:1 ratio) and guard

2.25.12 Alternative Drive Bill of Material (see figure 2.37)

1 motogear unit and inline reducer (Class II service), 100 rpm output, complete with 30 hp, 1750 rpm totally enclosed motor, 230/460 V, 3 ph, 60 Hz
1 roller-chain drive (2.18:1 ratio) and guard of Class II service, with a chain and sprocket combination of 22/48 ratio and hp of 25.5.

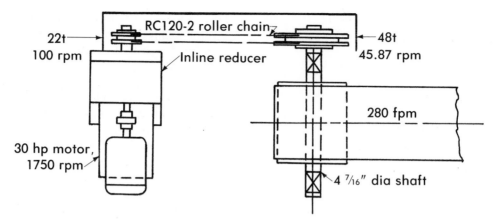

Figure 2.37 Diagram of a belt conveyor alternative drive.

Head shaft rpm $= \dfrac{280}{\pi \times 2'}$

$\phantom{\text{Head shaft rpm}} = \dfrac{280}{6.28}$

$\phantom{\text{Head shaft rpm}} = 44.58$ required

Reducer output $= 44.58 \times \dfrac{48}{22}$

$\phantom{\text{Reducer output}} = 97.26$

Refer to section 9 for the drive design.

2.26 BELT CONVEYOR CALCULATION (using Jeffrey Formula)

2.26.1 Data

Consider a belt conveyor that is to handle 30 tph of sulfur weighing 75 lb/ft.[3] The belt conveyor is 30 in. wide, 30 ft long, and rises 8 ft ($= 16° \angle$). The conveyor speed is 72 fpm.

2.26.2 Horsepower Calculation

Horsepower at the conveyor shaft is figured by

$$\frac{F(1.07L + 50)(C + 0.03QS) + CH}{1000}$$

where

$F =$ coefficient of friction for idlers: 0.03 for anti-friction idlers or 0.025 for well maintained units under ideal conditions
$L =$ length of conveyor: 30 ft
$C =$ capacity: 30 tph
$H =$ vertical rise: 8 ft
$S =$ conveyor speed: 72 ft/min
$Q =$ combined factor

$= 2B + \dfrac{W_c}{S_c} + \dfrac{W_r}{S_r}$

$= (2 \times 6.3) + \dfrac{38.1}{4} + \dfrac{32.8}{10}$

$= 12.6 + 9.52 + 3.28$

$= 25.4$

where $B =$ weight of the belt in lb/ft, 6.3 (30-in. wide, 4-ply, 3/16-in. cover); $W_c =$ weight of the rotating parts of the carrying idlers, 38.1 lb (6-in. diameter idlers); $S_c =$ the average spacing of the carrying idlers, 4 ft; $W_r =$ weight of the rotating parts of the return idlers, 32.8 lb; and $S_r =$ average spacing of the return idlers, 10 ft.

Therefore, the hp at the conveyor shaft is

$$0.03(1.07 \times 30 + 50)$$

$$\frac{\times (30 + 0.03 \times 25.4 \times 72) + (30 \times 8')}{1000}$$

$$= \frac{(0.03 \times 82.1 \times 84.86) + 240}{1000}$$

$$= 0.449$$

Figure the total hp required at the motor as

hp at conveyor shaft	$= 0.449$
10% of chain drive loss	$= 0.044$
5% reducer loss	$= \underline{0.022}$
	$0.515*$

Considering the heavy industrial service this conveyor will be subjected to, it was felt that a conservative selection of nothing less than a 3-hp motor should be made. Since the plant standardizes on 1200-rpm motors,

---

*If a vertical takeup is used, add 0.06 for takeup; 0.02 for high-tension bend pulleys; and 0.01 for low-tension bend pulleys; making a total of 0.09. Total horsepower at the headshaft = 1.09 × basic hp if hp is positive, or 0.91 basic hp if hp is negative.

a 3-hp, 1200-rpm tefc explosion-proof motor was selected.

2.26.3 Belt Selection

If antifriction journal bearings are used on the headshaft, the factor 0.96 becomes 0.98, and the 0.07 factor in the basic hp formula becomes 0.035.

Effective belt tension: T_e = hp pull

$$= \frac{0.96 \times hp \times 33,000}{S}$$
$$= \frac{0.96 \times 3 \times 33,000}{72}$$
$$= 1320 \text{ lb}$$

Slack side tension: T_s

$$= T_e \times R$$
$$= 1320 \times 1.25$$
$$= 1650 \text{ (for bare pulley)}$$

where R (mfr catalog) is the pulley wrap and takeup factor.

Tight side tension: T_T

$$= T_s + T_e$$
$$= 1650 + 1320$$
$$= 2970 \text{ lb}$$

(For an inclined conveyor, T_{si} should be used instead of T_s if T_{si} is greater.)

Corrected value for slack side tension: T_{si} (Jeffrey catalog)

$$= 0.9 \text{ BH}$$
$$= 0.9(6.3)(8)$$
$$= 45.4 \text{ lb}$$

is smaller than T_s. Use T_s.

Table 2.4 recommends an MP-50, 5-ply belt as a minimum for a 30-in. belt handling this material. Table 2.6 recommends a 4-ply, 36 HDNF belt. This belt, using table 2.4, would have an allowable pull of $5 \times 30 \times 50 = 7500$ lb, which is well above the 2970 lb required. For this material, a $\frac{1}{8}$ in. top rubber cover and a $\frac{1}{32}$ in. bottom rubber cover appear satisfactory. A skim coat should be provided between plies for better wear under flexing conditions.

2.26.4 Head Pulley Selection

Tables 2.22 and 2.24 should be followed as a minimum; anything larger would be better.

The percent of full tension stress of the belt is $2970/4000 = 74\%$, say, 75%. The allowable belt tension on 30-in. wide, 4-ply, $\frac{3}{16}$ in. cover is 4000 lb. The

pulley size is $0.75 \times 25 = 18.75$. Table 2.22 shows an 18-in. minimum pulley. Use a 24-in. head pulley as a conservative size for a 30-in. belt.

2.26.5 Tail Pulley Selection

The conveyor manufacturers recommend providing 4 in. of pulley diameter per ply for a 27-oz. belt, if the belt is stressed to full tension. If not, the proportional reduction used in determining the tail pulley should be used. The pulley size is 4-in. per ply $\times$ 5 plies $\times$ 75% = 15 in. Use a 20-in. tail pulley as a conservative size for a 30-in. belt.

Graphic charts give very close results to the analytical method, are faster, and can be used for general industrial work, if caution is exercised.

2.26.6 Design of Drive (see figure 2.38)

Use a 1.0 service factor for sprocket selection and use 19t and 60t sprockets. A 19t RC-100 sprocket has a 4-hp rating at 38 rpm and a 60t RC-100 sprocket has a 5-hp rating at 11.5 rpm. Therefore, the reducer ratio should be $1150/38 = 30:1$. The reducer hp required is 3×1.25 (the service factor) = 3.75 hp. Select an inline reducer with a ratio of 31:4 ($\pm 4\%$ variation). An output of 37 rpm at 5 hp would be close enough for any commercial work.

2.27 MINIMUM PULLEY DIAMETERS FOR BELTS (Minimum Pulley Sizes Required by CEMA)

Tables 2.22, 2.23, 2.24, and 2.25 are taken from the CEMA publication, "Belt Conveyors," and are reproduced here by courtesy of CEMA (refer to paragraph 2.25.5).

2.28 BEND, SNUB, TAKEUP, AND TAIL PULLEYS

The tail pulley minimum is the same as is given for the 40–60% classification, except those cases where the calculated tension is in a higher range. Then use the applicable table 2.22, 2.23, 2.24, or 2.25 (refer to paragraph 2.25.5).

The takeup pulley minimum is the same as is given for the 40–60% classification, except in those cases where the calculated tension is in a higher range. Then

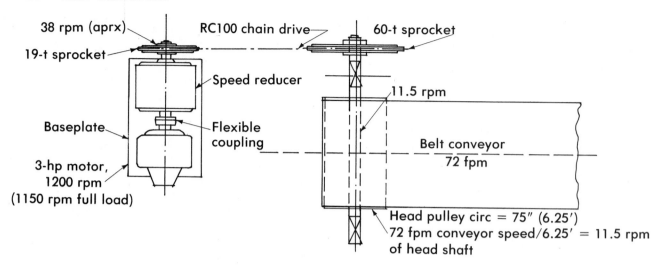

Figure 2.38   Jeffrey design belt conveyor, drive diagram.

use the applicable table 2.22, 2.23, 2.24, or 2.25 (refer to paragraph 2.25.5).

The slack side-bend pulleys, where belt wrap is 6 in. or less, may be 6 in. less than the diameters recommended above at the tensions encountered, but not less than 10 in. in diameter. Select high-tension bend and snub pulleys from the table at applicable percent of belt tensions, but never less than 40% under recommendation.

2.29   ENCLOSURES FOR BELT CONVEYORS

Figure 2.39 shows an enclosure for a belt conveyor built in a northern location. Note provision for a service walkway. The enclosure (gallery) ran over areas of the plant with clearance provided for traffic. Servicing can be done in any type of weather, and the conveyor can be supported to clear any obstacles.

2.30   IDLER SELECTION

The load carried by the idler has to be adjusted for the four items covered in tables 2.27 through 2.30. The adjusted load, *AL*, can be found by

$$AL = IL \times K_1 \times K_2 \times K_3 \times K_4$$

where *IL*, the actual load, equals $(W_b + W_m) \times S_i$. If the product of K_1, K_2, K_3, and K_4 is less than one, a value of one should be used. For return belt idlers, W_b (weight of belt) must be used for load. Refer to paragraph 2.16 and table 2.9.

Tables 2.31 through 2.35 show the load rating for CEMA equal-length-roll idlers A, B, C, D, and E. These

ratings are based on 90,000 hours minimum useful bearing life at 500 rpm. Note that these load ratings are minimum ratings for CEMA-rated idlers. Actual figures on load ratings supplied by the manufacturers may be higher.

To illustrate the use of tables 2.27 through 2.35, assume that the load on the idler in paragraph 2.25 must be checked.

$$IL = (5.5 + 35.7) \times 4.0$$

$$= 165 \text{ lb}$$

$K_1 = 1.0$ (table 2.27)
$K_2 = 1.10$ (table 2.28) for moderate conditions and
   fair maintenance
$K_3 = 1.0$ (table 2.29)
$K_4 = 0.85$ (table 2.30)

$$AL = IL \times K_1 \times K_2 \times K_3 \times K_4$$

$$= 165 \times 1.0 \times 1.10 \times 1.0 \times 0.85$$

$$= 165 \times 0.94$$

Because the product of $K_1 \times K_2 \times K_3 \times K_4$ is less than one, one is used.

$$= 165 \times 1.0$$

$$= 165 \text{ lb}$$

The load rating of a B idler, troughed 20°, and a 24-in. belt is 410 lb (table 2.32). The return load is 5.5 × 4.0 = 22, and the allowable load is 190. If table 2.9 is followed, this check can be dispensed with.

The type of idler recommended for various conditions is shown in table 2.9. If the spacing of the idlers has to be varied or if special loading conditions are involved, the load on the selected idler should be checked in tables 2.31 through 2.35.

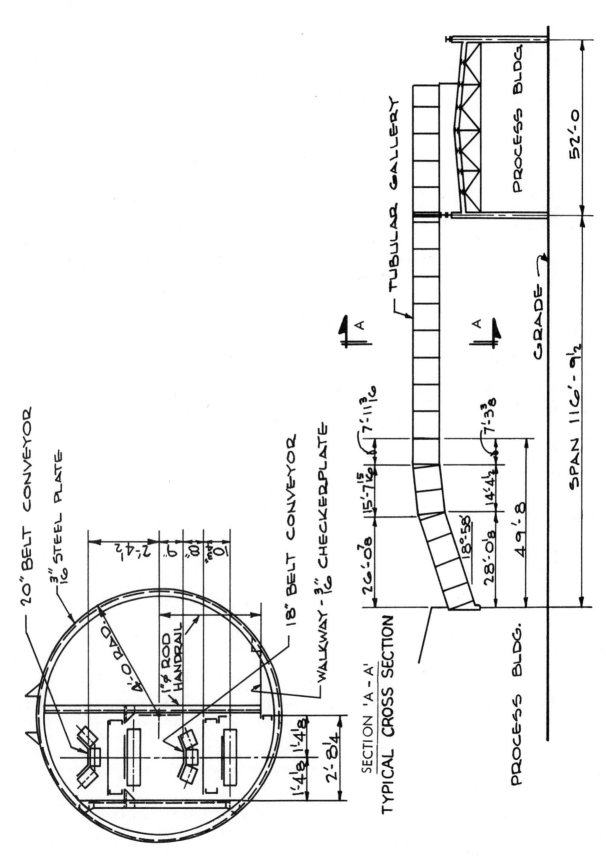

20" BELT CONVEYOR

$\frac{3}{16}$" STEEL PLATE

4'-0" RAD

$1"\phi$ ROD HANDRAIL

18" BELT CONVEYOR

WALKWAY - $\frac{3}{16}$" CHECKERPLATE

2'-4$\frac{1}{2}$"

9"

10"

$10\frac{1}{4}$"

$1'-4\frac{1}{8}$" $1'-4\frac{1}{8}$"

$2'-8\frac{1}{4}$"

SECTION 'A - A'
TYPICAL CROSS SECTION

TUBULAR GALLERY

PROCESS BLDG.

GRADE

PROCESS BLDG.

52'-0

SPAN 116'-9$\frac{1}{2}$

$7'-11\frac{3}{16}$

$15'-7\frac{15}{16}$

$26'-0\frac{3}{8}$

$7'-3\frac{3}{8}$

$14'-4\frac{1}{2}$

$28'-0\frac{3}{8}$

18°58'

49'-8

Figure 2.39   An 8-ft tubular steel gallery supporting a belt conveyor.

Table 2.27 K_1 Lump Adjustment Factor

Maximum Lump Size (in.)	Material Weight (lb/ft³)						
	50	75	100	125	150	175	200
4	1.0	1.0	1.0	1.0	1.1	1.1	1.1
6	1.0	1.0	1.0	1.1	1.1	1.1	1.1
8	1.0	1.0	1.1	1.1	1.1	1.2	1.2
10	1.0	1.1	1.1	1.2	1.2	1.2	1.2
12	1.0	1.1	1.1	1.2	1.2	1.2	1.3
14	1.1	1.1	1.1	1.2	1.2	1.3	1.3
16	1.1	1.1	1.2	1.2	1.3	1.3	1.4
18	1.1	1.1	1.2	1.2	1.3	1.3	1.4

Source: Courtesy CEMA

Table 2.28 K_2 Environmental and Maintenance Factors

Environmental Conditions	Maintenance		
	Good	Fair	Poor
Clean	1.00	1.08	1.11
Moderate	1.06	1.10	1.13
Dirty	1.09	1.12	1.15

Source: Courtesy CEMA

Table 2.29 K_3 Service Factor

Operation	Factor
Less than 6 hours per day	0.8
6 to 9 hours per day	1.0
10 to 16 hours per day	1.1
Over 16 hours per day	1.2

Source: Courtesy CEMA

Table 2.30 K_4 Belt Speed Correction Factor

Belt Speed (fpm)	Roll Diameter (in.)			
	4	5	6	7
100	0.80	0.80	0.80	0.80
200	0.83	0.80	0.80	0.80
300	0.90	0.85	0.83	0.81
400	0.95	0.91	0.88	0.85
500	0.99	0.95	0.92	0.88
600	1.03	0.98	0.95	0.92
700	1.05	1.01	0.98	0.95
800	—	1.04	1.00	0.97
900	—	1.06	1.03	1.00
1000	—	—	1.05	1.02

Source: Courtesy CEMA

Table 2.31 Load Ratings for CEMA A Idlers (lb)

Belt Width (in.)	Trough Angle			
	20°	35°	45°	Return
18	300	300	300	150
24	300	300	289	125
30	300	280	270	100
36	275	256	248	75

Source: Courtesy CEMA

Table 2.32 Load Ratings for CEMA B Idlers (lb)

Belt Width (in.)	Trough Angle			
	20°	35°	45°	Return
18	410	410	410	220
24	410	410	410	190
30	410	410	410	165
36	410	410	396	155
42	390	363	351	140
48	380	353	342	130

Source: Courtesy CEMA

Table 2.33 Load Ratings for CEMA C Idlers (lb)

Belt Width (in.)	Trough Angle			
	20°	35°	45°	Return
18	900	900	900	475
24	900	900	900	325
30	900	900	900	250
36	900	837	810	200
42	850	791	765	150
48	800	744	720	125
54	750	698	675	*
60	700	650	630	*

Source: Courtesy CEMA
*Use CEMA D return idlers.

2.31 CONVEYOR BELT TROUBLES

Table 2.36 is from the "Handbook of Conveyor & Elevator Belting" and is reproduced by courtesy of The Goodyear Tire & Rubber Company. Many of the difficulties encountered with conveyor belting have been listed, together with their causes and suggested remedies.

2.32 SELECTED NOMENCLATURE FOR SECTION 2

A_1 Belt tension, or force, required to overcome frictional resistance and rotate idlers (lb/idler)

Table 2.34 Load Ratings for CEMA D Idlers (lb)

Belt Width (in.)	Trough Angle			
	20°	35°	45°	Return
24	1200	1200	1200	600
30	1200	1200	1200	600
36	1200	1200	1200	600
42	1200	1200	1200	500
48	1200	1200	1200	425
54	1200	1116	1080	375
60	1150	1070	1035	280
72	1050	977	945	155

Source: Courtesy CEMA

Table 2.35 Load Ratings for CEMA E Idlers (lb)

Belt Width (in.)	Trough Angle			
	20°	35°	45°	Return
36	1800	1800	1800	1000
42	1800	1800	1800	1000
48	1800	1800	1800	1000
54	1800	1800	1800	925
60	1800	1800	1800	850
72	1800	1800	1800	700
84	1800	1674	1620	550
96	1750	1628	1575	400

Source: Courtesy CEMA

B Weight of belt (lb/ft)
C Capacity (tph)
C_s Skirtboard friction factor
C_w Wrap factor
F Coefficient of friction
H Vertical distance material is lifted or lowered (ft)
H_d Lift to drive pulley
H_x Vertical distance from tail pulley to point X (ft)
h_s Depth of material touching skirtboard (in.)
K_t Ambient temperature correction factor
K_x Factor used to calculate frictional resistance of idlers and sliding resistance between belt and idler rolls (lb/ft)
K_y Factor used to calculate combination of resistance of belt and resistance of load to flexure as belt and load move over idlers
L Length of conveyor (ft)
L_b Skirtboard length (one skirtboard) (ft)
L_x Distance from tail pulley to point X along conveyor (ft)
Q Combined factor (lb/ft)
R Pulley wrap and takeup factor

S Conveyor speed (ft/min)
S_c Average spacing of carrying idlers (ft)
S_i Idler spacing (ft)
S_r Average spacing of return idlers (ft)
T Belt tension to overcome skirtboard friction of two parallel skirtboards (lbs)
T_{ac} Total of tensions from friction of conveyor accessories (lb)
T_{am} Tension required to accelerate material (lb)
T_b Tension required to lift or lower belt (lb)
T_{cx} Tension in belt at point X on carrying run (lb)
T_e Effective belt tension at drive (lb)
T_{fcx} Tension in belt at point X on carrying run, resulting from friction (lb)
T_{frx} Tension in belt at point X on return run, resulting from friction (lb)
T_{hp} Tension at head or discharge pulley (lb)
$T_{\max}$ Maximum tension in belt (lb)
$T_{\min}$ Minimum tension in belt (lb)
T_o Minimum allowable sag tension in belt for a definite idler spacing (lb)
T_p Tension from belt flexure around pulleys plus pulley bearing friction (lb)
T_{pl} Tension from friction of plows (lb)
T_{rx} Tension in belt at point X on return run (lb)
T_s Slack side tension (lb)
T_{sb} Tension from skirtboard friction (lb)
T_{si} Slack side tension for inclined conveyor (lb)
T_t Tension in belt at tail pulley (lb)
T_{wcx} Tension in belt at point X on carrying run resulting from weight of belt plus material carried (lb)
T_{wrx} Tension in belt at point X on return run, resulting from weight of empty return belt (lb)
T_x Tension from friction of carrying and return idlers (lb)
T_y Carrying idler friction
T_{yr} Tension from belt flexure as belt rides over return idlers (lb)
T_1 Tension in belt at tight side of driving pulley (lb)
T_2 Tension in belt at slack side of driving pulley (lb)
TU Takeup
t Time (sec)
V Design belt speed (ft/min)
V_o Initial velocity of material fed onto belt (ft/min)
W Weight of material (lb/ft^3)
W_b Weight of belt (lb/ft of length)
W_c Weight of rotating parts for carrying idlers (lb)
W_m Weight of material conveyed (lb/ft of belt length)
W_r Weight of rotating parts for return idlers (lb)
w Weight per ft of belt and revolving parts (lb)

Table 2.36 Causes and Corrections of Conveyor Belt Troubles

Causes	Corrections
I. CONVEYOR RUNNING TO ONE SIDE AT GIVEN POINT ON STRUCTURE	
One or more idlers immediately preceding trouble point, not square (at right angles) to longitudinal axis of belt	Advance, in the direction of belt travel, the end of the idler to which the belt has shifted
Conveyor frame or structure crooked	Stretch string along edge to determine extent and make correction
One or more idler stands not centered on belt	Stretch string along edge to determine extent and make correction
Sticking idlers	Improve maintenance and lubrication
Belt runs off on terminal pulley	Check terminal pulley alignment; check alignment of idlers approaching terminal pulley
Buildup of material on idlers	Improve maintenance; install scrapers or other cleaning device
Structure not level and belt tends to shift to low side	Level structure
II. PARTICULAR SECTION OF CONVEYOR BELTS RUNS TO ONE SIDE AT ALL POINTS ON CONVEYOR	
Belt not joined squarely	Refasten, cutting ends square
Bowed belt	If belt is new, this condition should straighten out as soon as the belt has operated under full load tension and becomes broken in; in rare instances, belt must be straightened or replaced; avoid bad storage conditions such as telescoped rolls or one edge close to damp ground or wall; use self-aligning idlers, particularly on return run, approaching tail pulley to get central loading
III. CONVEYOR BELT RUNS TO ONE SIDE FOR LONG DISTANCE ALONG BED	
Load being placed on belt off-center	Adjust chute and loading conditions so as to place load in center of belt
See also causes and corrections under I	
IV. BELT IS ERRATIC (FOLLOWS NO PATTERN OF PERFORMANCE)	
Belt too stiff to train	Use self-aligning idlers; tilt troughing idlers forward but not more than 2 deg; use more troughable belt
Combination of causes under I and II with off-center loading	Correct loading first; then other causes can be identified

(continues)

TABLE 2.36 69

Table 2.36 (*Continued*)

Causes	Corrections
V. SEVERE WEAR ON PULLEY SIDE OF CONVEYOR BELT	
Slippage on drive pulley	Increase tension through screw takeup or more counterweight; lag drive pulley (grooved lagging if wet); increase arc of contact on drive pulley with snub
Spillage of material that is ground between belt and pulley or that builds up at loading point until belt is dragging	Improve loading conditions with chutes; if belt is loaded too full, increase belt speed or decrease feed onto belt; install decking between top and return runs; install plows or scrapers in front of tail pulley on return run; prevent leakage of abrasive fines at fasteners by changing to plate fasteners or vulcanized splice
Sticking idlers	Improve maintenance and lubrication
Bolt heads protruding above lagging	Tighten bolts; replace worn lagging; use vulcanized-on lagging
Excessive tilt to troughing idlers	Adjust to not more than 2 deg from line perpendicular to belt; straighten as many as possible to vertical.
VI. EXCESSIVE STRETCH IN CONVEYOR	
Tension too high	Increase speed, keeping tonnage same; reduce tonnage at same speed; reduce friction with better maintenance and replacing of wornout and frozen idlers; decrease tension by improving drive with lagging and/or increased arc of contact; use counterweight takeup of minimum amount; replace with belt of lower elongation
VII. BELT SHRINKS	
Absorbing moisture	Put in extra piece, and install with takeup half-way down
VIII. GROOVING, GOUGING, OR STRIPPING OF TOP COVER	
Skirt board seals too stiff and pressed against belt	Use more pliable seals (do not use old belting)
Excessive space between belt and skirt seals	Adjust seals to minimum clearance
Metal sides of chute or skirts too close to belt and gap not increasing in direction of belt travel	Adjust to at least 1-in. gap between metal and belt and have gap increasing in direction of travel so as to prevent material jamming at this point
Belt spanks down under impact at loading point allowing material to be trapped under skirts	Install cushion idlers to hold belt up against skirts
Material hanging under back panel of chute	Improve loading to prevent spillage or install baffles
Jamming of material in chute	Widen chute

Table 2.36   (*Continued*)

Causes	Corrections
IX. SHORT BREAKS IN CARCASS PARALLEL TO BELT EDGE AND STAR BREAKS IN CARCASS	
Impact of lumps falling on belt	Reduce impact; use cushion idlers
Material trapped between belt and pulley	Install plows or scrapers on return run ahead of tail pulley
X. TRANSVERSE BREAKS AT BELT EDGE	
Belt edges folding up on structure at or near pulleys	See I, II, III, and IV; install limit switches to stop belt in cases of extreme shifting; provide more lateral clearance
Mildew	Replace with mildew-resistant carcass
Final idler before head pulley located too close or too high with respect to head pulley	Adjust idler position in accordance with recommendations in Section 12
Inadequate convex curve	Adjust curve radius according to recommendations in Section 12
XI. COVER BLISTERS OR SAND BLISTERS	
Cover cuts or very small cover punctures allow fine particles to work under cover and cut cover away from carcass	Make spot repair with vulcanizer or self-curing repair material; in severe and repeating cases, refer all details to Goodyear for analysis
See XIV	
XII. EXCESSIVE TOP COVER WEAR, UNIFORM AROUND BELT	
Dirty, stuck, or misaligned return rolls	Install cleaning devices; wash belt; use rubber disc return idlers; repair, replace, and realign return rolls
Cover quality too low	Replace with belt of heavier cover gauge or higher quality rubber
Pileup of spilled material at tail pulley	Improve housekeeping and loading
Side loading or poor loading (speed of delivery of material too slow)	Redesign chute to make load feed onto belt in same direction as belt runs and at approximately same speed as belt
Excessive sag between idlers cause load to work and shuffle on belt as it passes over idlers	Increase tension if it is unnecessarily low; reduce idler spacing
XIII. BELT CENTER RAISES OFF TROUGHING IDLERS, BOWING TOWARD LOAD	
Oil in material	Remove source of oil, if possible; to complete life of present belt, cut longitudinal grooves in cover with tire grooving tool to relieve transverse pressure exerted by swelling rubber; replace with proper ORS belt

(continues)

a manner that the paint will adhere to the surface and be weather resistant. All painting shall be done in a neat and professional manner. The equipment shall not be loaded for shipment until the paint is thoroughly dry.

3. All polished parts, shafts, gearing, bearings, and so on shall be coated with rust-preventive compound and crated, if necessary, to prevent damage in shipment, and all equipment shall be thoroughly protected from the weather.

4. Services of an engineer, if required at the time of installation, shall be furnished at a rate stated in the proposal.

5. The successful bidder shall furnish three sets of prints for approval of the general design of the equipment and, later, _____ complete sets of certified prints to: _____ .

2.34 SPECIFICATION FOR 24" SHUTTLE BELT CONVEYOR

One shuttle belt conveyor required complete.

Capacity: 80–100 tph, $1\frac{1}{2}$" lime and coke mixture; 120°F temperature; weight of 40 lb/ft^3.

Conveyor length between centers of head and tail pulleys: 22'9".

Belt size: 24" wide, Imperial Sahara Belt or approved equal; provide lacing.

Troughing idlers: spacing 4' on centers, three-roll antifriction with 5" steel rolls; Alemite hydraulic fittings lubricated from one side of conveyor.

Return idlers: spacing 8' on centers, 5" single-pulley Link Belt 7517-30* or approved equal.

Head shaft: $2\frac{15}{16}$" diameter; angle pillow blocks, series 1500 babbitted bearings; Alemite hydraulic fittings.

Head pulley: 20" diameter by 26" dbcf.

Tail pulley: 20" diameter by 26" dbcf.

Tail shaft: $2\frac{7}{16}$".

Takeups: style DS, series 2800 babbitted bearings or approved equal, 12" adjustment, protected screw type.

Necessary safety collars: keys and safety setscrews; necessary safety guards for moving parts to meet safety laws of the state of _____ .

Conveyor supporting framework: steel, to be provided as shown on attached drawing.

Decking: 16 gauge.

Discharge spout from shuttle conveyor to bins: $\frac{3}{16}$", to be provided as shown on attached drawing.

Plate liners: $\frac{1}{4}$" steel, renewable, on _____ wearing surfaces.

_____

*Link-Belt catalog numbers are for guidance only.

Rails: furnished by manufacturer; Owner will furnish rail supports.

Trucks: single-flange, rigid, babbitted bearings with Alemite hydraulic fittings.

Truck centers: 9'0".

Wheels: 12" diameter.

Track: 3'9¾" gauge.

Rail: 16 lb.

Motor drive: the manufacturer shall furnish the drive system including motor, enclosed herringbone-gear-type speed reducer, located as shown on the attached drawing; flexible coupling; baseplate to mount reducer and motor and the necessary RO-40 roller chain drive from reducer to elevator head shaft. This arrangement should drive the belt at 50 fpm. Purchaser will specify power requirements for motor.

The drives shall consist of flint rim, plate center, driver and driven sprockets.

1. In general, the manufacturer shall design, furnish, fabricate, and deliver all equipment and accessories specified or implied herein, necessary to form a complete, workable unit.

2. All equipment shall be thoroughly cleaned and painted with a high grade of machinery paint in such a manner that the paint will adhere to the surface and be weather resistant. All painting shall be done in a neat and professional manner. The equipment shall not be loaded for shipment until the paint is thoroughly dry.

3. All polished parts, shafts, gearing, bearings, and so on shall be coated with rust-preventive compound and crated, if necessary, to prevent damage in shipment, and all equipment shall be thoroughly protected from the weather.

4. Services of an engineer, if required at time of installation, shall be furnished at a rate stated in the proposal.

5. The successful bidder shall furnish three sets of prints for approval of the general design of the equipment and, later, _____ complete sets of certified prints to: _____ .

2.35 SPECIFICATION FOR 30" REVERSIBLE BELT CONVEYOR

One conveyor required complete.

Capacity: 60–80 tph, $1\frac{1}{2}$" lime and coke mixture; weight of 40 lb/ft^3.

Conveyor length between centers of head and tail pulley: 63'0".

Belt: 30" wide, 5-ply, 28 oz duck with $\frac{1}{2}$" rubber top surface and $\frac{1}{32}$" rubber bottom surface or equal; provide lacing.

Troughing idlers: spacing 4′ on centers, three-roll anti-friction, Link-Belt 7507-30* or approved equal with 5″ steel rolls; Alemite hydraulic fittings lubricated from one side of conveyor.

Return idlers: spacing 8′ on centers, single pulley, Link-Belt 7517-30 or approved equal; Alemite hydraulic fittings.

Head shaft: $2^{15/16}$″ diameter; angle pillow blocks, series 1500 babbitted bearings or equal; Alemite hydraulic fittings.

Head pulley: 24″ diameter by 32″ dbcf.

Tail pulley: 24″ diameter by 32″ dbcf.

Tail shaft: $2^{7/16}$″ diameter.

Takeups: style DS, series 2800 babbitted bearings or approved equal; Alemite hydraulic fittings, 12″ adjustment, protected screw type.

Necessary safety collars: keys and safety setscrews; necessary safety guards for moving parts to meet requirements of the safety laws of the state of _____.

Supporting framework: to be provided as shown on attached drawing.

Guards, chutes, and spouts: to and from the conveyor, will be furnished by the Owner.

Decking: 16 gauge.

Motor drive: The owner will supply a 3-hp, 900-rpm totally enclosed motor for 440-V, 3-ph, 60-Hz service, together with magnetic starter and pushbutton control. The manufacturer shall furnish the rest of the drive, including herringbone-gear-type speed reducer and enclosed positive variable-speed reducer with 3 : 1 ratio, including remote control located as shown on the attached drawing; flexible coupling; baseplate to mount reducer and motor; and the necessary RO-40 roller chain drive from the reducer to the conveyor head shaft.

The drives shall consist of flint rim, plate center, driver and driven sprockets.

1. In general, the manufacturer shall design, furnish, fabricate, and deliver all equipment necessary to form a complete workable unit.
2. All equipment shall be thoroughly cleaned and painted with a high grade of machinery paint in such manner that the paint will adhere to the surface and be weather resistant. All painting shall be done in a professional manner. The equipment shall not be loaded for shipment until the paint is thoroughly dry.
3. All polished parts, shafts, gearing, bearings, and so on shall be coated with rust-preventive compound and crated, if necessary, to prevent damage in shipment, and all equipment shall be thoroughly protected from the weather.

---

*Link-Belt catalog numbers are for guidance only.

4. Services of an engineer, if required at time of installation, shall be furnished at a rate stated in the proposal.
5. The successful bidder shall furnish three sets of prints for approval of the general design of the equipment, and, later, _____ complete sets of certified prints to: _____.

2.36   SPECIFICATION FOR 24″ BELT CONVEYOR

One conveyor required complete.

Capacity: 80–100 tph, $1\frac{1}{2}$″ lime and coke mixture; belt speed about 300 fpm; 120°F temperature; weight of 40 lb/ft³.

Conveyor length between centers of head and tail pulley: 329′0″.

Belt: 24″ wide, 150°F temperature-resistant belt; provide lacing.

Troughing idlers: spacing 4′ on centers, three-roll anti-friction, Link-Belt 7507-30* or equal, with 5″ steel rolls; Alemite hydraulic fittings lubricated from one side of conveyor, 7507-24 self-aligning, spaced every tenth troughing idler.

Return idlers: spacing 8′ on centers, single-pulley, Link-Belt 7517-30 or equal; Alemite hydraulic fittings.

Head shaft: $2^{15/16}$″ diameter; angle pillow blocks, series 1500 babbitted bearings or equal; Alemite hydraulic fittings.

Head pulley: 24″ diameter by 26″ dbcf.

Tail pulley: 24″ diameter by 26″ dbcf.

Tail shaft: $2^{7/16}$″ diameter.

Takeups: style DS, series 2800 babbitted bearings or approved equal; Alemite hydraulic fittings, 24″ adjustment, protected screw type.

Necessary safety collars: keys and safety setscrews; necessary safety guards for moving parts to meet safety laws of the state of _____.

Supporting framework: to be provided as shown on attached drawing.

Guards: will be furnished by the Owner.

Decking: 16 gauge.

Tripper: one automatic type belt-propelled and self-reversing right-hand assembly, complete with supports and 16-pound rails; type 51, Link-Belt catalog 1050 (p. 523)*, or approved equal.

Motor drive: the Owner will supply a 5-hp, 900-rpm totally enclosed motor for 440-V, 3 ph, 60-Hz service, together with magnetic starter and pushbutton control. The manufacturer shall furnish the rest of the drive, including enclosed herringbone-gear-type or helical speed reducer, located as shown on attached drawing; flexible coupling; baseplate to mount reducer and motor; and necessary RO-40 roller chain drive from reducer to elevator head shaft.

The drives shall consist of flint rim, plate center, driver and driven sprockets.

1. In general, the manufacturer shall design, furnish, fabricate, and deliver all equipment and accessories, specified or implied herein, necessary to form a complete, workable unit.

2. All equipment shall be thoroughly cleaned and painted with a high grade of machinery paint in such manner that the paint will adhere to the surface and be weather resistant. All painting shall be done in a neat and professional manner. The equipment shall not be loaded for shipment until the paint is thoroughly dry.

3. All polished parts, shafts, gearing, bearings, and so on shall be coated with rust-preventive compound and crated, if necessary, to prevent damage in shipment, and all equipment shall be thoroughly protected from the weather.

4. Services of an engineer, if required at time of installation, shall be furnished at a rate stated in the proposal.

5. The successful bidder shall furnish three sets of prints for approval of the general design of the equipment, and, later, _____complete sets of certified prints to: _____ .

Bucket Elevators

3.1 GENERAL

A bucket elevator consists of a series of uniformly fed buckets mounted on an endless chain or belt which operates over head and foot wheels. The buckets are used to elevate (usually vertically) pulverized, granular, or lumpy materials. The material is received at the boot, raised and then discharged by passing over the head wheel at the top, into a discharge chute. Generally this mechanism is enclosed in a casing, especially the head and foot sections. Some elevators are self-supporting, but more often they are supported by, or at least braced against, a structural steel frame. Bucket elevators will be discussed under the seven headings listed below. Inclined elevators, which were seldom enclosed, were popular for handling crushed stone, but because of OSHA regulations, will probably go out of use (see figure 3.1).

3.1.1 Centrifugal-Discharge Bucket Elevator

In centrifugal-discharge bucket elevators, the material to be elevated is dug out of the boot and discharged by centrifugal force. They are comparatively high-speed elevators, used where the percentage and size of lumps are at a minimum.

3.1.2 Continuous Bucket Elevator

In continuous bucket elevators, buckets closely spaced on chain or belts are designed so that material is loaded directly into the buckets, usually through a loading leg, instead of being scooped up in the boot. Discharge over the head wheel is accomplished by transfer of material from the discharging bucket to the front of the preceding one, which acts as a moving chute to the fixed discharge chute. Sometimes styled Super Capacity, large-capacity continuous-bucket elevators are made with specially designed steel buckets attached at their sides to double-strand long-pitch steel chain.

3.1.3 Positive-Discharge Bucket Elevator

The positive-discharge bucket elevator should be considered where materials tend to stick in the buckets or where fluffy materials are handled. With its buckets at intervals on double-strand chain, this elevator picks up its load in the boot (as does the centrifugal-discharge type), but because of its lower speed, does not depend upon centrifugal force to discharge material from the buckets. The buckets are completely inverted by snubbing the chains after they have passed over the head wheels, giving them opportunity for complete discharge at relatively slow speed.

3.1.4 Combination Elevator–Conveyor

A combination elevator–conveyor is a gravity-discharge unit that elevates and conveys materials of a nonabrasive free-flowing nature, such as bituminous coal or material having similar characteristics. Operating vertically or on a steep incline, it functions as a bucket elevator

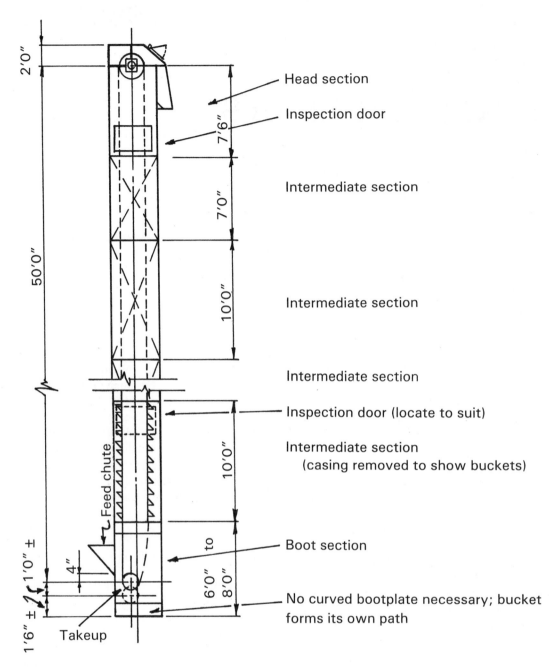

2'0"

50'0"

7'6"

7'0"

10'0"

10'0"

6'0" to 8'0"

Feed chute

1'6" ±

1'0" ±

4"

Takeup

Head section

Inspection door

Intermediate section

Intermediate section

Intermediate section

Inspection door (locate to suit)

Intermediate section
(casing removed to show buckets)

Boot section

No curved bootplate necessary; bucket
forms its own path

Figure 3.1   A typical elevator.

and horizontally as a scraper flight conveyor. Steel buckets, rigidly attached to double-strand steel bar-link, long-pitch roller chain, travel along a continuous steel trough on the horizontal loading run, picking up the material en route. At the lower corner upturn, a special steel corner trough is used to fill the buckets before starting their vertical run. At the upper corner, another curved corner piece is provided to transfer the load to the upper horizontal run, from where the material can be discharged at intervals through openings provided with slide gates. The discharge of the bucket is steep enough to empty as it passes over the discharge opening. It is seldom used, because it is slow and expensive.

3.1.5   Full Volume or Bulk-Flo Elevator–Conveyor Unit

The full volume or Bulk-Flo elevator–conveyor unit carries material in full volume through a dust and weathertight enclosure, completely filling its cross-section. The conveyor medium is an endless chain, upon which are mounted flights at suitable intervals. Flights are made of either the solid or the open-finger type. They are self-feeding and discharge through an opening in the enclosure in a manner depending on the design of the manufacturer.

3.1.6 Pivoted-Bucket Carrier

The pivoted-bucket carrier receives its name from the method of suspending the carrying bucket from double-strand long-pitch chains. Buckets are always in a load-carrying position on horizontal or inclined runs. Discharge is effected by trippers, engaging cams or similar devices located on the bucket ends. This is done in such a manner as to upend the buckets sufficiently to completely empty them.

3.1.7 Internal-Discharge Elevator

The internal-discharge elevator works well in continuously gently handling small bulk articles such as bolts or small castings. Buckets are loaded internally in casing from a chute extending through one side of the casing. Because of their infrequent use, no further space is devoted to them.

3.1.8 Bucket Elevator on Incline

A bucket elevator can operate on an incline, if the chain is guided. If the angle of inclination is over 45° from the vertical, it is better to use an apron or pan conveyor.

3.1.9 Centrifugal vs Continuous-Bucket Elevators

Normally a centrifugal-discharge bucket elevator can handle lumps up to 1 in. if they are not more than 10% of the required capacity. When the lumps are more than 10%, the continuous buckets should be used. Because of its lower speed and methods of loading, the continuous-bucket elevator will cause less breakage of fragile materials.

Belts should be used for corrosive material. If chain is used, it should be heat treated, and Everdur bronze pins and stainless-steel S-shaped cotter pins should be specified. Some material may require alloy buckets.

If the material is damp or wet, even if the capacity is small, a double-strand super-capacity continuous-bucket elevator, equipped with flat bottom buckets, or a positive-discharge bucket elevator should be used. If there

are fines present, there is the possibility of the fines sticking to the bottom.

3.1.10 Preliminary Selection of Type

Table 3.1 tabulates the properties of the various types of elevators.

3.2 CASING

The elevator is usually enclosed in a steel casing, to provide a means of support and as a matter of safety and dust retention. A casing can be made dust-tight, either by using a sealing medium, or continuously welding the corner angles to the plate. Figure 3.2 shows details of dusttight construction. These casings are regularly made with inspection cleanout doors. For free-standing elevators, structural considerations, such as strength of the sections and the size and number of anchor bolts to resist wind, often will dictate the narrow dimension of the casing and its composition.

Steel plates with corner angles provide a substantial support to the complete unit. Elevators are usually self-supporting up to 30 ft, with some even up to, say, 60 ft above the boot. Above these heights, the casing should be structurally supported, and braced against the building or silo for heights over 30 ft.

In some cases, the head shaft supporting the complete chain and buckets is mounted on building steel, to take the load off the casing, which then acts simply as a cover, carrying only its own weight. Above 100-ft centers, the drive (motor and speed reducer) should be supported directly on the building steel, rather than on the elevator casing.

3.2.1 Drives

A V-belt drive from the motor to speed reducer is recommended. Action is similar to that of a shear pin: the belt comes off the sheaves should anything become jammed.

Intermediate sections of elevators are usually made of no. 12 gauge steel, in sections 8–10-ft long. The casing acts as a column, and can support a very heavy vertical load. For easy inspection inside the casing, large inspection doors are usually placed in the intermediate section above the boot, or about 3 ft above the floor in a section that passes through a floor.

The head sections of elevators usually are designed with the hood or top cover split, so the two parts can be easily removed for inspection. A hinged door can also be located on top of the cover, or on the inclined portion

Table 3.1 Preliminary Selection of Elevator Type

Material	Elevator Type	Speed of Elevator Chain or Belt (fpm)	Single Strand of Chain with Spaced Buckets; Dig in Boot	Single Strand of Chain or Belt, Continuous Buckets, Fed in Boot	Handles Material −2 in. to Dust, Dry and Free-Flowing	Double Strand of Chain and Large-Capacity Buckets; Material Fed at Boot	Double Strand of Chain, Large Buckets; Material Scooped up in Boot	Double Strand of Long-Pitch Malleable Chain; Large-Capacity Pivoted Buckets
Grain, coal, sand, sugar, salt, chemicals, petroleum, coke, limestone dust, gypsum, sulfur, cement[a]	Centrifugal-discharge	185–300	Yes		Yes			
Sand, coke, cement, coal[b]	Continuous-bucket	100–150		Yes	Yes			
Coke, limestone, gypsum, cement, clinker, bauxite, coal, dolomite[b]	Super-capacity continuous-bucket	80–100			Yes; also material 6 in. and under	Yes		
Coal, carbon black, powdered bakelite[c]	Positive-discharge	120			Yes		Yes	
Grain, fine free-flowing material[d]	Bulk-Flo, full-volume, en masse elevator-conveyor	up to 80			Fine materials 1/4 in. and under [b,d]			
Crushed stone, coal, ashes, burnt lime, coke, cement, chemicals, fuller's earth[e]	Pivoted-bucket carrier	50 (max)			Well-built unit but expensive			Yes

Note: Centrifugal-discharge and continuous-bucket elevators constitute about 90% of all installations, but the other types are in use to meet the special requirements for which they are designed.

[a]Handles better when dry but can discharge damp material (not sticky).

[b]Material should be dry and free-flowing.

[c]Material should be dry.

[d]Material cannot be damp, abrasive, or corrosive.

[e]A large-capacity machine that can handle wet or dry material.

FIGURE 3.2   81

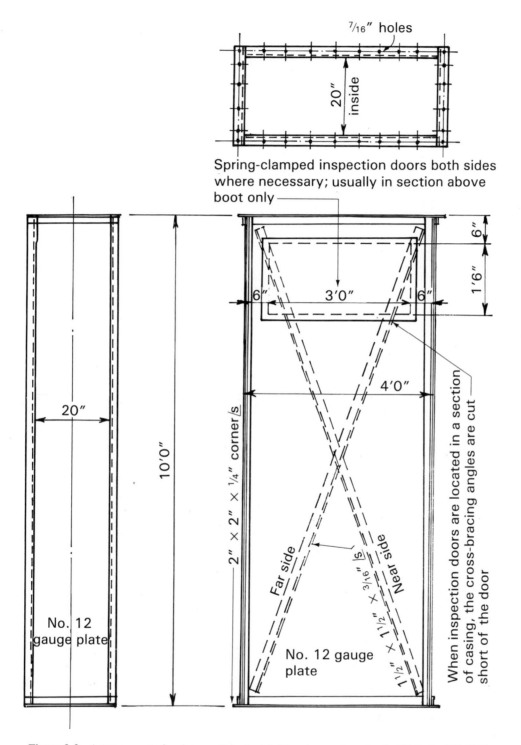

Figure 3.2 Arrangement of an intermediate dust-tight casing section. Apply 3M Sealer EC-1168 before spot welding any casing joint. The casing is Class IV dust-tight construction. Paint all inside and outside surfaces with one coat of grey lead and oil. *Note:* The cross bracing is not standard; furnished when requested.

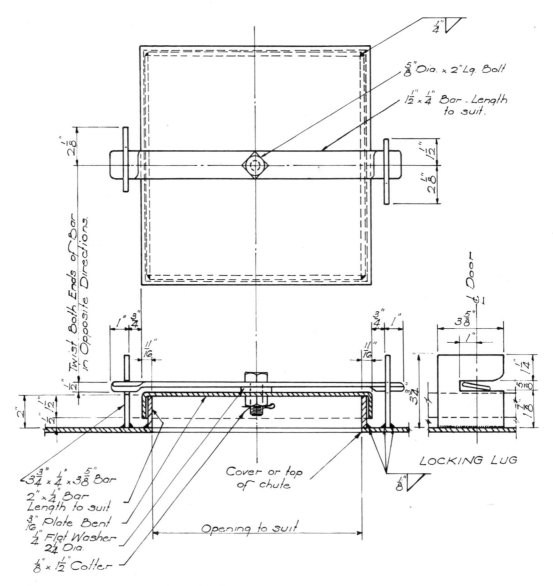

Figure 3.3   Detail of a typical inspection door.

forming part of the chute, both for inspection and when necessary for watching the discharge of material.

For servicing the elevator, a casing should be at least 6 in. wider than the bucket. For tall elevators, say over 75 ft, a bigger allowance should be made to prevent the buckets from slamming against the sides of the casing.

For handling explosive materials, refer to paragraph 14.13.

3.2.2   Inspection Doors

A very practical inspection door is shown on figure 3.3. When handling fine and dusty material, it is a necessity. This can be placed anywhere in the elevator casing, particularly in the head and boot sections. Figure 3.4 shows

a venting type of inspection door. Most manufacturers have their own design, and should be allowed to use it for economy.

3.3   ELEVATOR PITS

Elevators generally are placed in pits, although this should be avoided wherever possible. If pits must be used, ample space should be provided in both length and width to allow for maintenance. A good rule is to provide a minimum clearance of 24 in. on each side of the elevator. The feed inlet point of the continuous-type elevator is somewhat higher than that of the centrifugal elevator, necessitating a deeper pit when located below ground level.

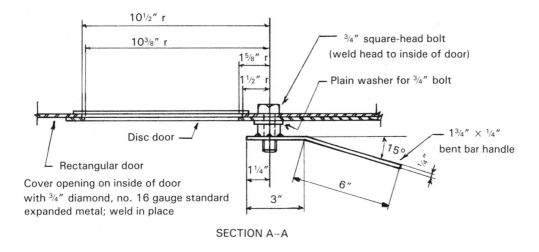

SECTION A–A

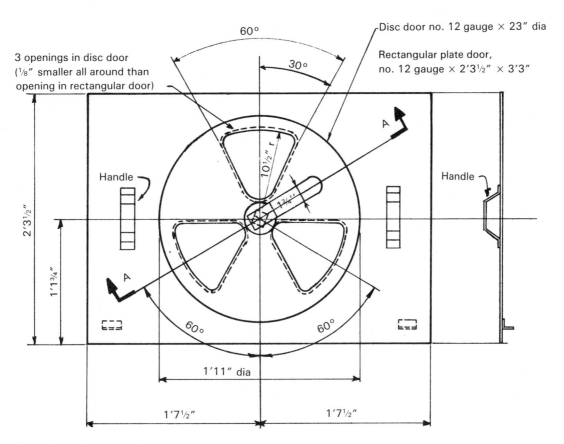

Figure 3.4 Venting-type inspection door.

3.4 BOOT SECTION

Steel elevator boot sections should be made of not less than 3/16-in. steel for elevators under 30 ft and 1/4 in. for elevators above 30 ft. In elevators, say, 30 ft and under, the boot section supports either part or all of the entire unit.

Removable doors and side plates can be installed in boot sections, to make it easier to clean out the boot by hand, when and if required. In industries where products cannot be mixed or contaminated, the boots have to be cleaned out after each operation or run. In some cases, the entire sides of the casing are made removable for cleaning.

Normally, the location of the point of inlet in a boot occurs between 4 in. and 6 in. for centrifugal types, and 20–26 in. for continuous types, above the centerline of the boot or shaft in its highest position. When using take-ups in the boot, and when handling materials that are mixed, say ¼ in. with 10% of lumps not exceeding 2 in., this point of inlet can be lowered to the centerline of the boot shaft without harm, but the capacity of the bucket may be slightly reduced, depending on the material handled.

An allowance of at least 6 in. below the buckets, with the takeup in the lowest position, should be made for cleaning-out purposes.

3.5   HEAD SECTION

Figure 3.5 shows an elevator head section with head takeup and one method of supporting the drive mechanism. This is a self-supporting casing. Normally, the point of discharge is located as shown on the figure; that is, 6 in. below centerline of the head shaft, projected on a 45° line downward. An adjustable throat plate in the bottom of the discharge spout is usually used to prevent materials from falling down the casing to the boot. When handling very fine and dry materials, the 6 in. vertical dimension should be made 12 in. This provides more

time for the buckets to discharge the fine material. In some cases, the head shaft supporting the chain and buckets is mounted on building steel. The casing then acts as only a cover, with no machinery load on it.

The head sections of elevators are made with either fixed shaft or takeup shaft. Covers should be made split where possible so that the two parts can be easily removed for inspection and maintenance. A door opening can be located on top of the cover or the inclined part of the discharge spout, both for inspection and for checking the discharge of material (see figure 3.6). For dust takeoff, one connection can be made in the boot section just above the loading hopper, and one at the discharge chute or at the top of the elevator. Provide pipe connection at the top of each.

In handling dusty material, a good head shaft dust seal should be used to prevent the dust from coming out of the head section of the elevator casing. A design for such a seal is shown on figure 3.7. It is inexpensive to make, and should be attached in the shop.

3.6   PLATFORMS AND LADDERS

On vertical elevators of any height where the head shaft cannot be easily reached by maintenance personnel, it is necessary to include a standard steel ladder attach-

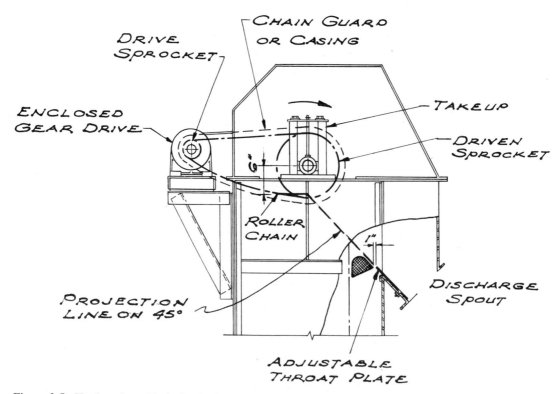

Figure 3.5   Head section with the head takeup drive supported on the casing. (*Courtesy FMC*)

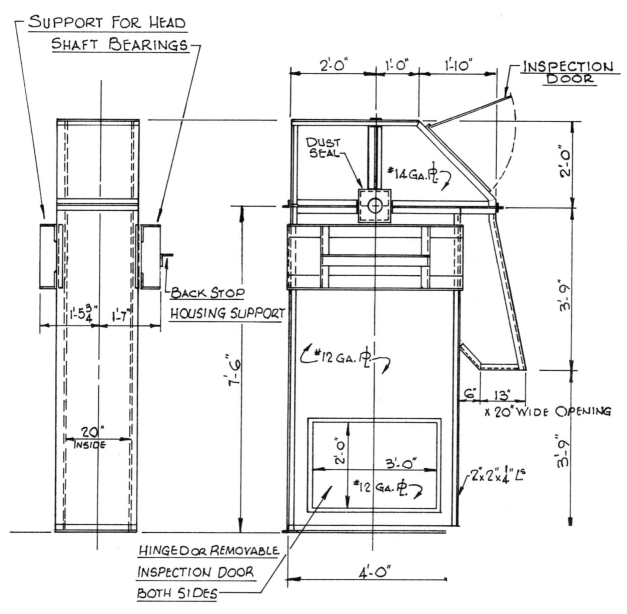

Figure 3.6 Dust-tight construction of the head section and hood arrangement.

ment to the casing, including a steel safety guard begin-ning 7 ft from lower floor level, and extending to a steel platform. This platform should be of ample size for working on, with the floor of expanded metal, grating, or diamond floor plates. In areas where considerable snow falls and the elevator is located outdoors, an open grating should be used to rid the platform of ice and snow and to prevent slipping. Intermediate platforms should be provided every 30 ft or so.

Handrails should be of standard design, made of an-gles or piping as approved by safety regulations. Steel toe plates about 6-in. high must be included to prevent a person's feet from moving off the platform. A hoist

beam can be provided about four feet above the top of the elevator casing on line of the head shaft, to assist in maintenance work. A V-belt drive from motor to speed reducer is preferred by some because the belt will come off the sheaves should anything become jammed (figure 3.8).

3.7 BUCKETS

Malleable iron buckets, either continuous type or type AA, have a Brinell hardness of about 120. Promal buck-ets are heat-treated, malleable-iron buckets with a Bri-

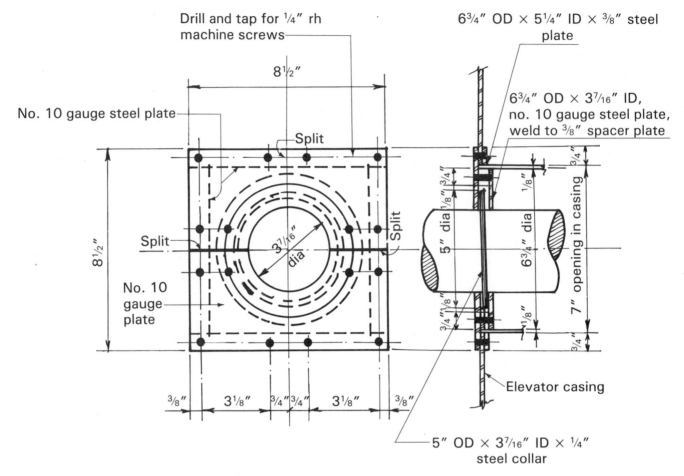

Figure 3.7 Head shaft dust seal.

nell hardness of 190. Buckets should be at least four times the size of the lumps, to get required capacity and avoid spill. For a width of bucket greater than 16 in., two strands of chain (or a belt) must be used. Charcoal, especially, requires wide buckets on two strands of

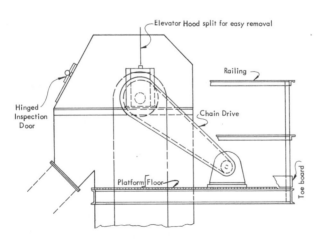

Figure 3.8 · Elevator platform.

chain. All steel buckets today are made of welded construction, either spot welded or continuously welded, depending on the fineness of material handled. For abrasive material, heat-treated, malleable-iron or cast buckets should be used.

The DuraBucket division of National Oats Co., Inc., and the Screw Conveyor Corporation make a high-density polyethylene bucket that is used primarily in the handling of grain, feed, cottonseed oil, salt production, soybean oil processing, and similar products. The buckets are rustproof, shatterproof, sparkproof, and self-cleaning. They weigh a third the weight of steel buckets and a fifth of malleable iron. DuraBucket also makes Low-Profile (LP), designed to increase the overall capacity by closing up the spacing on the belt (refer to paragraph 3.14 and table 3.9).

In light fluffy material, four or five holes, about ¼ in. in diameter in the bottom and ½ in. diameter in the sides near the top, are placed to break the suction or vacuum created by the speed of the bucket in picking up the load in the boot. Without these holes, the light materials usually stay in the bucket and go down to the

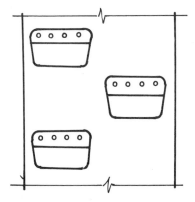 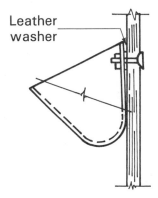

Figure 3.9 Alternating buckets.

boot again, often piling up and causing an undue strain for the buckets to pick up and, incidentally, increasing the horsepower required. Very little capacity is lost through the holes. In handling hot cement or gypsum, at 200–300°F, the holes in the bottom of the bucket help cool the material. The holes can be used for material at higher temperatures, and the cooling can be augmented by introducing outside air into the casing.

Alternating buckets are used on wide belts to obtain a better pickup in the boot of the elevator, and to prevent any possible flooding of the boot with an avalanche of material coming to it. With a single bucket extending across a wide belt and spaced apart at varying intervals, a slight void space could result as the buckets are turning around the foot pulley. With a continuous feed, the material would tend to pile up and finally stall the elevator (see figure 3.9).

Buckets are discussed further under each type of elevator.

3.8 CHAIN

Malleable chains are made with a Brinell hardness of about 120. If necessary, the chain can be made more tough by processing the malleable iron. When handling abrasive materials such as sand, gravel, stone, or alloys, toughened malleable iron should be used (refer to paragraph 3.8.3).

Rexnord calls its tough malleable chain "Z" metal. It rates a Brinell hardness of approximately 190, and an increase in strength of about 25%.

PTC, under the trade name Link-Belt, produces a tough malleable chain called Promal (*pro*cessed *mal*leable). In their process, the malleable links are reheated to approximately 1200°F and immediately quenched in oil. The result is an increase in hardness to Brinell 190, and a 25% gain in ultimate strength. It is almost impossible to get a Brinell hardness much above

190 by heat treatment. The malleable structure becomes disarranged, and gradually changes back to white iron, from which malleable iron starts.

Jeffrey Manufacturing Company calls their treated chain Super Mal. They heat the malleable iron and then plunge it into a bath of cyanide. This puts a coat about ¹⁄₆₄ in. in thickness on the chain. If this skin is broken, the material changes back to malleable iron.

For high elevators, say, 75 ft or over, it is also advisable to use two strands of chain. In that case, the specifications should require the strands to be matched and tagged right- or left-hand, although they are not actually right- or left-hand design. The two strands of chain are rigidly attached to the bucket and no two chains stretch alike during operation. To meet this specification, the manufacturer will shop-assemble these strands of chain, to make sure that the attachments are opposite each other and to tag each strand properly. If the erection is properly done, the buckets will be straight, even if the attachments are slightly off.

On double-strand chains, some preference has been expressed for the use of 6-in. pitch instead of the standard 12-in. pitch, with the bucket attachments every other pitch. This is done to run the chain more smoothly going over the head sprockets. While it is true that by shortening the pitch, the chain will follow more closely to the circumference of the sprocket, it is doubtful that the extra expense can be justified. Where the pitch is 18 in., a 9-in. pitch may be justified.

The selection of the type of chain, that is, malleable or steel combination, is dependent on the type of material to be handled, capacity required, type of duty (continuous or intermittent), and height of elevator. Class C combination chains are economical for general elevator service. SBS bushed chains are widely used for heavy duty and high elevators or those handling abrasive materials.

A variety of chain is available and shown in manufacturers' catalogs, and an effort is being made to stan-

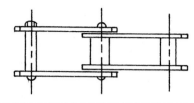

Table 3.2   Properties of SBS Class Chains

Chain Number	Average Pitch (in.)	Allowable Chain Pull[a] (lb)	Average Ultimate Strength (lb)	Average Weight Each Attachment (lb)	Weight Plain Chain (lb/ft)	Total Weight with Attachments[b] (lb/ft)
SBS 188	2.609	2,750	25,000	1.1 (K1)	3.8	4.0 (2)
SBS 110	6.000	6,300	40,000	4.3 (K2)	6.3	7.45 (1)
SBS 102B	4.000	6,500	40,000	3.0 (K2)	6.9	8.3 (2)
SBS 102½	4.040	7,900	50,000	4.8 (K3)	9.4	12.7 (2)
SBS 111	4.760	8,850	50,000	7.2 (K2)	10.2˙	12.3 (1)
SBS 856	6.000	14,000	100,000	11.5 (K2)	16.5	28.0 (1)
SBS 150 Plus	6.050	15,100	100,000	11.5 (K2)	16.6	28.0 (1)

Note: All steel bushed rollers.
[a]Ratings based on a service factor of 1.0. Apply service factors from paragraph 3.8.1.
[b]The number of attachments per foot is indicated in parentheses.

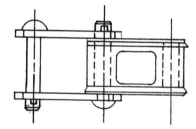

Table 3.3   Properties of Class C Combination Chain

Chain Number	Average Pitch (in.)	Allowable Chain Pull[a] (lb)	Average Ultimate Strength[b] (lb)	Weight (lb/ft) A	Attachments No.	Weight (lb/ft) B	Weight (lb/ft) C
C 188	2.609	1,950	14,000	3.6	K1	4.0	3.9
					K2	4.7	4.3
C 102B	4.000	4,000	24,000	6.8	K2	8.1	7.8
C 110	6.000	4,000	24,000	6.0	K2	7.1	7.3
C 102½	4.040	5,550	36,000	9.7	K3	11.8	—
C 111	4.760	5,950	36,000	9.8	K2	11.2	11.3
C 132	6.050	8,330	50,000	14.5	K2	16.3	16.1

Notes: Center links, cast; sidebars, steel.
Total weight per foot = weight of A + B or A + C.
B = center link attachment every second link.
C = steel sidebar attachment every second link.

[a]For malleable and Promal center links. Ratings are based on a service factor of 1.0. Apply service factors from paragraph 3.8.1.
[b]For malleable iron center links. Promal is approximately 25% stronger.

dardize the types. Those most commonly used are shown in tables 3.2 and 3.3. Basically, however, the SBS-110 and SBS-102B are easy to obtain and have proved themselves satisfactory in most installations. SBS and C combination chains are used in elevators. For handling gritty and abrasive materials and in high vertical elevators, SBS chain is preferred. Class C combination chain is less expensive, and is used on high vertical elevators handling aggregates, cement, and similar products. For other chains refer to table 1.3.

3.8.1 Service Factor

Here are summarized service factors for only those items that are normally involved in elevator and conveyor work:

Uniformly loaded: 1.0
Not uniformly loaded: 1.3
Reciprocating conveyors: 1.5
Multiple Strand Factor
2 strands: 1.7

3.8.2 Recommended Speeds

The recommended speeds for various types of elevators are shown in table 3.1. For the maximum speeds of all conveyor and elevator chains based on the number of teeth in the driving sprocket, refer to tables 1.5 and 1.6.

3.8.3 Chain versus Belt

Choosing between chain and belt as an elevating medium depends upon the characteristics of material handled. Where the temperature of the belt is likely to exceed about 250°F, it is safer to use chain, and select the best quality obtainable for the service. Hot materials up to 450°F (232°C) can be handled in continuous buckets mounted on standard chain. If the temperature goes to 600°F (315°C), special steel buckets mounted on Promal or Z chain should be used. Most malleable or steel chains will stand up to 600°F (315°C). Above 600°F, heat-treated alloy chain must be used. There are high-temperature belts on the market which may be used under certain conditions. It is advisable to consult with the belt manufacturer.

The Screw Conveyor Company, Sprout-Waldron Division of Koppers Company, and FMC sell a Wing-Type Pulley. These pulleys are installed on the boot or foot shaft and are usually self-cleaning, offering maximum protection from belt damage as a result of lumps or foreign material under belt. In this case, it is usually better to use chain. Whatever type of equipment is used, due

consideration should be given to the lift factor. The lowest initial cost frequently becomes the more expensive in the long run. Most materials with lumps up to 2 ½ in. can be handled with chain.

Abrasive materials, such as sand and abrasive grain, should be handled by belt instead of chain because the fine particle size could easily get into the chain joints and cause rapid wear. If any of this class of material should be damp or wet, the belt may slip on the head pulley unless lagged with a herringbone-cut-groove rubber cover. Belts should also be used for corrosive material. If chain is used, it should be heat-treated. Corrosive materials may require alloy buckets, and Everdur bronze pins and stainless steel S-shaped cotter pins should be specified.

Usually when belts are used, the diameter of the head pulley is larger to prevent slipping and, with a belt on a continuous elevator, especially outdoors, the pulley must be lagged, or covered with a rubber covering known as "rough-top brand" or with herringbone grooves cut into it, to get good contact with the belt. Larger pulley diameters at the same rpm as the head shaft give additional speed in feet per minute. When selecting a belt as an elevating medium, materials that pack and tend to build up between the belt and pulley, as well as rough or jagged particles that damage the belt by becoming lodged between buckets and belt, should be avoided. To some extent, these difficulties are alleviated through the use of wing pulleys on the foot shaft. In handling lumps with sharp edges, it is usually better to use chains. These lumps may become lodged between buckets and belt, resulting in damage to the belt as the material is picked up in the boot, or as the belt passes over the head pulley.

3.9 TRACTION WHEELS AND SPROCKETS

For general-purpose installations where there are no frequent shock loads, the arm or spoke-type sprocket is used. Plate-center sprockets (arm sprockets filled in to make a solid center) are used where shock loads are anticipated or where the maximum allowable chain pull on heavy-duty chains is required. Split sprockets can be furnished in arm or plate-center sprockets to facilitate mounting or removing them from the shaft without disturbing the bearings or the shaft itself (figure 3.10).

Hunting-tooth sprockets have an odd number of teeth, with the pitch of the teeth one-half that of the chain. Because of the odd number of teeth, the chain barrels contact the intermediate teeth after each revolution of the sprocket. Therefore, each tooth has one-half the number of contacts that it would have on a regular full-pitch sprocket over any period of time, thus increasing the life of the wheel on high-speed shafts.

Figure 3.10 Detail of a split sprocket.

the elevator clogs, the traction wheel will slip. It should be used for elevators 50 ft and over, and sometimes over 35 ft when handling abrasive materials. Some prefer to use the same sprockets at foot and head so that, in an emergency, the foot sprocket can be used at the head, if the teeth are not too badly worn.

Some manufacturers of sprockets have developed a method of casting chrome–nickel inserts into the rim of the sprocket to provide great strength, toughness, and abrasion-resisting qualities. Split (two-piece) sprockets, bolted together at the rims and at the hubs, also help to reduce labor costs.

Traction wheels (without teeth) and sprocket wheels also are made with cast-iron solid-hub centers, and with sectional bolted rims that can be removed without disturbing the hubs in any way, and replaced quickly with a minimum of down time. There is a growing tendency to use traction wheels at the bottom instead of sprockets. Traction wheels cannot be used at the foot unless a takeup is provided at that point, as the chain will slip off. This is especially true if the links are short. There is always traction at the head because of the load. When

3.10 TAKEUPS

Normally, elevators have the screw-type takeup on the foot or boot shaft unless space does not permit. If it is necessary to place the screw-type takeup on head shaft, the centers of the bucket elevator should not exceed 90 ft, because the total weight of chain (or belt) plus buckets and load in buckets on up or carry side, is hanging on the takeup screw in tension (see figure 3.11). Wherever a head takeup is used, the next larger sized head shaft from that recommended in the catalog should be

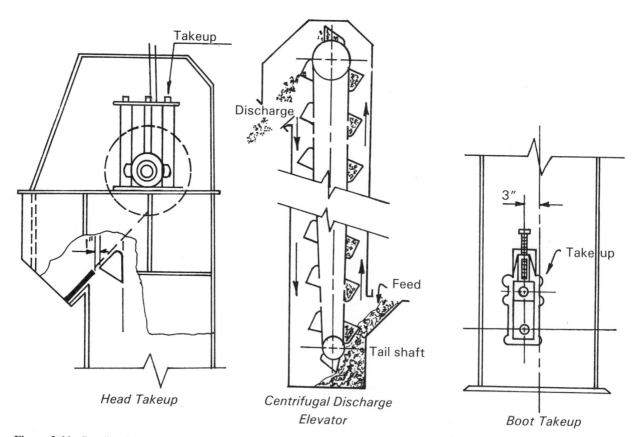

Figure 3.11 Details of a centrifugal-discharge elevator.

used, as the vibration is transferred to the head shaft through the pickup in the boot.

Gravity takeups are used on many elevators, particularly on powdery or aerated material such as cement, lime, and gypsum. A softening effect is encountered at the pickup which must be absorbed by this floating takeup. The frame supporting the shaft and wheel simply rides up and down in angle or channel guides, attached to the inside of the casing. Usually there is enough weight in the complete takeup to keep it in position but, if necessary, additional weight can be placed on the movable steel or cast steel frame (see figures 3.12 and 3.13). The sprocket, or traction wheel, runs loose on the shaft and is kept in place by safety collars on each side of the hub. The diameter of the hub is much larger than necessary, so that in the event the bore of the wheel

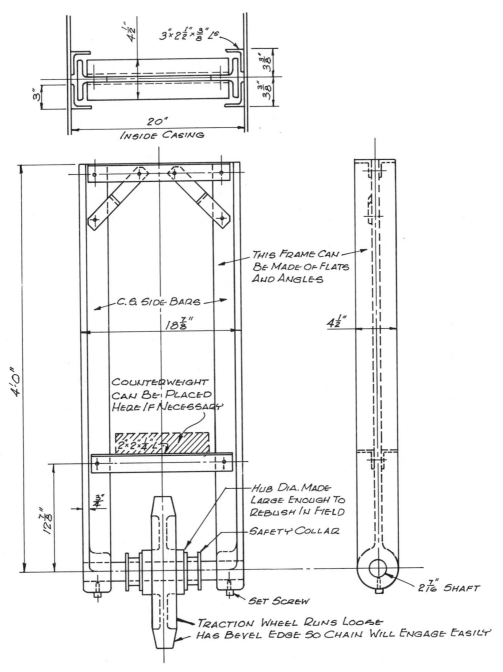

Figure 3.12 Arrangement of a gravity takeup inside a 20-in. wide casing. Note that the shaft does not go through the casing.

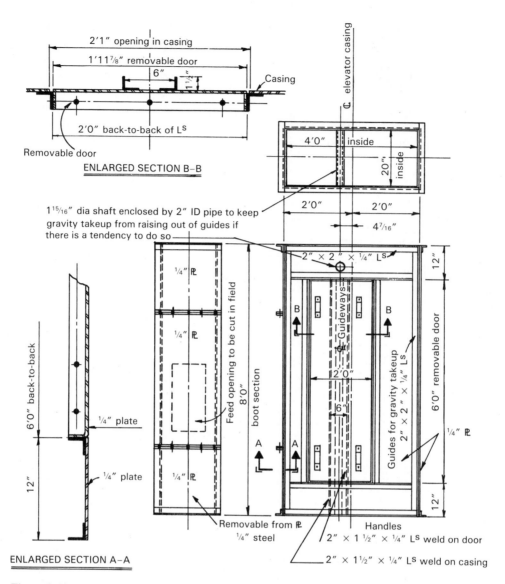

Figure 3.13 Arrangement of an elevator boot for a gravity takeup.

becomes sloppy, instead of discarding the wheel, it can be bushed. No lubrication is provided since the shaft is pinned and does not rotate.

Figure 3.14 shows a head shaft equipped with a differential band brake (back-stop) that is used to prevent the up, or carrying, run from running backward in the event of a power interruption. The backward drift is expected to be less than 2 ft and, upon resumption of power, the brake is immediately released.

Stephens-Adamson Mfg. Co. makes an enclosed roller-type holdback that is very effective, particularly on large horsepower drives. A ratchet-and-pawl back-stop has the disadvantage of the pawl or the teeth in the wheel failing under shock loads.

3.11 HORSEPOWER

There are several formulas in use for computing horsepower. In general, they are based on two principles:

1. The weight of the material in the loaded buckets. The weight of the chain or belt and the weight of the buckets on the up-run is balanced by the weight of the chain or belt and the buckets on the down-run.
2. An allowance is made for the extra load at the boot, and for boot pulley friction.

The various formulas will be illustrated under the various types of elevators. In addition, methods used in old Link-Belt and Jeffrey catalogs will be shown, in or-

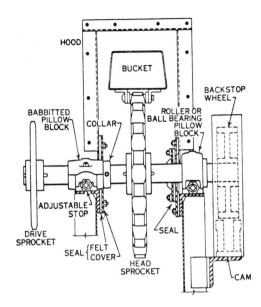

Figure 3.14 Headshaft equipped with a backstop. (*Courtesy FMC*)

der to help in review of old designs. It is to be noted that all formulas have given satisfactory results when good judgment, based on experience, was used.

All the calculations as presented are adequate for standard industrial elevators. For large, high-capacity, high-lift, engineered class elevators, a more elaborate case-by-case evaluation is generally warranted. It might be well to note that the CEMA method of computations is now the standard for the industry.

3.11.1 Tension in Chain or Belt (from Goodyear's *Handbook*, slightly modified)

a. If no foot tension is required, total tension is determined by

$$T_b + T_B + T_m + T_H$$

where

T_b = tension due to weight of buckets (lb)
   = $(12H \times w_b)/s$
H = height of elevator (ft), center to center of head and boot shafts plus diameter of head wheel (ft)
w_b = weight of each bucket (lb)
s = spacing of buckets (in.)
T_B = tension due to weight of belt or chain
   = $H \times w_c$

w_c = weight of belt or chain per foot (lb)
T_m = tension due to weight of material in buckets
   = $(12H \times w_m)/s$
w_m = weight of material in each bucket (lb)
T_H = tension due to pickup load and foot wheel friction
   = $(12H_o \times w_m)/s$
H_o = height factor (ft) to take care of pickup force
   = 10 for continuous-bucket elevators
   = 30 for centrifugal-discharge (spaced) bucket elevators

b. If foot tension is required, total tension is determined by

$$(1 + K) \times 12 \times w_m \left[(H + H_o)/S\right]$$

where

K = 0.97 for bare drive pulley or chain sprocket with screw takeup
   = 0.80 for lagged drive pulley with screw takeup
   = 0.64 for bare drive pulley or chain sprocket with counterweight takeup
   = 0.50 for lagged drive belt with counterweight takeup
S = speed of buckets (fpm)

3.11.2 Horsepower Allowance

The allowance for extra load at the boot has been a matter of judgment. From tests, the value has been assumed to be the equivalent of 250–500 lb of load. The value of H_o in paragraph 3.11.1 helps to place a more definite value on it. Thus, for continuous buckets, it will be $(10 \times 12 \times w_m)/s$, and $(30 \times 12 \times w_m)/s$ for centrifugal-discharge (spaced) bucket elevators.

Tests to check the formulas have been very difficult to make, primarily because the actual weight in the buckets at any time has been difficult to determine, due to both the volume in the buckets and the unit weight of the material at the time. This formula (where S is the speed of buckets in fpm) in use,

$$\text{hp} = \frac{w_m \times (H + H_o) \times 12 \times S}{33,000\,s} \times 1.5$$

where 1.5 is the efficiency factor, has resulted in motors being too large, since it is assumed that all buckets are filled to 100% of capacity, along with the maximum unit weight of the material. Applying a factor of 0.8 to w_m, that is, reducing the 1.5 value to 1.2, has given good results in installations where it was tried.

(*Text continues on page 100.*)

Table 3.4 Centrifugal-Discharge Bucket Elevator Selection/Specifications, Type 1—Chain

| Elev No.[a] | ft^3/hr[b] | Buckets[c] | | Chain | fpm | Max Lump Size | | Casing | Headshaft Sprkt | | | Footshaft Sprkt | | |
		Size	Space			100%	10%		Teeth	Pitch Dia	rpm	Teeth	Pitch Dia	Dia
102	280	6 × 4	13	C188	225	½	2½	9¾ × 35	24	20	43	18	15	1½
107	612	8 × 5	16	C102B	260	¾	3	11¾ × 42	19	24¼	41	14	18	2
108		8 × 5	16	SBS102B	260	¾	3	11¾ × 42	19	24¼	41	14	18	2
112	960	10 × 6	18	C110	268	1	3½	13¾ × 48	13	25	41	11	21¼	2
113		10 × 6	18	SBS110	268	1	3½	13¾ × 48	13	25	41	11	21¼	2
117	1536	12 × 7	18	SBS110	268	1¼	4	15¾ × 48	13	25	41	9	17½	2
128	2112	14 × 7	18	SBS110	306	1¼	4	17¾ × 54	16	30¾	38	12	23¼	2⁷⁄₁₆
134	3120	16 × 8	18	SBS110	306	1½	4½	19¾ × 54	16	30¾	38	11	21¼	2⁷⁄₁₆

[a]Bucket elevator assemblies include head shaft machinery with either ball or roller bearing pillow blocks, chain, buckets, casing, Style 1 or Style 2 discharge spout, stub inlet and internal gravity takeup with hard iron bearings. (Internal gravity takeup is available with cement mill type sleeves and bearings when handling highly abrasive materials.) Drives with backstops, service platforms, and ladders with safety cages can be furnished.

[b]Based on buckets filled to 75% of theoretical capacity.

[c]Style AA malleable iron buckets.

Elev No.	Aprx Wt. (lb) Terminals	Wt (lb/ft) Centers
102	686	58
107	906	82
108	887	83
112	1035	91
113	1140	92
117	1139	98
128	1525	107
134	1734	120

TABLE 3.4 95

| | Material Weight[d] | | | | | | | | | | | | | | |
| | 35 lb/ft³ | | | 50 lb/ft³ | | | 60 lb/ft³ | | | 75 lb/ft³ | | | 100 lb/ft³ | | |
Elev No.	Ctrs	Hd Shft	hp	Ctrs	Hd Shft	hp	Ctrs	Hd Shft	hp	Ctrs	Hd Shft	hp	Ctrs	Hd Shft	hp
102	Up to 87	$1\frac{15}{16}$	1	Up to 73	$1\frac{15}{16}$	1	Up to 58	$1\frac{15}{16}$	1	Up to 43	$1\frac{15}{16}$	1	Up to 28	$1\frac{15}{16}$	1
	88 to 100	$2\frac{7}{16}$	1	74 to 83	$1\frac{15}{16}$	$1\frac{1}{2}$	59 to 80	$1\frac{15}{16}$	$1\frac{1}{2}$	44 to 73	$1\frac{15}{16}$	$1\frac{1}{2}$	29 to 50	$1\frac{15}{16}$	$1\frac{1}{2}$
				84 to 100	$2\frac{7}{16}$	$1\frac{1}{2}$	81 to 95	$2\frac{7}{16}$	$1\frac{1}{2}$	74 to 100	$2\frac{7}{16}$	2	51 to 69	$1\frac{15}{16}$	2
							96 to 100	$2\frac{7}{16}$	2				70 to 97	$2\frac{7}{16}$	3
107 108	Up to 38	$1\frac{15}{16}$	1	Up to 20	$1\frac{15}{16}$	1	Up to 13	$1\frac{15}{16}$	1	Up to 20	$1\frac{15}{16}$	$1\frac{1}{2}$	Up to 10	$1\frac{15}{16}$	$1\frac{1}{2}$
	39 to 67	$2\frac{7}{16}$	$1\frac{1}{2}$	21 to 35	$1\frac{15}{16}$	$1\frac{1}{2}$	14 to 30	$1\frac{15}{16}$	$1\frac{1}{2}$	21 to 30	$1\frac{15}{16}$	2	11 to 20	$1\frac{15}{16}$	2
	68 to 93	$2\frac{7}{16}$	2	36 to 61	$2\frac{7}{16}$	2	31 to 47	$2\frac{7}{16}$	2	31 to 61	$2\frac{7}{16}$	3	21 to 41	$2\frac{7}{16}$	3
	94 to 100	$2\frac{15}{16}$	3	62 to 88	$2\frac{7}{16}$	3	48 to 82	$2\frac{7}{16}$	3	62 to 80	$2\frac{7}{16}$	5	42 to 72	$2\frac{7}{16}$	5
				89 to 100	$2\frac{15}{16}$	3	83 to 100	$2\frac{15}{16}$	5	81 to 100	$2\frac{15}{16}$	5	73 to 82	$2\frac{15}{16}$	5
													83 to 100	$2\frac{15}{16}$	$7\frac{1}{2}$
112 113	Up to 13	$1\frac{15}{16}$	1	Up to 15	$1\frac{15}{16}$	$1\frac{1}{2}$	Up to 19	$1\frac{15}{16}$	2	Up to 10	$1\frac{15}{16}$	2	Up to 10	$1\frac{15}{16}$	3
	14 to 27	$1\frac{15}{16}$	$1\frac{1}{2}$	16 to 22	$1\frac{15}{16}$	2	20 to 41	$2\frac{7}{16}$	3	11 to 28	$2\frac{7}{16}$	3	11 to 15	$2\frac{7}{16}$	3
	28 to 50	$2\frac{7}{16}$	2	23 to 54	$2\frac{7}{16}$	3	42 to 62	$2\frac{7}{16}$	5	29 to 56	$2\frac{7}{16}$	5	16 to 41	$2\frac{7}{16}$	5
	51 to 73	$2\frac{7}{16}$	3	55 to 66	$2\frac{7}{16}$	5	63 to 85	$2\frac{15}{16}$	5	57 to 63	$2\frac{15}{16}$	5	42 to 47	$2\frac{15}{16}$	$7\frac{1}{2}$
	74 to 88	$2\frac{15}{16}$	3	67 to 100	$2\frac{15}{16}$	5	86 to 100	$2\frac{15}{16}$	$7\frac{1}{2}$	64 to 100	$2\frac{15}{16}$	$7\frac{1}{2}$	48 to 74	$2\frac{15}{16}$	$7\frac{1}{2}$
	89 to 100	$2\frac{15}{16}$	5										75 to 100	$2\frac{15}{16}$	10
117	Up to 15	$1\frac{15}{16}$	$1\frac{1}{2}$	Up to 12	$1\frac{15}{16}$	2	Up to 21	$2\frac{7}{16}$	3	Up to 13	$2\frac{7}{16}$	3	Up to 21	$2\frac{7}{16}$	5
	16 to 27	$2\frac{7}{16}$	2	13 to 29	$2\frac{7}{16}$	3	22 to 38	$2\frac{7}{16}$	5	14 to 33	$2\frac{7}{16}$	5	22 to 26	$2\frac{7}{16}$	$7\frac{1}{2}$
	28 to 48	$2\frac{7}{16}$	3	30 to 42	$2\frac{7}{16}$	5	39 to 48	$2\frac{15}{16}$	5	34 to 62	$2\frac{15}{16}$	$7\frac{1}{2}$	27 to 41	$2\frac{15}{16}$	$7\frac{1}{2}$
	49 to 96	$2\frac{15}{16}$	5	43 to 62	$2\frac{15}{16}$	5	49 to 81	$2\frac{15}{16}$	$7\frac{1}{2}$	63 to 73	$2\frac{15}{16}$	10	42 to 62	$2\frac{15}{16}$	10
	97 to 100	$3\frac{7}{16}$	$7\frac{1}{2}$	63 to 87	$2\frac{15}{16}$	$7\frac{1}{2}$	82 to 100	$3\frac{7}{16}$	10	74 to 89	$3\frac{7}{16}$	10	63 to 100	$3\frac{7}{16}$	15
				88 to 100	$3\frac{7}{16}$	$7\frac{1}{2}$				90 to 100	$3\frac{7}{16}$	15			
128	Up to 24	$2\frac{7}{16}$	3	Up to 27	$2\frac{7}{16}$	5	Up to 22	$2\frac{7}{16}$	5	Up to 12	$2\frac{7}{16}$	5	Up to 17	$2\frac{15}{16}$	$7\frac{1}{2}$
	25 to 34	$2\frac{7}{16}$	5	28 to 62	$2\frac{15}{16}$	$7\frac{1}{2}$	23 to 47	$2\frac{15}{16}$	$7\frac{1}{2}$	13 to 17	$2\frac{7}{16}$	$7\frac{1}{2}$	18 to 32	$2\frac{15}{16}$	10
	35 to 58	$2\frac{15}{16}$	5	63 to 92	$3\frac{7}{16}$	10	48 to 57	$2\frac{15}{16}$	10	18 to 32	$2\frac{15}{16}$	$7\frac{1}{2}$	33 to 38	$2\frac{15}{16}$	15
	59 to 73	$2\frac{15}{16}$	$7\frac{1}{2}$	93 to 100	$3\frac{7}{16}$	15	58 to 72	$3\frac{7}{16}$	10	33 to 49	$2\frac{15}{16}$	10	39 to 62	$3\frac{7}{16}$	15
	74 to 100	$3\frac{7}{16}$	$7\frac{1}{2}$				73 to 100	$3\frac{7}{16}$	15	50 to 92	$3\frac{7}{16}$	15	63 to 76	$3\frac{7}{16}$	20
										93 to 100	$3\frac{15}{16}$	20	77 to 92	$3\frac{15}{16}$	20
													93 to 100	$3\frac{15}{16}$	25
134	Up to 32	$2\frac{15}{16}$	5	Up to 15	$2\frac{15}{16}$	5	Up to 25	$2\frac{15}{16}$	$7\frac{1}{2}$	Up to 15	$2\frac{15}{16}$	$7\frac{1}{2}$	Up to 15	$2\frac{15}{16}$	10
	33 to 46	$2\frac{15}{16}$	$7\frac{1}{2}$	16 to 35	$2\frac{15}{16}$	$7\frac{1}{2}$	26 to 33	$2\frac{15}{16}$	10	16 to 26	$2\frac{15}{16}$	10	16 to 35	$3\frac{7}{16}$	15
	47 to 61	$3\frac{7}{16}$	$7\frac{1}{2}$	36 to 55	$3\frac{7}{16}$	10	34 to 42	$3\frac{7}{16}$	10	27 to 55	$3\frac{7}{16}$	15	36 to 44	$3\frac{7}{16}$	20
	62 to 82	$3\frac{7}{16}$	10	56 to 71	$3\frac{7}{16}$	15	43 to 65	$3\frac{7}{16}$	15	56 to 82	$3\frac{15}{16}$	20	45 to 55	$3\frac{15}{16}$	20
	83 to 90	$3\frac{15}{16}$	10	72 to 96	$3\frac{15}{16}$	15	66 to 76	$3\frac{15}{16}$	15				56 to 76	$3\frac{15}{16}$	25
							77 to 100	$3\frac{15}{16}$	20						

Source: Courtesy FMC

Note: All dimensions, including centers (ctrs) and head shaft (hd shft) in inches.

[d]Based on buckets filled to 100% of theoretical capacity. If exact material weight is not shown, select drive and headshaft using the next heavier material weight.

Table 3.5 Centrifugal-Discharge Bucket Elevator Selection/Specifications, Type 1—Belt

| Elev No.[a] | ft³/hr[b] | Buckets[c] | | Belt Width | Belt Rating PIW | fpm | Max Lump Size | | Casing | Headshaft | | Footshaft | |
		Size	Space				100%	10%		Pulley Dia	rpm	Pulley Dia	Dia
141	280	6 × 4	13	7	160	225	½	2½	11¾ × 35	20	43	16	1½
143	609	8 × 5	16	9	160	258	¾	3	13¾ × 42	24	41	18	2
145	1045	10 × 6	16	11	240	258	1	3½	15¾ × 48	24	41	20	2
147	1698	12 × 7	18	13	240	298	1¼	4	17¾ × 54	30	38	24	2⁷/₁₆
149	2056	14 × 7	18	15	240	298	1¼	4	19¾ × 54	30	38	24	2⁷/₁₆
152	3039	16 × 8	18	18	320	298	1½	4½	22¾ × 54	30	38	24	2⁷/₁₆

[a]Bucket elevator assemblies include head shaft machinery with either ball or roller bearing pillow blocks, belt, buckets, casing, Style 1 or Style 2 discharge spout, stub inlet and internal gravity takeup with hard iron bearings. (Internal gravity takeup is available with cement mill type sleeves and bearings when handling highly abrasive materials.) Drives with backstops, service platforms, and ladders with safety cages can be furnished.

[b]Based on buckets filled to 75% of theoretical capacity.

[c]Style AA malleable iron buckets.

Elev No.	Aprx Wt (lb) Terminals	Wt (lb/ft) Centers
141	870	53
143	1080	71
145	1271	81
147	1669	93
149	1788	97
152	2023	110

TABLE 3.5   97

Elev No.	Material Weight[d] 35 lb/ft³			50 lb/ft³			60 lb/ft³			75 lb/ft³			100 lb/ft³		
	Ctrs	Hd Shft	hp	Ctrs	Hd Shft	hp	Ctrs	Hd Shft	hp	Ctrs	Hd Shft	hp	Ctrs	Hd Shft	hp
141	Up to 100	$1\frac{15}{16}$	1	Up to 72	$1\frac{15}{16}$	1	Up to 57	$1\frac{15}{16}$	1	Up to 42	$1\frac{15}{16}$	1	Up to 27	$1\frac{15}{16}$	1
				73 to 100	$1\frac{15}{16}$	1½	58 to 94	$1\frac{15}{16}$	1½	43 to 72	$1\frac{15}{16}$	1½	28 to 49	$1\frac{15}{16}$	1½
							95 to 100	$1\frac{15}{16}$	2	73 to 100	$1\frac{15}{16}$	2	50 to 72	$1\frac{15}{16}$	2
													73 to 100	$1\frac{15}{16}$	3
143	Up to 38	$1\frac{15}{16}$	1	Up to 20	$1\frac{15}{16}$	1	Up to 13	$1\frac{15}{16}$	1	Up to 20	$1\frac{15}{16}$	1½	Up to 10	$1\frac{15}{16}$	1½
	39 to 67	$1\frac{15}{16}$	1½	21 to 41	$1\frac{15}{16}$	1½	14 to 30	$1\frac{15}{16}$	1½	21 to 34	$1\frac{15}{16}$	2	11 to 20	$1\frac{15}{16}$	2
	68 to 97	$1\frac{15}{16}$	2	42 to 61	$1\frac{15}{16}$	2	31 to 48	$1\frac{15}{16}$	2	35 to 61	$1\frac{15}{16}$	3	21 to 41	$1\frac{15}{16}$	3
	98 to 100	$1\frac{15}{16}$	3	62 to 100	$1\frac{15}{16}$	3	49 to 82	$1\frac{15}{16}$	3	62 to 90	$1\frac{15}{16}$	5	42 to 73	$1\frac{15}{16}$	5
							83 to 100	$1\frac{15}{16}$	5	91 to 100	$2\frac{7}{16}$	5	74 to 82	$2\frac{7}{16}$	5
													83 to 100	$2\frac{7}{16}$	7½
145	Up to 11	$1\frac{15}{16}$	1	Up to 13	$1\frac{15}{16}$	1½	Up to 17	$1\frac{15}{16}$	2	Up to 25	$1\frac{15}{16}$	3	Up to 13	$1\frac{15}{16}$	3
	12 to 29	$1\frac{15}{16}$	1½	14 to 25	$1\frac{15}{16}$	2	18 to 37	$1\frac{15}{16}$	3	26 to 48	$1\frac{15}{16}$	5	14 to 35	$1\frac{15}{16}$	5
	30 to 46	$1\frac{15}{16}$	2	26 to 49	$1\frac{15}{16}$	3	38 to 57	$1\frac{15}{16}$	5	49 to 57	$2\frac{7}{16}$	5	36 to 67	$2\frac{7}{16}$	7½
	47 to 78	$1\frac{15}{16}$	3	50 to 64	$1\frac{15}{16}$	5	58 to 77	$2\frac{7}{16}$	5	58 to 98	$2\frac{7}{16}$	7½	68 to 92	$2\frac{7}{16}$	10
	79 to 100	$2\frac{7}{16}$	5	65 to 98	$2\frac{7}{16}$	5	78 to 100	$2\frac{7}{16}$	7½	99 to 100	$2\frac{7}{16}$	10	93 to 100	$2\frac{15}{16}$	15
				99 to 100	$2\frac{7}{16}$	7½									
147	Up to 15	$2\frac{7}{16}$	2	Up to 17	$2\frac{7}{16}$	3	Up to 34	$2\frac{7}{16}$	5	Up to 22	$2\frac{7}{16}$	5	Up to 28	$2\frac{7}{16}$	7½
	16 to 36	$2\frac{7}{16}$	3	18 to 46	$2\frac{7}{16}$	5	35 to 65	$2\frac{7}{16}$	7½	23 to 46	$2\frac{7}{16}$	7½	29 to 46	$2\frac{7}{16}$	10
	37 to 78	$2\frac{7}{16}$	5	47 to 84	$2\frac{7}{16}$	7½	67 to 85	$2\frac{7}{16}$	10	47 to 69	$2\frac{7}{16}$	10	47 to 50	$2\frac{7}{16}$	15
	79 to 100	$2\frac{7}{16}$	7½	85 to 97	$2\frac{7}{16}$	10	86 to 96	$2\frac{15}{16}$	10	70 to 100	$2\frac{15}{16}$	15	51 to 84	$2\frac{15}{16}$	15
				98 to 100	$2\frac{15}{16}$	10	97 to 100	$2\frac{15}{16}$	15				85 to 100	$2\frac{15}{16}$	20
149	Up to 25	$2\frac{7}{16}$	3	Up to 33	$2\frac{7}{16}$	5	Up to 23	$2\frac{7}{16}$	5	Up to 13	$2\frac{7}{16}$	5	Up to 18	$2\frac{7}{16}$	7½
	26 to 60	$2\frac{7}{16}$	5	34 to 64	$2\frac{7}{16}$	7½	29 to 49	$2\frac{7}{16}$	7½	14 to 33	$2\frac{7}{16}$	7½	19 to 33	$2\frac{7}{16}$	10
	61 to 91	$2\frac{7}{16}$	7½	65 to 72	$2\frac{7}{16}$	10	50 to 62	$2\frac{7}{16}$	10	34 to 49	$2\frac{7}{16}$	10	34 to 64	$2\frac{15}{16}$	15
	92 to 100	$2\frac{15}{16}$	7½	73 to 95	$2\frac{15}{16}$	10	63 to 74	$2\frac{15}{16}$	10	50 to 54	$2\frac{15}{16}$	10	65 to 79	$2\frac{15}{16}$	20
				96 to 100	$2\frac{15}{16}$	15	75 to 100	$2\frac{15}{16}$	15	55 to 95	$2\frac{15}{16}$	15	80 to 95	$3\frac{7}{16}$	20
										96 to 100	$2\frac{15}{16}$	20	96 to 100	$3\frac{7}{16}$	25
152	Up to 31	$2\frac{15}{16}$	5	Up to 13	$2\frac{15}{16}$	5	Up to 24	$2\frac{15}{16}$	7½	Up to 13	$2\frac{15}{16}$	7½	Up to 13	$2\frac{15}{16}$	10
	32 to 61	$2\frac{15}{16}$	7½	14 to 34	$2\frac{15}{16}$	7½	25 to 41	$2\frac{15}{16}$	10	14 to 27	$2\frac{15}{16}$	10	14 to 34	$2\frac{15}{16}$	15
	62 to 91	$2\frac{15}{16}$	10	35 to 55	$2\frac{15}{16}$	10	42 to 76	$2\frac{15}{16}$	15	28 to 55	$2\frac{15}{16}$	15	35 to 47	$2\frac{15}{16}$	20
	92 to 100	$2\frac{15}{16}$	15	56 to 92	$2\frac{15}{16}$	15	77 to 100	$3\frac{7}{16}$	20	56 to 65	$2\frac{15}{16}$	20	48 to 55	$3\frac{7}{16}$	20
				93 to 97	$3\frac{7}{16}$	15				66 to 83	$3\frac{7}{16}$	20	56 to 76	$3\frac{7}{16}$	25
				98 to 100	$3\frac{7}{16}$	20				84 to 100	$3\frac{7}{16}$	25	77 to 89	$3\frac{7}{16}$	30
													90 to 97	$3\frac{15}{16}$	30

Source: Courtesy FMC

Note: All dimensions, including centers (ctrs) and head shaft (hd shft) in inches.

[d]Based on buckets filled to 100% of theoretical capacity. If exact material weight is not shown, select drive and headshaft using the next heavier material weight.

Table 3.6 Continuous-Bucket Elevator Selection/Specifications, Type 7—Chain

Elev No.[a]	ft³/hr[b]	Buckets[c] Size	Space	Chain	fpm	Max Lump Size 100%	10%	Casing	Headshaft Sprkt Teeth	Pitch Dia	rpm	Footshaft Sprkt Teeth	Pitch Dia	Dia
766	590	8 × 5 × 7¾	8	C102B	125	¾	2½	11¾ × 39	16	20½	23.4	11	14¼	1½
767	590	8 × 5 × 7¾	8	SBS102B	125	¾	2½	11¾ × 39	16	20½	23.4	11	14¼	1½
768	750	10 × 5 × 7¾	8	C102B	125	¾	2½	13¾ × 39	16	20½	23.4	11	14¼	1½
769	750	10 × 5 × 7¾	8	SBS102B	125	¾	2½	13¾ × 39	16	20½	23.4	11	14¼	1½
770	1010	10 × 7 × 11⅝	12	C110	125	1	3	13¾ × 48	13	25	19.1	10	19½	2
771	1010	10 × 7 × 11⅝	12	SBS110	125	1	3	13¾ × 48	13	25	19.1	10	19½	2
776	1550	12 × 8 × 11⅝	12	C110	125	1¼	4	15¾ × 48	13	25	19.1	9	17½	2
777	1550	12 × 8 × 11⅝	12	SBS110	125	1¼	4	15¾ × 48	13	25	19.1	9	17½	2
781	2090	16 × 8 × 11⅝	12	SBS110	125	1½	4½	19¾ × 48	13	25	19.1	9	17½	2⁷⁄₁₆
783	2340	18 × 8 × 11⅝	12	SBS110	125	1½	4½	21¾ × 48	13	25	19.1	9	17½	2⁷⁄₁₆

[a]Bucket elevator assemblies include head shaft machinery with either ball or roller bearing pillow blocks, chain, buckets, casing, Style 1 or Style 2 discharge spout, stub inlet and internal gravity takeup with hard iron bearings. (Internal gravity takeup is available with cement mill type sleeves and bearings when handling highly abrasive materials.) Drives with backstops, service platforms, and ladders with safety cages can be furnished.

[b]Based on buckets filled to 75% of theoretical capacity.

[c]Style MF, medium front, continuous steel buckets.

Elev No.	Aprx Wt (lb) Terminals	Wt (lb/ft) Centers
766	867	83
767	862	83
768	827	94
769	816	94
770	1130	99
771	1223	103
776	1250	115
777	1462	121
781	1700	148
783	1642	137

	Material Weight[d]														
	35 lb/ft³			50 lb/ft³			60 lb/ft³			75 lb/ft³			100 lb/ft³		
Elev No.	Ctrs	Hd Shft	hp	Ctrs	Hd Shft	hp	Ctrs	Hd Shft	hp	Ctrs	Hd Shft	hp	Ctrs	Hd Shft	hp
766	Up to 25	$1\frac{15}{16}$	1	Up to 22	$1\frac{15}{16}$	1	Up to 20	$1\frac{15}{16}$	1	Up to 16	$1\frac{15}{16}$	1	Up to 10	$1\frac{15}{16}$	1
767	26 to 44	$2\frac{7}{16}$	1	23 to 28	$2\frac{7}{16}$	1	21 to 40	$2\frac{7}{16}$	1½	17 to 30	$2\frac{7}{16}$	1½	11 to 20	$2\frac{7}{16}$	1½
	45 to 62	$2\frac{7}{16}$	1½	29 to 50	$2\frac{7}{16}$	1½	41 to 55	$2\frac{7}{16}$	2	31 to 44	$2\frac{7}{16}$	2	21 to 31	$2\frac{7}{16}$	2
	63 to 75	$2\frac{15}{16}$	1½	51 to 58	$2\frac{7}{16}$	2	56 to 93	$2\frac{15}{16}$	3	45 to 51	$2\frac{7}{16}$	3	32 to 45	$2\frac{7}{16}$	3
	76 to 100	$2\frac{15}{16}$	2	59 to 71	$2\frac{15}{16}$	2	94 to 100	$2\frac{15}{16}$	5	52 to 73	$2\frac{15}{16}$	3	46 to 52	$2\frac{15}{16}$	3
				72 to 100	$2\frac{15}{16}$	3				74 to 100	$2\frac{15}{16}$	5	53 to 91	$2\frac{15}{16}$	5
													92 to 100	$3\frac{7}{16}$	7½
768	Up to 18	$1\frac{15}{16}$	1	Up to 15	$1\frac{15}{16}$	1	Up to 13	$1\frac{15}{16}$	1	Up to 10	$1\frac{15}{16}$	1	Up to 14	$2\frac{7}{16}$	1½
769	19 to 32	$2\frac{7}{16}$	1	16 to 20	$2\frac{7}{16}$	1	14 to 29	$2\frac{7}{16}$	1½	11 to 21	$2\frac{7}{16}$	1½	15 to 22	$2\frac{7}{16}$	2
	33 to 49	$2\frac{7}{16}$	1½	21 to 37	$2\frac{7}{16}$	1½	30 to 42	$2\frac{7}{16}$	2	22 to 33	$2\frac{7}{16}$	2	23 to 32	$2\frac{7}{16}$	3
	50 to 56	$2\frac{15}{16}$	1½	38 to 44	$2\frac{7}{16}$	2	43 to 71	$2\frac{15}{16}$	3	34 to 38	$2\frac{7}{16}$	3	33 to 39	$2\frac{15}{16}$	3
	57 to 80	$2\frac{15}{16}$	2	45 to 54	$2\frac{15}{16}$	2	72 to 83	$2\frac{15}{16}$	5	39 to 55	$2\frac{15}{16}$	3	40 to 69	$2\frac{15}{16}$	5
	81 to 94	$2\frac{15}{16}$	3	55 to 87	$2\frac{15}{16}$	3	84 to 100	$3\frac{7}{16}$	5	56 to 77	$2\frac{15}{16}$	5	70 to 73	$3\frac{7}{16}$	5
	95 to 100	$3\frac{7}{16}$	3	88 to 100	$3\frac{7}{16}$	5				78 to 100	$3\frac{7}{16}$	5	74 to 100	$3\frac{7}{16}$	7½
770	Up to 13	$1\frac{15}{16}$	1	Up to 10	$2\frac{7}{16}$	1	Up to 17	$2\frac{7}{16}$	1½	Up to 11	$2\frac{7}{16}$	1½	Up to 12	$2\frac{7}{16}$	2
771	14 to 19	$2\frac{7}{16}$	1	11 to 22	$2\frac{7}{16}$	1½	18 to 27	$2\frac{7}{16}$	2	12 to 19	$2\frac{7}{16}$	2	13 to 19	$2\frac{7}{16}$	3
	20 to 37	$2\frac{7}{16}$	1½	23 to 35	$2\frac{7}{16}$	2	28 to 33	$2\frac{7}{16}$	3	20 to 27	$2\frac{7}{16}$	3	20 to 24	$2\frac{15}{16}$	3
	38 to 45	$2\frac{7}{16}$	2	36 to 60	$2\frac{15}{16}$	3	34 to 48	$2\frac{15}{16}$	3	28 to 36	$2\frac{15}{16}$	3	25 to 49	$2\frac{15}{16}$	5
	46 to 54	$2\frac{15}{16}$	2	61 to 81	$2\frac{15}{16}$	5	49 to 75	$2\frac{15}{16}$	5	37 to 67	$2\frac{15}{16}$	5	50 to 55	$2\frac{15}{16}$	7½
	55 to 90	$2\frac{15}{16}$	3	82 to 100	$3\frac{7}{16}$	5	76 to 89	$3\frac{7}{16}$	5	68 to 100	$3\frac{7}{16}$	7½	56 to 80	$3\frac{7}{16}$	7½
	91 to 100	$3\frac{7}{16}$	5				90 to 100	$3\frac{7}{16}$	7½				81 to 100	$3\frac{7}{16}$	10
776	Up to 20	$2\frac{7}{16}$	1½	Up to 11	$2\frac{7}{16}$	1½	Up to 14	$2\frac{7}{16}$	2	Up to 11	$2\frac{7}{16}$	3	Up to 12	$2\frac{15}{16}$	3
777	21 to 26	$2\frac{7}{16}$	2	12 to 19	$2\frac{7}{16}$	2	15 to 28	$2\frac{15}{16}$	3	12 to 20	$2\frac{15}{16}$	3	13 to 28	$2\frac{15}{16}$	5
	27 to 32	$2\frac{15}{16}$	2	20 to 35	$2\frac{15}{16}$	3	29 to 44	$2\frac{15}{16}$	5	21 to 38	$2\frac{15}{16}$	5	29 to 49	$3\frac{7}{16}$	7½
	33 to 55	$2\frac{15}{16}$	3	36 to 49	$2\frac{15}{16}$	5	45 to 55	$3\frac{7}{16}$	5	39 to 69	$3\frac{7}{16}$	7½	50 to 60	$3\frac{7}{16}$	10
	56 to 58	$2\frac{15}{16}$	5	50 to 68	$3\frac{7}{16}$	5	56 to 82	$3\frac{7}{16}$	7½	70 to 87	$3\frac{15}{16}$	10	61 to 69	$3\frac{15}{16}$	10
	59 to 100	$3\frac{7}{16}$	5	69 to 89	$3\frac{7}{16}$	7½	83 to 89	$3\frac{15}{16}$	7½				70 to 72	$3\frac{15}{16}$	15
				90 to 100	$3\frac{15}{16}$	7½	90 to 98	$3\frac{15}{16}$	10						
777							90 to 100	$3\frac{15}{16}$	10	70 to 96	$3\frac{15}{16}$	10	70 to 100	$3\frac{15}{16}$	15
777										97 to 100	$3\frac{15}{16}$	15			
781	Up to 11	$2\frac{15}{16}$	1½	Up to 11	$2\frac{15}{16}$	2	Up to 17	$2\frac{15}{16}$	3	Up to 12	$2\frac{15}{16}$	3	Up to 18	$3\frac{7}{16}$	5
	12 to 20	$2\frac{15}{16}$	2	12 to 23	$2\frac{15}{16}$	3	18 to 23	$2\frac{15}{16}$	5	13 to 17	$2\frac{15}{16}$	5	19 to 31	$3\frac{7}{16}$	7½
	21 to 35	$2\frac{15}{16}$	3	24 to 28	$2\frac{15}{16}$	5	24 to 38	$3\frac{7}{16}$	5	18 to 28	$3\frac{7}{16}$	5	32 to 48	$3\frac{15}{16}$	10
	36 to 66	$3\frac{7}{16}$	5	29 to 47	$3\frac{7}{16}$	5	39 to 50	$3\frac{7}{16}$	7½	29 to 42	$3\frac{7}{16}$	7½	49 to 60	$3\frac{15}{16}$	15
	67 to 72	$3\frac{15}{16}$	5	48 to 56	$3\frac{7}{16}$	7½	51 to 63	$3\frac{15}{16}$	7½	43 to 48	$3\frac{15}{16}$	7½	61 to 79	$4\frac{7}{16}$	15
	73 to 100	$3\frac{15}{16}$	7½	57 to 77	$3\frac{15}{16}$	7½	64 to 84	$3\frac{15}{16}$	10	49 to 68	$3\frac{15}{16}$	10	80 to 91	$4\frac{7}{16}$	20
				78 to 92	$3\frac{15}{16}$	10	85 to 100	$4\frac{7}{16}$	15	69 to 74	$3\frac{15}{16}$	15			
				93 to 100	$4\frac{7}{16}$	10				75 to 100	$4\frac{7}{16}$	15			
783	Up to 17	$2\frac{15}{16}$	2	Up to 19	$2\frac{15}{16}$	3	Up to 14	$2\frac{15}{16}$	3	Up to 10	$2\frac{15}{16}$	3	Up to 15	$3\frac{7}{16}$	5
	18 to 28	$2\frac{15}{16}$	3	20 to 41	$3\frac{7}{16}$	5	15 to 32	$3\frac{7}{16}$	5	11 to 24	$3\frac{7}{16}$	5	16 to 22	$3\frac{7}{16}$	7½
	29 to 32	$3\frac{7}{16}$	3	42 to 45	$3\frac{7}{16}$	7½	33 to 39	$3\frac{7}{16}$	7½	25 to 32	$3\frac{7}{16}$	7½	23 to 29	$3\frac{15}{16}$	7½
	33 to 54	$3\frac{7}{16}$	5	46 to 68	$3\frac{15}{16}$	7½	40 to 55	$3\frac{15}{16}$	7½	33 to 42	$3\frac{15}{16}$	7½	30 to 42	$3\frac{15}{16}$	10
	55 to 63	$3\frac{15}{16}$	5	69 to 76	$3\frac{15}{16}$	10	56 to 69	$3\frac{15}{16}$	10	43 to 59	$3\frac{15}{16}$	10	43 to 69	$4\frac{7}{16}$	15
	64 to 88	$3\frac{15}{16}$	7½	77 to 95	$4\frac{7}{16}$	10	70 to 77	$4\frac{7}{16}$	10	60 to 96	$4\frac{7}{16}$	15	70 to 79	$4\frac{7}{16}$	20
	89 to 100	$4\frac{7}{16}$	7½	96 to 100	$4\frac{7}{16}$	15	78 to 100	$4\frac{7}{16}$	15						

Source: Courtesy FMC

Note: All dimensions, including centers (ctrs) and head shaft (hd shft) in inches.

[d]Based on buckets filled to 100% of theoretical capacity. If exact material weight is not shown, select drive and headshaft using the next heavier material weight.

3.12   STANDARD DESIGNS

Standard designs of elevators are given in tables 3.4, 3.5, and 3.6. It is not advisable to use the tables for final design. The weight of the buckets and their capacity are always subject to change, because of the abrasiveness of the material, the weight of the material, the size and percentage of lumps, the fluidity of the material, the rate of delivery of the material to the elevator, the moisture content of the material, and the speed of the buckets.

3.13   BELTS

The widths of the belts for various size buckets is given in table 3.5. It is good practice to use no fewer than four plies, even for the lightest loads. For design purposes, assume the weight of the belt and attachments to be 6½ lb/ft of belt width.

3.14   CENTRIFUGAL-DISCHARGE BUCKET ELEVATORS (Vertically Spaced)

3.14.1   General

A centrifugal-discharge elevator is designed to operate at a high speed, usually from 185 ft to 300 ft or more per minute, picking up material in the boot, as it is fed to it, and discharging the material by centrifugal force out of the buckets, as they pass over the head sprocket into a chute attached to the elevator casing. For very fine materials, similar to gypsum and cement (-10 mesh to 200 mesh), experience has shown that the speed can be reduced to about 185 fpm. The head wheel diameter will vary between 20 in. and 31 in. Speed is critical. Travelling slower than recommended may not allow material to be discharged by centrifugal force, and material may come back on the return run. Travelling faster

than recommended may cause material to hit the hood and bounce back down the return run.

The size of the head wheel (D, in ft) and the rpm of the head shaft may vary, but the speed of the elevator (fpm = $D \times$ rpm) must be maintained in order to avoid backlegging (return of material on the down run), regardless of the required capacity. For grain, cottonseed, wood chips, and other lightweight materials, however, the double-leg casing elevator, having buckets mounted on a belt and travelling at higher speeds, is frequently used. Bucket speeds for such units range between 350 fpm and 750 fpm, the head wheel diameter ranges from 24 in. to 84 in., and the spacing of the buckets will vary from 16 in. to 24 in. The capacity will vary between 14 tph and 152 tph for material weighing 100 lb/ft^3. This type of elevator usually is enclosed in a steel casing to provide a means of support, and as a matter of safety and dust retention.

On a centrifugal-discharge chain elevator, inclined about 30° from the vertical, the single strand of chain can be supported on the up, or carrying side, on single flanged rollers spaced 6–8 ft on centers. The return run can sag if there is plenty of clearance; if not, the return run can be supported by having the buckets slide on two angles, forming a track, to keep the return run in the proper path (see figure 3.15).

A centrifugal-discharge elevator will handle almost any kind of fine or small lump materials. It operates well when handling dry and free-flowing products such as grain, coal, petroleum coke, sand, sugar, salt, chemicals, limestone dust, gypsum, sulfur, and cement. The size of the lumps should be 2 in. and under, with the greater part of the volume under 1 in. If there are many 2-in. lumps, the buckets should be at least 12-in. wide, regardless of capacity. This type of elevator should not be used for materials containing over 10% to, say, 15% of lumps, because of the possibility of plugging in the boot, and the difficulty in retaining lumps in the buckets as they travel upward. It should not be used where breakage of material is to be avoided. Continuous-bucket elevators should be used instead.

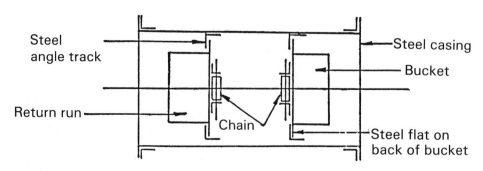

Figure 3.15   Centrifugal-discharge chain elevator.

3.14.2 Buckets

The size of the buckets ranges from 6 in. × 4 in. (6-in. long with 4-in. projection) to 16 in. × 8 in. The spacing of A or AA buckets can be between 13 in. and 24 in., depending on capacity. The A buckets are similar to AA, but are built lighter and of smaller capacity. They are seldom used. The buckets may be malleable or cast iron, steel, or plastic. The buckets are normally attached to a single strand of chain or belt with what are known as K attachments, spaced at intervals. The K-1 attachment has 2 holes, and the K-2 attachment has four holes. The manufacturers' catalogs give the type of attachment to be used for fastening to the back of the bucket, and the punching required for the belt. It depends on the minimum and maximum size of the bucket (refer to table 3.7).

Since the plastic buckets are not adapted for hard dig-ging in the elevator boots, it is advisable to place one style AA bucket (malleable iron, reinforced digging edge) on every sixth to eighth attachment. This bucket will clean a path through the caked material in the boot of the elevator housing, so that the plastic buckets do not have to dig, only elevate. This malleable iron bucket should be inspected at regular intervals for wear and corrosion.

Budd plastic buckets are listed in table 3.8; Dura plastic buckets, in table 3.9; AA-RB buckets, in table 3.10; and screw conveyor plastic buckets, in table 3.11. Screw conveyor buckets are made of high-density poly-ethylene. They remain flexible at ambient temperatures from −50°F to +220°F. The standard buckets are yel-low, but are available in white and other colors on spe-cial order. They are inert to most chemicals. Dura buck-ets are also made of high-density polyethylene. Due to the special shape of LP buckets, the capacities along

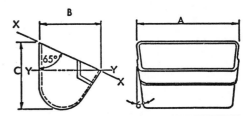

Table 3.7 AA Buckets

Nominal Size (in.)	Weight (lb)	Capacity[a] (ft³)		Dimensions (in.)		
		X-X	Y-Y	A	B	C
4 × 2	1.24	0.01	0.006	4¼	3	3
6 × 4	2.70	0.03	0.018	6¼	4⅜	4¼
7 × 4	3.70	0.05	0.030	7⅜	4¾	5
8 × 5	4.60	0.07	0.042	8¼	5⅜	5½
9 × 6	6.50	0.11	0.066	9¼	6⅜	6¼
10 × 6	7.70	0.12	0.072	10¼	6⅜	6¼
11 × 6	8.60	0.13	0.078	11¼	6⅜	6¼
12 × 6	9.40	0.14	0.084	12¼	6⅜	6¼
12 × 7	11.5	0.19	0.114	12⅜	7½	7¼
14 × 7	14.7	0.23	0.138	14⅜	7½	7¼
14 × 8	18.5	0.30	0.180	14⅜	8⅝	8½
14 × 8	22.0	0.23	0.138	14⅜	8⅝	8½
15 × 7	14.8	0.25	0.150	15⅜	7½	7¼
16 × 7	15.9	0.27	0.162	16⅜	7½	7¼
16 × 8	20.9	0.34	0.204	16⅜	8⅝	8½
18 × 8	23.3	0.39	0.235	18⅜	8⅝	8½
18 × 10	35.0	0.61	0.366	18½	10⅝	10½
20 × 8	25.7	0.43	0.258	20⅜	8⅝	8½
24 × 8	30.5	0.51	0.306	24⅜	8⅝	8½

[a]Actual capacity depends on the angle of repose of the material and the inclination of the elevator.

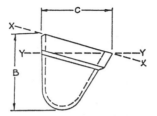

Table 3.8   Budd Cast Nylon Style AA Buckets

Nominal Size (in.)	Weight (lb)	Capacity[a] (ft³)		Dimensions[b] (in.)		
		X-X	Y-Y	A	B	C
6 × 4	0.47	0.028	0.018	6.47	4.32	4.06
8 × 5	1.03	0.057	0.037	8.41	6.53	5.00
9 × 3	0.61	0.024	0.009	9.53	3.41	3.30
10 × 6	1.54	0.102	0.066	10.59	6.25	6.09
12 × 7	2.78	0.174	0.113	12.40	7.33	7.53
14 × 7	3.13	0.216	0.140	14.72	7.42	7.59
14 × 8	3.68	0.256	0.162	14.53	8.47	8.06
15 × 5	2.78	0.121	0.081	15.75	5.68	5.44
16 × 8	4.78	0.290	0.190	16.62	8.56	8.06
18 × 8	4.52	0.309	0.210	17.78	8.47	8.03
18 × 10	8.92	0.486	0.312	18.84	10.56	10.19
19 × 5	3.20	0.132	0.085	19.88	5.16	5.25
24 × 8	9.06	0.361	0.258	24.06	8.69	8.00
24 × 13	18.58	1.00	0.548	24.00	12.72	12.25

[a]Actual capacity depends on the angle of repose of the material and the inclination of the elevator.
[b]Variations: dimension A +0.20 in.; dimensions B and C +0.15 in.; capacity +5%; weight +5%.

lines X-X and Y-Y would serve no useful purpose, so capacities as given by manufacturers are provided.

3.14.3   Inclined Elevators

A centrifugal-discharge elevator, equipped with either chain or belt, can operate on an incline at the same speeds as vertical elevators by welding or attaching steel flats on back of buckets, to ride on steel angle track attached to elevator casing sides, for both carrying and return runs.

The use of this steel angle track on return run prevents sagging of chain and bucket line, saving much space. A belt can be used instead of a chain, when necessary. Open inclined continuous bucket elevators are used in spite of the difficulties during rainy weather when the materials hang in the buckets and do not discharge properly. These elevators use either chain or belt. Normally, units of this kind are inclined 30° from the vertical, allowing the return run to sag (usually clearance permits this), and are not covered in any way or protected from the weather. The machinery for these elevators can be mounted on structural frames.

3.14.4   Centrifugal-Discharge Elevator Buckets

The edges of all these buckets are reinforced for digging, and the bottom of these buckets is rounded. The various types of buckets used in centrifugal discharge elevators are described below.

1. AA buckets are made of malleable iron for chain or belt mounting. They have a reinforced lip for digging. They are the most common type in use for centrifugal discharge elevators.
2. AA-RB buckets are the same as the AA buckets, except that the edges are thicker. They are used for heavy service, and for abrasive materials.
3. AC buckets (table 3.12) are made for chain mounting. The hooded back allows a closer spacing of buckets.
4. B buckets (table 3.13) are cast malleable iron buckets for chain or belt mounting. They are used on inclined elevators for handling coarse materials such as stone. They will produce a clean discharge at low speed.

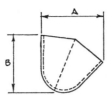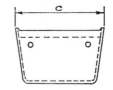

Table 3.9A Dura Buckets, Style SS, Plastic

| Bucket Size (in.) | Dimensions (in.) | | | Weight (lb) | Capacity (in³) |
	Projection A	Depth B	Width C		
4 × 3	3⅛	3	4¼	0.20	15.86
5 × 4	4⅛	3¾	5¼	0.31	38.45
6 × 4	4⅛	3¾	6¼	0.37	45.76
7 × 4	4⅛	3¾	7¼	0.40	55.23
6 × 5	5¼	4⅞	6⅜	0.50	77.41
7 × 5	5¼	4⅞	7⅜	0.63	90.62
8 × 5	5¼	4⅞	8⅜	0.70	105.87
9 × 5	5¼	4⅞	9⅜	0.74	118.99
8 × 6	6⅜	6¼	8⅜	1.08	103.53
9 × 6	6⅜	6¼	9⅜	1.16	189.78
10 × 6	6⅜	6¼	10⅜	1.22	202.59
11 × 6	6⅜	6¼	11⅜	1.34	223.96
12 × 6	6¾	6½	12⅜	1.70	265.76
13 × 6	6¾	6½	13⅜	2.00	291.69
10 × 7	7⅜	7¼	10⅜	1.75	286.81
11 × 7	7⅜	7¼	11⅜	1.90	315.42
12 × 7	7⅜	7¼	12⅜	2.10	324.64
13 × 7	7⅜	7¼	13⅜	2.25	373.47
14 × 7	7⅜	7¼	14⅜	2.50	411.30
15 × 7	7⅜	7¼	15⅜	2.59	427.17
16 × 7	7⅜	7¼	16⅜	2.67	433.88
12 × 8	8⅜	8¼	12⅜	2.50	416.18
14 × 8	8⅜	8¼	14⅜	2.80	507.72
16 × 8	8⅜	8¼	16⅜	3.20	594.37
18 × 8	8⅜	8¼	18⅜	3.50	629.77

5. C buckets (table 3.14) are cast malleable iron for chain or belt mounting, and are used for finely pulverized or wet materials that tend to stick to buckets.

3.14.5 Design of Centrifugal-Discharge Elevator Handling Sand

3.14.5.1 Data

Material: sand weighing 100 lb/ft³.
Buckets: style AA (reinforced lip), spaced 16 in. on centers.

Length: 72 ft (70 ft c to c of head and foot shafts, plus 2-ft diameter of head sprocket).
Capacity required: 30 tph. Manufacturers' catalogs show that 8 in. × 5 in. buckets traveling at 260 fpm will handle a capacity of 30.6 tph. The capacity of 8 in. × 5 in. buckets filled to water level is 0.042 ft³; filled full, 0.07 ft³; filled to 75%, 0.75 × 0.07 = 0.0525 ft³. In figuring capacity, the 75% value can be used. To figure hp, use 0.07 ft³.

$$w_m = 0.07 \times 100 = 7 \text{ lb/bucket}$$

Chain: SBS-102B (L-B) or 6102-BM (Jeffrey), steel se-

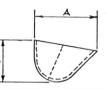 

Table 3.9B Dura Buckets, Style LP, Plastic

Bucket Size (in.)	Dimensions (in.)			Weight (lb)	Capacity (in³)
	Projection A	Depth B	Width C		
4 × 3	3⅛	3	4	0.16	16
5 × 4	4⅛	3	5⅛	0.24	31
6 × 4	4⅛	3	6⅛	0.30	38
7 × 4	4⅛	3	7⅛	0.32	43
6 × 5	5¼	3¾	6⅛	0.46	56
7 × 5	5¼	3¾	7⅛	0.48	70
8 × 5	5¼	3¾	8⅛	0.62	81
9 × 5	5¼	3¾	9⅛	0.65	90
8 × 6	6⅜	4½	8⅛	0.78	116
9 × 6	6⅜	4½	9⅛	0.84	141
10 × 6	6⅜	4½	10⅛	0.91	151
11 × 6	6⅜	4½	11⅛	0.95	165
12 × 6	6¾	5	12⅛	1.42	206
13 × 6	6¾	5	13⅛	1.66	233
10 × 7	7⅜	5¼	10⅛	1.17	215
11 × 7	7⅜	5¼	11⅛	1.26	238
12 × 7	7⅜	5¼	12⅛	1.52	248
13 × 7	7⅜	5¼	13⅛	1.67	281
14 × 7	7⅜	5¼	14⅜	1.88	303
15 × 7	7⅜	5¼	15⅛	1.90	325
16 × 7	7⅜	5¼	16⅛	2.14	331
12 × 8	8⅜	5¾	12⅛	1.84	307
14 × 8	8⅜	5¾	14⅛	1.97	363
16 × 8	8⅜	5¾	16⅛	2.64	414
18 × 8	8⅜	5¾	18¼	2.98	474

lected for heavy duty. Alloy heat-treated pins for handling sand.

3.14.5.2 Check of Capacity

$$\frac{260 \times 12}{16} = 195 \text{ buckets/min discharging}$$

$$\frac{195 \times 0.0525 \times 100 \times 60}{2000} = 30.7 \text{ tph}$$

These results check with the value in the catalogs, which is 30.6 tph.

3.14.5.3 Load on Up Run

Buckets are spaced 16 in. on centers = 1.33 ft (refer to table 3.4).

$$\frac{72}{1.33} = 54 \text{ buckets at 4.6 lb } = \quad 248 \text{ lb}$$

Chain = 72 ft × 8.3 lb/ft =   598 lb
Sand = 54 × 0.07 × 100   =   378 lb
                                    1224 lb

3.14.5.4 Load on Down Run

Chain   = 598 lb
Buckets = 248 lb
          846 lb

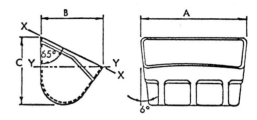

Table 3.10 AA-RB Buckets

Nominal Size (in.)	Weight (lb)	Capacity[a] (ft³)		Dimensions (in.)		
		X-X	Y-Y	A	B	C
8 × 5	5.0	0.070	0.042	8½	5⅜	5½
10 × 6	8.0	0.120	0.072	10½	6⅜	6¼
11 × 6	9.6	0.130	0.078	11½	6⅜	6¼
12 × 6	10.4	0.140	0.084	12½	6⅜	6¼
12 × 7	13.8	0.190	0.114	12⅝	7⁹⁄₁₆	7¼
14 × 7	16.5	0.230	0.138	14⅝	7⁹⁄₁₆	7¼
14 × 8	22.0	0.300	0.180	14¾	8¹¹⁄₁₆	8½
16 × 7	18.5	0.270	0.162	16⅝	7⁹⁄₁₆	7¼
16 × 8	26.0	0.340	0.204	16¾	8¹¹⁄₁₆	8½
18 × 8	30.0	0.390	0.230	18¾	8¹¹⁄₁₆	8½
18 × 10	45.0	0.610	0.366	18⅞	10⁹⁄₁₆	10¾
20 × 8	34.0	0.430	0.258	20¾	8¹¹⁄₁₆	8½
20 × 10	51.4	0.680	0.406	20⅞	10⅝	10
24 × 10	64.5	0.81	0.488	24⅞	10⅝	10
24 × 12	70.7	1.24	0.750	24⅞	12⅜	12

[a]Actual capacity depends on the angle of repose of the material and the inclination of the elevator.

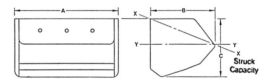

Table 3.11 SCC Polymer Buckets

Nominal Size (in.)	Average Weight (lb)	Dimensions (in.)			Capacity (in³)		Minimum Spacing on Belt (in.)
		A	B	C	Y-Y	X-X	
6 × 4	0.53	6⅜	4⅜	4⅛	46	60	6
7 × 4	0.61	7⅜	4⅜	4⅛	51	71	6
7 × 5	1.0	7½	5½	5⅛	74	108	7
8 × 5	1.15	8½	5½	5⅛	91	124	7
9 × 5	1.26	9½	5½	5⅛	103	137	7
9 × 6	1.61	9½	6½	6⅛	156	207	8
10 × 6	1.75	10½	6½	6⅛	168	224	8
11 × 6	1.80	11½	6½	6⅛	185	248	8
12 × 6	1.88	12½	6½	6⅛	200	278	8
11 × 7	2.25	11½	7⅝	7⅛	232	343	9
12 × 7	2.35	12½	7⅝	7⅛	255	383	9
14 × 7	2.75	14⅝	7⅝	7⅛	306	437	9
16 × 7	3.13	16⅝	7⅝	7⅛	361	513	9
16 × 8	5.0	16⅝	8¾	8⅛	516	734	10
18 × 8	5.5	18⅝	8¾	8⅛	576	814	10

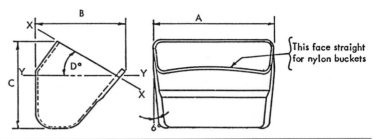

Table 3.12   AC Buckets

Nominal Size (in.)	Weight (lb)	Capacity[a] (ft^3)		Dimensions[b] (in.)		
		X-X	Y-Y	A	B	C
Steel AC Buckets						
12 × 8	28.0	0.28	0.21	12	8	8
16 × 8	34.0	0.38	0.28	16	8	8
18 × 10	52.0	0.62	0.49	18	11	11
24 × 10	72.0	0.85	0.68	24	11	11
Budd Cast Nylon AC Buckets						
18 × 10	10.45	0.654	0.459	19.0	11.06	11.38
24 × 10	15.46	0.781	0.572	25.0	10.75	11.06

[a]Actual capacity depends on the angle of repose of the material and the inclination of the elevator.
[b]Variations for nylon buckets: capacity and weight, +5%; dimension A, +0.20 in.; dimensions B and C, +0.15 in.

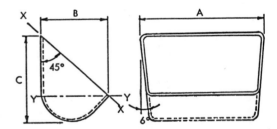

Table 3.13   B Buckets

Nominal Size (in.)	Weight (lb)	Capacity[a] (ft^3)		Dimensions (in.)		
		X-X	Y-Y	A	B	C
4 × 1	1.0	0.004	0.0014	4¼	1¾	2¼
7 × 3	2.2	0.03	0.007	7⁵⁄₁₆	3¾	5
8 × 3	2.3	0.04	0.008	8⁵⁄₁₆	3¾	5
10 × 4	4.0	0.06	0.014	10⅜	4¼	5¾
12 × 5	6.5	0.14	0.035	12⅜	5¾	8
14 × 6	12.0	0.21	0.039	14⅜	6¾	9⅜
16 × 6	13.5	0.24	0.044	16⁷⁄₁₆	6¾	9⅜
18 × 6	13.2	0.29	0.110	18⅜	6⅜	9⅛
24 × 5	13.0	0.28	0.070	24⁷⁄₁₆	5⅞	8

[a]Actual capacity depends on the angle of repose of the material and the inclination of the elevator.

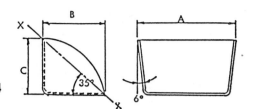

Table 3.14
C Buckets

Dimensions (in.)			Weight	Capacity (ft^3)[a]
A	B	C	(lb)	X-X
6	4½	4	2.0	0.026
7	4½	4	2.4	0.030
8	4½	4	2.8	0.035
10	5	4	4.0	0.052
12	5	4	4.8	0.061
14	7	5½	8.5	0.138
16	7	5½	10.5	0.158

[a]Actual capacity depends on the angle of repose of the material and the inclination of the elevator.

3.14.5.5 Horsepower

The sand weighs 378 lb, assume the pick up load in the boot to be 250 lb, which can vary. The maximum value is about 500 lb. Since the sand is free flowing and has no lumps, use 250 lb as the minimum value.

$$hp = \frac{(378 + 250) \times 260}{33,000} = 4.95$$

Or, use the formula for chain elevators

$$T_h = \frac{12H_o w_m}{s} = \frac{12 \times 30 \times 7}{16 \text{ in.}}$$

$$= 158 \text{ lb}$$

where $H_o = 30$ for centrifugal discharge elevators, w_m = weight (lb/ft³) of sand in the buckets, and s = spacing of buckets (in.).
 Thus,

$$hp = \frac{(378 + 158) \times 260}{33,000} = 4.22$$

3.14.5.6 Link-Belt (Catalog 1000)

As checked on page 320 of the catalog, this elevator is type 108. For material weighing 100 lb/ft³,

hp terminals   = 0.86
hp height = 0.041 × 72 = <u>2.95</u>
                              3.81

3.14.5.7 Jeffrey (Catalog PT-222A)

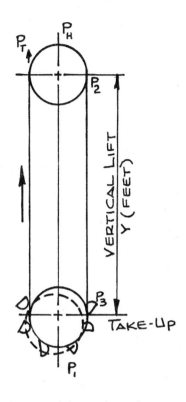

M = weight per foot of conveyor or elevator, including buckets (lb)

  = $\dfrac{\text{no. buckets} \times W_b}{d} + W_B$

  = $\dfrac{12 \times 4.6}{16} + 7$

  = 10.5 lb

d = bucket spacing (in.)
  = 16

W_b = weight of one bucket (lb) = 4.6

W_B = weight of material in one bucket (lb)
  = 7

S = speed (ft/min) = 260

T = capacity (tph) = 30.7

W = weight per ft of material to be conveyed (lb)

  = $\dfrac{12 \text{ in.}}{\text{bucket centers (in.)}} \times W_B$

  = $\dfrac{12}{16} \times 7 = 5.25$ lb

Y = vertical lift (ft)
  = 70 + 2 = 72 ft

Z = an empirical corrective factor with the following values for stated conditions:

  = 1.0 for centrifugal-discharge elevators handling coarse lumpy material

= 0.67 for centrifugal-discharge elevators handling fine, free-flowing material

= 0.50 for continuous bucket elevators

P_1, P_2, P_3 = total chain pull at various points on conveyor or elevator

P_1 = takeup tension, for screw-type takeups, will be as near zero as possible with proper adjustment; for gravity takeups, it will be equal to the tail carriage machinery plus any resultant force from added weights (see sketch)

= 110 lb (catalog)

$P_2 = MY = 10.5 \times 72$

= 756 lb

$P_3 = \dfrac{P_1}{2}$ (least tension)

= $\dfrac{110}{2}$ = 55 lb.

$P_Q = \dfrac{12\,W_B Z}{d} \times$ boot sprocket dia (in.)

= $\dfrac{12 \times 7 \times 0.67}{16} \times 24$

= 84.42 (use 84)

P_T = total maximum chain pull (lb.)

= $Y(W + M) + P_Q + \dfrac{P_1}{2}$

= $72(5.25 + 10.5) + 84 + \dfrac{110}{2}$

= 1134 + 84 + 55

= 1273

P_H = effective chain pull at head shaft (lb) (used in hp formula)

= $P_T - P_2$

= 1273 − 756

= 517

hp = $\dfrac{1.1\,P_H S}{33{,}000} = \dfrac{1.1 \times 517 \times 260}{33{,}000}$

= 4.5

As a check with Jeffrey Catalog 418,

hp = 3.7 − 0.035 × (88 − 72)

= 3.7 − 0.6

= 3.1

Using the formula below for chain elevators:

hp = $E \times \dfrac{15WH}{10{,}000}$

= $1.2 \times \dfrac{15 \times 30 \times 72}{10{,}000}$

= 3.89

where W = capacity (tph), H = height (ft, c to c of sprockets + diameter of one sprocket), and E = efficiency and drive factor = 1.2 (refer to paragraph 3.11.2). Therefore, use a 5-hp motor.

3.14.5.8 Chain

Tension = 1134 lb, without foot tension

= 1134 + 250 = 1384 lb, with foot tension

The allowable tension on SBS-102B chain is 6500 lb. The average pitch is 4 in. The weight per ft is 6.9 lb. For K_2 attachments, use 8.3 lb/ft (see table 3.2).

3.14.6 Design of Centrifugal-Discharge Elevator Handling Salt

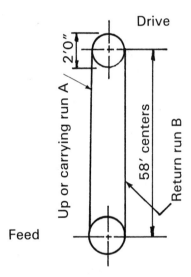

Material: 60 tons salt = 120,000 lb or 1600 ft³.

Length: 60 ft (58 ft c to c of head and foot shafts, plus 2-ft diameter of head sprocket.

Chain: C-111, with K2 attachments = 11 lb/ft (aprx).

Buckets: MI style A, 14 in. × 7 in., say, 13 lb each.

Capacity required: 60 tph. The capacity of buckets filled to water level (about 10 lb salt) is 0.138 ft³; filled full (about 17 lb salt), 0.23 ft³; filled to 75% (about 14 lb salt), 0.1725 ft³. When figuring for capacity, always figure the buckets 75% full on centrifugal-discharge elevators. When figuring for hp, use the full capacity of each bucket, because it is possible to get a surge load feed to the boot from another conveyor, or by another means.

3.14.6.1 Load on Up Run A

Chain = 60 ft × 11 lb/ft	=	660 lb
Buckets = 40 buckets × 13 lb	=	520 lb
Salt = 40 buckets × 17 lb	=	680 lb
		1860 lb

3.14.6.2 Load on Return Run B

Chain = 660 lb
Buckets = 520 lb
1180 lb

3.14.6.3 Capacity

The buckets are traveling at the rate of 260 fpm and are spaced every 19 in., resulting in approximately 164 buckets/min discharged, or

$$164 \text{ buckets} \times 14 \text{ lb} = 2296 \text{ lb/min}$$

$$2296 \text{ lb/min} \times 60 \text{ min} = 137{,}760 \text{ lb/hr}$$

$$137{,}760 \text{ lb/hr} \div 2000 \text{ lb/t} = 68 \text{ tph (aprx)}$$

based on solid material in the buckets. There will be some voids in a bucket because the salt does not have time to pack solid in a bucket traveling at 260 fpm. Therefore, from experiments in the field, it is best to take about 93% of the capacity for safe figuring, resulting in 63 tph.

3.14.6.4 Figuring Theoretical Horsepower

For figuring theoretical hp at the head shaft, assume (in a vertical elevator) that the weight of the chain and empty buckets balance. Then, the load to be moved is the load or amount of salt in each bucket on the up or carrying run, with the buckets full and solid with salt, plus an additional bucket pickup load in the boot of 250–500 lb (refer to paragraph 3.14.5), depending on the type of material. Handling fine, dry, free-flowing (if not allowed to stand in the boot and cake up) salt, use 350 lb (or $[12 \times 30 \times 17]/19 = 322$ lb).

On the up, or carrying run, there will be approximately 40 buckets $\times$ 17 lb in each full bucket = 680 lb + 350 lb (pickup in boot) = 1030 lb. The 350-lb pickup also takes care of starting the elevator loaded on the up run from a rest condition in the event of interruption of power.

$$hp = \frac{1030 \text{ lb} \times 260 \text{ fpm}}{33{,}000} = 8.1 \text{ (aprx)}$$

$$\frac{8.1}{0.8} = 10.1 \text{ hp (refer to paragraph 3.11.2)}$$

Use a 10-hp motor.

3.14.6.5 Determining Size of Head Shaft of Single-Strand Centrifugal-Discharge Elevators

The loads will be

Up run A = 1860 lb
Return run B = 1180 lb
Pickup in boot = 350 lb

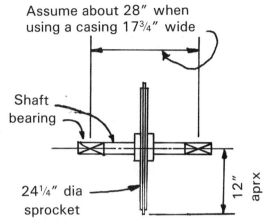

Assume about 28″ when using a casing 17¾″ wide

Shaft bearing

24¼″ dia sprocket

12″ aprx

Figure 3.16 Detail at a sprocket.

Weight of shaft and
sprocket (assumed) = 250 lb
Total load on head
shaft = 3640 lb
(say 3700 lb)

Total torsional moment is found by

$$1030 \text{ lb} \times 12 \text{ in.} = 12{,}360$$

where 1030 lb is the up run load (680 + 350) and 12 in. is one-half the diameter of the sprocket shown on figure 3.16.

Total bending moment is found by

$$\frac{WI}{4} = \frac{3700 \times 28 \text{ in.}}{4} = 25{,}900 \text{ in.-lb}$$

Using figure 9.3, with a factor of 1.5 in bending and 1.0 for torsion, a $3^{7}/_{16}$-in. shaft is required. A $3^{15}/_{16}$ shaft was used to reduce the number of stock parts at the particular plant.

The torsion will be increased by the difference in tension of the chain or belt up and down runs and, making an allowance for these, would require a $3^{15}/_{16}$-in. shaft.

3.14.7 Specification for Vertical Centrifugal-Discharge Elevator

Required: one vertical centrifugal-discharge elevator complete.
Capacity: 60 tons (uniform feed) of dry, fine salt (similar to common table salt).
Elevator boot and head shaft centers: 60′0″.
Buckets: 14″ × 7″ malleable iron, style AA, enameled, attached to chain by cut and bronze lockwashers and stainless steel bolts.
Chain: C-111 combination chain, or equal.
Attachments: K2 every fourth link (19″).

Headshaft: 3^{15}/$_{16}$" diameter with standard style GS-3100 takeup babbitted bearings, or equal.

Head sprockets: flint rim, split, 24.40" pd, 16T, 4.76" pitch, bore 3^{15}/$_{16}$" KS and SS.

Automatic differential band brake: backstop, 24" diameter on head shaft.

Boot shaft: 2^{3}/$_{16}$" diameter bearings, babbitted flanged type.

Boot traction wheel: flint rim, 18^{3}/$_{4}$" pd (actual od = 17^{3}/$_{8}$"), bore of 2^{3}/$_{16}$" KS and SS.

Casing: no. 12 gauge steel plate and angle construction for intermediate sections; dust-tight construction, 17^{3}/$_{4}$" × 48" inside dimensions, with two large inspection doors above boot section opposite each other, spring clamped. Boot section to be ^{3}/$_{16}$" steel with removable end panel and clean out doors at bottom, also ^{1}/$_{4}$" curved boot plate to clear buckets by ^{1}/$_{2}$".

Head section: no. 12 gauge steel.

Hood section: no. 14 gauge steel, split in two pieces, with hinged inspection doors front and back.

Discharge chute: ^{3}/$_{16}$" steel with no. 10 gauge steel removable liner, and ^{3}/$_{16}$" steel adjustable throat plate, acting as a liner in bottom of chute, adjusted so it comes within 1" of returning buckets.

Motor drive: the owner will supply a 10-hp motor at 1200 rpm for 440V, 60 Hz, 3-ph service, together with magnetic starter and push button control. The manufacturer shall furnish the rest of the drive, including enclosed herringbone or helical-gear type drive, flexible coupling, steel baseplate to mount reducer and motor also necessary, finished roller-chain drive from reducer to elevator head shaft. (Some prefer to have manufacturer specify size of motor.)

Coupling and chain guard: furnished by owner.

Platform, ladder, and safety guard: a standard ladder is required from boot of elevator to platform, serving driving machinery mounted on it just below head shaft. This platform shall be large enough for two men to service machinery when necessary. Standard handrail of angles or pipe will be acceptable; toe guards to be provided 6" high around platform. The floor of platform should be open nonslip grating. Complete platform to be supported from elevator casing. A safety guard, in accordance with state of _____ safety laws, must be included, to start 7' from bottom of ladder and attach to platform.

In general, the manufacturer shall design, furnish, fabricate, and deliver (freight allowed to plant) all equipment and accessories specified or implied herein, necessary to form a complete workable unit.

Boot and head section shall be completely assembled for ease in erection.

All equipment shall be thoroughly cleaned and painted with a high grade of machinery paint, in such a manner that the paint will adhere to the surface and be weather resistant. All painting shall be done in a neat and professional manner. The equipment shall not be loaded for shipment until the paint is thoroughly dry.

All polished parts, shafts, bearings, etc., shall be coated with rust preventive compound and crated, if necessary, to prevent damage in shipment, and all equipment shall be thoroughly protected from the weather.

Services of an engineer, if required, at time of installation, shall be furnished at the rate stated in the Proposal.

The successful bidder shall furnish three sets of prints for approval of the general design of the equipment and later, _____ sets of certified as-built prints to _____ .

3.15 CONTINUOUS-BUCKET ELEVATORS

3.15.1 General

In the continuous-bucket elevator, buckets closely spaced on chain or belt are designed so that material is loaded directly into the buckets through a loading leg, instead of being scooped up in the boot. It is designed to operate at a low speed. The low operating speed and the method of loading and discharging minimizes breakage of fragile materials. These elevators are thus especially well adapted where degradation of the material is to be minimized and where extreme dust conditions are to be avoided. They will handle efficiently almost any kind of dry, fine, or small lump material that is not damp. Where lumps are over 2 in., or where the 2-in. lumps are over 10% of the capacity, super-capacity continuous-bucket elevators should be used. Feed inlet point of the continuous-type elevator is somewhat higher than that of the centrifugal elevator, necessitating a deeper pit when located below ground level. This elevator handles limestone, lime, cement, dry chemicals, and ferroalloys. Such materials as fine salt, sand, clay, and many chemicals, dry or damp, should not, generally speaking, be handled by continuous-type elevators, as the fine particles get into the bottom of the bucket because of the V-shape. The particles clog there, will not discharge and finally, the material piles upon itself and gets hard in the bucket until very little capacity is left.

3.15.2 Speed

The speed of a continuous-bucket elevator on chain preferably should not exceed 150 fpm. If the material is not entirely free-flowing, the speed should be reduced to 100–125 fpm. When mounted on belts, and when inclined, the bucket speed may be increased up to about 200 fpm. When belts are used, the diameter of the head pulley is generally larger to prevent slipping. Where highly abrasive materials are handled, reduced speeds are advisable. Generally, continuous-bucket elevators are equipped with chain. Larger pulley diameters, at the same rpm of head shaft, give additional speed in feet per minute. The use of belts is preferred where dusty, abrasive material can get to the chain joints.

3.15.3 Buckets

The buckets are not designed or intended to scoop material from the boot. Discharge of material over the head sprocket or wheel is accomplished by transfer of material from the discharging buckets to the front or bottom of the preceding bucket, which thus acts as a moving chute to the fixed discharge chute attached to the elevator casing (refer to paragraph 3.15.5).

The steel buckets on this type of elevator do not have a so-called round bottom like the type AA used on centrifugal elevators. The V shape of the bucket will fill up fast and the material will get hard, if damp or wet material is handled, thus reducing the actual capacity of the buckets. Some manufacturers install a filler plate in the bottom of the ∨ or ∀, either flat or curved. This may help a little, depending on the character of the material handled, but the possibility of material packing in this restricted area still remains.

The loading leg is used to direct material to buckets and is attached to the casing. It fits closely around the path of the buckets to prevent as much spillage of any fines in the material as possible, although some fine material will go to the bottom of the boot where it can be cleaned out at intervals. Any fines accumulating in the boot generally do not tend to hamper the operation of the elevator, and are usually scooped up by buckets if material does not get packed or hard. (see figure 3.17). For material +1 in., a loading leg should always be used.

The continuous-bucket elevator is quieter than the centrifugal discharge, especially when handling lumpy material.

3.15.4 Belts

Usually, when belts are used, the diameter of the head pulley is larger to prevent slipping. With a belt on a continuous elevator outdoors and not encased, the pulley must be lagged, or covered with a rubber covering known as "rough top brand," or with herringbone grooves cut into it, to get good contact with the belt.

When selecting a belt as an elevating medium, materials that pack and tend to build up between the belt and pulley, as well as rough or jagged particles which damage the belt by becoming lodged between buckets and belt, should be avoided. To some extent, these difficulties are alleviated through the use of wing pulleys on the foot shaft.

3.15.5 Types of Continuous Buckets

1. Type MF is a medium-front bucket that is not overlapping. It is the type most frequently used for continuous bucket elevators. Small flat or curved filler pieces are welded into buckets as shown. Refer to table 3.15 for capacity and weight.
2. Type HF buckets are made with a high front that is not overlapping. These buckets are used for higher capacities than medium-front buckets. Refer to table 3.16 for capacity and weight.
3. Type LF buckets are low-front, not overlapping buckets, designed for inclined bucket elevators or to handle fine or damp materials that would stick or pack in buckets of other styles (refer to table 3.17).
4. Type HFO buckets are high-front overlapping buckets. They are similar to type HF high-front buckets but are made overlapping to prevent leakage between the buckets (refer to table 3.18).
5. Type D buckets are generally used for crushed stone plants or concrete plants located at construction sites (refer to table 3.19).
6. Budd cast nylon type D buckets data are given in table 3.20.

In summary, the MF, HF, and LF buckets are not the overlapping type, and are spaced on the chain or belt with about ½ in. between them so they will not foul each other because of poor assembly in the field. Such spacing is considered good practice and allows little leakage. The overlapping buckets are designed to actually fit into each other to prevent any leakage when discharging.

All of the buckets described are currently made of welded construction; either spot welded or continuously welded, depending on the fineness of the material handled.

3.15.6 Design of Continuous-Bucket Elevator

3.15.6.1 Data

Required: a continuous bucket elevator to lift 25 tons of an abrasive material.
Height: 112 vertical ft.
Operating conditions: outdoors, exposed to weather, commercially dust-tight.

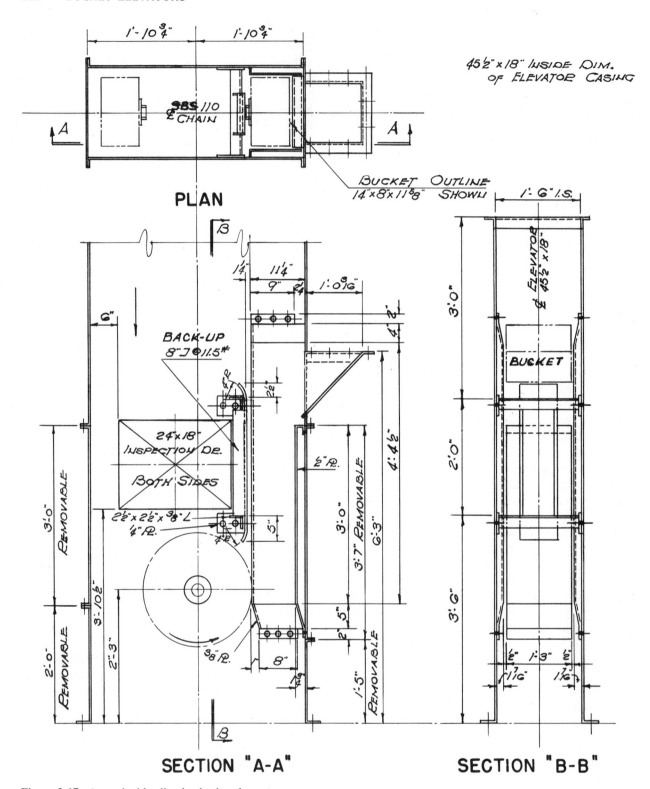

Figure 3.17   A standard loading leg bucket elevator.

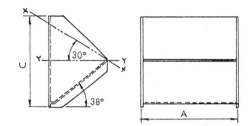

Table 3.15 Type MF Buckets

Bucket Size (in.)			Weight (lb)				Capacity (ft^3)	
A	B	C	12 ga	10 ga	$^3/_{16}$ in.	$^1/_4$ in.	X-X	Y-Y
8	5	7¾	5.1	6.3	8.7		0.070	0.040
9	6	9¼	6.7	8.6	11.9		0.118	0.068
10	5	7¾	5.9	7.4	10.2		0.090	0.050
10	6	9¼	7.2	9.2	12.7		0.130	0.075
10	7	11⅝	9.3	11.9	16.5		0.180	0.103
10	8	11⅝	9.9	12.8	17.8	23.2	0.235	0.135
11	6	9¼	7.7	9.9	13.6		0.145	0.081
12	6	9¼	8.1	10.5	14.5		0.155	0.091
12	7	11⅝	10.4	13.4	18.6		0.218	0.125
12	8	11⅝	11.2	14.4	20.0	26.1	0.275	0.163
14	7	11⅝	11.6	14.9	20.7		0.253	0.145
14	8	11⅝	12.4	16.0	22.2	29.1	0.325	0.190
16	8	11⅝	13.7	17.6	24.5	32.0	0.375	0.220
16	12	17⅝		29.9	40.6	54.8	0.852	0.490
18	8	11⅝	14.9	19.2	26.7	35.0	0.420	0.250
18	10	15		25.9	36.1	47.3	0.662	0.379
20	8	11⅝	16.1	20.8	29.0	38.0	0.470	0.270
20	12	17⅝		34.8	48.5	63.9	1.075	0.620
24	10	11⅝		27.4	38.2	50.0	0.850	0.512
24	12	17⅝		39.8	55.4	73.1	1.295	0.745

Material: abrasive, weighing 40-50 lb/ft³ with a 200°F temperature.

Buckets: style MF, ¼-in. plate. Because of the size of the material (some 2-in. lumps), try a 16 in. × 8 in. × 11⅝ in. bucket at a speed of 125 fpm. (This is either as a result of a previous test or as suggested by manufacturers' catalogs.) Bucket capacity is 0.220 ft³ (table 3.15) and buckets are spaced at 12 in. or 1 ft 0 in. The number of buckets/min is 125 (125/1.0).

3.15.6.2 Capacity

For capacity

$$\frac{0.220 \times 75\% \times 125 \times 60 \times 40 \text{ lb}}{2000} = 24.75 \text{ tons}$$

For hp

$$\frac{0.220 \times 80\% \times 125 \times 60 \times 50 \text{ lb}}{2000}$$

$$= 24.75 \times \frac{5}{4} \times \frac{80}{75}$$

$$= 33 \text{ tons}$$

The capacity is acceptable but close. Increasing the speed to 130 fpm should be about the maximum needed to get clean bucket discharge for MF buckets. The 80% value was used here for hp because of experience with this material at this location.

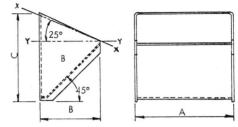

Table 3.16 Type HF Buckets

Bucket Size (in.)			Weight (lb)				Capacity (ft^3)	
A	B	C	12 ga	10 ga	$^3/_{16}$ in.	$^1/_4$ in.	X-X	Y-Y
8	5	7¾	4.9	6.2	8.5		0.080	0.052
10	5	7¾	5.7	7.3	10.0		0.100	0.065
10	6	9¼	7.2	9.1	12.6		0.145	0.098
10	7	11⅝	9.1	11.6	16.0	20.9	0.190	0.130
12	6	9¼	8.3	10.4	14.4		0.175	0.115
12	7	11⅝	10.3	13.2	18.2	23.9	0.240	0.155
12	8	11⅝	11.3	14.3	20.0	26.0	0.295	0.205
14	7	11⅝	11.5	14.8	20.4	26.7	0.280	0.184
14	8	11⅝	12.6	16.0	22.4	28.1	0.350	0.240
16	8	11⅝	13.9	17.7	24.7	32.2	0.395	0.275
16	12	17⅝		30.3	41.9	55.0	0.900	0.635
18	10	15		26.2	36.1	47.7	0.720	0.485
20	12	17⅝		35.1	49.1	64.6	1.150	0.800
24	12	17⅝		40.5	56.3	74.3	1.335	0.960

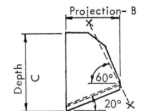
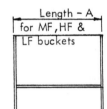

Table 3.17 Type LF Buckets

Bucket Size (in.)			Weight (lb)				Capacity (ft^3)	
A	B	C	12 ga	10 ga	$^3/_{16}$ in.	$^1/_4$ in.	X-X	Y-Y
10	6	9¼	6.8	8.8	12.1		0.168	0.035
10	7	11⅝	8.5	10.8	15.1		0.242	0.050
12	6	9¼	7.8	10.0	13.8		0.201	0.042
12	7	11⅝	9.6	12.3	17.1		0.302	0.060
12	8	11⅝	11.2	14.4	20.1		0.347	0.075
14	7	11⅝	10.7	13.7	19.1		0.345	0.070
16	8	11⅝	13.6	17.4	24.3		0.463	0.101
16	12	17⅝		29.3	40.7	53.6	1.093	0.229
18	10	15		25.4	35.0	46.5	0.940	0.183
20	8	11⅝	15.9	20.5	28.5		0.573	0.126
20	12	17⅝		33.9	47.1	62.0	1.365	0.287
24	12	17⅝		38.5	53.5	70.5	1.643	0.346

[a]Actual capacity depends on the angle of repose of the material and the inclination of the elevator.

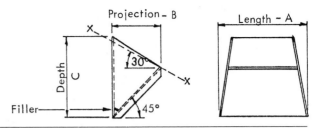

Table 3.18 Type HFO Buckets

Bucket Size (in.)			Weight (lb)				Capacity (ft^3)	
A	B	C	12 ga	10 ga	$^3/_{16}$ in.	$^1/_4$ in.	X-X	Y-Y
8	5	8½	5.1	6.5	8.9		0.089	0.059
10	5	8½	5.9	7.6	10.5		0.112	0.077
10	6	10	7.5	9.5	13.1		0.162	0.108
10	7	12½	9.6	12.3	16.7		0.227	0.150
12	6	10	8.6	10.8	15.0		0.193	0.126
12	7	12½	10.8	14.0	19.0		0.275	0.182
12	8	12½	11.8	15.0	20.5	27.1	0.320	0.200
14	7	12½	12.1	15.7	21.3		0.333	0.224
14	8	12½	13.1	16.8	22.9	30.4	0.386	0.246
16	8	12½	14.5	18.6	25.2	33.6	0.425	0.265
16	12	18⅝		31.1	43.0	56.8	0.962	0.605
20	12	18⅝		36.4	50.4	66.6	1.203	0.755
24	12	18⅝		41.7	57.8	76.4	1.444	0.905

[a]Actual capacity depends on the angle of repose of the material and the inclination of the elevator.

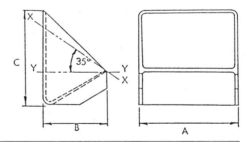

Table 3.19
Type D Buckets

Bucket Size (in.)			Weight (lb)	Capacity (ft^3)	
A	B	C		X-X	Y-Y
8	5	7¾	8.37	0.068	0.028
12	5	7¾	11.00	0.106	0.045
12	7	11¾	19.43	0.219	0.099
16	7	11¾	20.00	0.288	0.129
20	9	11¾	32.00	0.573	0.251

Note: These buckets are generally used for crushed stone plants or concrete plants located at construction sites.

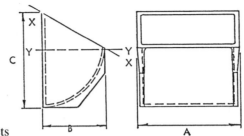

Table 3.20
Type D Budd
Cast Nylon Buckets

Bucket Size (in.)			Weight	Capacity[a] (ft^3)	
A	B	C	(lb)	X-X	Y-Y
8.12	5.06	7.75	1.70	0.083	0.057
10.06	5.09	7.78	2.06	0.102	0.069
12.25	8.12	11.68	5.50	0.301	0.193
14.18	8.12	11.68	6.00	0.347	0.219
16.18	8.12	11.75	6.10	0.399	0.251

[a]Variations: capacity and weight, +5%; dimension A, +0.20 in.; dimensions B and C, +0.15 in.

3.15.6.3 Load on Up Run

Chain = 6.3 lb × 112   = 706 lb
Buckets = 112 each at 32 lb   = 3584 lb
Material = 112 at 11 × 0.8   = 986 lb
$(12wH_o)/s = (12 × 11 × 10)/12$ = 110 lb
                                                      5386 lb

3.15.6.4 Load on Down Run

Chain   = 706 lb
Buckets = 3584 lb
               4290 lb

3.15.6.5 Horsepower

For belt elevators, the pull, P, required to pick up the load, overcome friction, and raise the material in the buckets is given by

$$P = \frac{12w(H_o + H)(1 + K)}{s}$$

$$= \frac{12 × 8.8(10 + 112)(1 + 0.97)}{12}$$

$$= 2115 \text{ lb}$$

where

w = weight of material in a single filled bucket (lb)
  = 0.220 × 50 × 0.8 = 8.8
H_o = height factor to take care of pickup force (ft)
  = 10 for continuous buckets and 30 for all spaced buckets (centrifugal)

H = height of elevator (ft)
  = 110 + 2 = 112 (c to c of head and boot shaft + diameter of one pulley)
K = factors depending on drive pulley (refer to paragraph 3.11.1 for values)
  = 0.97
s = center to center of buckets (in.) = 12

Therefore,

$$\text{hp} = \frac{PS}{33,000} = \frac{2115 × 130}{33,000}$$

$$= 8.33$$

where S = speed of buckets in fpm = 130. With a drive efficiency of 80%,

$$\frac{8.33}{0.8} = 10.4 \text{ hp}$$

Consequently, use a 15-hp motor for a belt elevator.

For chain elevators, the allowable chain pull for SBS-110 chain is 6300 lb. The maximum pull is 5386 lb. Using the formula

$$\frac{15WH}{10,000} × 2 \text{ (service factor)}$$

$$= \frac{15 × 33 \text{ tons} × 112}{10,000} × 2$$

$$= 11.09 \text{ hp}$$

The service factor, $(1 + K)$, in this case is 1.97, so use 2. With a drive efficiency of 80%,

$$\frac{11.09}{0.8} = 13.86 \text{ hp}$$

Therefore, use a 15-hp motor for a chain elevator.

Where elevators are over 80-ft high, takeups probably will be required at the head and boot sections.

3.15.6.6 Drive

Horsepower: 15.

Speed of motor: 1800 rpm no load, 1750 rpm full load

$$\frac{1750}{31.4} = 55.7 \text{ (say 56) rpm.}$$

Speed of reducer: Falk Motoreducer type 151-70FZ, or Link-Belt Motogear no. 15 FDBL 2; ratio: 31.4:1; slow-speed shaft, 56 rpm using 1750-rpm motor speed.

Selection of speed reducer: say 25-in. pitch diameter of head sprocket; assumed elevator speed = 125 fpm; required rpm = $125/2.08\pi = 19+$.

Drive sprocket selected: 14T, 8.988-in. pitch diameter.

Driven sprocket: 40T, 25.491-in. pitch diameter; $40/14 \times 19 = 54.3$ rpm required at head shaft (output shaft of speed reducer).

Selections of RC chain and sprockets for the 15-hp motor and shaft bores are RC-160 chain with 14T, 8.988-in. pd, type B drive sprocket, and 40T, 25.491-in. pd, type B driven sprocket. The ratio of reduction for the sprockets is 2.86:1.

3.15.6.7 Tail Shaft (figure 3.18)

N = pull for making chain taut (lb)
  = $(6.3 \times 112)2 = 1411$ lb.
Bending load = $F_1 = 1411/2 = 706$ lb.
Bending moment = 706×16 in. = 11,296 in.-lb.
Torsional moment = 1411×12 in. = 16,932 in.-lb.

Shaft size = $3^{7}/_{16}$ in. at 6000 lb shear (see figure 9.3 and paragraph 9.6). Using 10,000 lb shear = $3^{7}/_{16} \times 0.8433 = 3.437 \times 0.8433 = 2.90$, say $2^{15}/_{16}$-in. diameter.

3.15.6.8 Head Shaft (figure 3.19)

Shaft: refer to section 9.
Bending moment, at point 3 on the figure = 4838×16 = 77,408 in.-lb.
Torsional moment = $1096 \times 12 = 13,152$ in.-lb.
Shaft size = $4^{7}/_{16}$ in. For $K_b = 1.5$ and $K_t = 1.0$, figure 9.3 shows a $4^{15}/_{16}$-in. shaft with 6000 psi shear (shafts with key seats). At 6000 psi, a $4^{7}/_{16}$-in. shaft is slightly light. It is generally good policy to play safe on the shaft. In this case, however, the overhanging load will reduce the bending moment, and the available shaft steel had an allowable shear stress of 8000 psi.

3.15.6.9 Bill of Materials

Chain: single strand SBS-110 with K_2 attachments; 12″ spacing, riveted construction.

Buckets: 16″ × 8″ × 11⁵⁄₈″, style MF, ¼″ steel continuous welded to the back plate with necessary bolts, nuts, and washers.

Head shaft terminal
  shaft: 4⁷⁄₁₆″ diameter with necessary keyways, keys
  takeups: spherical roller bearing, 4⁷⁄₁₆″ diameter
  backstop: enclosed one-way cam and roller type preferred for maximum load conditions
  traction wheel: 24″ diameter, 25¼″ pd, split type, flint rim, 4⁷⁄₁₆″ diameter bore for SBS-110 chain

Foot shaft assembly
  shaft: 2¹⁵⁄₁₆″ diameter with necessary keyway, key, SS, and collars

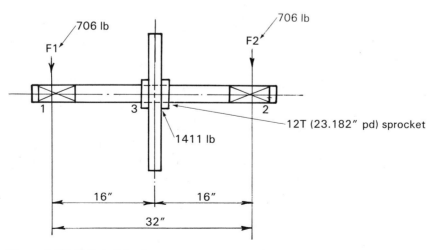

Figure 3.18 Tail shaft loads.

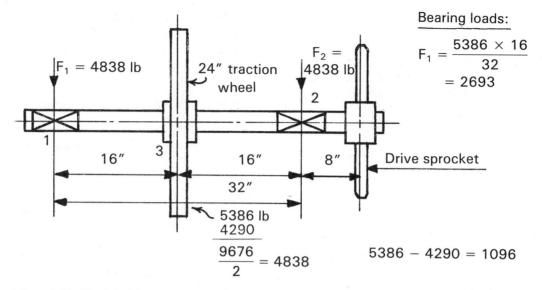

$$F_1 = \frac{5386 \times 16}{32} = 2693$$

Bearing loads:

$F_1 = 4838$ lb   24″ traction wheel $F_2 = 4838$ lb   Drive sprocket

16″   16″   8″

32″

5386 lb

$$\frac{4290}{9676} \atop 2 = 4838$$

$$5386 - 4290 = 1096$$

Figure 3.19 Head shaft loads.

bearings: spherical roller bearing pillow blocks, $2^{15}/_{16}$″ diameter bore, to be shelf-mounted to boot section

sprocket: 12T, 23.182″ pd, $2^{15}/_{16}$″ diameter bore, chilled tooth

Drive: Motogear unit with 25.6:1 ratio, 68-rpm slow-speed shaft, complete with 15-hp, 1800-rpm, 220-V, 60-Hz, 3-ph, TEFC motor; flexible coupling; safety guard over coupling and adjustable baseplate for the unit; drive to be based on buckets being 100% loaded.

Drive chain and sprockets

chain: RC-160, necessary length (to suit center)

drive sprocket: 14T, 8.988″ pd, hub to one side, $3^5/_8$″ diameter bore

driven sprocket: 40T, 25.491″ pd, hub on one side, $4^7/_{16}$″ diameter bore

safety guard

necessary keyways, keys, and SS

Casing: intermediate sections, $^3/_{16}$″ diagonally braced, commercially dust tight and waterproof.

Head section: no. 12 gauge steel, split, style 1 discharge spout lined with $^1/_4$″ steel plate; $^3/_{16}$″ steel, well braced to support head bearings, drive, windload, etc. Drive shall be mounted on a steel bracket, rigidly attached to the head section.

Hood section: no. 12 gauge steel, split, with hinged dust-tight inspection door.

Boot section: $^1/_4$″ steel, complete with removable clean-out panels; hinged, full-width doors both sides of boot, for assembly and inspection of buckets and chain. Bottom of feed chute to be three buckets high, upright loading position from center of boot shaft, loading leg, and curved bottom plate.

Service platform: supported from casing, including hand

railings, toe plates, subway grating deck, caged safety ladder, hoist beam and supports off casing and platform.

Painting: shall be seller's standard shop and finish coat, using rust resistant paint; casing shall be painted inside and outside.

Lubrication: shall be by means of hydraulic-type Ale-mite fittings.

Shop assembly: all units shall be shop assembled prior to shipment, matched, marked, and shipped in convenient pieces or subassembled for field erection.

3.15.6.10 Specification for Vertical Continuous-Bucket Elevator

Required: one vertical continuous-bucket elevator complete.

Capacity: 120 tph limestone, 3 in. $\times$ D, weighing 85–90 lb/ft³.

Elevator boot and head shaft centers: 72 ft $1^3/_8$ in.

Boot shaft: height (in highest position) above the bottom of boot section is 4 ft 0 in.

Buckets: continuous steel 20 in. $\times$ 12 in. $\times$ $17^3/_4$ in., made of $^1/_4$-in. plate with $^1/_4$-in. reinforced lip and filler plate bottom of bucket. Welded construction, with continuous inside and outside welds. Attached to chain by bolts, with locknuts peened set after tightening.

Chain: double strand of SS-1833 or approved equal; 18-in. pitch steel thimble roller chain with chilled iron rollers; ultimate strength of 42,000 lb.

Head shaft: $5^7/_{16}$-in. diameter extended to receive head-shaft-driven sprocket and differential band break backstop.

Head sprockets: two split chilled rim, 36.12-in. pd, 6T, $5^7/_{16}$-in. bore, KS and SS.

Backstop: Marland one-way backstop, or equal.

Bearings: three 5$^{7}/_{16}$ in. babbitted pillow blocks.

Head shaft bearing supports: provided independent of elevator casing.

Boot shaft: 2$^{15}/_{16}$-in. diameter.

Bearings: 2$^{15}/_{16}$ in. self-aligning adjustable, with 12 in. of takeup.

Boot traction wheels: two split chilled rim, 36-in. pd, 2$^{15}/_{16}$-in. bore, securely fastened to boot shaft.

Casing: no. 10 gauge steel plate on long sides, 12 gauge on short sides and angles; dust-tight construction; 33 in. × 66 in. inside dimensions; maximum length of sections: 10 ft 0 in.

Manufacturer shall furnish 10-ft boot section made of ¼-in. plate, with receiving hopper and the necessary loading leg, as is normally required for that type elevator, with guide angles on the loading side only.

Inspection doors: openings 4 ft 6 in. wide by 3 ft 0 in. high with clamped removable doors both sides of elevator; lower edge of each opening 18 in. above the boot shaft in its upper position.

Additional inspection doors: two additional doors (one each side) of the same size as those previously specified shall be provided in each elevator, placed so bottoms of doors are 3 ft above ground level.

Removable panel cleanouts: shall be provided on front and back of boot section, 24-in. high, equipped with spring holding clamps.

Safety features: necessary safety collars, keys, safety setscrews, etc.; safety guards for shafts, sprockets, chain, couplings, etc.

Hopper and chute (attached to boot section, to receive discharge from belt conveyor): to be provided by the manufacturer as shown on attached drawing.

Chutes or spouts (from the discharge of elevator to belt conveyor): shall be provided by the manufacturer, as shown on purchaser's drawing.

Lubrication: all bearings shall be provided with Alemite or equal industrial hydraulic-type grease fittings; extension piping shall be provided where necessary or desirable, for easy access or safety.

Drive system: shall include a gear-reducer unit, flexible coupling, and drive motor mounted as a unit on a steel baseplate. The gear reducer shall be an enclosed herringbone or helical-gear-type drive, according to AGMA specifications, with a service factor of (specify the required service factor). The drive system shall also include a steel roller chain drive of flint-rim plate center sprockets and chain, from speed reducer to elevator head shaft.

Motor: the drive motor shall be totally enclosed, 20 hp, 440 V, 60 Hz, 1800 rpm, 3-ph. (This motor specification shall be made to satisfy customer's requirements).

The manufacturer shall design, furnish, fabricate, and deliver, freight allowed to destination, all equipment and accessories, specified or implied herein, necessary to form a complete workable unit. Parts shall be shop-assembled for minimum field assembly work.

All equipment shall be thoroughly cleaned and painted with a high grade of machinery paint, in such a manner that the paint will adhere to the surface and be weather resistant. All painting shall be done in a neat and professional manner. The equipment shall not be loaded until the paint is thoroughly dry.

All polished parts, shafts, gearing, bearings, etc, shall be coated with rust-preventive compound and crated, if necessary, to prevent damage in shipment, and all equipment shall be thoroughly protected from the weather.

A platform for support of the drive system and head shaft bearings of the elevator will be furnished by the Purchaser.

Services of an engineer, if required at time of installation, shall be furnished at rate stated in the Proposal.

The successful Bidder shall furnish three sets of prints for approval of general design of the equipment and later, _____ complete sets of certified as-built prints to: _____, Consulting Engineers.

3.16 SUPER-CAPACITY CONTINUOUS BUCKET ELEVATOR

3.16.1 General

Large-capacity (super-capacity) elevators are made with specially designed steel buckets attached at their sides to double-strand long-pitch steel roller chain. Super-capacity buckets are generally made up to 36 in. long by 12$^{5}/_{8}$ in. projection by 17$^{5}/_{8}$ in. deep. Other sizes are given in table 3.21. These elevators are slow-operating, travelling anywhere between 75 fpm and 120 fpm.

3.16.2 Materials Handled

These elevators are of the continuous-bucket type, for high lifts and to handle large capacities of friable or abrasive materials from fines to large or heavy lumps. Material is fed to the buckets through a loading leg. Materials usually handled are dry and free-flowing. Dry, sized limestone or lime, dry gypsum rock, lump coal and ferro alloys are successfully handled. The flat bottom of these buckets helps prevent material from lodging in the bottom.

These elevators are ruggedly built to stand extreme

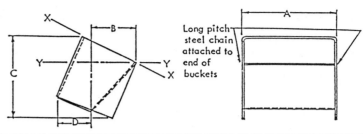

Table 3.21 SC Buckets

Bucket Size (in.)				Weight (lb)				Capacity (ft^3)	
A	B	C	D	10 ga	$^3/_{16}$ in.	$^1/_4$ in.	$^5/_{16}$ in.	X-X	Y-Y
12	8¾	11⅝	4⁹⁄₁₆	22	29	39	49	0.54	0.35
14	8¾	11⅝	4⁹⁄₁₆	23	31	41	51	0.63	0.41
16	8¾	11⅝	4⁹⁄₁₆	25	34	45	56	0.72	0.46
16	12⁷⁄₁₆	17⅜	6½	43	58	76	95	1.55	1.11
18	8¾	11⅝	4⁹⁄₁₆	27	36	48	60	0.81	0.52
20	8¾	11⅝	4⁹⁄₁₆	29	39	52	65	0.90	0.58
20	12⁷⁄₁₆	17⅜	6½	49	67	58	110	1.94	1.40
24	12⁷⁄₁₆	17⅜	6½	55	75	104	130	2.33	1.68
30	12⁷⁄₁₆	17⅜	6½	65	88	117	146	2.91	2.11
36	12⁷⁄₁₆	17⅜	6½	73	99	132	165	3.49	2.53

Note: The flat bottom of these buckets helps prevent material from lodging in them.

service and, given proper maintenance, will give a low cost per ton of material handled. They must be operated at a speed in the range noted above because of being equipped with long-pitch steel chains (9 in. and over), which actually cause some vibration when passing over the head sprocket.

From the open quarries, overburden (ground and claylike material) must be removed from the surface before reaching the gypsum rock or stone deposits, as these products are mined in the open. They are loaded on railroad cars or ships, delivered to the processing plant, at times not clean or free of overburden. Particularly in wet weather, the overburden gets sticky or tacky. When in this condition, it should not be handled in this or any other type of bucket elevator (if there is any way to avoid it), as the buckets fill up with this sticky mass, which gets as hard as concrete when the machinery is shut down overnight or weekend, and the buckets will not discharge their load. Wherever possible, a belt conveyor should be used to elevate this type of material. If the gypsum rock or stone products are clean of dust or overburden, they can be handled satisfactorily in the super-capacity elevator.

Charcoal, produced in northern Pennsylvania, as received by chemical plants, weighs only 18–25 lb/ft³. It is shipped in sizes ranging from 4 in. to 6 in., like branches of trees. Charcoal pieces tangle together and, in handling it in a super-capacity bucket elevator, result in only a few pieces actually getting into the bucket. As a result, there is much void space in the buckets, and it is wise to figure on actual loading of only 40% of given catalog capacity. This type of bucket elevator with flat bottom buckets works well handling charcoal, but the machinery parts are usually much heavier than need be, due to the character of the charcoal. A skip hoist bucket arrangement could also be used if space permits, but these units usually run higher in cost.

Damp or wet material should not be handled in this type of elevator. High capacities can be obtained with material −6 in. and under, at low operating speeds of 75–100 fpm (maximum speed: 120 fpm), if material is dry. When handling wet or damp material becomes necessary, a filler plate should be welded in the bottom of the bucket, thus: ∀ or ∨. This is not a solution, but it may help. The fines may still stick to the bottom of the buckets. But, even if the capacity is small, it is better to use a double-strand super-capacity continuous-bucket elevator, equipped with flat bottom buckets.

Materials above 2-in. lumps can be handled easily, using as a rough guide that the length of the bucket should be four times the size of the largest lumps.

The super-capacity elevators can be inclined. Most

of these elevators are furnished vertical, wherever possible, and are fully enclosed in steel casing.

3.16.3 Guide Angles

Sometimes this type of vertical elevator is equipped with internal guide angles for the chain travel and to keep it in place, particularly on high elevators (50-ft centers and above), and when handling ores or stone products.

Sometimes the guide angles are placed above the boot or foot sprockets for about 10 ft, then midway in the casing for 10 ft, and below head sprockets for 10 ft, turning the ends of the angles out so chain will not foul entering or leaving guides.

Because of the weight of the machinery parts, the head shaft bearings are located on the building steel to take the load off the casing. This usually can be done in new construction but, in existing construction, the building steel has probably not been designed to carry these heavy loads, so the steel casing head must be reinforced to take the load.

3.16.4 Takeups

Gravity takeups are used at times. Traction wheels can be used instead of sprockets. The foot shaft on a gravity takeup does not go through the boot steel and simply rides in steel guides attached to inside of boot section. This arrangement keeps the pit below floor level clean.

3.16.5 Specification for Super-Capacity Elevator

Required: One vertical continuous elevator, complete.
Capacity: 120 tph of $1\frac{1}{2}$-in. lime.
Elevator boot and head shaft centers: 72 ft $1\frac{3}{8}$ in.
Boot Shaft: height (highest position) above the bottom of the boot section is 4 ft 0 in.
Buckets: continuous steel 20 in. $\times$ 12 $\times$ $17\frac{3}{4}$ in. made of $\frac{1}{4}$-in. plate with $\frac{1}{4}$-in. reinforced lip and filler plate bottom of bucket. Welded construction with continuous inside and outside welds. Attached to chain by bolts with locknuts peened set after tightening.
Chain: SBS-1833 or approved equal; 18-in. pitch steel thimble roller chain with chilled iron rollers; ultimate strength of 42,000 lb.
Head shaft: $5\frac{7}{16}$-in. diameter extended to receive head-shaft-driven sprocket and differential band brake backstop.
Head sprockets: two, split chilled rim, 36.12-in. pd, 6 T, $5\frac{7}{16}$-in. bore, KS and SS.

Automatic differential band brake backstop: 30 in. on head shaft.
Bearings: Three $5\frac{7}{16}$ in. babbitted pillow blocks.
Head shaft bearing supports: provide independent of the elevator casing.
Boot shaft: $2\frac{15}{16}$-in. diameter.
Bearings: $2\frac{15}{16}$ in. self-aligning adjustable with 12 in. of takeup.
Boot traction wheels: two, split chilled rim, 36 in. pd, $2\frac{15}{16}$-in. bore, securely fastened to the boot shaft.
Casing: no. 10 gauge steel plate on long sides, 12 gauge on short sides and angles; dust-tight construction; 33 in. $\times$ 66 in. inside dimensions; maximum length of sections: 10 ft 0 in.

Manufacturer shall furnish 10-ft boot section made of $\frac{1}{4}$-in. plate, with receiving hopper and the necessary loading leg, as is normally required for that type elevator, with guide angles on the loading side only. Also, furnish the structural steel for headshaft supports and motor-drive supports and necessary guards.
Platform: furnish a platform for support of the drive system and head shaft bearing of the elevator (in some cases furnished by the purchaser).
Inspection doors: openings 4 ft 6 in. wide $\times$ 3 ft 0 in. high with clamped removable doors both sides of elevator; lower edge of each opening 18 in. above the boot shaft in its upper position.
Additional inspection doors: two additional doors (one each side) of the same size as those previously specified shall be provided in each elevator, placed so bottoms of doors are 3 ft above ground level.
Removable panel cleanouts: shall be provided on front and back of boot section, 24-in. high, equipped with spring holding clamps.
Safety features: necessary safety collars, keys, safety setscrews, etc.; necessary safety guards for shafts, sprockets, chain, couplings, etc.
Hopper and chute (attached to boot section): to receive discharge (from belt conveyor); to be provided by the manufacturer as shown on purchaser's drawing.
Chutes or spouts (from the discharge of the elevator to the belt conveyor): shall be provided by the manufacturer, as shown on purchaser's drawing.
Lubrication: All bearings shall be provided with Alemite or equal industrial hydraulic-type grease fittings; extension piping shall be provided where necessary or desirable for easy access or safety.
Motor drive: furnish 20-hp, 900-rpm totally enclosed motor for 440 V, 60 Hz, 3-ph service, together with magnetic starter and pushbutton control. The manufacturer shall furnish the rest of the drive, including enclosed herringbone-gear type speed reducer, located as shown on purchaser's drawings; flexible coupling,

baseplate to mount reducer and motor and the necessary flint-rim plate center sprockets and RC-40 roller chain drive, from reducer to elevator head shaft.

In general, the manufacturer shall design, furnish, fabricate, and deliver all equipment and accessories, specified or implied herein, necessary to form a complete workable unit. Parts shall be shop-assembled for minimum field assembly.

All equipment shall be thoroughly cleaned and painted with a high grade of machinery paint in such a manner that the paint will adhere to the surface and be weather resistant. All painting shall be done in a neat and professional manner. The equipment shall not be loaded for shipment until the paint is thoroughly dry.

All polished parts, shafts, gearing, bearings, etc shall be coated with rust-preventive compound and crated, if necessary, to prevent damage in shipment, and all equipment shall be thoroughly protected from the weather.

3.16.6   Design of Super-Capacity Elevator

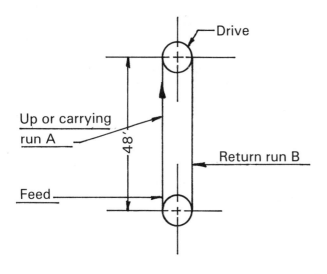

3.16.6.1   Data

Capacity required: 200 tph crushed stone.
Material: 6 in. down to 2½ in. and smaller, weighing 100 lb/ft³.
Height: 48 ft + 2 ft (sprocket) = 50 ft.
Buckets: 20 in. × 11⅝ in. × 8¾ in; 10 ga, weighing 29 lb each (table 3.6); *s* (c to c) is 12 in. The capacity of each bucket is 0.90 ft³ (table 3.6), which allows for 90 lb of material per bucket (0.90 × 100 lb). To allow for lumps of the size described, use 72 lb (90 × 0.8).

3.16.6.2   Capacity

Using 12-in. pitch chain, with a bucket spaced every pitch, and the chain traveling at 100 fpm, 100 buckets/ min would be passing the discharge point, resulting in a capacity of

100 buckets × 90 lb stone/bucket = 9000 lb

9000 lb × 60 min = 540,000 lb/hr

540,000 lb/hr ÷ 2000 lb/ton = 270 tph

Figuring 75% capacity gives

270 tph × 0.75 = 202.50 tph

which agrees with manufacturers' catalogs.

3.16.6.3   Load on Up Run A

Chain (two strands)

$= 50 \text{ ft} \times 2 \times 16.5 \text{ lb}$ $= 1650 \text{ lb}$

Buckets $= 50$ each at 29 lb $= 1450$ lb

Material $= 50 \times 72$ lb $= 3600$ lb

$(12 \, w_m H_o)/s = 72 \times 10$ $= \underline{720 \text{ lb}}$

7420 lb

3.16.6.4   Load on Return Run B

Chain (two strands)

$= 50 \text{ ft} \times 2 \times 16.5 \text{ lb}$ $= 1650 \text{ lb}$

Buckets $= 50$ each at 29 lb $= \underline{1450 \text{ lb}}$

3100 lb

3.16.6.5   Size of Head Shaft (figure 3.20)

For an unbalanced load:

	Loads on Head Shaft	
	Torsional	*Bending*
Weight of material	3600 lb	7420 lb
Pickup shock passing		
loading leg	720 lb	3100 lb
Totals	4320 lb	10,520 lb

Torsional moment = 4320 lb × 16 in. = 69,120 in.-lb.

Bending load = 10,520 lb + 350 lb (weight of shaft) + 400 lb (weight of two sprockets) = 11,270 lb.

Bending moment = $(W/2) \times 7$ in. = 39,445 in.-lb.

Shaft size = 5¹⁵⁄₁₆ in. shown on figure 9.3 as required

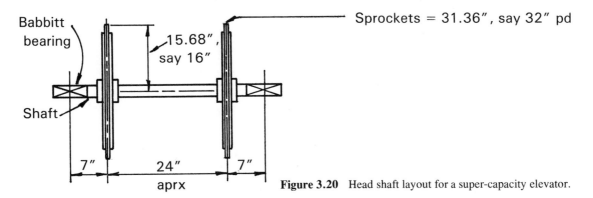

Figure 3.20 Head shaft layout for a super-capacity elevator.

assuming 24-hr operation and using a service factor of 1.5 in torsion and bending, and if allowable shear is 6000 psi. For 8000 psi shear, diameter = 0.843 × 5.9375 = 5.0 in. Therefore, use a $5\frac{7}{16}$-in. diameter shaft.

3.16.6.6 Horsepower

$$\text{theoretical hp} = \frac{4320 \times 100 \text{ fpm}}{33,000}$$

$$= 13.09$$

$$\text{hp} = 13.09 \times 1.50$$

$$= 19.6$$

Therefore, use 25 hp.

Using the chain formula,

$$\text{hp} = \frac{15 \ WH}{10,000} \times \text{service factor}$$

$$= \frac{15 \times 200 \times 50}{10,000} \times 1.50$$

$$= 22.5$$

Again, use 25 hp.

3.17 POSITIVE-DISCHARGE BUCKET ELEVATOR

3.17.1 General

The positive-discharge bucket elevator is considered a special type, similar to the centrifugal-discharge elevator, except that the spaced buckets are end-mounted between two strands of chain. The buckets are spaced 18–24 in. apart, depending on the size of the bucket, which can vary from 8 in. × 5 in. up to 24 in. wide × 8 in. projection. The buckets are attached to the chains, either fixed with bolts to the ends of the buckets (G attachment) or swivel (wing attachment). Attachments engage sprockets located below headshaft. The buckets (such as type AA malleable iron) operate at a low speed, about 120 fpm, allowing material to drop out of a bucket, even if it is slightly damp or wet. Buckets are usually welded. Chain is either no. 730 pintle or no. 1130 roller chain, heat-treated. Connection is of swivel design so that, in picking up large lumps in the boot, there is some play in either strand of chain. If the chain were rigidly attached to the bucket, unusual strains may develop in the bucket attachment bolts. The chains are usually guided and ride on angle track, whether the elevators are vertical or inclined.

This elevator picks up its load in the boot (as does the centrifugal-discharge type) but, because of its lower speed, does not depend upon centrifugal force to discharge material from the buckets. The buckets are completely inverted by snubbing the chains under the head sprockets after they have passed over the head wheels, giving them opportunity for complete discharge at relatively slow speed.

Since the positive-discharge bucket elevator is a low-capacity unit for the cost involved, it does not replace the centrifugal-discharge or continuous-bucket elevator where they are applicable. It will solve some difficult handling problems, however, where the others will not. It is an inexpensive unit for handling run-of-mine coal, compared with double strand super-capacity continuous-bucket elevators.

Takeups on the head shaft are not used on this type of elevator. Gravity takeups can be used in place of regular type screw takeups, the same as for centrifugal-discharge bucket elevators. No angle chain guides are necessary.

Positive-discharge bucket elevators (sometimes called perfect discharge) should be used in handling material

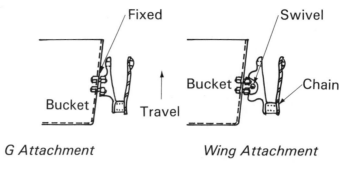

Figure 3.21 Bucket attachments.

that has a tendency to stick to steel, or where materials are light, fluffy, or fragile and where the required capacity is small. They are slow operating and, in general, should be used only on dry and free-flowing material. Fine material to 6-in. lumps can be handled. Experiments have indicated that enameling the bucket helps considerably, but this should not be done if a food product is being handled.

Because of the difficulty of having two strands of chain stretch the same amount during operation, there is a tendency to put an unusual strain on one chain, causing the buckets to get out of line, and eventually to break the G-attachment bolts. The wing, or swivel attachment, makes the better arrangement (figure 3.21). One part of this combination is rigidly attached to the bucket, and is attached to the chain through a bolt, forming a swivel joint. If any stretch occurs in the chain during operation, the bucket usually remains in line, and no parts are damaged.

A unique modification of this type uses a single strand of chain (figure 3.22), with buckets on both sides of chain rather than centrally, thus permitting the snub wheel to engage the chain after passing the head wheel.

3.17.2 Perfect-Discharge Bucket Elevator Handling Run-of-Mine Bituminous Coal (ROM)

Figure 3.23 shows the arrangement of a track hopper installed on a dead-end siding with absolute minimum clearance of 6 ft 6 in. from centerline of track to building wall. This clearance was used on some dead-end sidings, with the approval of the railroad and provision that no locomotive cross over it. This hopper, in plan, is 8 ft 0 in. wide. The railroad now insists on 8 ft 0 in. to 8 ft 6 in. from centerline of track to building wall, to get ample clearance for cars to pass with a man possibly hanging on the side of the car.

Safety laws also demand that a protective grating be placed over hopper opening, with the maximum size of opening 4 in. × 4 in. so an operator's leg cannot go down through it. When handling run-of-mine coal, this grating can be removed before car is placed over hopper, but after car is emptied, grating must be replaced. A large 24 in. × 24 in. undercut counterweighted-gate hand chain, operated as indicated, will allow the coal to flow to boot of inclined (can also be vertical) bucket

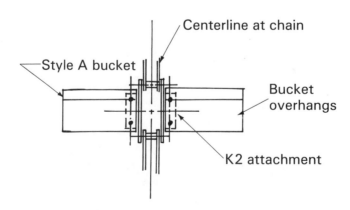

Figure 3.22 Two buckets on a single strand of chain.

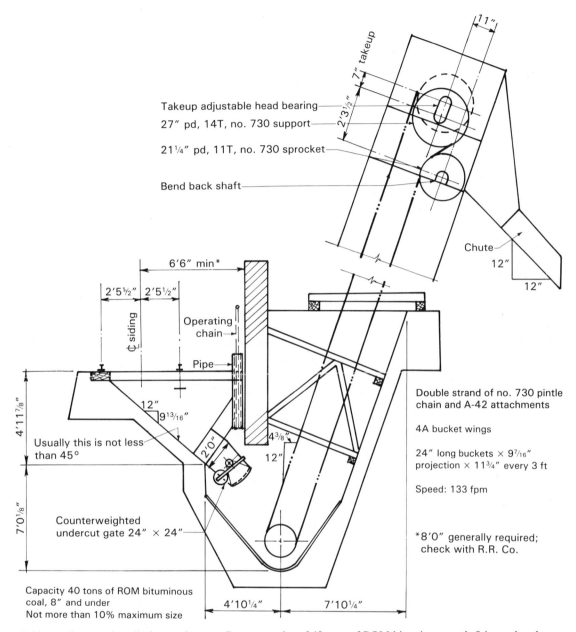

Figure 3.23 Inclined perfect-discharge elevator. Data: capacity of 40 tons of ROM bituminous coal, 8 in. and under; not more than 10% maximum size; double strand of no. 730 pintle chain; 4A bucket wings; 24-in. long buckets × 9⁷/₁₆ in. projection × 11³/₄ in. every 3 ft; speed is 133 fpm.

elevator, where the perfect-discharge buckets, spaced 3 ft 0 in. apart, will pick up coal, operating at a slow speed of 133 fpm, resulting in a capacity of 40 tph with buckets 24-in. long.

This type of elevator is an inexpensive unit to handle ROM coal compared with the double-strand super-capacity continuous bucket elevators that are usually recommended.

Figure 3.23 also shows a slide gate control, which is the method used in this installation. Under normal conditions, a vibrating feeder would be used.

3.17.3 Specification for Positive-Discharge Bucket Elevator-Conveyor

Required: One vertical positive-discharge bucket elevator.*

Capacity: 20 tph bituminous coal 1¼ in. and under, weighing 50 lb/ft³.

---

*Link-Belt catalog references are for guidance only.

Elevator boot and head shaft centers: 70 ft 0 in.

Buckets: 16 in. × 8 in. × 8½ in. malleable iron, style A, with attachment bolts and lockwashers.

Chain and attachments: Double strand of no. 730 pintle chain with A42 attachment and 6A bucket wing every 24 in. each strand assembled right and left hand, matched and tagged (see figure 3.24).

Head Shaft: $2^{15}/_{16}$-in. diameter, with babbitted series 2-1300 pillow blocks or equal, at 15 rpm.

Head sprockets: 30¾-in. pd, 16T, flint rim, bore $2^{15}/_{16}$ in., KS in line and SS.

Snub sprockets: 17½-in. pd, 9T, flint rim, bore $1^{15}/_{16}$ in. SS only, with series 2-1200 babbitt bearings.

Automatic differential band brake: 18-in diameter on head shaft.

Boot shaft: $2^{3}/_{16}$-in. diameter, with series MS-2200 babbitt takeups with dust seals.

Boot sprockets: 25.07-in. pd, 13T, flint rim, bore $2^{3}/_{16}$ in., one SS and one plain.

Casing: no. 12 gauge steel plate and angle construction for intermediate sections; dust-tight construction; 30½ in. × 42 in. inside dimensions, with two large inspection doors above boot section opposite each other, spring clamped.

Boot section: $^{3}/_{16}$-in. steel with removable end panel and cleanout doors at bottom.

Head section: no. 14 gauge steel, split in two pieces.

Discharge chute: $^{3}/_{16}$-in. steel with no. 10 gauge steel removable liner.

Motor Drive: 5-hp, 1200 rpm motor for 440 V, 60 Hz, 3-ph service, together with magnetic starter and push-button control.

The Manufacturer shall furnish the rest of the drive, including enclosed herringbone- or helical-gear type reducer, flexible coupling, steel baseplate to mount reducer and motor, also necessary finished roller-chain drive from reducer to elevator head shaft. Include coupling and chain guard.

Platform, ladder, and safety guard: a standard ladder is required from boot of elevator to platform serving driving machinery mounted on it just below head shaft. This platform shall be large enough for two people to service machinery when necessary. Standard handrail of angles or pipe is acceptable, 6-in. high toe guards to be provided around platform. Floor of platform shall be open, nonslip grating. Complete platform to be supported from elevator casing. A safety guard, in accordance with state of _____ safety laws, must be

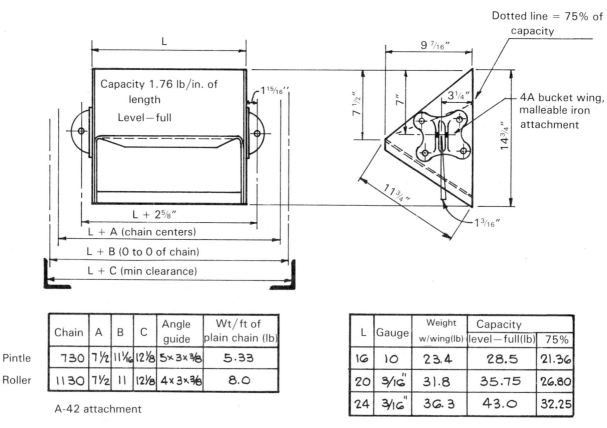

	Chain	A	B	C	Angle guide	Wt/ft of plain chain (lb)
Pintle	730	7½	11¹⁄₁₆	12⅛	5×3×⅜	5.33
Roller	1130	7½	11	12⅛	4×3×⅜	8.0

A-42 attachment

L	Gauge	Weight w/wing(lb)	Capacity level—full(lb)	75%
16	10	23.4	28.5	21.36
20	3/16"	31.8	35.75	26.80
24	3/16"	36.3	43.0	32.25

Figure 3.24 Perfect-discharge bucket on a double strand of chain. All dimensions are in inches except for no. 10 gauge.

included, to start 10 ft from bottom of ladder and attach to platform.

In general, the manufacturer shall design, furnish, fabricate, and deliver (freight allowed to plant) all equipment and accessories specified or implied herein, necessary to form a complete, workable unit.

All equipment shall be thoroughly cleaned and painted with a high grade of machinery paint in such a manner that the paint will adhere to the surface and be weather resistant. All painting shall be done in a neat and professional manner. The equipment shall not be loaded for shipment until the paint is thoroughly dry.

All polished parts, shafts, bearings, etc., shall be coated with rust-preventive compound and crated, if necessary, to prevent damage in shipment, and all equipment shall be thoroughly protected from the weather.

Services of an engineer, if required, at time of installation, shall be furnished at rate stated in the Proposal.

The successful Bidder shall furnish three sets of prints for approval of the general design of the equipment and later, _____ sets of certified as-built prints to _____.

3.18 GRAVITY-DISCHARGE ELEVATOR–CONVEYOR

3.18.1 General

A gravity-discharge elevator–conveyor (figure 3.25) is a unit that elevates and conveys materials of a nonabrasive, free-flowing nature, such as bituminous coal or material having similar characteristics. Operating vertically or on an incline, it functions as a bucket elevator and horizontally, as a scraper flight conveyor. Steel buckets spaced at 24 in. or 36 in. on centers (table 3.22)

are rigidly attached to double-strand steel bar link, long-pitch (12 in. or 18 in., no swivel joints) roller chain, and travel along a continuous steel trough on the horizontal loading run, picking up the material enroute. At the lower corner upturn, a special steel corner trough is used to fill the buckets before starting their vertical run. At the upper corner, another curved corner piece is provided to transfer the load to the upper horizontal run, from where the material can be discharged at intervals through openings provided with slide gates. The discharge angle of the bucket is steep enough to empty as it passes over the discharge opening.

The elevator–conveyor can handle bituminous coal with maximum size lumps of, say 6 in., not more than 10% of capacity. Based on coal at 50 lb/ft^3, it can provide a capacity from 23.8 tph to 95.5 tph. Where coal is stored in several silos and reclaimed, this is an economical method of doing the work. It is not good for handling ashes, as scraping the ashes in the horizontal run soon wears the troughs, and acid corrosion will rust the metal. Because of its low operating speed of 100 fpm, this unit gently picks up and discharges material and is, therefore, usually selected where breakage or degradation of conveyed material is an important consideration.

Some typical arrangements of this unit are shown on figure 3.26.

3.18.2 Design of Gravity-Discharge Elevator–Conveyor

3.18.2.1 Data

Arrangement: 1 (figure 3.26)
Material: bituminous coal weighing 50 lb/ft^3
Capacity required: 60 tph
Horizontal runs, loaded: 20 ft, lower; 100 ft, upper

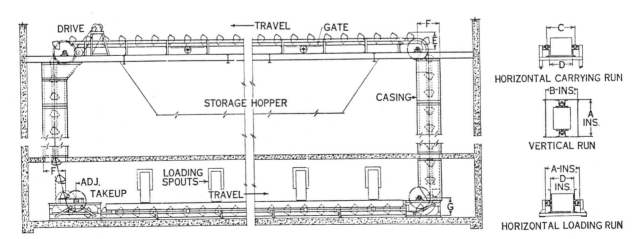

Figure 3.25 Gravity-discharge elevator. (*Courtesy FMC*)

Table. 3.22 Gravity-Discharge Conveyor-Elevator Capacity and Horsepower

Length (in.)	Width (in.)	Depth (in.)	Style	Thickness	Spacing (in.)	Max. Lumps (in.)	Capacity[a] (ft^3)	Vertical Loaded	Horizontal Empty	Horizontal Loaded	Chain Number	Empty	Loaded	Head Shaft rpm[c]
12	12	6	A	12 ga	18	4	0.238	0.30	0.014	0.032	SS-927	36.2	44.2	16.6
16	12	6	A	12 ga	18	4	0.318	0.04	0.014	0.038	SS-927	37.5	48.1	16.6
16	15	7	A	10 ga	24	6	0.463	0.045	0.014	0.040	SS-1227	35.3	46.9	12.5
20	15	7	A	10 ga	24	6	0.582	0.055	0.014	0.047	SS-1227	36.8	51.3	12.5
24	15	7	A	10 ga	24	6	0.699	0.067	0.015	0.055	SS-1227	38.8	56.3	12.5
20	20	10	B	3/16 in.	36	8	1.045	0.067	0.014	0.054	SS-1827	36.6	54.0	8.33
24	20	10	B	3/16 in.	36	8	1.266	0.08	0.015	0.063	SS-1827	38.6	59.7	8.33
30	20	10	B	3/16 in.	36	8	1.887	0.10	0.018	0.078	SS-1827	46.0	72.4	8.33
36	20	10	B	3/16 in.	36	8	1.909	0.12	0.019	0.093	SS-1827	50.0	81.8	8.33

[a]Based on 80% loaded.

[b]Based on material weighing 50 lb/ft³, buckets 80% loaded and traveling at 100 fpm. Add 10% to hp at head shaft for corners.

[c]Based on 8T sprocket wheels and speed of 100 fpm.

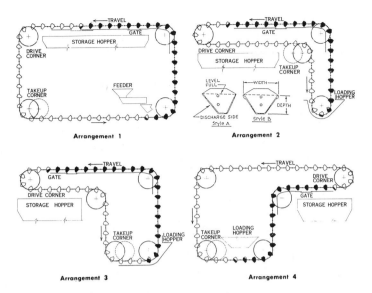

Figure 3.26 Gravity-discharge conveyor-elevator.

Shaft centers: horizontal, 110 ft; vertical, 40 ft
Buckets: 24 in. × 20 in., style B, at 36 in. oc (table 3.22)
Capacity, per bucket: 1.266 ft³ (table 3.22)
Capacity (lb/hr, 80% full):

$$1.266 \times \frac{12}{36} \times 50 \text{ lb} \times 100 \text{ fpm} \times 60 \text{ min}$$

$$= 126,600 \text{ lb} = 63.3 \text{ tph}$$

3.18.2.2 Horsepower

Speed required: 60/63.3 × 100 = 94.8 fpm
Weights:
    empty bucket per ft of run = 38.6 lb (table 3.22)
    bucket loaded 80% = 59.7 lb/ft of run (table 3.22)
    material in bucket (at 80%) = 59.7 − 38.6
$$= 21.1 \text{ lb/ft of run or}$$
$$= 1.266 \times 50 \times$$
$$(12/36)$$
$$= 21.1$$

For hp, it is best to figure on the bucket fully loaded; thus, the weight of the material will be 21.1/0.8 = 26.4 lb/ft of run.

The hp per ft of rise (weight of buckets balanced on up and down run) is

$$\frac{26.4 \times 100}{33,000} = 0.08 \text{ hp}$$

as shown in table 3.22. Therefore, hp is found by

$$40 \times 0.08 + 120 \times 0.063 + (2 \times 110 - 120) \times$$
$$0.015$$
$$= 3.2 + 7.56 + 1.5$$
$$= 12.26$$

Allowing 10% for corners, the hp would be 13.49. The hp at the head shaft is

$$\frac{13.49 \times 94.8}{100} = 12.79$$

Use 15 hp.

3.18.3 Specification for Gravity-Discharge Elevator-Conveyor

Required: one gravity-discharge elevator-conveyor.
Capacity: 60 tph bituminous coal, 6 in. and under, weighing 50 lb/ft³
Speed of conveyor: 100 fpm.
Shaft centers: 20 ft of lower run and 100 ft of upper run loaded; horizontal, 110 ft; vertical, 40 ft.
Buckets: 24 in. long × 20 in. wide × 10 in. deep; ³⁄₁₆ in. steel, spot-welded, style B, with attachment bolts and lockwashers.
Chain and attachments: double strand of SS-1827 steel strap roller chain with rigid attachments every 36 in.,

assembled right- and left-hand, matched and tagged, or approved equal.

Head shaft: to be determined by the manufacturer.

Corner shafts: head shaft bearings, Series 1500, babbitt type.

Corner bearings: series 2-1300, babbitt type.

Takeup bearings: series DS-2800, babbitt type.

Sprockets: all sprockets shall be 47.03-in. pd, 8T, flint-rim.

Casing for vertical runs: no. 10 gauge with corner angles and guide angles for chain.

Trough: trough at loading point to lower turn trough to be $3/16$-in. steel. No trough necessary from takeup corner to loading point. Upper trough turn and trough on upper horizontal run to be $3/16$-in. steel; rack-and-pinion slide gates with handwheels to be spaced every 15 ft.

Motor Drive: supply a 15-hp, 900 rpm motor (theoretical hp at head shaft: 12.79) for 440 V, 60 Hz, 3 ph operation, together with magnetic starter and pushbutton control.

The manufacturer shall furnish the rest of the drive including, if necessary, a countershaft with cut-tooth spur gears, herringbone or helical reducer with finished roller-chain drive to countershaft and, if necessary, flexible coupling. Steel baseplate to mount reducer and motor. All drive parts to be supported on Owner's steel supports. Coupling and chain guard by Owner.

Ladders, stairs, handrails, walkways and safety guards by Owner.

In general, the Manufacturer shall design, furnish, fabricate, and deliver all equipment and accessories specified or implied herein, necessary to form a complete, workable unit.

All equipment shall be thoroughly cleaned and painted with a high grade of machinery paint in such manner that the paint will adhere to the surface and be weather resistant. All painting shall be done in a neat and professional manner. The equipment shall not be loaded for shipment until the paint is thoroughly dry.

All polished parts, shafts, gearing, bearings, etc, shall be coated with rust-preventive compound and crated, if necessary, to prevent damage in shipment, and all equipment shall be thoroughly protected from the weather.

Services of an engineer, if required at time of installation, shall be furnished at the rate stated in the Proposal.

The successful Bidder shall furnish three sets of prints for approval of general design of the equipment, and later ＿＿＿＿＿＿ complete sets of certified as-built prints to ＿＿＿＿＿＿＿＿＿＿, Consulting Engineer.

Enclosure; Dwg ＿＿＿＿＿＿

3.19 FULL VOLUME OR EN MASSE ELEVATOR-CONVEYOR UNIT

The en masse elevator carries material in full volume, through an enclosure that can be made dust- and weather-tight, completely filling its cross-section. The conveyor medium is an endless chain, upon which are mounted flights at suitable intervals. Flights are made of either the solid or the open-finger type. The flights are buried in the material and move it without trying to compress it. In the conveying run, the friction against the side walls is primarily that of the material against the walls, rather than the flights. The unit can move material horizontally and vertically without need of transfer of material. It can take material from hopper or bin, and carry it along to discharge at several points. The feed loading should stop earlier to clear the casing of any material. With open-type flights, the casing may not clear completely.

This elevator–conveyor is well adapted to handling granular or pulverized free-flowing, dry materials, which are not abrasive or corrosive. Intimate contact between material and casing limits its application. Casings can be made out of abrasion-resistant steel and bolted for easy replacement. It will handle lumps one inch and under, if the lumps are less than 10% of capacity. This unit is not well adapted for handling materials containing a large percentage of lumps, where degradation is a factor. Coal must be dry and fine. Sawdust must be free from slivers and sticks. It should not be used for handling food for human consumption. It is very difficult to sterilize or clean when the occasion requires, even if made up of bolted sections. Various designs are available for handling fine powders, which aerate, and for handling somewhat sticky materials.

This unit is quite flexible in its application and takes up little space compared to other equipment. Speed of travel is usually limited to about 80 fpm, although slower speeds are preferred.

These units, made by several manufacturers, look similar but actually vary considerably in design of various parts; yet they all accomplish about the same result in capacity desired.

Stephens-Adamson Co. produces the Redler conveyor (figure 3.27), which finds much use in industry. They also produce a high-speed machined-link type Redler (Lagos Redler) which has T-bar flights that vary in width with the application. With this machined link, higher sprocket speeds are available, and smaller units can be used.

Another type is their side-pull Redler, which has unique applications for distribution of materials. It has

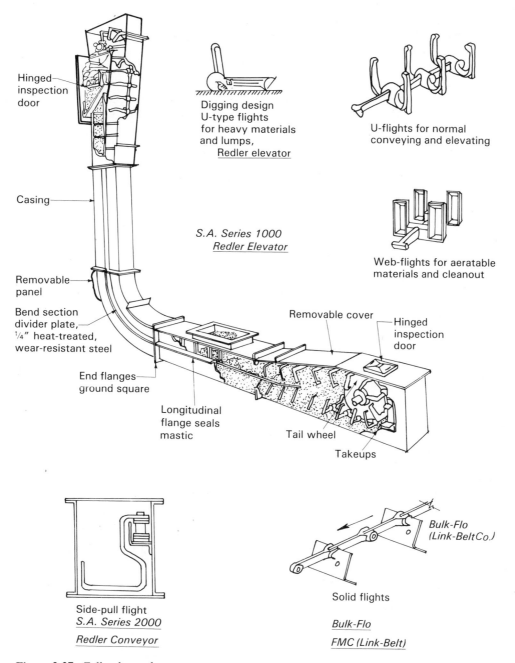

Figure 3.27 Full-volume elevator.

a recirculation feature that eliminates the requirement for a separate conveying system to return the unused material. On this side-pull Redler, the metal-to-metal wear is confined to a chain track that is isolated from the material being conveyed. These vertical closed circuit Redlers are particularly useful where it is necessary to fill several bins, or reclaim from one or more bins and reload into others with a minimum of handling,

breakage, dust, and power required. Only one drive is required to elevate and convey in two directions.

The regular Redler conveyors are available in widths up to 23 in. The side-pull Redler is available in units as large as 19 in. The Lagos Redler is made in sizes up to 630 millimeters (24¾ in.) wide. FMC's unit is called Bulk-Flo.

The Screw Conveyor Corporation makes a conveyor

with the shape of the flights a little different. It is called Super-Flo. These units can operate on a straight line only, whether horizontal or inclined.

Jeffrey Manufacturing Company makes the Tube-Flo conveyor (figure 3.28) in 4-, 6-, and 8-in. diameter tubes, with net capacities of 0.066 ft^3 of conveyor length for the 4 in., 0.162 ft^3 for the 6 in., and 0.297 ft^3 for the 8 in. Discs pulled through the tubes push the material forward. The Jeffrey catalog lists the maximum particle size, percent of conveyor loading, material friction factor, and properties of various materials that it recommends for handling by the Tube-Flo conveyor. It gives typical calculations for capacity, speed of travel, and hp required. Their catalog has the most complete engineering data.

For the sizes, capacities, and methods of figuring horsepower, it is best to consult individual manufacturers' catalogs.

Note: Redler, Super-Flo, and Bulk-Flo are registered trademarks.

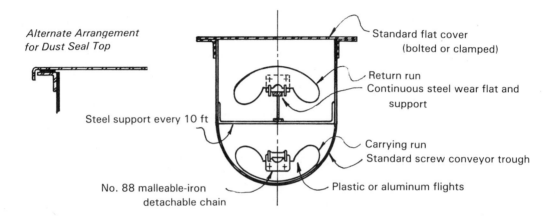

Cross-section of 12-in. Trough Flight Conveyor

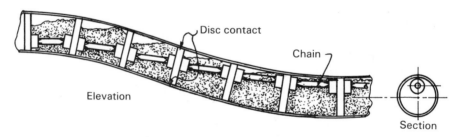

Tube-Flo Conveyor

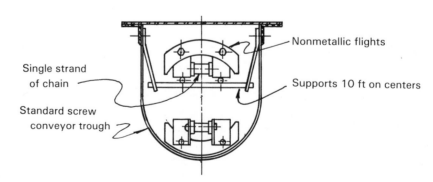

Cross-section of 12-in. Clean-Flo Conveyor

Figure 3.28 Cross-section of en masse elevators.

3.20 PIVOTED BUCKET CARRIER

3.20.1 General

The pivoted bucket carrier derives its name from the method of suspending the carrying bucket from two-strand long-pitch chains. Speed of travel is low, 40–50 fpm, seldom over 80 fpm, and horsepower requirements are low (see figure 3.29).

Buckets are always in a load-carrying position on horizontal or inclined runs, except when discharging.

Discharge is effected by trippers, engaging cams or similar devices located on the bucket ends. This is done in such a manner as to up-end the buckets sufficiently to empty them completely. The tripper may be of the fixed or travelling type.

While this type of equipment is probably the costliest of any, it will accomplish results not attainable by other types, because of its ability to both elevate and convey such materials as coal, ash, coke, ore, lime, crushed stone, and many other products, both abrasive and non-abrasive, hot or cold. In cold weather areas, coal that

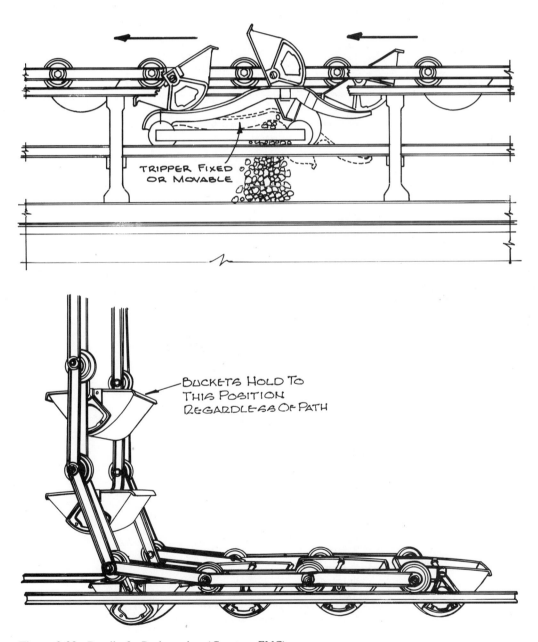

Figure 3.29 Detail of a Peck carrier. (*Courtesy FMC*)

arrives damp may freeze and form into lumps. In this case, a crusher should be used to squeeze the lumps enough to break the ice crystals, and to reduce the material to proper size before feeding the buckets. If stored outdoors, the top surface may freeze to a depth of 12 in.

or more, where the top layer can be broken into lumps with a shovel.

The buckets are made of malleable iron, steel, or a combination of both, depending on the material handled. The chains are preferably of malleable iron, long-

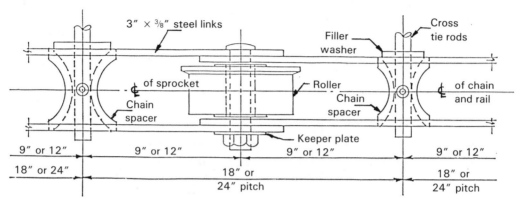

Stephens-Adamson Company

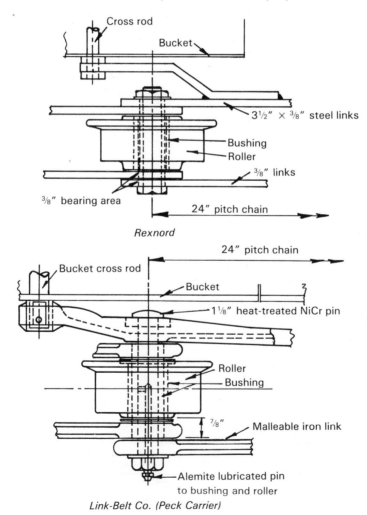

Rexnord

Link-Belt Co. (Peck Carrier)

Figure 3.30   Strap chains for pivoted-bucket carriers.

Table 3.23 Peck Carrier Capacity and Weight

Pitch (in.)	Width (in.)	Capacity of Bucket^a Level-Full (ft^3)		Weight (lb/lin ft)		Recommended Max Speed (fpm)
		per lin ft	per bucket	A	B	
18	15	0.50	0.74	50	176	
	18	0.60	0.89	53	182	40
	21	0.70	1.04	56	188	
24	18	0.78	1.55	90	228	
	24	1.04	2.08	103	242	50
	30	1.28	2.55	112	257	
30	24	1.46	3.65	120	269	
	30	1.82	4.55	140	294	60
	36	2.20	5.47	146	307	
36	36	2.66	8.00	240	413	80

Notes: A includes chain and buckets, and B, the upper run, includes cross channels, walkway, hand rails, track rails, roll chairs, chain, and empty buckets.
^{a}All capacities should be figured for 80% of tabulated values.

pitch links. Steel or forged steel (on special projects), with case-hardened pins and bushings, and large-size flanged rollers, either Alemite lubricated through the pin and bushing, or Alemite greased directly to the roller, are also used. Steel chain will stretch, and the punched holes in the links will become elongated and loose. The long-pitch chain will eventually try to climb the teeth of the drive sprocket wheel. The chain will then have to be replaced. Malleable iron chains are designed with ample bearing area to provide for the stretchout wear in the pin joints.

Because buckets overlap on all horizontal runs to

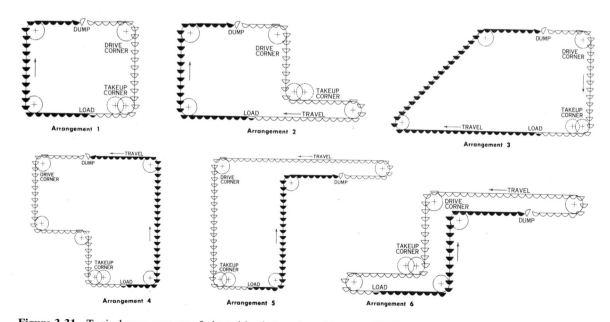

Figure 3.31 Typical arrangements of pivoted-bucket carriers. (*Courtesy FMC*)

form a continuous carrier line, any location on lower horizontal run may be used as a loading point, and moving buckets may be filled uniformly by an uninterrupted flow of material, without spillage. Buckets can discharge anywhere along the upper horizontal run. In transition from horizontal to vertical or inclined travel, buckets separate smoothly without tipping and, when horizontal travel is resumed, they are brought together again automatically. During the entire circuit, buckets remain suspended in normal position. They cannot come into the loading zone inverted, nor can they discharge at any point except when emptied by a tripper. This positive coordination is produced by suspending the buckets from projecting ends of the chain sidebars, so that as the chain pivots at the joints when entering or leaving the turns, each point of bucket suspension describes an arc that frees the overlapping bucket lips or brings them together, as required, without interference.

3.20.2   Strap Chains

Some details of the strap chains used by various manufacturers are shown on figure 3.30. The pitch of the chain can be 18 in., 24 in., 30 in., or 36 in.

3.20.3   Travel Paths

The variety of paths of travel permissible, combined with the fact that buckets may be loaded or discharged on horizontal runs, often enables a single carrier to perform functions which would require several separate conveyors of other types. A typical example is a boiler plant installation in which ashes are loaded directly into the carrier from the cleanout doors and discharged into the ash bin, and the same carrier used at a different time to distribute coal to the bunker. Contamination of the product is avoided, except where material is damp and is likely to stick to the bucket.

Some of the more common arrangements are illustrated on figure 3.31.

3.20.4   Peck Carrier Capacity and Weight

Capacity and weight information for the Peck carrier is given in table 3.23. For example, the capacity in tph for a 24-24 carrier handling material weighing 50 lb/ft³,

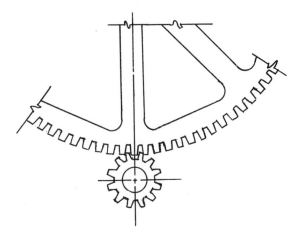

Figure 3.32   Cast-tooth equalizing gears.

allowing for 80% of level-full capacity, and traveling at 45 fpm, will be

$$\frac{0.8 \times 1.04 \times 45 \, (\text{fpm}) \times 60 \, (\text{min}) \times 50 \, (\text{lb})}{2000}$$

$$= 56.16 \, \text{tph}$$

3.20.5   Cast-Tooth Equalizing Gears

Cast-tooth equalizing gears connect parallel shafts in drives for conveyors and elevators employing long-pitch chains, to compensate for the pulsating motion caused by the difference between the true pitch circle and the chordal pitch of the polygonal sprocket wheels. This results in a smooth peripheral speed (see figure 3.32).

An equalizing gear set consists of a wave-line spur gear and an eccentrically bored cast-steel pinion. The number of elevations and depressions of the gear, or the ratio of the gear set should equal, or be a multiple of, the number of teeth in the conveyor driving-sprocket wheel. These gears eliminate the strains set up by the repeated accelerations and decelerations that occur with the use of long-pitch chains.

3.21   SELECTED ABBREVIATIONS AND LETTER SYMBOLS USED IN SECTION 3

HF   Elevator bucket, high front
HFO   Elevator bucket, high front overlapping
H_o Height factor to take care of pickup force (ft)
LF   Elevator bucket, low front
MF   Elevator bucket, medium front

MP Multiple-ply belt

$P_1, P_2,$

$\quad P_3$ Chain pull at various points on elevator

$\quad P_H$ Effective chain pull at head shaft (lb)

PIW Pounds per inch width of belt

$\quad P_t$ Total maximum chain pull (lb)

$\quad S$ Speed of bucket elevator (rpm or fpm)

$\quad s$ Spacing of elevator buckets (in.)

SC Elevator bucket, super capacity

T Capacity (tons/hour)

T_B Tension due to weight of belt or chain

T_b Tension due to weight of buckets

T_h Tension due to pickup of load and foot-wheel friction

T_m Tension due to weight of material in buckets

W_c Weight of belt or chain (lb/ft)

W_b Weight of each bucket (lb)

W_m Weight of material in each bucket (lb)

SECTION *4*

Screw Conveyors

4.1 GENERAL

Screw conveyors consist of metal flights, each formed as a helix and mounted on a pipe or solid shaft when necessary. The helicoid type is made by cold rolling a continuous strip to form a helix of desired diameter, pitch, and thickness, to fit a given size pipe or shaft. The flights are made hard and smooth in the rolling process. The flighting is secured to the pipe by end lugs welded to the back face of the flights and by intermediate spaced welds (see figure 4.1). For more information and data on larger sizes, refer to CEMA Book No. 350.

In the sectional flight type, the flights are made in sections equal to one pitch of the helix, and are secured to each other by butt welding the sections together.

A screw conveyor takes up very little space aside from that necessarily occupied by the moving stream of material. This advantage derives from the disappearance of the return run of a conventional chain and belt conveyor. As a result of this advantage, spiral or screw conveyors are applied in hundreds of locations where no other type can be considered.

A good installation of screw conveyors permits other machines to be located in the most advantageous positions for maximum production, per cubic foot of space. This advantage is obtained not only from the compactness of the conveyor itself, but also from the convenient methods of driving the conveyor and the location of inlet and discharge chutes.

4.2 ADVANTAGES OF SCREW CONVEYOR

The screw conveyor is among the simplest and most versatile types of material-handling equipment. It can:

1. Convey horizontally, on an incline, or vertically.
2. Accommodate multiple inlet and discharge openings; the latter fitted with control gates.
3. Provide troughs designed for dust-tight construction.
4. Provide for a variety of flight and trough thicknesses.
5. Provide flights made one-half, two-thirds, full, or double pitch, depending upon the problem encountered.
6. It can be used as a feeder, either as an independent unit or as part of a conveyor.
7. Convey in either direction in one unit, by means of right- and left-hand flights.

4.3 HELICOID VERSUS SECTIONAL CONVEYORS

The cross-section of a helicoid is thicker at the base than at the outer edge, while the sectional flights are the same thickness throughout.

With sectional flight conveyors, if a piece of flighting is damaged, it can be removed and replaced by a new section without removing the length of screw from the trough. With a helicoid, the length of screw would have

139

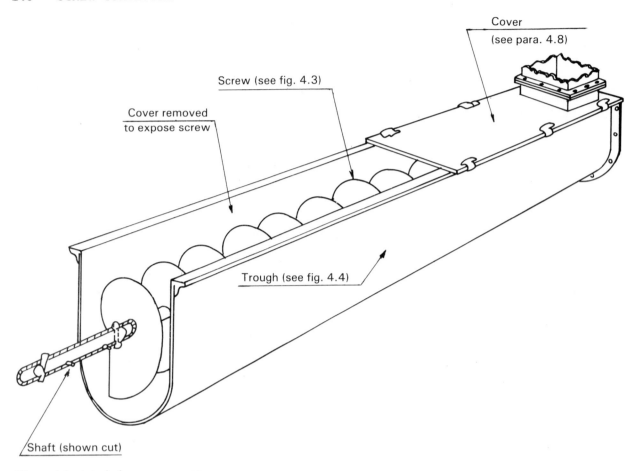

Figure 4.1 A typical screw conveyor.

to be moved out to be repaired; however, helicoid screw sections are very popular.

The helicoid conveyor has the advantage over the sectional flight type in that the continuous flights add strength, and there are no crevices or voids where material might lodge. The helicoid is applied particularly to nonabrasive materials and grains.

The sectional flight conveyor has the advantage where special features are required, such as: odd diameters, large diameters, extra thickness of flights, hardened flights, or flights of special materials.

4.4 TYPES OF FLIGHTS
(see figure 4.2)

For special problems involving corrosion, temperature, etc, flights and pipe can be made of monel, stainless steel, aluminum, brass, or special alloy steels. They also can be cast in sections. For resisting abrasion, steel flights may be tipped with stellite or similar materials. Rubber also has been used to cover the flights to resist chemical action or wear. Flighting can be tack welded or continuously welded, on one or both sides of pipe. It can be machined true to a desired diameter to give very close clearance between the flights and the inside of the trough for the entire length of the pipe. This gets into a special, and usually expensive, job.

Ribbon conveyors are a modification of the helicoid and are used on sticky materials where the solid flights will not clear themselves. These are rated at about 75% of the capacity of solid screws.

Paddle conveyor screws are used where more than one material is carried and it is desired to mix the feed. The angle of the paddle can be varied. Their carrying capacity is rated at about 50% of that of a solid screw.

Cut flight conveyor screws can be either helicoid or sectional. They add a slight mixing action, and their capacity is rated at 60% of that of solid screws.

4.5 PITCH OF CONVEYOR SCREW
(see figure 4.3)

The pitch of the screw is usually made equal to its diameter; thus, a 9-in. diameter screw is made with 9-in. pitch. This is called for as standard pitch.

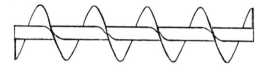

Sectional conveyor

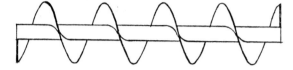

Helicoid conveyor

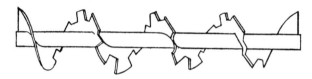

Cut flight conveyor

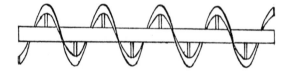

Ribbon conveyor

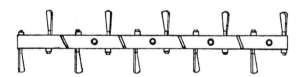

Paddle conveyor

Figure 4.2 Types of conveyor screws.

Variable-pitch screws are sometimes used for the same purpose as stepped-pitch screws. For lumpy materials ½-in. and up, use short pitch and tapered screws.

In short-pitch conveyor screws, the pitch is less than the diameter. They are used as feeders to other equipment or as parts of standard screws. They are recommended to prevent flooding and for use on screw conveyors inclined over 20°.

Both helicoid and sectional conveyors are standard in sizes 6, 9, 12, 14, and 16 in. Sectional

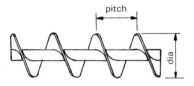

Standard pitch (dia = pitch)

Variable pitch

Short pitch

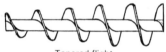

Tapered flight

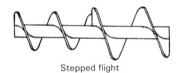

Stepped flight

Stepped pitch

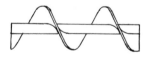

Long pitch

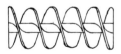

Double flight

Figure 4.3 Types of screw conveyor pitch.

conveyors also are standard in sizes 18, 20, and 24 in.

Tapered-flight screw conveyors are used as feeders for handling friable, lumpy material from bins or hoppers.

Stepped-pitch screws have flights progressively increasing in pitch. They provide for a uniform feed of free-flowing material over the entire length of the feed. It is a modified form of the short-pitch conveyor.

Stepped-flight screws are used as feeders to regulate the flow of the material. The smaller diameter usually operates in a smaller trough.

Long-pitch screws are used occasionally for rapid conveying of very free-flowing materials or for agitating liquids.

Double-flight screws are used to provide smooth flow for free-flowing materials.

4.6 SHORT FEEDER SECTIONS (refer to paragraph 7.10)

Frequently, short feeder sections are made part of a screw conveyor line, where material is to be fed from hoppers or bins and conveyed some distance. Feeder sections should be made no longer than a standard length section to avoid the use of hangers. The cross-section of the trough is filled completely with the material being handled. Fine mesh materials which hang up in bins and suddenly let go tend to flush past standard-pitch screws. In such installations, short-pitch flights are recommended, to prevent flushing. Conveyors usually are loaded about 15% for very abrasive material, 30% for abrasive, and 45% for mildly abrasive.

4.7 SUPPORTING HANGERS

Standard spacings of conveyor supporting hangers are 10 ft for up to and including 10-in. diameter screws; for 12–24-in. diameter screws, the spacing is 12 ft.

In some instances, the length of a screw can exceed the standard lengths specified by a manufacturer by 2 or 3 ft on the same size pipe, but it is not recommended. The published standard lengths of screws have been developed so that, at the hanger, there will be ½ in. clearance between the flights and the inside of the trough; at the longitudinal center, the weight of the length of screw may theoretically cause a sag of ¼ in., leaving ¼-in. clearance between the flights and the inside of the trough. Sometimes, if the problem dictates, larger diameter or heavier pipe is used to prevent sagging. Table

4.1 gives deflections of pipe shaft for various sizes of schedule 40 pipe. Table 4.2 gives the properties of schedule 40 and schedule 80 pipe, with the allowable torque values for each size used in screw conveyor work.

The clearance of the flights at any point is decreased by the pipe deflection and increased by the trough deflection. The trough deflection can be disregarded because of the stiffness of the trough. It will seldom exceed 0.015 in.

Each manufacturer of screw-conveyor parts can furnish standard hangers that are, in a general way, of the same construction as those of other manufacturers. These designs have been in use for many years. Several of them cannot be located in a trough where one section joins another, because of the narrow top steel support bent down with two attachment holes in each bent-down part. This would put these holes directly through the connection or splice, which is not good practice. To overcome this objection, the steel supporting plate is made wider, so that the attachment holes are parallel with the trough, and this design allows hangers such as CEMA style 216 or 226 to be attached to the trough at the splice. Thus, by using a standard length screw section with a 10–12-ft long trough, erection or maintenance work is simplified.

The hangers shown on figure 4.4 are for mounting inside the conveyor trough. For mounting on top of the trough angles or flanges, the top member is a flat plate. No. 226 has more clearance for passage of material. No. 270 has self-aligning ball bearings, resulting in quieter operation. When handling materials that may have a temperature of 150°F or over, an expansion type hanger, such as No. 326, should be used. The top bars can slide on the angle guides that are fastened to the trough. Ball bearings are not to be used for corrosive material, nor for finely ground material. Hard iron bearings have a Brinell hardness of approximately 450. Depending on the material handled, the life of the hanger bearing can be several years.

4.8 TROUGHS

For the most part, the steel troughs are made U-shaped in cross-section and in 10 or 12-ft lengths. They can be furnished in various gauge thicknesses, depending upon the service requirements. Several special-purpose troughs are available. Troughs are flanged or angle flanged. Flanged should be used where possible, because they are more dust-tight and more economical.

Troughs generally are provided with covers which can be made dust-tight, when necessary. Inlet or feed and discharge openings are located to suit conditions.

Table 4.1 Deflections of Screw Conveyor Pipe

Conveyor Size	Pipe Size† (in.)	Wt (lb/ft)	I (in.⁴)	Deflection for Length of Pipe = Δ*					
				10-0	11-0	12-0	13-0	14-0	15-0
6H-304	2.0	5.1	0.67	0.059	0.086	0.122	0.169	0.227	0.299
6H-308	2.0	6.5	0.67	0.075	0.110	0.156	0.215	0.289	0.381
6H-312	2.0	7.9	0.67	0.091	0.134	0.190	0.261	0.351	0.463
9H-306	2.0	6.8	0.67	0.079	0.115	0.163	0.225	0.303	0.399
9H-312	2.0	9.9	0.67	0.115	0.168	0.238	0.327	0.440	0.580
9H-406	2.5	8.8	1.53	0.045	0.065	0.093	0.127	0.171	0.226
9H-412	2.5	11.8	1.53	0.060	0.088	0.124	0.171	0.230	0.303
9H-414	2.5	12.8	1.53	0.065	0.095	0.135	0.185	0.249	0.329
10H-306	2.0	7.9	0.67	0.091	0.134	0.190	0.261	0.351	0.463
10H-412	2.5	12.7	1.53	0.064	0.094	0.134	0.184	0.247	0.326
12H-408	2.5	12.5	1.53	0.063	0.093	0.131	0.181	0.244	0.321
12H-412	2.5	16.0	1.53	0.081	0.119	0.168	0.232	0.312	0.411
12H-508	3.0	14.0	3.02	0.036	0.053	0.075	0.103	0.138	0.182
12H-512	3.0	17.2	3.02	0.044	0.065	0.092	0.126	0.170	0.224
12H-614	3.5	20.3	4.79	0.033	0.048	0.068	0.094	0.126	0.166
14H-508	3.0	16.0	3.02	0.041	0.060	0.085	0.117	0.158	0.208
14H-614	3.5	22.3	4.79	0.036	0.053	0.075	0.103	0.139	0.183
16H-610	3.5	21.1	4.79	0.034	0.050	0.071	0.098	0.131	0.173
16H-614	4.0	27.1	7.23	0.029	0.043	0.060	0.083	0.112	0.147

*Δ is in inches = $(5/384) \times (WL^3/EI)$, where W = total weight (lb), L = length (in.), $E = 29 \times 10^6$, and I = moment of inertia (in.).
†All pipe is schedule 40.

Table 4.2 Properties of Pipe

Nominal size (in.)	Schedule	r_o	r_i	Area of Metal (in.²)	I_p (in.⁴)	T
1¼	40	0.83	0.69	0.668	0.3894	3,140
2	40	1.1875	1.0335	1.0745	1.3315	7,500
2½	40	1.4375	1.2345	1.7041	3.0591	14,250
2½	80	1.4375	1.1615	2.2535	3.8485	17,900
3	40	1.75	1.534	2.2285	6.0343	23,100
3	80	1.75	1.450	3.0159	7.7886	29,800
3½	40	2.00	1.774	2.6795	9.5754	32,100
3½	80	2.00	1.682	3.6784	12.5602	42,100
4	40	2.25	2.013	3.1740	14.4652	43,100
4	80	2.25	1.913	4.4074	19.2210	57,200

r_o = outside radius (in.)
r_i = inside radius (in.)
S = allowable shear = 6700 psi
I_p = polar moment of inertia (in.)
T = safe torque (in.-lb)
N = rpm
v = ft/min
$= \dfrac{2\pi r_o}{12} \times \dfrac{N}{60}$

$I_p = \dfrac{\pi}{2}\left(r_o^4 - r_i^4\right)$

$T = \dfrac{S \times I_p}{r_o} = \dfrac{6700\, I_p}{r_o}$

safe hp $= \dfrac{T \times 2\pi \times N}{12 \times 33,000} = \dfrac{TN}{63,025}$

HANGERS

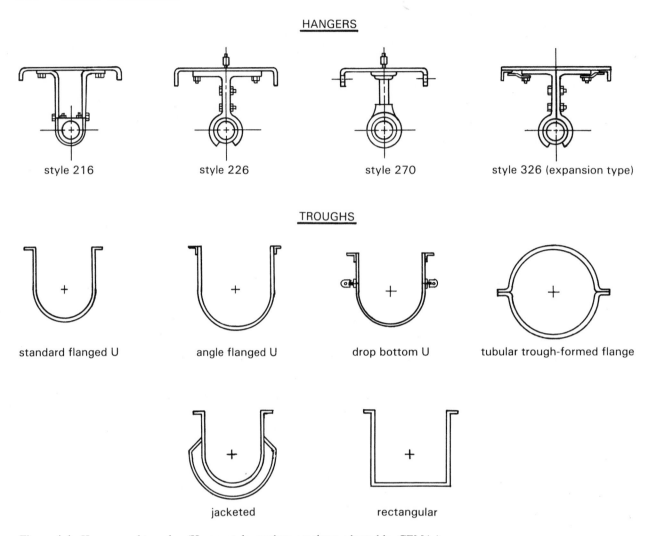

style 216   style 226   style 270   style 326 (expansion type)

TROUGHS

standard flanged U   angle flanged U   drop bottom U   tubular trough-formed flange

jacketed   rectangular

Figure 4.4 Hangers and troughs. (Hanger style numbers are those adopted by CEMA.)

Where necessary, they are equipped with hand-operated slide gates or rack-and-pinion type gates (see figure 4.5).

Covers are of light gauge, usually no. 14 or 16 gauge, either bolted or spring-clamped to screw casing. Screw Conveyor Corp., Hammond, IN, furnishes the patented "Tite Seal" cover. This cover is commercially dust-tight and eliminates the necessity of bolt and nut fasteners. The cover is held securely in place by U-edging along both sides of the trough. A maintenance man has to use care in removing or installing the U-edging so as not to damage it or get it out of shape. Attempts have been made to locate the screw conveyor several feet above the floor, and make arrangements for the screw casing to be turned upside down. The cover can then be removed and the casing cleaned. This method, however, is not as satisfactory, and may be much more expensive, especially with larger sizes. Most troughs are covered with a flat, clamped or bolted arrangement. Where extreme dust conditions are encountered, special attention should be given to the details of the cover.

The hip-top cover (figure 4.6) is designed for outdoor, weather-tight applications. The cover is formed

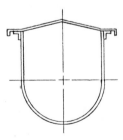

Figure 4.6 Hip-top cover.

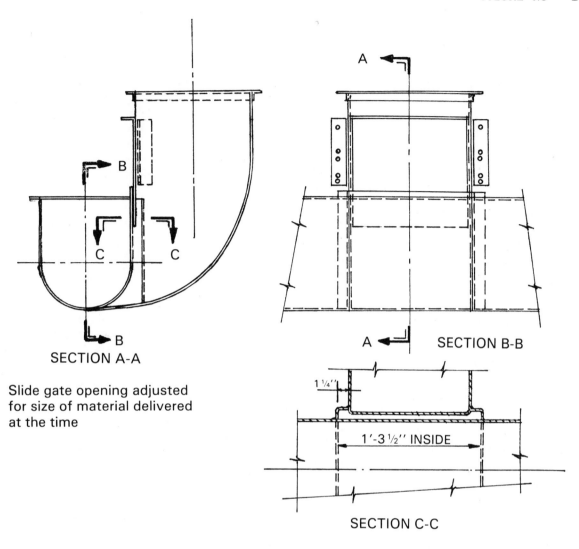

SECTION A-A

Slide gate opening adjusted
for size of material delivered
at the time

SECTION B-B

1 1/4''

1'-3 1/2'' INSIDE

SECTION C-C

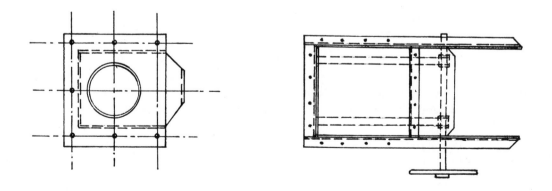

RACK AND PINION SLIDE GATE

Figure 4.5 One possible design of a slide gate at the side inlet of a 9-in. screw conveyor.

FIGURE 4.5 *145*

with a raised center section and is flanged down along the sides to fit over the standard conveyor trough side angles. This prevents an accumulation of moisture on the cover or water from entering the trough.

4.9 NOMENCLATURE

The first unit (a one- or two-digit number) in the nomenclature of a screw conveyor gives the size (diameter) of the conveyor in inches; the second (a letter), the type of conveyor; the third (a single digit), the coupling diameter (2 = 1 in., 3 = 1½ in., 4 = 2 in., 5 = 2⁷⁄₁₆ in., 6 = 3 in., 7 = 3⁷⁄₁₆ in.); and the fourth and fifth (two digits), the flight-tip thickness (in 64ths of an inch). Thus, 12H408 signifies a 12-in. diameter helicoid conveyor, with a 2-in. coupling and a ⅛-in. flight-tip thickness. Likewise, 12S616 would indicate a 12-in. screw sectional conveyor, with a 3-in. coupling, and ¹⁶⁄₆₄ in. (or ¼-in.) thick flight tip.

4.10 HANDLING FOOD PRODUCTS

FDA regulations state that screw conveyors handling a food product for human or animal consumption must be kept clean to avoid contamination. Manufacturers of such products must fabricate of stainless steel all parts coming into contact with food products. These screw conveyors usually are washed down after every run. Parts should be ground, polished, and buffed.

Some screw conveyor casings are made with a hinged bottom which is dropped after each run and washed clean. Similarly, where a screw conveyor has to move different products, as in feed and grain mills, the bottom of the screw casing should be hinged so that it can be cleaned easily, before a new product is fed. No hidden corners or spaces where food particles can lodge are allowed. In designing screw conveyors for the food industry, the utmost control of cleanliness must be exercised. These always cost more than conveyors for nonfood commercial purposes because of the rigid requirements of health departments.

Any hangers are usually equipped with stainless-steel supports and oilless bearings. The bearings can be made of close-grained, dense, hardwood and impregnated with a lubricant (which will not oxidize up to 180°F and is acceptable in the food industry). The bearings should eliminate contamination, resist corrosion and abrasion, and prevent squealing. They should operate under dry, dusty conditions, or where subjected to moisture.

4.11 SHAFTS, COUPLINGS, AND COUPLING BOLTS

Drive shafts can be extended with a keyway through the bearing for any nonstandard drive, with or without provision for a trough end seal. On most shaft-mount reducers, manufacturers have developed a screw conveyor drive package which includes their own special drive shaft. Tail shafts are extended just through the tail bearing with or without provision for a trough end seal.

Coupling shafts are made of either cold rolled steel, hardened steel, (hardened only for bearing length) or stainless steel. Standard construction consists of two coupling bolts each side of hanger bearing. Body-bound coupling bolts have caused some problems. While it is a simple problem to fit a body-bound bolt in the shop, it is difficult to get maintenance men to take the necessary time and care to fit these bolts. Hence, the threads and nuts become damaged, and are useless for the work for which they were originally intended.

Couplings hold two conveyor screws together as well as provide a bearing surface for an internal bearing. Coupling bolts hold all shafts inside the pipe and collar of the conveyor screws.

4.12 LONG SCREW CONVEYOR DESIGN

Data: 20″ screw conveyor is 113 ft long. It handles 100,000 lb of abrasive material weighing 52 lb/ft³. It is driven by a 40-hp motor.
Design hp: 30
Speed: 40 rpm
Developed torque: $(30 \times 63,000)/40 = 47,250$ in.-lb

Use 4-in. Schedule 80 pipe (refer to table 4.2), which has an allowable torque (T) of 43,100 in.-lb.

A standard 3⁷⁄₁₆-in. coupling is rated at 42,470 in.-lb. Use hardened coupling that is good for 53,100 in.-lb (refer to table 4.3).

Using ⅞-in. bolts (2-bolt connection) allows $2(10,900) = 21,800$ in.-lb in bearing (refer to table 4.4) and $2(12,800) = 25,600$ in.-lb in shear (refer to table 4.5).

A 3-bolt connection allows 1.5 times the above values, or 32,700 and 38,400 in.-lb (refer to table 4.6). The drive shaft of the roller bearing pillow block should have a 3-bolt connection and be of A-325 steel.

The trough endplate should be made at least 1 in. thick to take any twisting action of the drive. A very

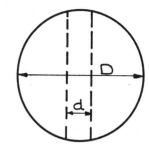

Table 4.3   Design Torque on Coupling Shafts

D Size of Shaft (in.)	d Size of Bolt Hole (in.)	I_{ps} (in.⁴)	Deduct for Hole (in.⁴)	I_p (in.⁴)	S (in.³)	T_1 @ 7000 psi	T_2 @ 8750 psi
1	¹³⁄₃₂	0.09817	0.03944	0.05873	0.1175	820	1030
1½	¹⁷⁄₃₂	0.4970	0.1681	0.3289	0.4385	3070	3840
2	²¹⁄₃₂	1.5708	0.4846	1.0862	1.0862	7600	9500
2⁷⁄₁₆	²¹⁄₃₂	3.4656	0.8494	2.6162	2.1466	15020	18800
3	²⁵⁄₃₂	7.9522	1.8770	6.0752	4.0501	28350	35400
3⁷⁄₁₆	²⁹⁄₃₂	13.7079	3.2808	10.4271	6.0667	42470	53100

I_{ps} = polar moment of inertia of solid shaft
I_p = net moment of inertia, after deducting for hole
The deduction is based on a rectangle with one side equal to the
 diameter of the solid shaft and the other side equal to the width of
 the opening. This results in deductions of slightly over 5% for a
 1-in. shaft to about 1% for a 3-in. shaft.

S = net polar section modulus of shaft
T_1 = torsional rating for allowable shear stress = 7000 psi (unhardened)
    = $S \times 7000$
T_2 = torsional rating for allowable shear stress = 8750 psi (hardened)
    = $S \times 8750$

Table 4.4   Allowable Torque on Bolts in Bearing

Coupling dia (in.)	Pipe size (in.)	Bushing ID (in.) (max)	Bushing ID (in.) (min)	Pipe OD (in.)	Bearing depth (in.)	Bolt dia (in.)	A	r	T (in.-lb)
1	1¼	1.016	1.005	1.660	0.644	0.375	0.242	0:68	958
1½	2	1.516	1.505	2.375	0.859	0.500	0.430	0.97	2500
2	2½	2.016	2.005	2.875	0.859	0.625	0.537	1.22	3930
2⁷⁄₁₆	3	2.458	2.443	3.500	1.042	0.625	0.650	1.48	5820
3	3½	3.025	3.005	4.000	0.975	0.750	0.730	1.75	7770
3	4	3.025	3.005	4.500	1.475	0.750	1.112	1.87	12500
3⁷⁄₁₆	4	3.467	3.443	4.500	1.033	0.875	0.915	1.98	10900

Bearing depth = Pipe OD − Bushing ID (max)
Projected area = Bearing depth × Bolt diameter = A
Allowable bearing = 6000 psi
Load radius (r) = (Coupling size/2) + (Bearing thickness/4)
Allowable Torque per bolt (T) = 6000 × A × r
T values used are from CEMA Book 350, although they vary slightly from values resulting from computations above.

Table 4.5 Allowable Torque of Coupling Bolts in Shear

Coupling Dia (in.)	Bolt Dia (in.)	Bolt Area (in.^{2})	S on 1 Bolt (lb)	r (in.)	Torque (in. lb)
1	3/8	0.1104	1369	0.50	690
1½	½	0.1963	2434	0.75	1830
2	5/8	0.3068	3804	1.0	3800
2⁷/₁₆	5/8	0.3068	3804	1.21875	4635
3	¾	0.4418	5478	1.5	8200
3⁷/₁₆	7/8	0.6013	7456	1.71875	12800

Allowable shear = 6200 psi in single shear, 12,400 psi in double shear.
S values are for one bolt in double shear.

dependable drive on large screw conveyors is the use of shelf-type trough endplates, where the bearing is supported on shelf and can be removed easily for inspection in any emergency, saving valuable downtime. The sectional butt-welded screw sections should be welded to the pipe continuously on one side to stiffen the flighting.

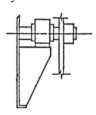

When the length of a screw conveyor exceeds, say,

75 ft., it is best to check various parts, as shown above. Some engineers would prefer to make a 113-ft long screw in two sections, and drive each conveyor with a separate drive, using standard parts. But, as shown, a conveyor can be made in one length, if the safe limits are not exceeded. (See figure 4.7 for a screw conveyor built in two sections, and figure 4.8 for one screw conveyor feeding another on an incline.)

4.13 INTERMEDIATE HANGER BEARINGS

CEMA lists four types of bearings: babbitted or bronze, self-lubricating, ball, and hard iron.

Table 4.6 Torque Capacities of Conveyor Pipe Couplings and Coupling Bolts with Conveyor Pipe Having Internal Collars (Also without Collars)

(in.)			Torque Capacity (in.-lb)								
			Pipe @ 6700 (psf)	Cplg Shaft		Bolts					
						@6200 (psf) Shear			@ 6000 (psi) Bearing Value		
Cplg Dia	Pipe Nom Dia	Cplg Bolt Size	Plain	Std @ 7000 (psi)	Hard Stl @ 8750 (psi)	1 Bolt	2 Bolts	3 Bolts	1 Bolt	2 Bolts	3 Bolts
1	1¼	3/8	3140	820	1030	690	1380	2070	958	1916	2874
1½	2	½	7500	3070	3840	1830	3660	5490	2500	5000	7500
2	2½	5/8	14250	7600	9500	3800	7600	11400	3930	7860	11790
2⁷/₁₆	3	5/8	23100	15020	18800	4635	9270	13905	5820	11640	17460
3	3½	¾	32100	28350	35400	8200	16400	24600	7770	15540	23310
	4		43100						12500	25000	37500
3⁷/₁₆	4	7/8	43100	42470	53100	12800	25600	38400	10900	21800	32700

Note: Tables 4.2, 4.3, 4.4, and 4.5 have been combined in this table for easier selection of standard combinations of parts.

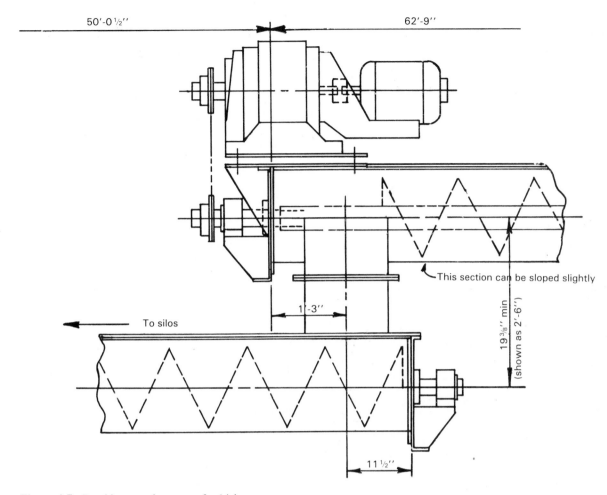

Figure 4.7 Breaking up a long run of a 14-in. conveyor.

Babbitted bearings have a maximum operating temperature of 130°F, lubricated bronze bearings a maximum of 225°F. They must not be used for food or other products where contamination by bearing wear or lubricants is possible.

Self-lubricated bearings are available in several types:

1. Oil-impregnated hard maple wood has a maximum operating temperature of 160°F.
2. Oil-impregnated sintered bronze has a maximum operating temperature of 200°F.
3. Plastic and reinforced fiber compounds are available in a wide variety of compositions and constructions, and can be obtained from many sources. They require no grease or oil lubrication and usually are run dry. They are best suited for use in conveyors handling a material wetted with water. Maximum operating temperatures vary with the composition and construction of the bearing. When appropriately used, the wear rate usually is low. (Consult bearing manufacturer for recommendations.)

4. Graphited bronze bearings have a maximum operating temperature of 500°F.
5. Commercial carbon bearings may be used for operating temperatures up to 700°F.

Ball bearings are preferred for use when handling granular or pelletized materials not containing any fine powder. Maximum operating temperature is 225°F with petroleum base lubricants, or 270°F with high-temperature synthetic lubricants. When appropriately used and sealed against loss of lubricant, ball bearings usually involve no contamination of the material conveyed.

Hard white-iron or chilled bearings are used with hardened coupling shafts for handling abrasive materials. Depending on circumstances manganese steel, stellite, or hardened nickel-iron may be used in place of hard-iron bearings. Hard-iron bearings are not normally lubricated. The maximum operating temperature is 500°F.

Conveyor screw speeds must be considered when using hard-iron bearings on hardened coupling shafts,

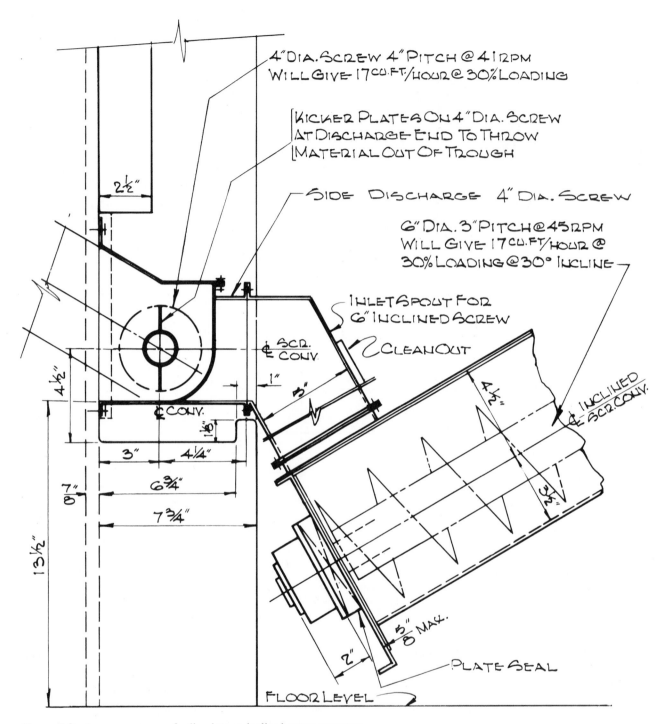

Figure 4.8 A screw conveyor feeding into an inclined screw conveyor.

in order to minimize wear and to reduce the squealing noise of dry metal on metal. The following formula gives maximum recommended operating speed:

$$N = \frac{120}{\text{shaft diameter in inches}}$$

where N = maximum operating rpm of screw.

Another material used for screw conveyor bearings, especially when handling food materials, is Lignumvitae*, a tree grown in the West Indies, in the coastal re-

*Data furnished by Forest Products Laboratory from text "Tropical Timbers of the World."

gions of tropical Mexico, and in some other nearby tropical areas. Its basic specific gravity is defined as

$$\frac{\text{oven dry weight}}{\text{green volume}} = 1.05$$

Air-dry density (12% moisture content) = 80 lb/ft^3 and its maximum crushing strength is 11,400 psi.

Heartwood is dark greenish-brown to almost black and sharply demarcated from the narrow pale yellow or cream-colored sapwood. The texture is very fine; its grain is strongly interlocked; a slight scent is evident when warmed or rubbed. It has a characteristic oily feel due to the resin content that may be as high as one-fourth of the air-dry weight. It is not recommended where the temperature runs higher than 150°F.

Lignumvitae is produced from logs (often 10–12 in. in diameter by 4–5 ft long). It is machinable into many shapes and forms.

4.14 END BEARINGS

The end bearings include a separate end plate and a separate flange bearing or pillow block. Thus, a bearing can be easily replaced with little downtime. A thrust ball or roller bearing, flanged bearing, or pillow block is used to take care of the developed thrust. The standard drive bearing is a double-row ball-bearing flanged block which takes the thrust developed in most applications.

Steel-plate outboard bearing trough ends can be furnished with babbitt, ball, or roller bearing pillow blocks. Dust-seal glands are provided to fit over shaft and against steel end plate to prevent leakage of fine material and to protect against dust or fumes.

A plain babbitt bearing is commonly used on one end of a screw conveyor and an antifriction bearing on the other end.

Details of a pedestal-type end bearing for a 16H616 helicoid spiral conveyor are shown on figure 4.9.

4.15 LIVE BOTTOM SCREW COVERINGS IN REFUSE BINS

In many of the new industrial retail shopping centers built around the cities, a power plant is used not only for heating, but for the disposal of wooden and paper corrugated boxes.

This scheme involves the dumping of such material on a short, flat-top apron feeder to act as a picking table, the top of which would be about 30 in. above the floor level. Any foreign pieces, similar to steel box bands, can be picked out. The material then is fed into a shred-

der, which produces very small pieces, about ¾ in. in size. A very successful unit is manufactured by Montgomery Industries of Jacksonville, Florida and called the "Montgomery Blo-Hog." The small pieces (down to dust) are blown into an overhead storage bin of desired capacity.

The bottom of the bin is equipped with live screw conveyors, usually 9 or 12 in. in diameter, and spaced close together. If 9-in. diameter screws are used, they are spaced on about 12-in. centers, and if 12-in. diameter screws are used, on about 20-in. centers. Some are made right-hand and some left-hand, operating at 2 rpm, delivering into a cross screw (see figure 14.30). This cross screw is driven separately and delivers material directly to boiler stoker.

Refuse does not pack together in a bin and, therefore, the screws should not be smaller than 9 in. There is a lot of void space, so capacities are generally figured about 50–60% of a solid space.

The cross screw handles material from the bin screws that has gone through the hog and normally reduced to minus ¾ in. in size. Sometimes, a piece of wood somehow gets through a hog in the form of shavings or a small block. In any event, the cross screw is made larger and operates at 20 rpm, because six or more screws usually discharge out of the bin to the cross screw at one time. It is also placed in the next larger size trough and out of center about 1½ in. to 2 in. (see figure 4.10).

Using 9-in. bin screws, the cross conveyor is usually 12 in. in diameter, operating in a 14-in. U-shaped trough. Using 12-in. bin screws, the cross conveyor is usually 14 in. in diameter, operating in a 16-in. U-shaped trough.

This same scheme (of handling hogged refuse) is used for handling long, stringy, wood shavings, say, 6 in. and under, from a lumber mill or furniture manufacturing plant.

An oversized screw trough is used with the screw offset, providing a large clearance on the carrying side. A steel baffle is welded on one side, so arranged that the clearance around the screw increases toward the carrying side as the screw revolves. The effect is to have the screw relieve itself, since any shavings or slivers which may get started around the flighting will gradually free themselves, due to the increase clearance.

4.16 INCLINED SCREW CONVEYORS
(see figure 4.11)

Figure 4.12, prepared by Mr. S. Snyder of the Link-Belt Company, is intended to give approximate values. The curves are not consistent, but they are as good as anything available to date.

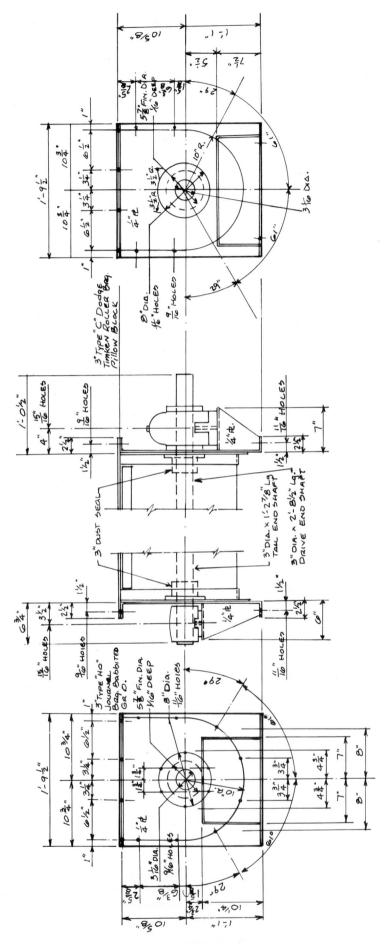

Figure 4.9 Detail of pedestal-type end bearings for 16H616 helicoid spiral conveyor.

152

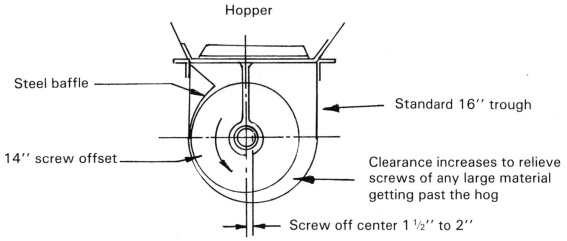

Figure 4.10 Cross conveyor.

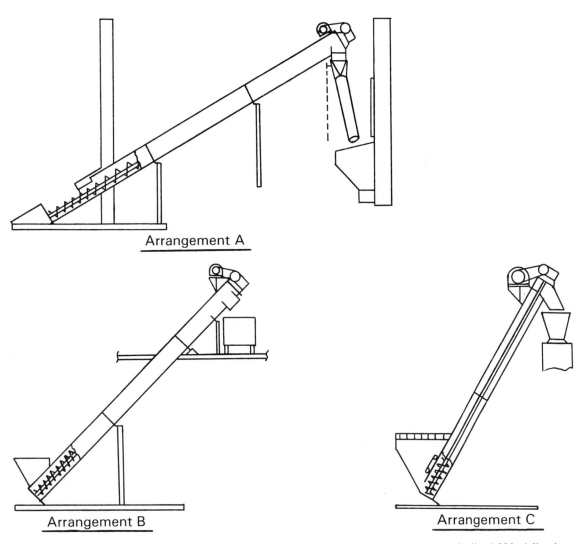

Figure 4.11 Inclined screw conveyors. *Arrangement A:* A typical coal-handling screw conveyor, inclined 20°, delivering stoker coal from a bin into two stoker hoppers. It consists of a short-pitch, tapered feeder section followed by a full-diameter, normal-pitch screw. The discharge is directed to either of the two stoker hoppers by a bifurcated spout with a flap gate. *Arrangement B:* A general-purpose 45° inclined screw conveyor. It consists of a full-diameter, short-pitch conveyor screw in a regular trough with a shroud cover plate. *Arrangement C:* A general-purpose 60° inclined screw conveyor. The short-pitch screw is in a split tubular casing and includes a feed hopper with a bar grating and an adjustable feed inlet gate.

153

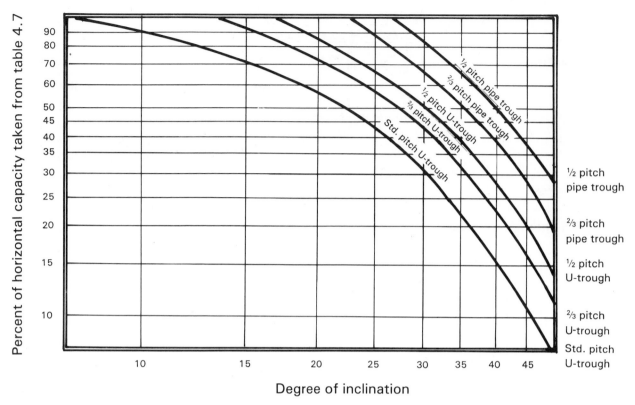

Figure 4.12 Inclined screw conveyor capacity chart.

To illustrate the use of the charts, assume 14-in. short-pitch screw, at 30% capacity, on 30° slope. Table 4.7 indicates capacity on horizontal to be 20.62 ft³/hr at 1 rpm. For half pitch, multiply capacity by 2.00. Thus, theoretical capacity on horizontal is then

$$20.62 \times 2.0 = 41.24 \text{ ft}^3/\text{hr at 1 rpm}$$

Enter figure 4.12 for standard pitch and 30° inclination, and read

$$31\% \text{ or } 0.31 \times 41.24 = 12.78 \text{ ft}^3/\text{hr at 1 rpm.}$$

Now, using the curve for half pitch, the curve reads 52%. The curve labelled "½ pitch U-trough" is intended to correct for ½ pitch, or $0.52 \times 20.62 = 10.72$. Thus, there is about 20% difference in taking the increased capacity in accordance with table 4.7, or taking the increased capacity given by figure 4.12. This is due, in some measure, to the factors of 1.5 and 2.0 for reduced pitch not being exact, but also to allow for the curve inaccuracy.

Inclined screw conveyors can be operated successfully in ways other than as ordinary horizontal or slightly inclined installations. With the proper flights, hangers, bearings, terminals, and housings, nearly all free-flowing, fine dry material (lumps above ½ in. usually cause

trouble because of the limited space between flighting) can be handled up steep inclines or lifted vertically. An inclined screw conveyor in a tubular housing is shown on figure 14.12.

Above 15° slope from the horizontal, short pitch (⅔ pitch or ½ pitch), screws should be used up to about 30°. This angle can vary a little, depending on the character of the material. The reason for making the screw short pitch (standard screw sections are made with flighting, having the pitch equal the diameter) is to make the helix angle of flighting more vertical, so the flighting can push the material instead of letting the material hesitate and fall back on itself.

Up to 35°, normally a U-shaped trough can be used. Above that, pipe should be used so the material cannot slide or flow backward under any condition. Figure 4.3 shows a short-pitch conveyor screw. If the inclined conveyor is under 20 ft in length and operating in a pipe, it is often better to make the screw in one piece. Or, if shorter lengths are used, they can be coupled end-to-end of pipe or welded together (no intermediate hangers used); on account of the length, the screw will bottom (due to its own weight) on the inside of pipe, causing some friction when operating empty or until carrying a full load in pipe.

Table 4.7 Capacity of Screw Conveyors

Screw Dia (in.)	Pipe Dia* (in.)	NA (ft²)	FC (1 rpm)		Capacity (ft³/hr)		
			1 turn FC₁	per hr FC₂	@ 45%	@ 30%	@ 15%
6	2	0.1656	0.0828	4.97	2.24	1.49	0.75
9	2½	0.3967	0.2975	17.85	8.03	5.36	2.68
12	3	0.7186	0.7186	43.12	19.40	12.93	6.47
14	3½	0.9817	1.1454	68.72	30.93	20.62	10.31
16	3½	1.309	1.7453	104.72	47.12	31.42	15.71
18	4	1.657	2.4850	149.10	67.10	44.73	22.37
20	4	2.071	3.4520	207.12	93.20	61.44	30.72
24	4	3.031	6.0623	363.74	163.68	109.12	54.56

Note: Capacity is reduced on inclined conveyors (see paragraph 4.16)

*Nominal size of pipe (OD figured for area)

$$NA = \text{net area (ft}^2) = \frac{(\text{area of screw} - \text{area occupied by pipe}) (\text{in.}^2)}{144}$$

$$FC_1 = 100\% \text{ capacity for one turn} = \text{net area} \times \frac{\text{screw dia}}{12} \text{ (ft}^3)$$

$$FC_2 = 100\% \text{ capacity} = \text{capacity for one turn} \times 60 \text{ (ft}^3/\text{hr)}$$

Capacity (ft³/hr) in schedule is for 1 rpm. To get full capacity, multiply by rpm of screw conveyor shaft. For short pitch, multiply values by 1.50; for half pitch, by 2.00; for long pitch, by 0.67.

To obtain the proper size of screw to use, divide the required capacity in tons/hr by the factor below. This will give theoretical capacity required.

Cut flight and mixing paddles

type:	45+%	30%	15%
cut flight screw	0.73	0.67	0.63
cut flight screw with mixing paddles	0.60	0.54	0.49
standard screw with mixing paddles	0.89	0.87	0.86

When the pipe is full of material, the tendency normally is for the material to lift the screw slightly, relieving the screw itself from scraping the inside of the pipe. Usually a screw in a pipe fits closely to the inside diameter of the pipe and there is not much clearance, but just enough to prevent so-called screw friction on the inside of the pipe.

4.17 SCREW CONVEYORS OVER 20 FEET LONG

If the screw is over 20 ft long and must have hangers between the sections in a pipe, it is difficult to provide a hanger of any kind that will not obstruct the flow of material. Most of the standard hangers are of such design that considerable space (in a pipe) is occupied by them. To support a coupling shaft connecting each section (figure 4.13), a successful method used consists of a round rod of steel, giving only point contact on coupling used. It is true that the rod wears, but it can be replaced easily, inexpensively, and effectively, and usually does not interfere with the movement of material.

This method, with screws operating in pipes, has worked well and with little maintenance when handling a fine granular cellulose product.

4.18 SCREW ELEVATOR

A screw elevator is designed to elevate vertically any free-flowing bulk material, normally ½ in. and under. The unit consists of a vertical helicoid screw with accurate pitch and exact diameter, supported and held in suspension.

The screw is enclosed in a split tubular housing rolled to exact inside diameter made dust-tight and watertight against weather. These units can be driven either from the bottom (free end) or the top.

A constant flow of material must be fed to the bottom of the rotor lift by a short horizontal screw feeder to cause a force feed, so the vertical screw can pick up the

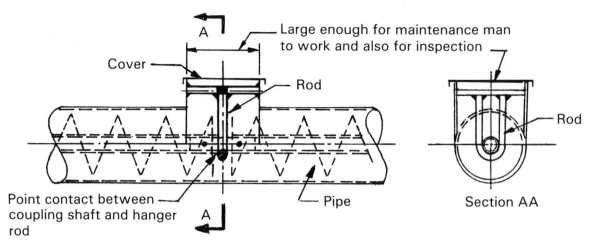

Figure 4.13 Support at a coupling.

load. Otherwise, if this feed is stopped, the vertical rotor lift will discharge only about two-thirds of the material in the housing. Most housings are made with large cleanout doors so that material can be removed, and in the case of food products, parts can be hosed down.

Road contractors use these units when handling cement. No attention is required during dry, warm weather, but in the event of several days of rain with the unit at rest, it must be cleared of cement before the cement picks up enough moisture to harden.

4.19 JACKETED SCREWS

Where there is a heat transfer or cooling problem, any standard U-shaped trough can be made with a jacket, welded continuously to trough to handle so much heat in degrees and pressure, or water pressure at so many psi. In most states, a pressure over 15 psi is subject to the Boiler Code and an Underwriters Lab certificate is

necessary (after tests) and posted on trough, showing that jacketed trough has met the proper specification.

4.20 OVERFLOW PROTECTION

Where no electrical equipment is provided to shut off a screw when the trough is filled, and the material tries to get out of last discharge opening and cannot (due to the discharge chute being full) some prefer to extend the trough another foot or so beyond the last discharge opening, and install an opening with a pipe leading to a receptacle to catch the overflow. A further precaution is to make the hand of the screw opposite to that of the main screw, for a distance between the edge of the last discharge opening and end of trough, thus preventing material from accumulating beyond last discharge opening (see figure 4-14). A successful method used in screw casings consists of a swinging baffle, as shown on figure 4.15. When an overflow of material comes in contact

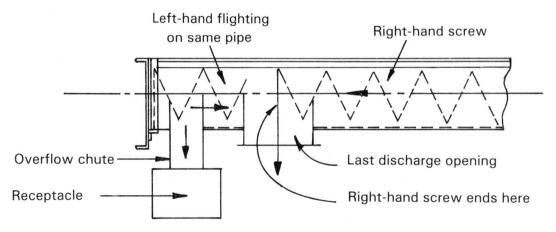

Figure 4.14 Overflow chute.

FIGURE 4.15   157

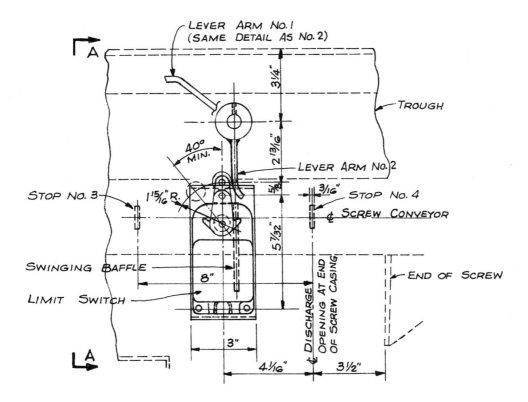

ELEVATION

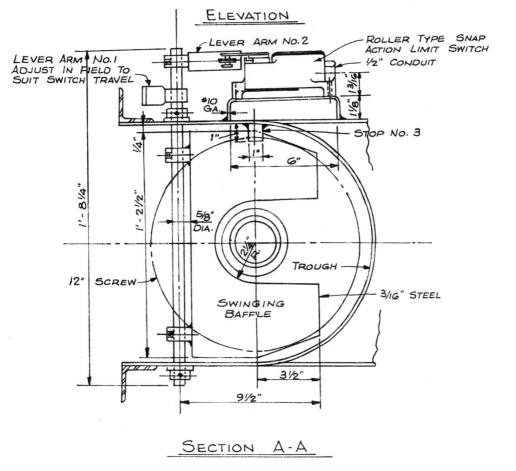

SECTION A-A

Figure 4.15 Protection against overflow.

with it (when discharge chute plugs up), Lever Arm no. 2 moves against roller on the Allen-Bradley roller-type snap-action limit switch, and moves it in an arc of 40° minimum. The swinging baffle in the meantime moves until it comes in contact with Stop no. 3. Finally, the limit switch operates, and cuts off power to the motor driving the screw, stopping it. Lever Arm no. 1 is the same detail as Lever Arm no. 2; it simply acts as a balance under normal operating conditions, allowing Lever Arm no. 2 to just rest against the switch roller. Stop no. 4 is located on centerline of discharge opening, as a precaution against swinging baffle moving in the wrong direction for some unknown reason. This electrical setup has proved very effective. A simple type overflow control is shown on figure 14.2.

4.21 SCREW CONVEYOR DRIVE (refer to section 9)

Many screw conveyors today are designed to be driven by a reduction unit mounted on drive shaft of conveyor,

to simplify the drive and save space. The motor may be supported directly on the reducer (this puts additional load on drive shaft) or by an adapter unit on end plate of screw casing (see figure 9.2).

Many shaft-mount reducers are on the market, all antifriction type with an effective lubrication system. Internal gears are usually of the helical design, made of precision-machined, hardened alloy steel, rated in accordance with AGMA standards. Precision ball bearings with ample capacities for maximum loading are used throughout the reducer. These bearings will easily handle the radial and thrust loads imposed by helical gears, V-belt drive, and screw conveyor.

4.22 RECLAIMING SCREW CONVEYORS HANDLING COAL

On figure 4.16, only the coal which gets through the opening below the adjustable plate will flow into the conveyor, and the conveyor must operate as shown to control the flow of coal into the trough.

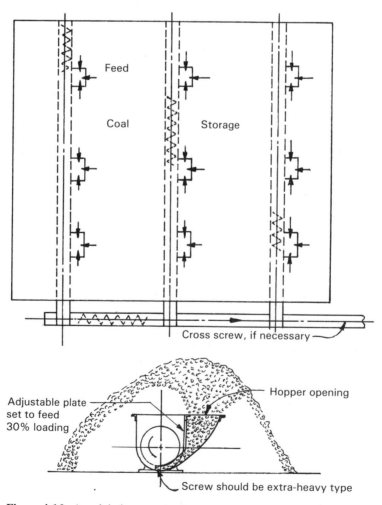

Figure 4.16 A reclaiming screw conveyor.

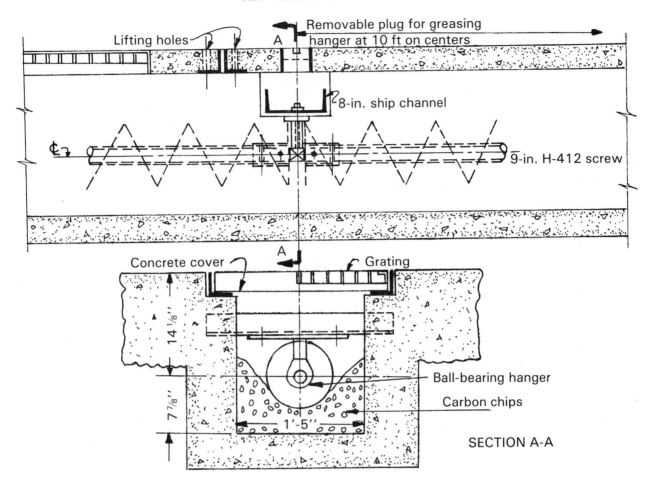

Figure 4.17 Carbon chip screw conveyor.

4.23 CARBON-CHIP REFUSE SCREW CONVEYORS

A number of 9-in. carbon-chip-reclaiming screw conveyors were installed in a concrete trough below the floor. The finishing machines were located over the conveyors, so chips and fine material would fall into the conveyor or the refuse could be swept or shoveled onto grating over trough (trench). The steel grating (removable) has 1-in. square openings and is 2 ft 0 in. long under each machine; the balance of the trench cover is a concrete slab made in 5-ft removable sections.

Actual capacity required for each conveyor was 1 ton of carbon chip refuse per hour, with possible lumps 2 in. $\times D$ (not over 10–15%). Large lumps are broken on grating. Refuse weighs 45 lb/ft^3.

Due to limited space, a standard motor could not be installed. A Vickers hydraulic-pump motor operates each conveyor.

As far as capacity is concerned, a 6-in. screw would take care of present requirements, but a 6-in. screw in one length should not exceed 75 ft, because the coupling shafts are only 1½ in. in diameter. If a small screw like this should become jammed for some reason, a small diameter coupling shaft could easily shear off. Therefore, it was decided to make the screws 9 in. in diameter, having a 2-in. diameter coupling shaft, and run the screw at a slow speed (59 rpm). The screw was mounted on 2½-in. ID pipe, collared at ends for 2-in. couplings, helicoid flighting ⅜-in. thick inner edge next to pipe, and outer edge 3/16 in. thick.

A concrete trough was designed 17 in. wide, and the center of the screw set 7 ⅞ in. from bottom. The carbon refuse fills in the corners and forms its own trough (see figure 4.17).

4.24 DESIGN OF A SCREW CONVEYOR HANDLING SULFUR

Maximum capacity to be handled: 30 tons plus 50% = 45 tons/hr (1200 ft^3).

For a screw conveyor designed to handle 30 tons per

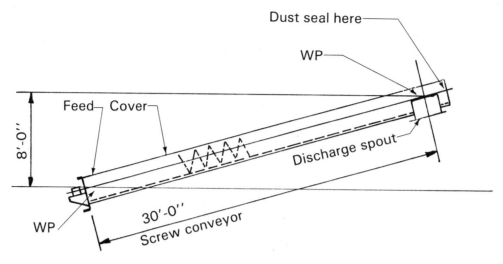

Figure 4.18 Sulfur screw conveyor.

hour of sulfur weighing 75 lb/ft^3, consisting of medium lumps mixed with fines, see figure 4.18. The maximum lump size is 2 in. (maximum of 25% of lumps in this size). An occasional overload of 50% is possible during operation. The conveyor is 30 ft long and rises 8 ft. The capacity of inclined screw conveyors is reduced. On a 25° slope, the reduction is about 40%, and on a 15° slope, about 25%. On slopes over 20°, it is best to check with a manufacturer.

To get a comparison of the various formulas used, the hp will be figured according to the Jeffrey, Stephens-Adamson, Link-Belt, and CEMA formulas.

4.24.1 Jeffrey Screw Conveyor, Catalog No. 951*

According to the Jeffrey catalog (page 7) the average weight of lumpy sulfur is 70–75 lb/ft^3, the loading classification is II, the horsepower factor is 0.6, and explosive dust is possible.

On page 11, according to the heading of Class II (30% full), a 12-in. screw is too small for handling 1200 ft^3/hr of sulfur. Because of the size of the lumps, a 14-in. screw is recommended.

The Capacity Table gives tons per hour at 1 rpm for 100 lb/ft^3 material as 1.05. The factor for material "E," (page 6) is 0.6, since a 14-in. screw was selected because of the size of the lumps.

Maximum rpm of screw = 85 (page 11)

---

*Jeffrey screw conveyors are now manufactured by the Goodman Company.

Conveyor capacity for 75 lb/ft^3 material at 30% full is

$$1.05 \text{ tph} \times \frac{75 \text{ lb/ft}^3}{100 \text{ lb/ft}^3}$$

$$= 0.787 \text{ tph at 1 rpm}$$

The required rpm equals

$$\frac{45 \text{ tph}}{0.787 \text{ tph at 1 rpm}} = 57 \text{ rpm}$$

This is well within the allowable maximum rpm.

Horsepower Calculation, Jeffrey Formula

At the conveyor shaft, the hp is

$$\frac{L(DWfN + 2TF) + TH}{1000}$$

where, for inclined conveyors (see paragraph 4.16)

L = length of conveyor (ft): 30 ft
W = weight of conveyor (lb/ft): 22 lb (screw section and coupling)
D = diameter of coupling or revolving parts of bearing (in.): 3 in.
f = factor for power to overcome friction for babbitted bearings: 0.01
N = rpm of conveyor: 57
T = capacity in tons per hour: 45 (max)
F = factor for material: 0.6
H = rise of conveyor (ft): 8 ft

Therefore, horsepower at the conveyor drive shaft is

$$\frac{30\left[(3 \times 22 \times 0.01 \times 57) + (2 \times 45 \times 0.6)\right] + (45 \times 8)}{1000}$$

$$= 3.11 \text{ hp}$$

Allow 5% for herringbone reducer losses, giving 0.16 hp, added to the 3.11 hp for a total of 3.27 hp at the motor. Therefore, use a 5-hp motor.

4.24.2 Stephens-Adamson Catalog No. 66

Refer to page 214 in the catalog for details on solving the problem as previously described.

$$\text{hp} = \frac{(30 \times 135 \times 57) + (1200 \times 75 \times 30 \times 0.8)}{1,000,000}$$

$$+ \frac{8 \times 45}{990}$$

$$= 2.75$$

Allowing 5% for reducer losses, giving 0.14 hp, the total hp at the motor is

$$2.75 + 0.14 = 2.89$$

4.24.3 Link-Belt Catalog 1000

According to the catalog, lumpy sulfur is class D-26 (1967 ed., p. 254); under the heading of 14-in. screw and weight of material at 75 lb/ft^3 (1967 ed., p. 256), the horsepower factor K is 86. The Link-Belt short formula (1967 ed., p. 252) gives:

$$A = \frac{C \times L \times K}{1,000,000}$$

$$= \frac{1200 \times 30 \times 86}{1,000,000}$$

$$= 3.10$$

where A is the calculated hp, C is capacity in ft^3/hr ($= 1200$), L is conveyor length in ft ($= 30$), and K is the hp factor (86). Consequently, the horsepower is found by

$$A \times G = 3.10 \times 1.07 = 3.32$$

where the value of G is found on page 289 (1958 ed.) of the catalog.*

---

*In the 1967 edition of the catalog, this calculation has been replaced by the use of a graph (p. 260) that combines the A and G factors.

Allowing 5% for reducer losses, giving 0.17 hp, the total hp for a horizontal conveyor is

$$3.32 + 0.17 = 3.49$$

For an inclined conveyor,

$$3.49 \times 1.25 = 4.36 \text{ hp}$$

where the value 1.25 has been determined from tests in the field for an incline up to 20°.

Whichever formula is used, a 5-hp motor is required and has performed satisfactorily.

4.24.4 CEMA Formula

CEMA classifies lumpy sulfur (3 in.) as type 83D$_3$35N (table 1.2), having a material factor (F_m) of 0.8 (table 1.2) and an actual weight of 75 lb/ft^3. Where hp$_f$ is the horsepower required to overcome friction,

$$\text{hp}_f = \frac{LNF_dF_b}{1,000,000}$$

With L being 30 ft, N being 57 rpm, F_b being 1.0 (babbitt bearing, table 4.8), and F_d being 78 (table 4.9):

$$\text{hp}_f = \frac{30 \times 57 \times 78 \times 1.0}{1,000,000}$$

$$= 0.13$$

Where

$C = 1200 \text{ ft}^3$
$W = 75 \text{ lb/ft}^3$
$F_f = 1.0 \text{ (table 4.10)}$
$F_m = 0.8 \text{ (table 1.2)}$
$F_p = 1.0 \text{ (table 4.11)},$

$$\text{hp}_m = \frac{CLWF_fF_mF_p}{1,000,000}$$

$$= \frac{1200 \times 30 \times 75 \times 1.0 \times 0.8 \times 1.0}{1,000,000}$$

$$= 2.16$$

The horsepower needed for raising the material 8 ft is found by

$$\frac{\text{height} \times \text{capacity (tons)}}{990}$$

$$= \frac{8 \times 45}{990}$$

$$= 0.36$$

Table 4.8   Hanger Bearing Factor, F_b

Bearings Component Groups	Bearing Type	Hanger Bearing Factor F_b
Group A	Ball	1.0
Group B	Babbit Bronze Graphite bronze* Canvas-base phenolic* Oil-impregnated bronze* Oil-impregnated woods*	1.7
Group C	Plastic* Nylon* Teflon*	2.0
Group D	Chilled hard iron* Hardened alloy sleeve*	4.4

Source: Courtesy CEMA
*Nonlubricated bearings or bearings not additionally lubricated.

Table 4.9   Conveyor Diameter Factor, F_d

Screw Dia (in.)	F_d
4	12.0
6	18.0
9	31.0
10	37.0
12	55.0
14	78.0
16	106.0
18	135.0
20	165.0
24	235.0

Source: Courtesy CEMA

Table 4.10   Flight Factor, F_f

Type of Flight	Conveyor Loading			
	15%	30%	45%	95%
Standard	1.0	1.0	1.0	1.0
Cut flight	1.10	1.15	1.20	1.3
Cut & folded flight	*	1.50	1.70	2.20
Ribbon flight	1.05	1.14	1.20	—

Source: Courtesy CEMA
*Not recommended

Table 4.11   Paddle Factor, F_p

Standard paddles per pitch set,
  paddles set at 45° reverse pitch

Number of paddles per pitch	none	1	2	3	4
Paddle factor, F_p	1.0	1.29	1.58	1.87	2.16

Source: Courtesy CEMA

If the overload factor, F_o, is 1.55 (see figure 4.20), the total hp is given by

$$0.13 + 1.55(2.16 + 0.36)$$
$$= 4.04$$

The amount to be added for the material tumbling because of its being raised can generally be neglected, because the motor selected is always larger than the minimum computations show.

4.24.5   Comparison of Results

Jeffrey:	3.27 hp
Stephens-Adamson:	2.89 hp
Link-Belt:	4.36 hp
CEMA:	4.04 hp

The CEMA formula should be used, since it represents the latest combined thinking of all the manufacturers. The other formulas are given for use in reviewing previous designs.

4.24.6   Final Design

When the design was completed, the conveyor appeared as shown on figure 4.19.

4.24.7   Specifications

Description: screw conveyor to handle 30 tph of sulfur weighing 75 lb/ft^3 and consisting of mixed fines and 2-in. lumps (25% lumps), with the conveyor inclined 15° and being 26 ft 2 in. long.

Material and details of construction: the conveyor shall be designed in accordance with the latest CEMA standards and using the following manufacturer's standards:

Flights: shall be 14-in. diameter carbon steel, right hand, helicoid $7/16$ in. by $7/32$ in. thick, mounted on $3\frac{1}{2}$

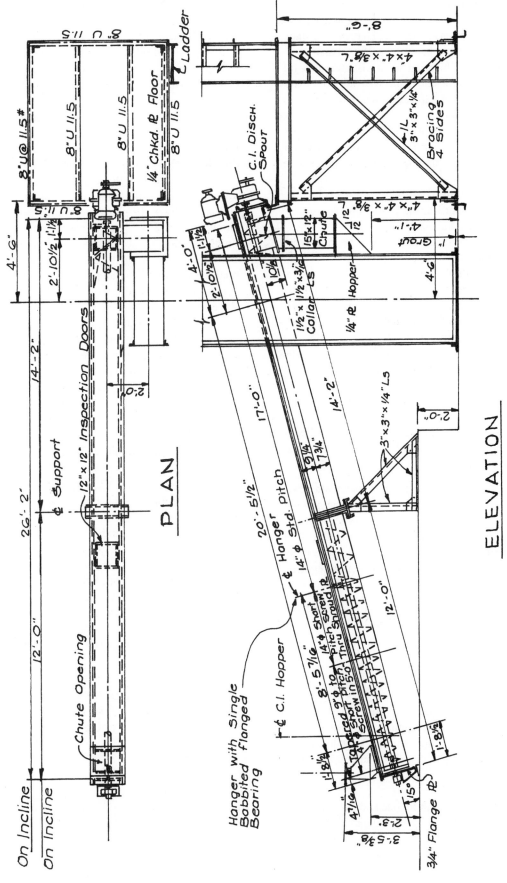

Figure 4.19 An inclined screw conveyor handling sulfur.

163

in. standard pipe with ends bushed for 3-in. diameter couplings; refer to the bill of material (paragraph 4.24.8) for short and tapered pitch at one end. To prevent a static spark on screw conveyors handling sulfur, many operators like to break the flighting for a distance of 4 in. This appears to be sufficient to prevent a static spark from jumping across the gap, and does not appear to stop the flow of material.

Hanger: style 226 with babbitted bearings. The use of white-metal hanger bearings is recommended for handling sulfur and clay, although babbitt bearings give fair service.

Shafts: tail, drive and coupling, carbon steel.

End bearings: roller-bearing pillow block.

End seal: split seal gland.

Trough end plate: tail end, outboard bearing type; drive end, for type "C" conveyor drive.

Trough: $^3/_{16}$-in. thick flanged type.

Cover: semiflanged with spring clamps and gaskets.

Connections: inlets and outlets shall have flanged openings. Openings shall be manufacturer's standard, unless otherwise indicated on the attached sketches.

Motor base: adjustable base shall be furnished. The drive shall be supported on the conveyor at the discharge end.

Motor: 230/460V, 3-ph, 60 Hz, totally enclosed, 1750 rpm.

Drive: dust-tight, explosive hazard screw conveyor, 30% full. Complete with V-belt drive, sheaves, and guard (OSHA approved).

Assembly of equipment: It is the buyer's standard practice to have equipment preassembled to the maximum extent permitted by shipping facilities in order to minimize field erection and assembly time. The seller shall clearly state in the proposal the extent of disassembly for shipment and the degree of interconnection required for erection.

Painting: All external carbon steel shall be protected with manufacturer's standard finish paint.

4.24.8 Bill of Material

Generally, bills of material should be revised to list manufacturer's catalog numbers to make it easier to order replacement parts. Other engineers prefer to leave bills of material in original bid form. Both methods have been used here. This project was furnished by Link-Belt Co.

Refer to figure 4.19 for an inclined screw conveyor, 14-in. diameter, 26 ft 2 in. long, running at 57 rpm with a 5-hp motor.

1 Length conveyor screw 8' 5$^7/_{16}$" long, helicoid flight rh, flights $^7/_{16}$" × $^7/_{32}$" mounted on 3½" sched 40 pipe, with ends bushed for 3" dia shafts, made as follows:
   5' 0" of tapered flighting 9" dia to 14" dia, half pitch 3' 5$^7/_{16}$" of 14" dia, half pitch

1 Length conveyor screw 16' 9" long, rh, std pitch no. 14H614

1 Style 226 hanger 14" × 3" with babbitted bearing no. 162-390E

1 Hardened coupling shaft 3" dia, no. 170-38-13

8 Coupling bolts ¾" dia × 5" long no. 126-527-J

1 Outboard bearing trough end 14" × 3", complete with split seal gland and 3" dia roller bearing

1 Special 3" dia × 19" long end shaft

1 Shroud 14" × 2' 4" long × no. 7 ga

1 Trough section 12' 0" long × $^3/_{16}$" flanged

1 Trough section 14' 2" long × $^3/_{16}$" flanged with discharge spout fitted and attached. Center line spout located 13½" from end

2 Cover plates, semiflanged × 12' 0" long with spring clamps on 2' 0" centers

1 Cover plate, semiflanged × 2' 4" long with spring clamps on 1' 0" centers, complete with relief door

1 Support foot no. 166-6-1

1 Intermediate conveyor support (see drawing for layout)

4.24.9 Drive Equipment

1 Electric motor 5 hp, 1750 rpm, totally enclosed, 230/460 V, 3 ph 60 Hz, frame 184T

1 Type C screw conveyor drive (class II service), complete with adapter kit and 3" dia shaft kit

1 Motor bracket assembly, complete with 184T motor plate

1 Driver sheave 2B5.0" pd, hub 1$^1/_8$" dia

1 Driven sheave 2B6.4" pd, hub 1$^3/_8$" dia

2 B60 V-belts (21.9" centers)

1 V-belt guard (OSHA approved)

1 Trough end for type C screw conveyor drive

4.25 DESIGN OF A SCREW CONVEYOR HANDLING PETROLEUM COKE

4.25.1 Data

Material: petroleum coke lumps 2–3 in. (max) in size, with an apparent density (W) of 40 lb/ft^3.

Conveyor: 63 ft long (*L*) with a capacity (*CW*) of 10 tph (= 20,000 lb), so

$$C = \frac{20,000}{40}$$

$$= 500 \text{ ft}^3/\text{hr}$$

and trough loading of 15%.

4.25.2 Size of the Conveyor

Because of the abrasiveness of the material, the capacity should be figured at 30%, or even at 15% when using hangers. On short conveyors, if hangers are not used, a selection of up to 45% might be possible.

Using 15% trough loading, a 14-in. screw at 1 rpm will have a capacity of 10.31 ft^3/hr (table 4.7). For a capacity of 500 ft^3/hr, the speed required is

$$\frac{500}{10.31} = 48.5 \text{ rpm}$$

The actual computed speed is 46.5 rpm (refer to paragraph 4.25.4).

Since the maximum material size is 2–3 in., and the speed is under 65 rpm, the 14-in. screw is satisfactory. An 18-in. trough was used because of the abrasive material.

4.25.3 Horsepower

For mechanical friction, where

L = conveyor length in ft (63),
F_b = 4.4 (table 4.8),

T_d = 78.0 (table 4.9), and
N = 46.5 rpm,

$$\text{hp}_f = \frac{LF_b F_d N}{1,000,000}$$

$$= \frac{63 \times 4.4 \times 78 \times 46.5}{1,000,000}$$

$$= 1.0$$

For the material, where

CW = capacity,
L = conveyor length in ft (63), and
F_m = 1.3 (table 1.1),

$$\text{hp} = \frac{CWLF_m}{1,000,000}$$

$$= \frac{20,000 \times 63 \times 1.3}{1,000,000}$$

$$= 1.64$$

Therefore, the total hp is $1.0 + 1.6 = 2.6$. Since 2.6 is less than 5 hp, the overload factor, F_o, is 1.45 (figure 4.20).

$$2.6 \times 1.45 = 3.8 \text{ hp}$$

With an 85% drive efficiency,

$$\frac{3.8}{0.85} = 4.5 \text{ hp}$$

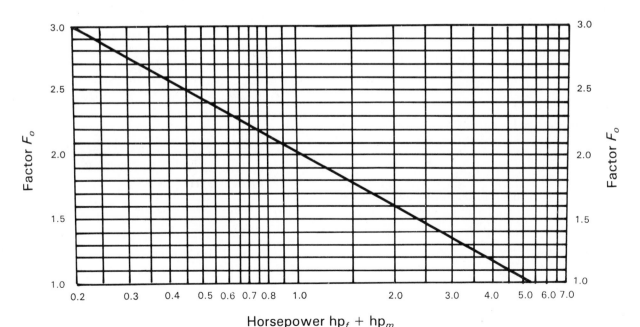

Figure 4.20 A chart for the values of factor F_o. For values of hp$_f$ + hp$_m$ greater than 5.2, F_o is 1.0. *(Courtesy of CEMA)*

For shock loading and 16-hr operation, apply a factor of 1.5 (refer to table 9.1):

$$4.5 \times 1.5 = 6.75 \text{ hp}$$

Consequently, a 7½ hp motor has been selected.

4.25.4 Drive

The desired speed of the conveyor is about 46 rpm. Using a shaft-mounted reducer for economy and to save space, the Falk Corporation catalog (all manufacturers have similar catalogs) lists a 1215-24 unit which serves for 56-33 output rpm; this one is selected.

Using V-belt and sheaves, the actual drive ratio (from catalog) is 22.93:1. The sheave ratio will then be $1750/22.93 = 76.32$. The minimum sheave diameter recommended by NEMA (refer to table 9.2) is 3 in. Where the driver and driven sheave diameters are not far apart, it is best to use a larger size. In this case, a 6.5-in. OD was chosen for the driver, and a 10.6-in. OD for the driven sheave. Both sizes are stock units (per catalogs). The rpm of the conveyor will be

$$\frac{76.32 \times 6.5}{10.6} = 46.8$$

Using RC-40 chain, with a 25-tooth driver sprocket and a 41-tooth driven sprocket, the rpm of the screw conveyor will be

$$76.32 \times \frac{25}{41} = 46.5$$

Using the FMC (Link-Belt) catalog, a 7.5 DTB2 Link-Belt motogear with a ratio of 31.4:1 will have output shaft rpm of

$$\frac{1750}{31.4} = 55.7.$$

A 10EDB2 unit with a ratio of 38.4:1 would result in

$$\frac{1750}{38.4} = 45.6 \text{ rpm},$$

but would not permit variation. The DTB2 unit connected to a drive sprocket of 26 teeth, with a driven sprocket of 32 teeth, will reduce the speed to

$$\frac{55.7 \times 26}{32} = 45.3 \text{ rpm.}$$

4.25.5 End Thrust

The end thrust in screw conveyors is very seldom a factor in the design of the shaft. The following formula suggested in CEMA Book no. 350, 1st edition, as a conservative calculation, is

$$\text{thrust (lb)} = \frac{252,000\,(\text{hp})K^*}{ND_s}$$

where $K = 1.0$ for standard pitch (diameter), 1.5 for two-thirds pitch, 2.0 for one-half pitch, 3.0 for one-third pitch, and 4.0 for one-fourth pitch; $N = $ rpm; and $D_s = $ conveyor screw diameter (in.).

In this case, $K = 1.0$, $N = 46.5$ rpm, and $D_s = 14$ in., so the thrust is

$$\frac{252,000 \times 7.5 \times 1}{46.5 \times 14} = 2903 \text{ lb}$$

From Link-Belt Catalog 1000 (1958 ed., pp. 181, 187, and 284), horsepower equals AG†, where AG is given by

$$\frac{C \times L \times K}{1,000,000}$$

$$= \frac{500 \times 63 \times 155}{1,000,000}$$

$$= 4.88$$

where C is the capacity in ft³/hr, L is the length in ft, and the value 155 is for class D-28 material for a 14-in. screw. The 4.88 value compares with 4.5 in the new formula. For class D-28, with material weighing 30–40 lb/ft³, the torque developed is

$$\frac{4.88 \times 63,000}{46.5} = 6612 \text{ in.-lb}$$

The class D-28 refers to a 16-in. screw conveyor carrying a maximum 3-in. lump size, at a recommended maximum speed of 40 fpm, with 15% loading. Link-Belt catalog 1000 recommended a 16-in. screw conveyor for 3-in. maximum size in the construction. Note that table 1.2 codes petroleum coke as 40D$_7$-37. The old FMC classification was used in the problem so as not to change the original computations.

4.26 SCREW CONVEYOR HANDLING PETROLEUM COKE, SCREW FREE OF HANGERS

4.26.1 Data

Material: petroleum coke, 2 in. and under, with a weight of 40 lb/ft³. It may be damp or wet.
Conveyor: 16 ft long with a capacity of 40 tph.

---

*This equation has been removed from the second edition because it is very conservative, and since there is no practical way to make an accurate determination of thrust—CEMA.

†In the 1967 edition of the catalog, the AG calculation has been replaced by the use of a graph appearing on page 260.

4.26.2 Size of the Conveyor

The material is generally considered very abrasive, but some consider it mildly abrasive because of a so-called light greasy feel. Because of the absence of hangers, 45% trough loading may be used.

The capacity of a 12-in. screw at 45% capacity is 19.40 ft³/hr at 1 rpm (table 4.7).

$$\text{rpm} = \frac{40 \text{ tons} \times 2000}{40 \text{ lb/ft}^3 \times 19.40}$$

$$= \frac{2000}{19.4}$$

$$= 103$$

Experience has shown that this is too fast and causes considerable maintenance.

The capacity of a 14-in. screw conveyor at 1 rpm = 30.93 ft³/hr (table 4.7). So,

$$\text{rpm} = \frac{2000}{30.93}$$

$$= 65$$

To improve the maintenance, the 14-in. screw was installed in a 16-in. trough. It could have been installed in a 16- or 18-in. square trough. The coke thus makes its own trough.

U-shaped troughs are generally made of ¼-in. steel.

4.26.3 Horsepower (compare paragraph 4.25.3)

For mechanical friction,

$$\text{hp}_f = \frac{16 \times 65 \times 78 \times 1.0}{1,000,000}$$

$$= 0.08$$

For the material,

$$\text{hp}_m = \frac{2000 \times 16 \times 40 \times 1.0 \times 1.6 \times 1.0}{1,000,000}$$

$$= 2.05$$

Because of the moisture content, 1.6 is used for F_m instead of 1.3.

Therefore, the total hp is

$$(2.05 + 0.08) \times 1.7 = 3.6 \text{ hp}$$

and

$$\frac{3.6}{0.85} = 4.26 \text{ hp}$$

This was considered too close to allow for surges, and a 7½ hp motor was used. The Link-Belt formula for class D-28 (40–50 lb/ft³) material (p. 287) and the formula on page 281 of Catalog 1000 (1958 ed.*) gives

$$\text{hp} = \frac{C \times L \times K}{1,000,000}$$

$$= \frac{2000 \times 16 \times 183}{1,000,000}$$

$$= 5.86$$

4.26.4 Deflection

Table 4.1 does not give the deflection for a 16-ft span, but the deflection for a 10-ft span is 0.036 in. For a 16-ft span,

$$\Delta = \left(\frac{16}{10}\right)^4 \times 0.036 = 0.236 \text{ in.}$$

---

*Compare the results of the calculation to the use of the graph on page 260 of the 1967 edition of the catalog.

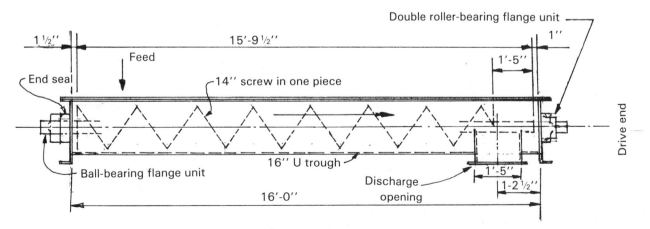

Figure 4.21 Elevation of a screw conveyor. This layout with a roller chain allows a change in the speed of the screw. Where the screw is direct-connected to a speed reducer, a change in speed can be made by connecting the reducer and motor with a V-belt drive.

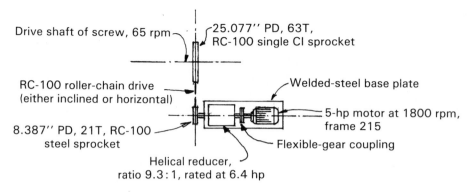

Figure 4.22 Drive machinery.

This is too close to the maximum of ¼ in. allowable, and schedule 80 pipe should be used. Since the outside diameter of the pipe is the same as that of schedule 40, no other changes have to be made.

4.26.5 Components

An elevation of the screw conveyor is shown on figure 4.21 and the drive is shown on figure 4.22.

4.26.6 Finishes

Details of continuous-weld finishes are given in table 4.12.

4.27 LUMP SIZE LIMITATIONS

If the lumps cannot be broken down in the screw conveyor, its size may be determined by the size and proportion of the lumps in the conveyed material. The size

Table 4.12 Continuous-Weld Finishes

	Class of finish				
Operation	*I*	*II*	*III*	*IV*	*V*
Weld spatter and slag removed	X	X	X	X	X
Rough grind welds to remove heavy weld ripple or unusual roughness (equivalent to a 40–50 grit finish)		X			
Medium grind welds—leaving some pits and crevices (equivalent to a 80–100 grit finish)			X		
Fine grind welds—no pits or crevices permissible (equivalent to a 140–150 grit finish)				X	X
Polish to a bright uniform finish					X

Source: Courtesy CEMA

Notes: "Grind smooth" is a general term and is subject to various interpretations. This table establishes CEMA-recommended classes of finishes, which should be used to help find the class required for an application. (Special weld finishes do not apply to standard stock conveyor screws, but should be specified where desired.)

Class I finish has the weld spatter and slag removed, but no grinding of the welds.

Class II finish is a refinement of the "as welded condition" with the welds rough ground to remove heavy weld ripple or unusual roughness.

Class III finish has the weld medium ground with some pits and crevices permitted. This finish is recommended for materials that do not tend to contaminate or hang up in pits or crevices.

Classes IV and V finishes have the welds ground fine with no pits or crevices. The only difference between the two finishes is the degree of polish. These finishes are recommended where sanitary regulations dictate exclusion of the materials being handled from the welded surface. The type of finish is dependent on the application and/or industry.

Table 4.13 Maximum Lump Size

Screw dia (in.)	Pipe OD (in.)	Radial Clearance (in.)	Class 1 10% lumps Ratio R = 1.75 Max lump (in.)	Class 2 25% lumps Ratio R = 2.5 Max lump (in.)	Class 3 95% lumps Ratio R = 4.5 Max lump (in.)
6	2⅜	2 5/16	1¼	¾	½
9	2⅜	3 13/16	2¼	1½	¾
9	2⅞	3 9/16	2¼	1½	¾
12	2⅞	5 1/16	2¾	2	1
12	3½	4¾	2¾	2	1
12	4	4½	2¾	2	1
14	3½	5¾	3¼	2½	1¼
14	4	5½	3¼	2½	1¼
16	4	6½	3¾	2¾	1½
16	4½	6¼	3¾	2¾	1½
18	4	7½	4¼	3	1¾
18	4½	7¼	4¼	3	1¾
20	4	8½	4¾	3½	2
20	4½	8¼	4¾	3½	2
24	4½	10¼	6	3¾	2½

of a lump is its maximum dimension. CEMA makes three classifications of lump sizes (CEMA Book 350, 1981).

Class 1. A mixture of lumps and fines in which not more than 10% are lumps ranging from maximum size to one half of the maximum; and 90% are lumps smaller than one half of the maximum size.

Class 2. A mixture of lumps and fines in which not more than 25% are lumps ranging from the maximum size to one half of the maximum; and 75% are lumps smaller than one half of the maximum size.

Class 3. A mixture of lumps only in which 95% or more are lumps ranging from maximum size to one half of the maximum size; and 5% or less are lumps less than one tenth of the maximum size.

Table 4.13 shows the recommended maximum lump size for each customary screw diameter and the three lump classes. The ratio, R, is included to show the average factor used for the normal screw diameters, which then may be used as a guide for special screw sizes and constructions. For example:

$$R = \frac{\text{radial clearance (in.)}}{\text{lump size (in.)}}$$

This ratio applies to such unusual cases as screws 16 in. in diameter mounted on 2-in. solid shafts; or 12-in. diameter screws mounted on 6-in. diameter pipes (the large pipe serving to reduce deflection of the screw).

Example: To illustrate the choice of screw size from table 4.13, say the material is ice, with Material Characteristics code number D15, 35–45 lb/ft^3 and with size distribution as follows: 4 in. × 2 in., 9%; 2 in. × 1 in., 41%; 1 in. × ⅜ in., 22%; minus ⅜ in., 28%.

This lump size distribution falls under Class 1; from table 4.13, the ratio (R) is 1.75, and the radial clearance is 4 × 1.75, or 7 in. This calls for an 18-in. diameter screw.

4.28 SAFETY (Courtesy CEMA)

All applicable OSHA regulations must be considered in the erection and operation of screw conveyors. Heavy gratings must be used in all loading areas and solid covers in other areas. Covers, guards, and gratings at inlet points must be such that personnel cannot be injured by the screw. Guards must be provided for all exposed equipment such as drives, gears, shafts, couplings, and so on.

For hazardous conditions or hazardous materials, consult manufacturer. For explosive materials, refer to paragraph 14.13. For handling foodstuffs, refer to paragraph 4.10.

Author's note: Goodman Conveyor Company and Thomas Conveyor Company have issued new catalogs but, regretfully, they arrived too late to be of use in this edition.

Apron and Mold Conveyors

5.1 GENERAL

Apron conveyors for handling bulk materials are usually made of metal pans, either overlapping or hinged at the articulation points, and mounted on a double strand of steel strap roller chain and riding on single-flanged rollers. They operate horizontally or at inclines up to about 25°, sometimes even 30°, from the horizontal, depending on material and size of lumps. They can be made in heavy- or light-duty designs, depending upon the severity of service required, and are usually made up in widths varying from 12 in. to 60 in., depending upon the capacity and size of lumps in the material handled. They are particularly useful as feeders (refer to section 7). In this application, they are generally provided with stationary side plates for increased capacities and handling of large, lumpy materials. Their operating speeds as feeders are slow, not much over 10 fpm, but as conveyors, they can run safely at up to 100 fpm.

5.2 NO-LEAK TYPES

The no-leak or leakproof type is well suited to handle materials with small particle sizes, because the side plates are made part of the pans, and special attention is paid to obtaining an overlap with the closest clearance possible to prevent leakage. It is designed to withstand the wear of abrasive materials, the chains being located under the pans to avoid direct fall of material on them. Carrying rollers are located outboard on the cross rods for easy accessibility for repairs. A double-beaded type of steel pan is used on the no-leak or leakproof aprons and on light service conveyors where short pitch, cheaper chains may be used. There are many variations of the overlapping pans, all shown in manufacturers' catalogs.

The popular type of apron conveyor pans are shown on figures 5.1, 5.2, 5.3, and 5.4. The pans of the single-beaded design, Style AC (Type 1), are available in several thicknesses in 9- and 12-in. pitch chains. This style of pan provides free discharge of material and, because of the pan shape, is suitable for use on conveyors inclined up to 25°. For handling any heavy lumps, the pan can be equipped with steel-capped wood filler blocks. The filler blocks must be capped to avoid contamination. The general dimensions of the pans, complete with double strand of chain, are shown in tables 5.1, 5.2, 5.3, 5.4, and in manufacturers' catalogs.

Shallow skirts are used for handling individual pieces, rather than bulk materials. Where the materials run to volume rather than weight, higher skirts are used. Where there is danger of material becoming entangled in the side wings, the guards project to the pans and overlap the side wings.

5.3 HEAVY ORES AND LARGE LUMPS

Heavy ores and large lumps are handled on apron feeders and conveyors designed for this class of service. Pans can be made of heavy steel, reinforced on the bottom by tee rails, or of cast manganese of heavy thickness.

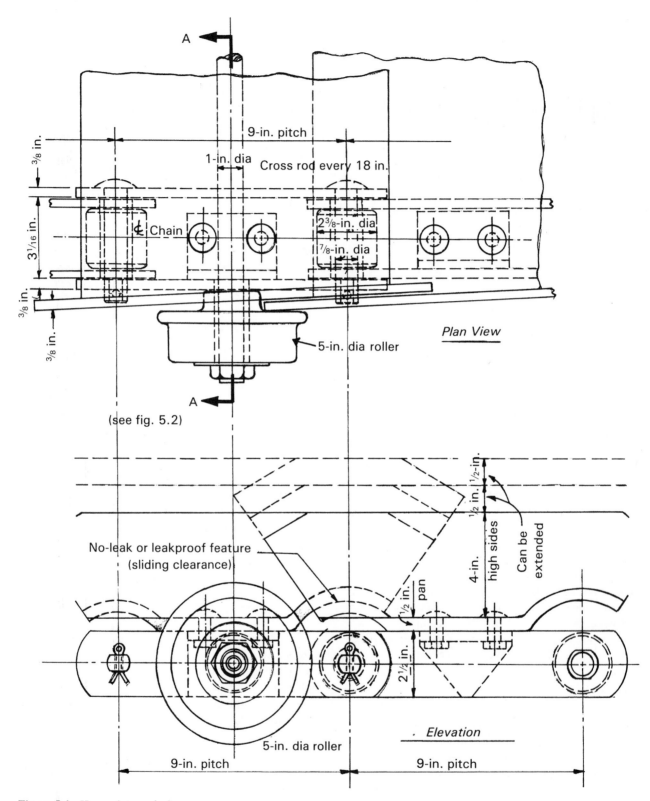

A

9-in. pitch
1-in. dia   Cross rod every 18 in.
3/8 in.
3 1/16 in.
2 3/8-in. dia
7/8-in. dia
¢ Chain
3/8 in.
3/8 in.
5-in. dia roller

Plan View

A

(see fig. 5.2)

No-leak or leakproof feature
(sliding clearance)

1/2 in. 1/2-in.

4-in. high sides

Can be extended

1/2 in. pan

2 1/2 in.

5-in. dia roller

Elevation

9-in. pitch   9-in. pitch

Figure 5.1 Heavy-duty no-leak apron.

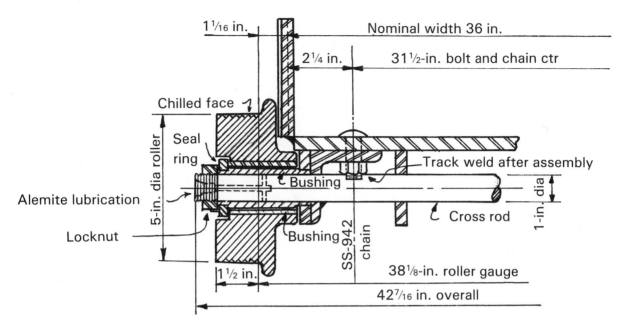

Figure 5.2 Section A-A (from figure 5.1) through the heavy apron.

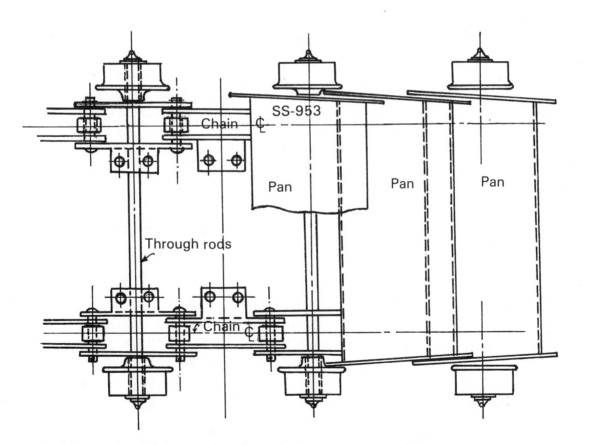

Figure 5.3 Assembly of chain-supporting rollers and pans on a standard no-leak apron.

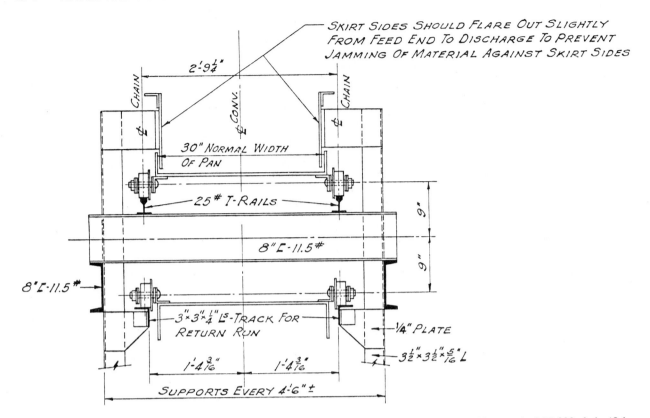

Figure 5.4 A 30-in. apron conveyor with ½-in. steel pans, 3 in. × ⅜ in. side plates on a double strand of SS-922 chain (9-in. pitch), and 3½-in. diameter single-flange roller.

Manganese chains are located under the pans for protection and are carried on rollers closely spaced. The manganese pans are well adapted as feeders for handling heavy ores, such as chrome, and/or where the service is especially severe.

5.4 DATA ON APRON CONVEYORS

Tables 5.1, 5.2, 5.3, and 5.4 give data on the popular types of apron conveyors. The tables give the chain pitch, type of chain, coefficient of rolling friction, minimum radius of upturn, and weights for these conveyors. The data are taken from Link-Belt Company's 1967 catalog 1000. For capacities, consult manufacturers.

5.5 SELECTION OF APRON

When handling any material, experience generally will govern the selection of the width and speed of the apron.

A rule often used is to make the width of the apron four times the lump size. Table 5.5 can be used as a guide.

5.6 DESIGN OF APRON CONVEYOR

5.6.1 General Data

Apron: 30-in. wide with type no. 1 pans.
Length: 75-ft centers.
Material: dry sulfur weighing 80 lb/ft³, with 25% of capacity—3-in. lumps—balance down to ¼ in., and fines screened out.
Capacity: 100 tph.
Conveyor: horizontal, fed uniformly by a vibrating feeder. Due to size and percentage of lumps, only 75% of cross-section filled by testing.

5.6.2 Capacity

Recommended skirt size is four times the size of the lumps being conveyed, which would then be in this ex-

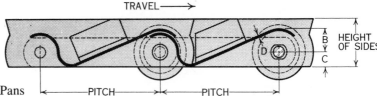

Table 5.1 Data for Aprons with Style AC Pans

Chain and Pan Pitch (in.)	Chain Number	Allowable Chain Pull per Strand (lb)	Factor C		Minimum Radius of Upturn (ft)	Weight per Foot (lb)[a]					
			Chain Not Lubricated	Chain Lubricated		For Width A (in.)				Each Extra 6-in. Width	A[b] (in.)
						Height of Sides (in.)					
						3½	4	5	6		
	SS 922	7,200	0.161	0.113	2.5	49	50	56	60	4	18
	SS 922	7,200	0.161	0.113	2.5	54	55	61	65	5	18
	SS 922	7,200	0.161	0.113	2.5	64	65	70	74	7.3	18
9.00	SS 933	9,200	0.156	0.109	2.5	63	65	69	73	5.8	18
	SS 933	9,200	0.156	0.109	2.5	73	75	79	83	8.3	18
	SS 940	9,200	0.125	0.078	6	65	67	71	77	5.8	18
	SS 940	9,200	0.125	0.078	6	76	78	82	88	8.3	18
	SS 1222	7,200	0.161	0.113	2.5	57	59	63	67	6.8	24
	SS 1222	7,200	0.161	0.113	2.5	69	70	74	78	9.5	24
	SS 1233	9,200	0.156	0.109	2.5	62	63	67	71	7	24
12.00	SS 1233	9,200	0.156	0.109	2.5	75	76	80	84	10.3	24
	SS 1240	9,200	0.125	0.078	6	62	65	69	73	7	24
	SS 1240	9,200	0.125	0.078	6	75	78	82	86	10.3	24
	SS 1244	12,700	0.120	0.075	4	81	83	87	91	11	24
	SS 1244	12,700	0.120	0.075	4	108	110	116	121	14	24

Source: Courtesy Link Belt (FMC)

Notes: Have dimensions certified for installation purposes. The minimum recommended sprocket is a 6-tooth one.
[a]Weight of cross rods is included, spaced every 36 in.
[b]Minimum apron width. Maximum recommended width is 48 in. for 9-in. pitch chains and 60 in. for 12-in. pitch chains.

ample 4 × 3 in. = 12 in. However, experience shows that 6 in. is satisfactory for this grading of material (see figure 5.5*a*). The suggested speed is 50 fpm, a good velocity to keep maintenance low. Capacity is obtained through the following calculations:

2.5 ft × 0.5 ft × 1.0 × 0.75 × 80 lb = 75 lb/lin ft

75 lb/lin ft × 50 fpm × 60 min = 225,000 lb/hr

or 112.5 tph

5.6.3 Horsepower

Weight of sulfur per lin ft   = 75 lb
(refer to table 1.2)

Weight of apron (18 in. wide
with 6-in. sides) per ft   = 88 lb

Correction for 30-in. width
= 2 × 8.3   =   17 lb (approx)

Total   = 180 lb/ft

With a chain not Alemite lubricated, and the oil simply being poured on by hand, use a factor of 0.11 in calculating the weights on carrying run A and return run B.

Carrying run A
75 ft × 180 lb × 0.11 = 1485 lb
Return run B
75 ft × 105 lb (apron only) × 0.11 = 867 lb

Table 5.2 Data for Aprons with Style AA Pans

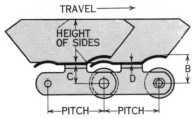

Style AA1 pans

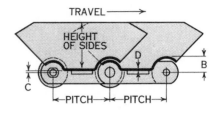

Style AA2 pans

Chain and Pan Pitch (in.)	Chain and Attachment Number	Allowable Chain Pull per Strand (lb)	Pan Style	Factor C		Minimum Radius of Upturn (ft)	Weight per Foot (lb)[a]					
				Chain Not Lubri-cated	Chain Lubri-cated		For Width A (in.)					
							Without sides	Height of Sides (in.)			Each Extra 6-in. Width	A[b] (in.)
								2	3	4		
2.97	MR 1½-A1[c]	2820	AA1	0.200	0.140	6	24	27	29	31	4	12
	MR 1½-K1[c]	2820	AA1	0.200	0.140	6	29	32	34	36	4	12
3.00	RS 3013-A1	2100	AA1	0.167	0.104	6	18	21	23	25	4	12
	RS 3013-K1	2100	AA1	0.167	0.104	6	20	23	25	27	4	12
4.00	1120-A63[c]	1080	AA2	0.125	0.088	2	18	22.5	24.8	27	4.5	12
	1124-A63[c]	1870	AA2	0.167	0.117	2	20	24.5	26.8	29	4.5	12
	RS 4019-A11	2450	AA2	0.192	0.120	2	22	26.5	28.8	31	4.5	12
4.04	1113-A4C	3220	AA1	0.172	0.120	2	31	37	40	42.3	7.5	12
	RS 60-A1	4950	AA1	0.212	0.133	2	36	42	45	47.3	7.5	12
	RS 60-A1	4950	AA1	0.212	0.133	2	43	46.8	49	52	11	12
	RS 60-K1	4950	AA1	0.212	0.133	2	38	41.8	44	47	7.5	12
	RS 60-K1	4950	AA1	0.212	0.133	2	45	48.8	51	54	11	12
6.00	1130-A135[c]	3750	AA2	0.150	0.105	3	42	46.5	48.5	51	7	18
	1130-A135[c]	3750	AA2	0.150	0.105	3	52	56.5	58.5	61	10	18
	RS 6238-A98[d]	5600	AA2	0.179	0.112	3	46	50.5	52.5	55	7	18
	RS 6238-A98[d]	5600	AA2	0.179	0.112	3	56	60.5	62.5	65	10	18
	1130-K1[c]	3750	AA1	0.150	0.105	3	44	48	50	52.5	7	18
	1130-K1[c]	3750	AA1	0.150	0.105	3	55	59	61	63.5	11	18
	RS 6238-K1[d]	5600	AA1	0.179	0.112	3	66	70	72	74.5	11	18
	RS 944-K1[d]	5900	AA1	0.240	0.150	2	45	49	51	53.5	11	18
	RS 944-K1[d]	5900	AA1	0.240	0.150	2	57	62.5	65	68	15	18

Source: Courtesy Link Belt (FMC)

Notes: Have dimensions certified for installation purposes. Minimum recommended sprockets: 10-tooth for 2.97- and 3.00-in. pitch chains; 9-tooth for 4.00- and 4.04-in. pitch chains; and 6-tooth for 6-in. pitch chains.

[a] Weight of cross rods not included.

[b] Minimum apron width. Maximum recommended width is 48 in.

[c] Not suitable for cross rods.

[d] Cross rods recommended every 4th pitch.

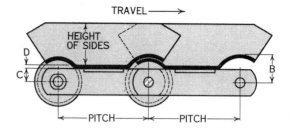

Table 5.3 Data for Aprons with Style AB Pans

Chain and Pan Pitch (in.)	Chain Number	Allowable Chain Pull per Strand (lb)	Factor C		Minimum Radius of Upturn (ft)	Weight per Foot (lb)[a]					
			Chain Not Lubricated	Chain Lubricated		For Width A (in.)				Each Extra 6-in. Width	A^b (in.)
						Height of Sides (in.)					
						3	4	5	6		
	SS 658	4650	0.156	0.110	5	46	50	53	56	5.3	18
6.00	SS 658	4650	0.156	0.110	5	54	58	62	66	7.3	18
	SS 658	4650	0.156	0.110	5	69	74	79	86	11.3	18
	SS 922	7200	0.161	0.113	5	65	68	71	74	7.5	24
9.00	SS 922	7200	0.161	0.113	7	82	86	90	95	10.5	24
	SS 933	9200	0.156	0.109	11	75	78	81	84	7.5	24
	SS 933	9200	0.156	0.109	15	92	96	101	106	10.5	24
	SS 1222	7200	0.161	0.112	5	58	61	63	66	7	24
	SS 1222	7200	0.161	0.112	7	74	77	81	86	10	24
12.00	SS 1233	9200	0.156	0.109	7	67	69	72	75	7	24
	SS 1233	9200	0.156	0.109	9	83	86	90	95	10.5	24
	SS 1233	9200	0.156	0.109	15	98	108	108	114	13.5	24

Source: Courtesy Link Belt (FMC)

Notes: Have dimensions certified for installation purposes. The minimum recommended sprocket is a 6-tooth one.
[a] Weight of cross rods included, spaced every 3rd pitch.
[b] Minimum apron width. Maximum recommended width is 60 in.

Carrying run pull = 1485 lb
Return run pull   = 867 lb
Total   = 2352 lb

For nonlubricated 6-in. pitch chain, use the factor 0.22 on a no-leak apron. On all other nonlubricated chain, use the 0.11 factor and on all lubricated chain, use 0.07.

The total pull on the head shaft sprockets is 2352 lb, but use 2400 lb for calculations. Therefore, the theoretical hp at the head shaft is

$$\frac{2400 \text{ lb} \times 50 \text{ fpm}}{33,000} = 3.7$$

Under normal conditions, a 7.5-hp motor would take

care of any friction developed by a roller-chain drive, speed reducer, and motor.

Many conveyor manufacturers have published formulas in their catalogs.

5.6.4 Size of Head Shaft

A diagram of the head shaft design is shown on figure 5.6. Assume 10 in. for the bearing-to-sprocket centerline distance, until the job is detailed. It may vary, but the results do not have to change due to a safety factor of 6, usually included.

The total pull equals 2400 lb, which can be divided by two strands of chain so that there is a working load or pull of only 1200 lb per strand. SS-922 chain has an

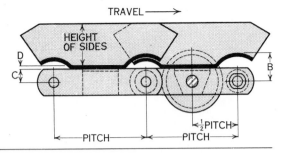

Table 5.4 Data for No-Leak Aprons with Style AB Pans

Chain and Pan Pitch (in.)	Chain Number	Allowable Chain Pull per Strand (lb)	Factor C Chain Not Lubricated	Factor C Chain Lubricated	Minimum Radius of Upturn (ft)	Weight per Foot (lb)[a] For Width A (in.) Height of Sides (in.) 3	4	5	6	Each Extra 6-in. Width	A[b] (in.)
	RS 953	5600	0.257	0.161	5	56	59	61	64	6	18
6.00	RS 953	5600	0.257	0.161	5	63	67	70	74	8	18
	RS 953	5600	0.257	0.161	5	77	83	88	94	12	18
	SS 942	9200	0.211	0.132	11	89	92	95	98	8	24
9.00	SS 942	9200	0.211	0.132	15	106	110	115	120	11.5	24
	SS 928	7200	0.240	0.150	5	71	74	78	81	7.5	24
	SS 928	7200	0.240	0.150	7	88	92	97	102	11	24
	SS 1242	9200	0.211	0.132	7	79	82	85	88	10	24
12.00	SS 1242	9200	0.211	0.132	9	95	99	103	108	13.5	24
	SS 1242	9200	0.211	0.132	15	111	116	121	127	16.5	24

Source: Courtesy Link Belt (FMC)

Notes: Have dimensions certified for installation purposes. Minimum recommended sprocket is a 6-tooth one.
[a] Weight of cross rods included, spaced every 2nd pitch.
[b] Minimum apron width. Maximum recommended width is 60 in.

Table 5.5 Maximum Lump Sizes and Speeds

Apron Width (in.)	Maximum Lump Size (in.)[a] Unsized	Sized	Recommended Conveyor Speed (fpm) Apron Pitch (in.) 3–6	9	12	Picking Table
18	6	4	100	100	—	75
24	8	5	100	95	—	70
30	12	6	90	90	85	65
36	14	7	90	85	80	60
42	16	8	80	75	70	55
48	19	10	80	70	65	50
54	21	11	—	—	60	45
60	24	12	—	—	60	40

Source: Courtesy Link Belt (FMC)
[a] Maximum size lumps for unsized material not to exceed 10% of total volume and at least 75% of total to be less than one-half the maximum lump size.

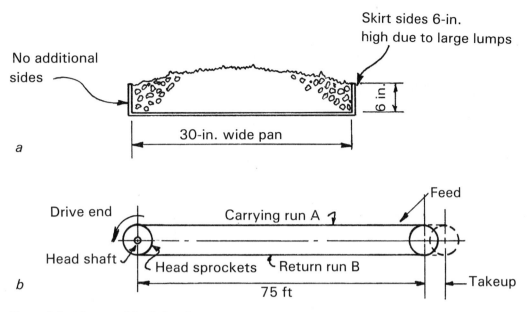

Skirt sides 6-in. high due to large lumps

No additional sides

6 in.

30-in. wide pan

a

Drive end

Carrying run A

Feed

Head shaft

Head sprockets

Return run B

Takeup

75 ft

b

Figure 5.5 Diagram of the designed conveyor.

allowable chain pull per strand of 7200 lb, which is far in excess of that needed for the work intended. This chain has a 9-in. pitch. A shorter-pitch chain, say 6 in., can be used, but a 6-in. pitch chain with pans to form an apron costs more because of more chain joints; and maintenance is somewhat higher because of more chain joints wearing. Also, in a 6-in. pitch chain, the chain pin usually is smaller in diameter. In apron conveyor work, a chain with a ³/₄-in. diameter pin is preferred. This has been determined by experience in the field and from tests made on all kinds of material to determine the wear qualities of various parts of chain, operating indoors or outdoors, and exposed to all kinds of weather.

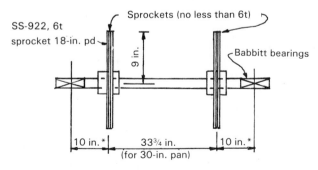

SS-922, 6t sprocket 18-in. pd

Sprockets (no less than 6t)

9 in.

Babbitt bearings

10 in.*

33³/₄ in. (for 30-in. pan)

10 in.*

Figure 5.6 Diagram of the head shaft of the designed conveyor.
*Assume 10 in. (a good figure to use until the job is detailed). It may vary, but the results do not have to change due to a safety factor of 6, usually included.

5.6.5 Drive Layout

SBS straight side-bar roller chain is used on slow-speed, heavy-duty conveyors (see figure 5.7). If no. 1113 malleable chain were used, the coefficient of friction would be 0.20, and the weight of a ³/₁₆-in. apron, with 10-gauge side plates, would be 65 lb/ft.

Weight of sulfur	=	75 lb/ft
Weight of apron	=	65 lb
Total weight on run A	=	140 lb/ft

Carrying run A:
  75 ft × 140 lb × 0.20 = 2100 lb
Return run B (empty):
  75 ft × 65 lb × 0.20 = 975 lb
Total = 3075 lb

The hp at the head shaft is

$$\frac{3075 \text{ lb} \times 50 \text{ ft}}{33,000} = 4.66$$

Thus, using no. 1113 chain requires 1 hp additional power. But if a 7.5-hp motor and drive, as shown, would still be satisfactory, the pull on the head shaft is over the 2400 lb, but still good. The cost might be 15% lower. In this case, the SS-992 chain was used because it requires less maintenance.

The total torsional pull on both sprockets is 2400 lb. The total bending pull on each sprocket is 2400/2 = 1200 lb. The torsional pull on two sprockets is not re-

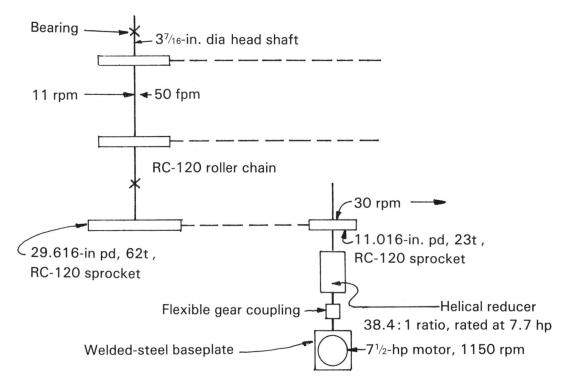

Figure 5.7 Schematic of an apron conveyor drive.

duced or split (for instance, one half on each sprocket) and the torsional moment is

2400 lb × 9 in. ($\frac{1}{2}$ sprocket dia) = 21,600 in.-lb

The bending pull is 2400 lb (the same amount as the torsional pull), but is split between the two wheels, so one-half of 2400 lb, or 1200 lb × 10 in. (distance from centerline of each wheel to centerline of each bearing) = 12,000 in.-lb. Referring to catalogs showing Shaft Diameter Calculations, it will be found that a 3$\frac{7}{16}$-in. diameter shaft is required, based on gradually applied loads (see figure 9.3).

The reason for not dividing the torsional pull between the two sprockets is that it is safer to take the total torsional pull on one wheel because of possible stretch in one strand of chain, or to allow for one chain wearing more quickly than the other. In other words, both chains should wear the same amount if care is taken when assembling the chain to the pans. This is not often the case during erection by contractors. The reason for dividing the bending pull so that each wheel takes one-half of the pull is that, normally, each wheel is close to a bearing and most of the load actually is taken up by the first one-third of the distance in the bearing bore (see figure 5.8).

For safety, it is best to use the center of the bearing as the point where the load is transmitted. In very large conveyor applications using babbitt bearings, the one-third distance point is considered.

If antifriction (roller or ball) bearings are used on the conveyor application just calculated, all loads are figured to centerline of bearings, and shaft sizes normally can be reduced, due to less friction. Generally, babbitt bearings are composed of cast-iron housings, solid or split. Babbitt bearings should not be used in temperatures over 160°F.

Bronze is used instead of babbitt where heavy bearing pressures, shock loads, high shaft speeds, and heat conditions up to 450°F occur.

For extremely heavy loads or heavy shocks, bearings are made with cast steel housings.

5.6.6 Inclined Conveyor

If this conveyor were inclined 15°, the capacity would not be affected in any way (25° is usually the maximum; if 30° is necessary, steel angle pushers usually are welded to each pan to keep the load from rolling backward).

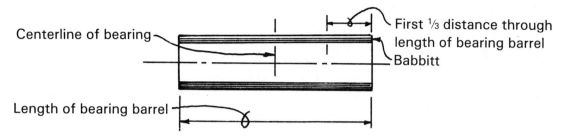

Figure 5.8 Bearing detail.

The hp required to drive a conveyor on a 15° incline can be determined by the following calculations:

Weight of sulfur per ft   =   75 lb
Assumed weight of apron per ft = 105 lb
Total   = 180 lb/ft

$$180 \times 75 = 13,500 \text{ lb}$$

$$105 \times 75 = 7875 \text{ lb}$$

Where P is the weight of the apron and sulfur, the total pull on carrying run A is

$$P \times [\sin 15° + (f \times \cos 15°)]$$

$$= 13,500 \text{ lb} \times [0.2588 + (0.11 \times 0.9659)]$$

$$= 4928 \text{ lb}$$

For the total pull on the return run B, where P_a is the weight of the apron only,

$$P_a \times [-\sin 15° + (f \times \cos 15°)]$$

$$= 7875 \text{ lb} \times [-0.2588 + (0.11 \times 0.9659)]$$

$$= 7875 \times -0.1525$$

$$= -1201 \text{ lb}$$

Total carrying run pull =   4928 lb
Total return run pull   = −1201 lb
Total   =   3727 lb

Therefore, theoretical hp at the head shaft is

$$\frac{3727 \times 50 \text{ ft}}{33,000} = 5.65$$

A 10-hp motor would be required to take care of any friction developed by a roller-chain drive, speed reducer, and motor.

The total torsional pull is 3727 lb and the total bending pull is 3727/2 = 1864 lb. The torsional moment is 1864 lb × 9 in. = 16,776 in.-lb, and the bending mo-

ment is 1864 lb × 10 in. = 18,640 in.-lb. A 3 7/16-in. diameter shaft is used (see figure 9.3).

The diagrams in tables 5.1 through 5.4 show the arrangement of corrugated pans that tend to hold any material in place, either on the horizontal or incline. On these aprons, 24-in. and wider, the chain pin becomes a through rod on every third pitch to hold the apron rigidly together and not let pans get out of line.

5.7 LEAKPROOF CONVEYORS

Although these conveyors were originally designed for handling dry, fine, foundry sand, they are suitable and economical for handling any other materials, such as coal, crushed stone, sand, chemicals, ferro-alloys, and coke. This apron conveyor is more expensive, however, than those equipped with a no. 1 (Style AC) apron. Because of the close-fitting contacts of the overlapping surfaces, leakage is held to the absolute minimum. The overlapping surfaces are so close they practically scrape each other, and prevent fine material from sifting through the joint. Pans are accurately formed to ensure close-fitting overlapping joints and are provided with staggered offset side plates welded integrally with the ends of the pans. This construction effectively prevents leakage at the pan ends, which may occur with normal construction. Two strands of conveyor chain are attached underneath and close to the ends of the pans, where they are protected from grit and abrasive materials that may spill over the conveyor sides. The steel strap, bushed roller chains have steel driving collars, hardened bushings and pins at the chain joints to minimize wear and friction, and this gives long and dependable service.

The weight of the conveyor and its load is supported on the conveyor rails by single, flanged rollers, assembled every second pitch on cross rods extending through the links of both conveyor chains at mid-pitch joints.

5.8 HANDLING HEAVY MATERIAL

When handling heavy material such as ferro-alloys, with many large lumps, very abrasive material, and sharp edges, weighing at times up to 200 lb/ft^3, special attention must be given to the size of the opening in the bottom of the hopper. Additional reinforcement must be added to the bottoms of pans to absorb the shock of dumping such material onto the apron (if the hopper or bin is empty when dumping begins). At times, the ferro-alloy may be fairly hot from the furnaces and this, too, must be taken into account.

5.9 APRON CONVEYOR WITH HORIZONTAL AND INCLINED SECTIONS

A 30-in. wide overlapping apron for handling ferro-alloys with a large percentage of lumps and very little fines was installed with pans ½ in. thick, with ⅞ in. dia chain pins, and SS-933 chain. The alloy material was fairly hot but not glowing. The material discharged directly onto the apron, which had to take the shock, and after many years it still worked well. Figures 5.9 and 5.10 show the design of this conveyor. On the bottom of each pan is a continuously welded tee rail that slides on the ¾ in. wear plates attached to the top of the 6-in. I beams, spaced diagonally as shown, so as to distribute the wear over the head of the tee rails and wear plate.

There is a ¼ in. clearance between the chain rollers and the top of the track. When a load of alloys is dumped, the shock can be absorbed by the tee rail attached to each pan and the ¾ in. wear plate, thus saving the chain joints from shock. Eventually, the hardened head of the tee rail attached to the pan wears and so will the ¾ in. wear plates, which are removable and thus can be replaced.

Because the alloys are fairly hot at this point, an allowance for possible additional expansion was provided by bending down the return angle track. This apron was not of the no-leak type; it was not necessary because of so many large lumps and little fines. A clearance of 3/16 in. should be provided when handling hot materials. The

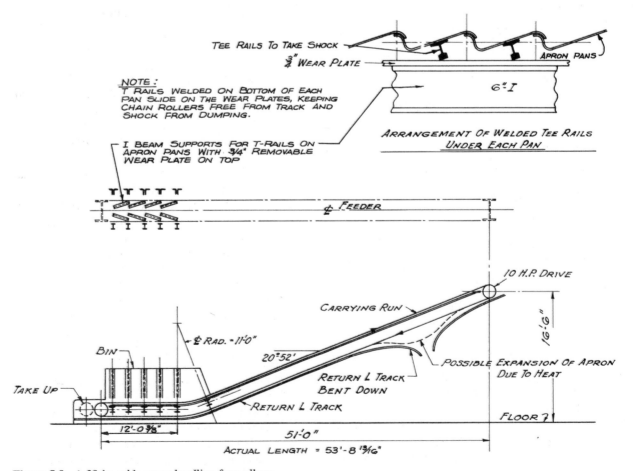

Figure 5.9 A 30-in. wide apron handling ferro-alloys.

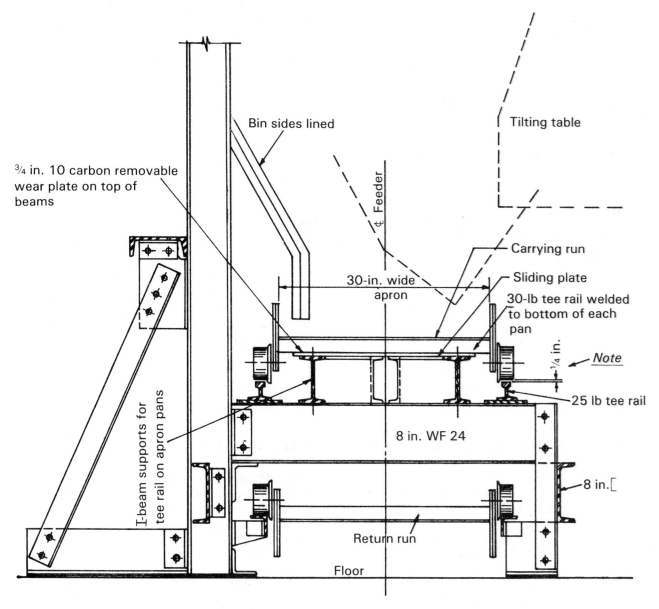

Figure 5.10 Cross-section of an apron at the loading bin. *Data:* pans, ½-in. plate; chain pins, ⅞-in. diameter; chain, SS-933; material, lumps 10 in. and smaller with little fines; weight, 250 lb/ft³ (30% voids); hot but not glowing.

small sliding friction has been neglected in the following computations.

5.10 DESIGN OF APRON CONVEYOR WITH HORIZONTAL AND INCLINED SECTIONS

5.10.1 Data

Apron conveyor: 30 in. wide, steel (see figure 5.11), 53 ft 8 ¹³⁄₁₆ in. centers, 12 ft ⅜ in. horizontal, then inclined as shown on figure 5.12.

Material: ferro-alloys, 10 in. and smaller, very abrasive, weighing 250 lb/ft³ with 60% voids.
Speed: 50 fpm.
Capacity: 185 tph.

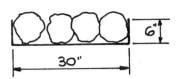

Figure 5.11 Loading sketch.

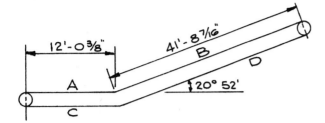

Figure 5.12 Diagram of the designed conveyor.

5.10.2 Capacity

2.5 ft × 0.5 ft × 1 ft (length of apron) = 1.25 ft³/ft

With 60% voids,

$$1.25 \times 0.40 = 0.50 \text{ ft}^3/\text{ft}$$

$$0.50 \times 50 \text{ fpm} \times 250 \text{ lb} = 6250 \text{ lb/min}$$

$$= 375,000 \text{ lb/hr}$$

$$= 187.5 \text{ tph}$$

5.10.3 Loads

The weight of the apron with a double strand of SS-933 chain and ½-in. steel pans that are 6 in. high with ⅜-in. steel sides is 110 lb/ft.

Where $f = 0.25$ (for steel pin on bored roller, greased), the rolling friction equals

$$f \times \frac{\text{dia of pin or bushing}}{\text{dia of roller}}$$

$$= 0.25 \times \frac{1.25}{4.00}$$

$$= 0.0781$$

say, 0.10 to allow for dust or material on the track.

The sliding friction of the T-rails on the wear plates can be neglected.

5.10.4 Motor Selection

Carrying run A

$$12 \times \frac{250}{2} \text{ (ferro-alloys)} = 1500 \text{ lb}$$

$$12 \times 110 \text{ (apron)} \quad = \underline{1320 \text{ lb}}$$

Total \qquad\qquad\qquad $= 2820 \text{ lb} \times 0.10 = 282 \text{ lb}$

Carrying run B

$$42 \text{ ft} \times 125 \text{ lb} = 5250$$
$$42 \text{ ft} \times 110 \text{ lb} = \underline{4620}$$
Total \qquad\qquad $= 9870$

$$9870 \text{ lb} \times [\sin 20°52' + (0.10 \cos 20°52')]$$

$$= 9870 \times (0.356 + 0.093)$$

$$= 4432 \text{ lb}$$

Return run C

$$12 \text{ ft} \times 110 \text{ lb} \times 0.10 = 132 \text{ lb}$$

Return run D

$$42 \text{ ft} \times 110 \text{ lb} = 4620 \text{ lb}$$

$$4620(-0.356 + 0.093) = -1215$$

Total pull on the shaft = 282 lb + 4432 lb + 132 lb − 1215 lb = 3631 lb. Adding 20% to that figure for curving upgrade gives 4357 lb, say 4400 lb. Therefore, the theoretical hp is

$$\frac{4400 \text{ lb} \times 50 \text{ fpm}}{33,000} = 6.67$$

Use a 10-hp motor.

5.10.5 Head Shaft

Use a 3 ¹⁵⁄₁₆-in. shaft to allow for slight overload on a gradually applied load.

The total torsional moment = 4400 × 9 in. = 39,600 in.-lb and the total bending moment = 4400/2 × 9 in. = 19,800 in.-lb.

5.10.6 Layout of Drive

The drive layout is shown on figure 5.13.

5.11 HINGED-STEEL BELT CONVEYORS

Mayfran International, Inc. makes a Piano-Hinged-Type Conveyor—die-formed hinged-steel belting, developed for handling scrap metal, stampings, forgings, and wet or dry chips from production machines (see figure 5.14). This apron is made from 4 in. to 96 in. wide, and the double strand of chain is available in 1¼-, 2½-, 4-, 6-, and 9-in. pitch. Hinged belts can be articulated through tight convex or concave bends and, with optional pusher cleats, can convey materials up steep inclines (15°, 30°, 45°, 60°, 75°, or 90° are standard).

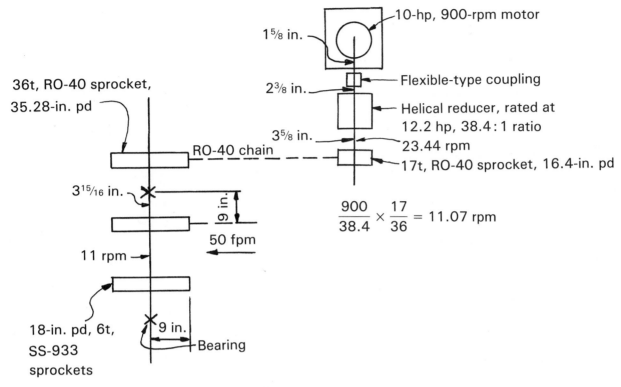

Figure 5.13 Apron conveyor drive diagram.

5.12 FLAT-PLATE CONVEYORS

Stearns Airport Equipment makes a flat steel conveyor for handling airport baggage, as shown on figure 5.15. The standard unit is 35 in. wide overall, and the over-

Belts can be perforated for drainage, or made with raised dimples to prevent surface adhesion of flat, oily parts. Belts can be reinforced to take heavy loads and possible shocks. Details of these conveyors are given in the manufacturers' catalogs.

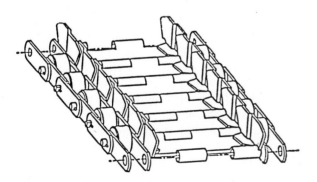

Figure 5.14 Mayfran's hinged-steel belt conveyor.

size unit is 41 in. wide overall. The side and top trim are stainless steel. Because it is a package-handling conveyor, it is described further in paragraph 17.13.

5.13 OPEN-TOP DEEP-BUCKET CARRIER

A typical design is shown on figure 5.16. This is a heavily built unit and is intended to carry material weighing close to 100 lb/ft^3 up a steep incline (up to 45°), in a limited space. The chain consists of 2½ in. × ⅜-in. wide steel flats. The roller is single flange, 4 in. diameter, Alemite-lubricated through 1-in. diameter pins and 1¼-in. OD steel bushings. Deep buckets are made not less than ¼ in. thick and are continuously welded on the inside.

On steep inclines, say 40° slopes, the return run plays an important part in figuring the horsepower, because of the weight of the empty buckets or pans with the two strands of chain, rolling down hill on a tee-rail track. The force exerted will help the carrying run, and the horsepower necessary to drive the head shaft will be affected more, compared to a conveyor of this type if the angle is small. This does not decrease the bending or torsional pulls on the head shaft.

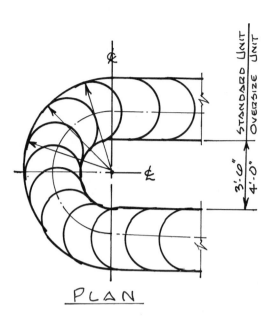

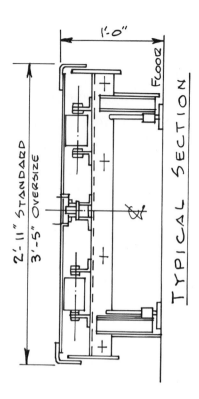

Figure 5.15   Stearns baggage conveyor.

In quarries, the operator must maintain complete control of an apron feeder to deliver rock to the crusher at uniform volume so that the balance of the equipment is not overloaded. All apron feeders must always be stopped before the hopper is completely empty to ensure a bed of material on the apron to cushion the impact of the next load. It is much better to have the apron at rest when dumping into a hopper so as to allow the large pieces to find their place and settle; otherwise, large pieces could get wedged together (if apron is moving), causing possible jamming against the feeder skirt sides. If this happens, there is much downtime until the rock is released by steel hand poles or other mechanical means, such as electrically operated tongs.

5.14 COOLING AND MOLD CONVEYORS

Carbide comes out of the furnace at about 2200°F when it is poured into cast-steel molds that must move after being filled. Figure 5.17 shows details of these pans. The pigs should be fed to the crusher at a temperature not over 1000–1200°F. To avoid gumming up the crusher, it should not be semiplastic when arriving at the crusher. In this climate (Buffalo, NY) it should travel

outdoors for at least one hour before being fed to the crusher. Since the outside cools first, the pigs retain their shape well.

Metals from furnace operations are cast in molds that generally are part of apron conveyors and move over a sufficient distance (time) to cool to a temperature at which they can be processed further. The molds may be attached securely to the chains and made portable.

The size or length of the device that takes away the pigs from the furnace must be such as to complete the tapping in about an hour. The conveyors must be kept moving to avoid damage because of the hot pigs. An expansion loop should be provided on the return run of the chain to avoid trouble at the takeups (see figure 5.9).

If the conveyors stand still with pigs in them, trouble may develop at the takeup, unless an expansion loop is provided on the return run, free of track or structure. The elongation in the conveyor chain causes the free-moving counterweight takeups to bind.

It is best to provide long runs on the conveyors to avoid spillage at junctions. Where one conveyor transfers its pigs to another at right angles, a guide must be provided to prevent the pigs from piling up.

For carbide, the rolling friction of the conveyor can be figured as 0.10, and the power used should allow 50% for surges. The live load on the conveyors should be figured at 125 lb/lin ft.

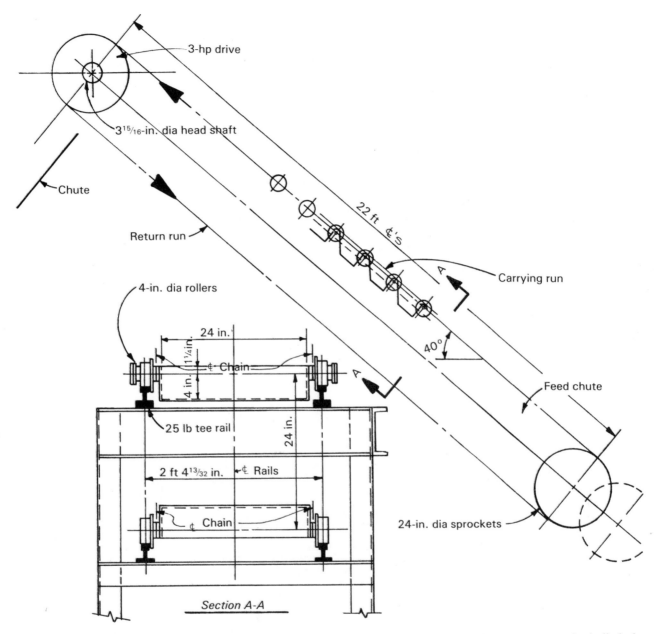

Figure 5.16 An open-top carrier for handling limestone, gypsum, and the like, 3 in. and under, up a steep angle, in limited space. Speed: 50 fpm.

Figures 5.18 and 5.19 show an apron conveyor used for cooling aluminum ingots. Figure 5.20 shows a conveyor carrying coil springs through an oil bath. The same type can be used for cleaning and treating such materials as coils of wire, cylinders, and so on. For another method of cleaning and coating such materials, see figure 12.8.

Various other devices are used to convey hot material, such as air-cooled conveyors and vibrating plates that raise the material as it moves, and allow air to cir-

culate around it. Figure 5.21 shows a 42-in. wide shuttle cast-steel apron conveyor carrying carbide pigs.

5.15 COAL-HANDLING SYSTEM

A coal-handling system, consisting of a 24-in. wide apron feeder with Style no. 1 pans, operating on an incline of 25° at 10 fpm, is shown on figure 5.22. It receives either slack (−¾ in. and under) or less expensive

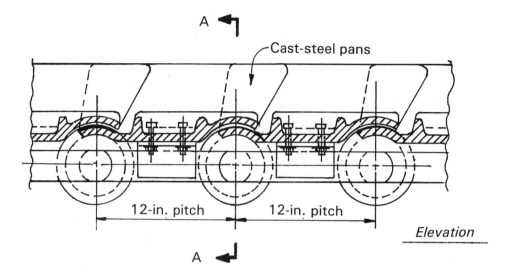

Cast-steel pans

12-in. pitch   12-in. pitch

Elevation

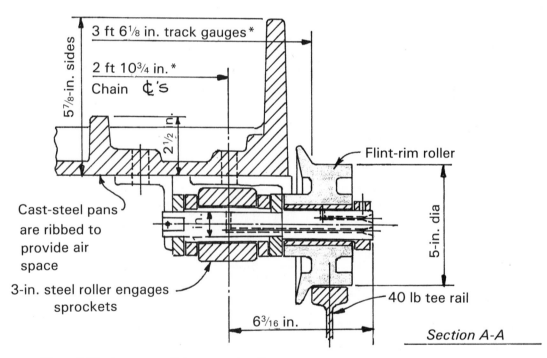

3 ft 6⅛ in. track gauges *

2 ft 10¾ in. *
Chain ₵'s

5⅞-in. sides

2½ in.

Flint-rim roller

Cast-steel pans
are ribbed to
provide air
space

1 in.

5-in. dia

3-in. steel roller engages
sprockets

40 lb tee rail

6³⁄₁₆ in.

Section A-A

Figure 5.17 An apron cooling conveyor handling carbide cakes. Speed: 12-24 fpm (through PIV units).
*Dimensions are given for a 36-in. conveyor. For a 42-in. conveyor, they are 4 ft ⅛ in. and 3 ft 4¾ in.

8-in.-and-under coal from the track hopper. The capacity is 30 tons per hour. A crusher receives the discharge of the apron feeder to break up the larger lumps. Figure 5.22 shows, in addition to the general arrangement of this installation, the bypass chute to transfer −¾ in. coal directly to a chute feeding the bucket elevator.

Considering the amount of coal handled yearly, plus

installation charges, and the small amount of money spent for repairs over a long period, the cost per ton handled (not including electric power, overhead, or fixed charges) will be found to be very reasonable. All it requires is the occasional placing of heavy oil on the chain joints. Reused engine oil is satisfactory.

This apron feeder operates on an angle of 25°, which

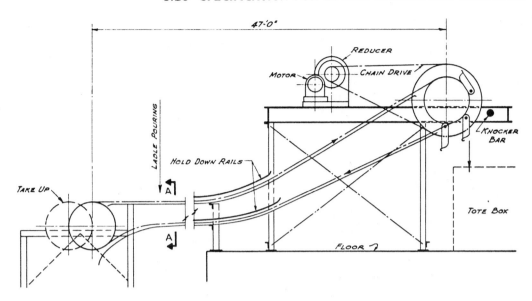

Figure 5.18 A mold conveyor handling aluminum alloy ingots from a furnace at the rate of 450 every 15 min. The molds are mounted on a double strand of steel strap chain. The operating speed is 1.88 fpm and a 12-min cooling period is required.

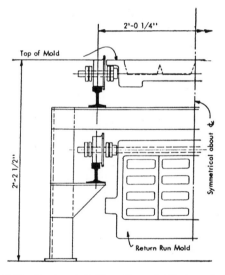

Figure 5.19 Section A-A (from figure 5.18) through the mold conveyor.

is maximum without steel angles or cleats being welded to each pan (to help up incline and prevent it from falling backward). Thirty degrees is maximum in most cases. The steel frame can be made of the open-truss construction or as shown in cross-section. The 18-in. sides were used because of the 8-in. coal.

5.16 SPECIFICATION FOR 24-INCH STEEL-APRON CONVEYOR

Description: One 24-in. wide no-leak-type apron conveyor complete.

Material: 6 in. × D calcium carbide weighing 70 lb/ft³.

Capacity: 60 tph.

Centers: approximately 23 ft.

Apron feeder: to be made of 24-in. wide overlapping double-beaded pans of 3/8-in. steel plate having 3-in. high, 1/4-in. thick, welded overlapping sides, secured to A-3 attachments to two equal strands of 3/8 in. × 2 in. × 6 in. pitch SS-953 steel bar link chain, fitted at articulations with 3/4-in. diameter heat-treated pins, case-hardened steel bushings and 1 3/4-in. steel driving collars.

This apron is to be supported every 4 ft with 3/4-in. diameter through rods and at every other pitch by 4-in. diameter single-flange hard-iron outboard supporting rollers turning on case-hardened steel sleeves on the ends of 3/4-in. diameter steel pins.

Alemite or equivalent lubricating fittings shall be provided at the ends for lubrication.

Terminal driving machinery: 2 15/16-in. turned and polished shaft with Alemite or approved equivalent lubricated angle pillow blocks, split safety collars.

Head and boot sprockets: 8-tooth 15.67-in. pd, case-hardened steel head sprockets with flame-hardened teeth.

Drive chain: RO-40 bushed roller chain with plate center sprocket wheels.

Foot machinery: 2 7/16-in. plain foot shaft, turned and polished with babbitted-type screw takeups for 12 in. adjustment and split safety collar.

The conveyor is to be complete with structural steel supports, including head and foot shaft supports and 20-

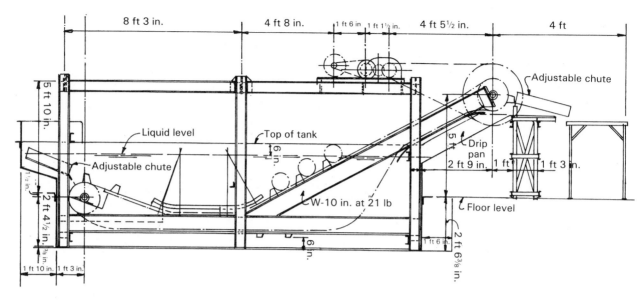

Figure 5.20 A 30-in. wide apron conveyor handling material for cooling or cleaning. Speed: 25 fpm.

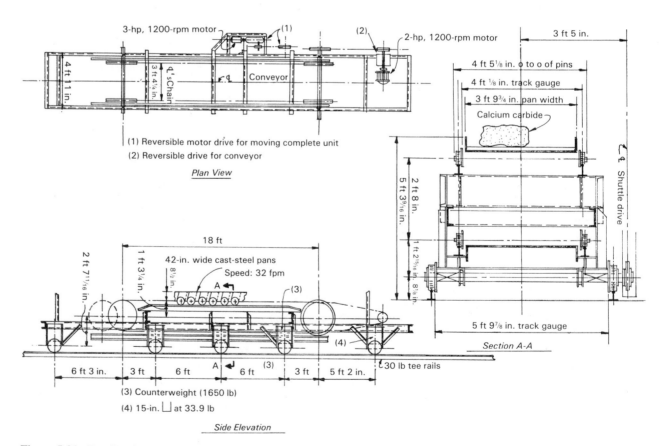

Figure 5.21 Details of a 42-in. wide shuttle cast-steel apron conveyor.

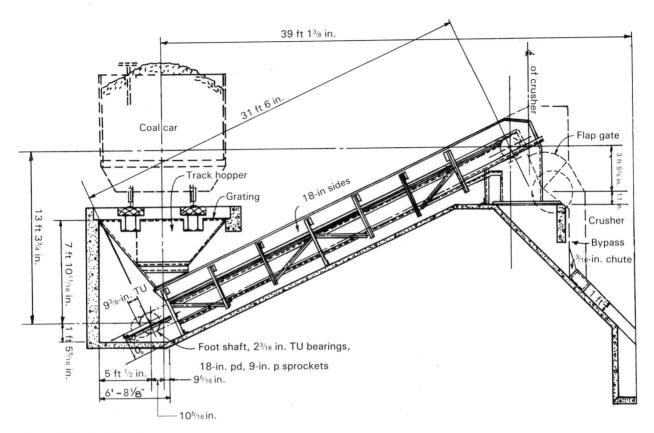

Figure 5.22   Coal-handling system.

lb tee-rail tracking for the apron on carrying and return runs.

Supports to floor to be suitable for inclined conveyor 2 ft 0 in. from floor to center of tail sprockets and 9 ft 10 in. from floor to center of head sprockets.

Drive systems: the manufacturer shall furnish the drive system, including a 5 hp, 720-rpm motor; herringbone-gear-type speed reducer; and enclosed positive variable-speed reducer, ratio 3 : 1, including remote electrical control and a baseplate to mount reducer and motor. The length of the RO-40 drive chain shall be suitable for a center distance of 10 ft 0 in. between the slow-speed shaft of the reducer and the conveyor head shaft. Price per foot of the drive chain shall be submitted for addition or deduction as required. Location of the above equipment shall be as shown on Purchaser's drawing. The conveyor speed shall be capable of regulation so that capacities may be varied from 20 tph to 60 tph. Power: 440 V, 60 Hz, 3 ph.

In general, the manufacturer shall design, furnish, fabricate, and deliver all equipment and accessories, specified or implied herein, necessary to form a complete, workable unit. All equipment shall be thoroughly cleaned and painted with a high grade of machinery paint in such a manner that the paint will adhere to the surface and be weather resistant. All painting shall be done in a neat and professional manner. The equipment shall not be loaded for shipment until the paint is thoroughly dry.

All polished parts, shafts, gearings, bearings, and so on, shall be coated with rust-preventive compound and crated, it necessary, to prevent damage in shipment, and all equipment shall be thoroughly protected from the weather.

Services of an engineer, if required at time of installation, shall be furnished at rate stated in the proposal.

Price per foot shall be submitted for conveyor as required for lengthening or shortening conveyor in final design.

The successful bidder shall furnish three sets of prints for approval of general design of the equipment, and later _____ complete sets of certified as-built prints to _____, the Consulting Engineer.

SECTION 6

Flight and Drag Conveyors

6.1 GENERAL

A scraper flight conveyor consists of a series of metal, wood, or plastic flights attached at uniform intervals to one or two strands of endless chain which moves the flights along in a trough, the chain or chains passing around sprockets at the ends of the trough. The flights push or slide the material along; therefore, abrasive materials should not be handled by this type of conveyor, unless the speed of the conveyor is very low. Material can be discharged at any point along the trough through gates or at the end of the trough. See figure 6.1 for typical flight conveyors. Operating speed should not exceed 100 fpm.

In a drag conveyor, the chain itself acts as a scraper.

6.2 USES

There are many variations of flight conveyors handling bulk materials such as coal and similar materials weighing about 50 lb/ft³, as well as dry sewage sludge, dry wood chips, and other nonabrasive, free-flowing and fairly dry materials. Every cannery uses a form of scraper flight conveyor for moving to refuse bin or boiler such materials as string beans, ears of corn before husking, and mush waste. Sugar-producing plants use a double strand of roller chain with steel flights to remove ''bagasse'' (refuse from sugar cane or grapes) from mill to furnaces.

Material such as coal is pushed along the trough by the flight, which actually bottoms itself in the trough (scrapes in the bottom of trough). The flights usually are made of malleable iron, cast with wear shoes, to prevent breakage of material and keep maintenance down. Many coal dealers reported that the steel trough would last between six and eight years, and that the chain usually operated for up to 10 years.

6.3 SINGLE-STRAND FLIGHT CONVEYORS

Single-strand flight conveyors with scraper flights were popular in retail anthracite coal outlets to distribute the coal to various bins. Those with shoe-suspended flights (figure 6.2) are similar, except the flight does not touch the trough and the wear is taken by the flight shoes sliding on an angle track. These provide inexpensive methods of handling certain materials if the dust problem is not a consideration. The single-strand flight conveyor with roller-suspended flights accomplishes the same result with less power. The flights do not scrape the trough but are attached to a through shaft, supported by single-flange rollers, which roll on their track, giving only rolling friction instead of sliding friction. This roller-suspended type handles free-flowing material with lumps not larger than 3 in. or 4 in. for the shoe-suspended type.

6.4 DOUBLE-STRAND FLIGHT CONVEYOR WITH ROLLER-SUSPENDED FLIGHTS

The double-strand flight conveyor with roller-suspended flights (figure 6.3) is used where the material contains

193

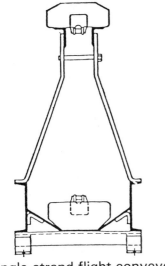

Single-strand flight conveyor
with scraper flights

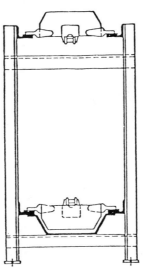

Single-strand flight conveyor
with shoe-suspended flights

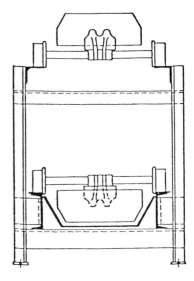

Single-strand flight conveyor
with roller-suspended flights

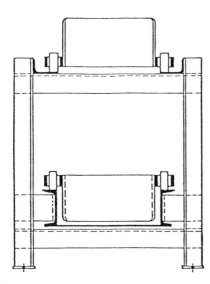

Double-strand flight conveyor
with roller-suspended flights

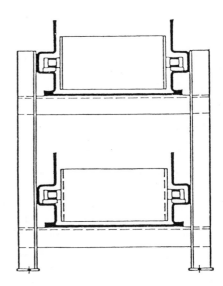

Double-strand flight conveyor
with sliding chain-suspended flights

Figure 6.1   Popular types of flight conveyors. (*Courtesy FMC*)

large lumps, or where the capacity might be, say, 75–125 tph. These units are ideal for distribution of coal in a bunker (bin) where the length of the bunker is up to 100 ft. Beyond that, it may be possible to use a belt conveyor with tripper. Most of these conveyors are installed to operate horizontally, but could start out horizontally and travel up an incline of about 30° maximum, if a radius of about 8 to 10 feet is provided between the horizontal and incline paths, and the number of rollers is held down going through the curve. The various manufacturers' catalogs provide engineering in-

formation. A 30-in.-wide flight can handle material with lumps up to six inches, and with lumps not to exceed 25% of capacity.

6.5 DRAG CONVEYORS

In drag conveyors, a special chain is used which acts as a scraper. The design of a typical drag conveyor carrying ashes is given in paragraph 6.8.

Figure 6.4 shows a typical drag conveyor handling

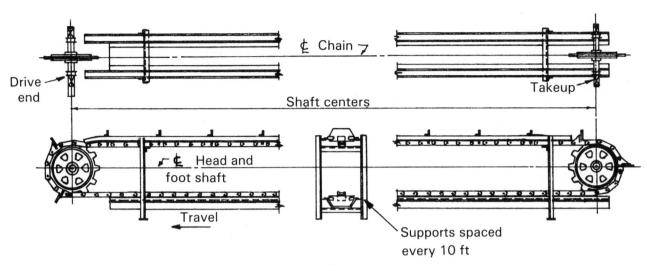

Figure 6.2 A single-strand flight conveyor with shoe-suspended flights.

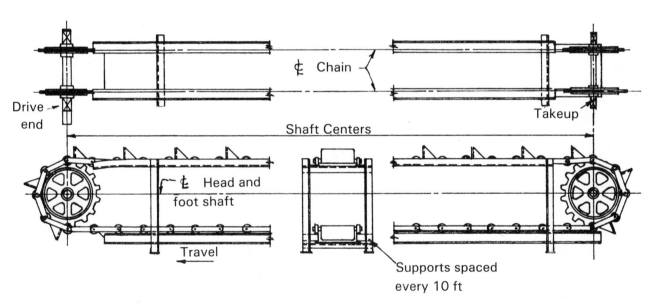

Figure 6.3 A double-strand flight conveyor with roller-suspended flights.

ashes, just below the power house floor level. Ashes can be pulled out of a boiler by hand onto a steel grating where any large lumps can be broken by hand shovel. The ashes falling through the grating to the carrying run of H-class chain, slide along in a hard white cast-iron trough (replaceable sections) set in a concrete pit. These ashes can be hot or glowing at times, but they cool rapidly or are quenched by water, and do not distort the trough or the chain, which normally can withstand temperatures up to 600°F or, if made of heat-treated metal, up to 1000°F.

Figure 6.5 shows an ashes conveyor in a tunnel be-

low the power house floor, where a man can walk along the conveyor, open a slide gate to allow ashes to flow onto the steel grating, and be conveyed by the top run of a conveyor chain operating in a hard white-iron trough. The concrete chute usually is installed at the front of the boiler, but ashes still must be pulled to an opening at the power house floor level and allowed to fall down the concrete chute. The return run of the conveyors travels on single-flange cast-iron rollers spaced every 10 feet. When lumps are broken up by hand on the steel grating and made small enough to go through grating, the ashes can be handled by a centrifugal dis-

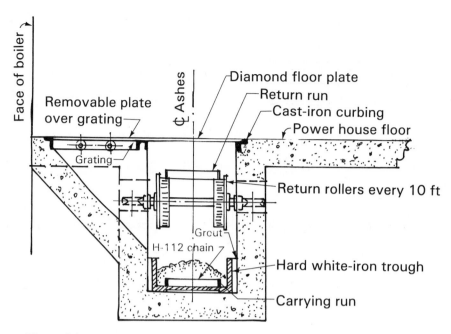

Figure 6.4 A bar scraper feeder carrying coal from the hopper to the elevator drag conveyor in a boiler plant.

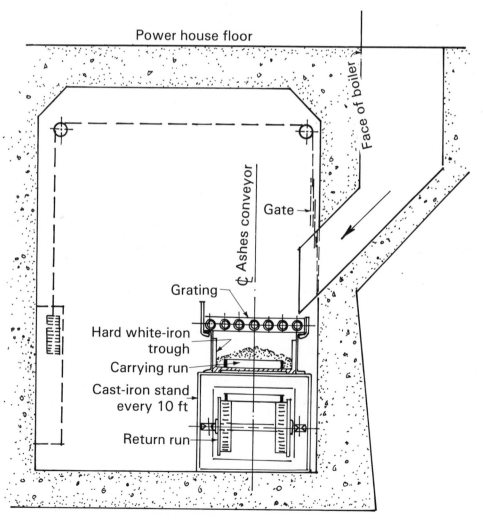

Figure 6.5 A drag conveyor handling ashes just below the floor in a boiler plant.

charge bucket elevator high enough to allow ashes to flow into a truck for disposal. Chute angle to truck should be at least 45°.

6.6 SELECTION OF A SCRAPER FLIGHT CONVEYOR

6.6.1 Data

Material: coal (slack bituminous, ¾ in. and smaller), weighing 50 lb/ft³.
Moisture: coal could be damp or partly frozen into small lumps, 3-in. maximum, but easily broken by movement along the trough or by the drop into the bin.
Capacity: 50 tph (2000 ft³/hr).
Centers of sprockets: 100 ft.

6.6.2 Size Selection of Flights

Because of the 3-in. lumps, assume the flights to be 15 in. long and 6 in. deep. The spacing of flights with wear shoes (shoe-suspended flights) could be assumed at 24 in. and conveyor operating at 100 fpm.

Refer to the portion of table 6.1 headed "Single-strand flight conveyor with shoe-suspended flights";

Table 6.1 Flight Conveyor Capacities

Flight					
Size (in.)		Spacing (in.)	Lump Size[a] (in.)		Capacity[b] (ft³/hr)
Length	Width		Maximum	Average	
Single-strand flight conveyor with shoe-suspended flights					
10	4	24	3	1½	932
12	5	24	3½	1¾	1386
12	5	24	3½	1¾	1386
15	6	24	4½	2½	2106
18	6	24	5	2¾	2541
Double-strand flight conveyor with roller-suspended flights					
16	8⅝	24	8	4	3186
20	10⅝	24	10	6	5243
24	10⅝	36	12	8	6093
30	11⅝	36	14	10	10860
36	12⅝	36	16	12	11561

Source: Courtesy FMC
[a] Maximum size lumps not to exceed 10% of total volume.
[b] Capacity based on conveyors with a maximum 30° incline, operating at 100 fpm with 75% of the area of trough cross-section filled.

with flights 15 in. long and 6 in. wide, a conveyor has a capacity of 2106 ft³/hr, operating at 100 fpm, with 75% of the cross-section of the trough filled. With flight conveyors, the same capacity can be had whether horizontal or inclined to 30° maximum. As the material settles and is pushed by each flight, it does not separate. The horsepower requirement is greater as the incline increases.

There are many catalogs dealing with coal handling. These give the drive and hp requirements. It is important to check the length of the conveyor, the speed of the conveyor, the material handled, and the slope of the conveyor used in preparing the tables.

6.7 DESIGN OF A DOUBLE-STRAND SCRAPER CONVEYOR (figures 6.6, 6.7, and 6.8)

6.7.1 General Data

Material handled: dry shakeout foundry sand.
Weight: 100 lb/ft³.
Center-to-center: 33 ft 4 in.

6.7.2 Considerations

Trough belt conveyors were considered in place of scraper conveyors, but the sand at this plant had to be dumped into the conveyor trough anywhere along the conveyor through the steel floor grating, without spillage anywhere.

A belt conveyor with continuous skirt sides could have been worked out, but it is always a problem to keep the rubber skirt-board liners close to the belt because of wear, and considerable labor would have been necessary to replace the rubber liners.

A belt conveyor must be supported by steel framing, and this usually requires a deeper pit and one wide enough to allow a man to clean up any spill. It is normally difficult to keep a belt-conveyor pit clean because, when the flasks are shaken out, small pieces of glowing hot iron may fall through the grating and injure a belt, unless it is a heat-resisting type. Also, scraper conveyors are very "forgiving" for hot, hard, and sharp materials, with the loss of a single flight not interfering with the continued operation of the conveyor.

When using a well designed scraper conveyor, any hot metal pieces falling onto the sand in the pit will turn over and over, getting cooler as they are pushed along to the discharge point.

When using table 6.2, the chain speed for very abrasive material should not exceed 8 fpm, and the

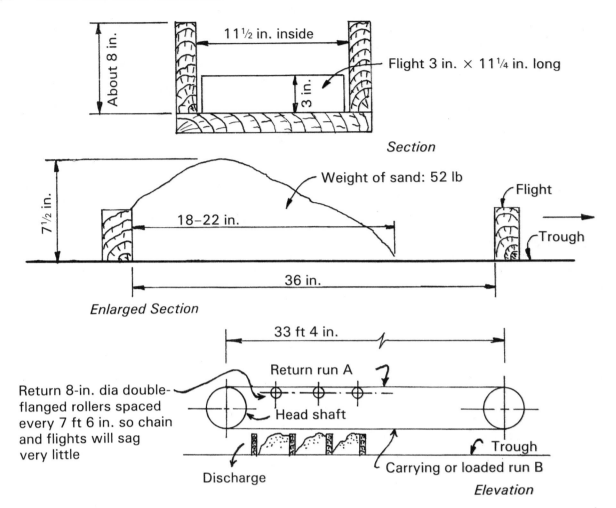

Figure 6.6 A double-strand scraper conveyor.

trough can be hard iron. For mildly abrasive material, the speed should not exceed 20 fpm, and the trough can be concrete. For nonabrasive material, the speed should not exceed 80 fpm, and the trough is generally made of steel. Capacity is based on a speed v of 10 fpm. For any other speed, capacity = $v/10$ × tabulated value.

SD-28 and 29 are for handling cold material; SD-19 and 21 should be used for handling hot material.

H drag chains (table 6.3) are used to convey sawdust, wood chips, and other bulk materials under moderate operating conditions. WD drag chains are used on heavy-duty drag conveyors. The chain links are welded and connected by large-diameter steel pins. Esco chains are of alloy steel and are popular for handling logs.

When using table 6.3, the chain speed for very abrasive material should be less than 10 fpm. For mildly abrasive material, the speed should not exceed 20 fpm; no material should exceed 100 fpm. Capacity is based on a speed v of 10 fpm. To obtain capacity at any other speed, multiply the tabulated value by $v/10$.

6.7.3 Capacity

An experiment was made using the dry shakeout sand. A wooden trough 11½ in. wide (inside) with wooden flights 11¼ in. wide by 3 in. high, spaced on 36-in. centers, was set up to determine how the sand would react if dumped into the trough on a conveyor traveling at about 30 fpm. It was found that the amount of sand that would settle in front of each flight was 52 lb (figure 6.6). The conveyor was level in one case and inclined 8 in. in 6 ft (about 11%) in another. It was found that the weight of sand was almost the same in both cases.

The capacity required was approximately 40 tph; therefore, if a flight or trough was made 30 in. wide, the result would be:

$$\frac{52 \ (\text{lb}) \times 30 \ (\text{in.})}{11\frac{1}{2} \ (\text{in.}) \ (\text{width of experimental trough})}$$

$$= 136 \ \text{lb sand}$$

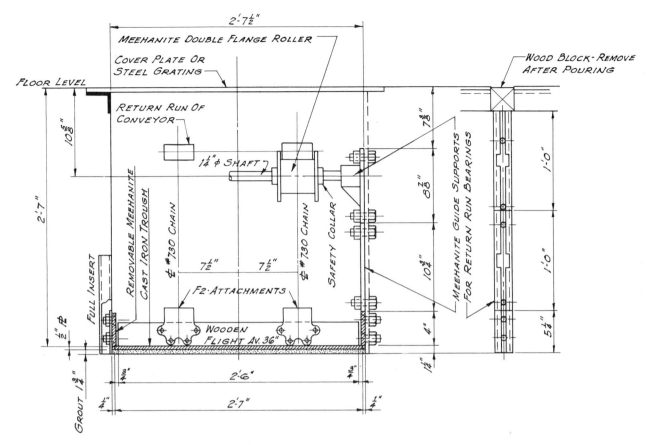

Figure 6.7 A section through the pit of a scraper conveyor.

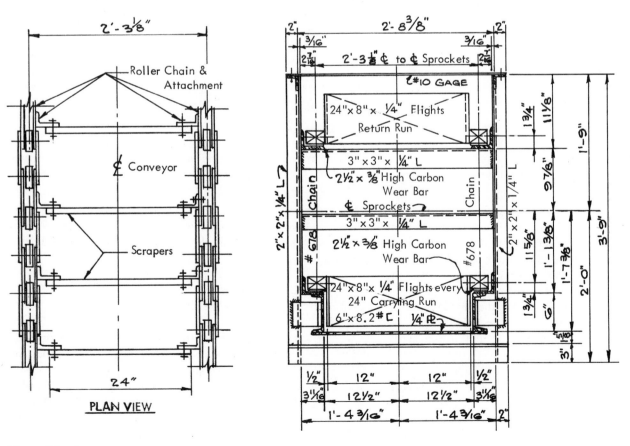

Figure 6.8 Scraper conveyors.

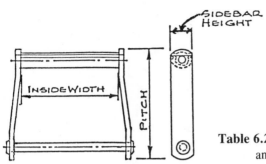

Table 6.2 Properties of SD and WD Drag Chains

Chain No.	Pitch (in.)	Avg Ult Strength (lb)	Allowable Chain Pull (lb)	Inside Width (in.)	Sidebar Height (in.)	Capacity (ft³/hr)
SD-19	6.000	100,000	16,700	3.63	2.00	124
SD-21	9.000	140,000	23,400	9.50	2.50	230
SD-27	9.000	115,000	15,200	4.25	2.50	124
SD-28	9.000	120,000	17,500	10.13	2.13	230
SD-29	9.000	120,000	17,500	6.13	2.13	186
WD-480	8.000	70,000	11,500	11.13	2.00	334
WDH-480	8.000	90,000	15,000	11.13	2.00	334

Note: Weight of chain includes attachments.

Table 6.3 Wide-Chain Drag Conveyors

Chain No.	Width of Material (in.)	Depth of Material (in.)	Capacity (ft³/hr)	Pitch (in.)	Ultimate Strength[a] (lb)	Allowable Chain Pull[b] (lb)
H-104	10½	4	124	6	28,000	4160
H-112	16	4	186	8	28,000	4160
H-116	19¾	4	230	8	28,000	4160
H-480	20	6	334	8	40,000	6670

Source: Courtesy L-B Co.

[a] For malleable iron chain; Promal chains are approximately 25% stronger.

[b] For malleable iron or Promal. Ratings are based on a service factor of 1.0. Apply service and correction factors from tables 1.5 and 1.6.

This would mean it could dump 136 lb of sand between flights spaced 36 in. apart, although the sand would not spread out beyond about 22 in., as shown on figure 6.6. For figuring capacity with the flights every 36 in. and sand spreading out to 22 in., there would be some void space beyond the sand to the next flight but, on average, there would be 136 lb divided by 3 ft (spacing of flights), or 45⅓ lb/lin ft of conveyor.

Then, operating the conveyor at 30 fpm would give 45⅓ lb × 30 fpm = 1360 lb/min or, 1360 × 60 = 81,600 lb say, 82,000 lb, or 41 tph. This satisfied the requirement of 40 tph with some leeway of void space between flighting, which could be filled in the event more capacity is required in the future. A belt conveyor would have required a deep and wide pit, a belt to withstand heat, a metal remover, and continuous cleaning.

6.7.4 Horsepower (see figure 6.6)

It was assumed that two strands of no. 730 pintle chain with attachments and flights weigh 20 lb/ft. The sand would be dragged along a trough, fairly smooth but still creating much friction. The chains would be attached to the flights, and the flights also would be dragged along, creating friction. The conservative method of figuring the total pull on the head shaft follows.

Return run B
chain and flights = 33.33 ft × 20 lb × 0.5
= 333 lb
sand = 45 lb/ft × 33.33 ft × 0.5 = 749 lb
(assumed friction for hard scraping)

Carrying run A
chain and flights only, returning on loose rollers
mounted on 1¼ in. shafts =
33.33 ft × 20 lb × 0.4 = 267 lb
(assumed friction for possible stalling of rollers due
to sand getting into bore or piling up in trough)

Return run pull = 1082 lb
Carrying run pull = 267 lb
Total = 1349 lb pull required.

Use 2000 lb to be on the safe side.
Therefore, the theoretical hp at the head shaft is

$$\frac{2000 \text{ lb} \times 30 \text{ fpm}}{33,000} = 2 \text{ hp (aprx)}$$

A 3-hp motor could be used on this conveyor safely because of the safety factors used throughout, and the incorporation of overload shear-pins.

6.8 DESIGN OF A DRAG CONVEYOR

6.8.1 Data

Material: ashes weighing 40 lb/ft³.
Length of run: 75-ft centers.
Capacity required: 3.7 tph.

6.8.2 Size of Trough (figure 6.9)

Try a 16-in. width (1.33 ft) and 3-in. depth (0.25 ft).
Then the cross-section is

$$1.33 \times 0.25 = 0.3325 \text{ ft}^2$$

For capacity, 3.7 tph equals 7400 lb/hr. Therefore,

$$\frac{7400 \text{ lb/hr}}{40 \text{ lb/ft}^3} = 185 \text{ ft}^3/\text{hr}$$

and

$$\frac{185}{0.3325 \times 60} = 9.3 \text{ fpm}$$

so use 10 fpm. Table 6.3 gives the capacity as 186 ft³/hr for 10 fpm.

The space occupied by the chain should be deducted from the 0.3325 ft³/ft length figured. The loss amounts to about 13.5 ft³/hr. Therefore, the actual capacity would be

$$185 \times \frac{10}{9.3} = 198.9 - 13.5 = 185.4 \text{ ft}^3/\text{hr}$$

So the size selected is satisfactory. The conveyor can be operated up to 20 fpm, or even faster, but excessive wear will result.

Type H chain is generally used for drag conveyors. H-112 chain is 12½ in. wide (o to o), and is the size commonly used for handling ashes. Type H chain has a space between the sidebars of each link that will get filled up with fine ashes, but this is not generally considered as part of the capacity. The chain is made of malleable iron (type H), combination malleable iron or steel (type HC), or welded steel (type WS) (see figure 6.10 and tables 6.2 and 6.3).

6.8.3 Head Shaft

$$0.3325 \times 40 = 13.3 \text{ lb, say } 14 \text{ lb}$$

The approximate weight of the chain is 11 lb/ft. The sliding factor for ashes, by experiment, is 0.53 and the

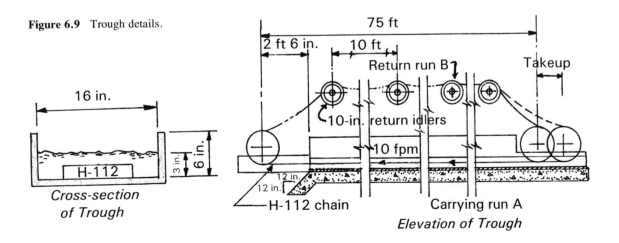

Figure 6.9 Trough details.

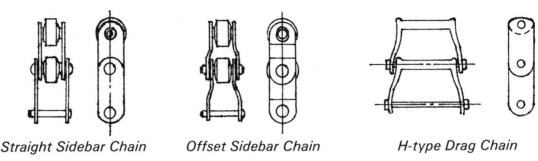

Straight Sidebar Chain   *Offset Sidebar Chain*   *H-type Drag Chain*

Figure 6.10 Details of steel chain for double-strand flight conveyor with roller-suspended flights (SS-1227 or SS-1827).

chain sliding factor on the carrying run is usually taken as $\frac{1}{3}$.

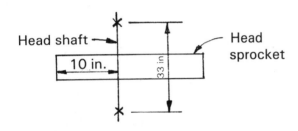

Carrying run A

75 ft × 11 lb × 0.333 = 275 lb chain pull

75 ft × 14 lb × 0.53 = <u>557 lb ashes pull</u>

Total pull   = 832 lb

Return run B

75 ft × 11 lb × 0.10 = 83 lb

Total bending pull = 832 lb + 83 lb = 915 lb.

The total bending moment is

$$\frac{915 \times 33}{4} = 7549 \text{ in.-lb}$$

and the total torsional moment is

$$915 \times 10 = 9150 \text{ in.-lb}$$

Figure 9.3, for gradually applied loads, shows that the shaft required has a diameter of $2^{15}/_{16}$ in.

6.8.4 Horsepower

$$\text{Theoretical hp} = 915 \times \frac{10}{33,000} = 0.3 \text{ hp}$$

Because ashes are likely to be extremely hard and fused together, the theoretical hp generally is multiplied by 3. Allowing for friction losses in the chain drive, reducer, and other moving parts that may not get oiled or greased regularly, a 2-hp motor should be used.

6.8.5 Design of Drive (figure 6.11)

6.8.6 Bill of Material*

1 Head shaft, $2^{15}/_{16}''$ dia × about 3′ 6″ long; ks, with keys 4½″ long

2 Series 2-1300 babbitted bearings (pillow blocks), $2^{15}/_{16}''$

2 Solid malleable iron safety collars, $2^{15}/_{16}''$

1 Flint-rim sprocket, 20.90″ pd, 8t, plate-center, standard hubs, bore, ks and ss.

1 Flint-rim sprocket, 29.93″ pd, 36t, plate-center, RO-882 with standard hubs, bore, ks and ss.

1 Drive chain, RO-882, pin and cotter type, 15 lin ft

1 Flint-rim sprocket, 10.90″ pd, 13t, plate-center, standard hubs, bore, ks and ss; for reducer low-speed shaft

1 Link-Belt FT-1 helical reducer, ratio 195:1, rated at 3.5 hp

1 SG-210 gear-type coupling

1 Motor, 2 hp, 1200 rpm, tefc, type K, frame 213

1 Baseplate for reducer and motor, welded steel

1 Plain foot shaft, $2^{3}/_{16}''$ dia × about 3′ long

2 Babbitt takeups, $2^{3}/_{16}'' × 12''$

2 Solid malleable-iron safety collars, $2^{3}/_{16}''$

1 Flint-rim sprocket, 20.90″ pd, 8t, plate-center, standard hubs (to rebore to $2^{15}/_{16}''$); bore $2^{3}/_{16}''$, to run loose on shaft (to prevent fine abrasive ashes from clogging any movement of the sprocket; it is better to run the sprocket loose than to keyseat it or even setscrew it, which would turn the shaft in the babbitted takeups in a location where one very seldom makes an inspection)

1 H-112 Promal chain (to resist abrasion), 157′ 4″

_____

*Numbers are from the Link-Belt Catalog.

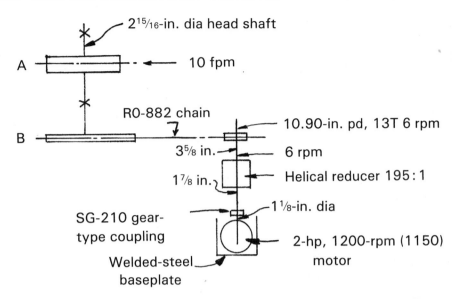

Figure 6.11 Drive diagram of drag conveyor. A = 20.9 in. pd, 8t, plate-center, flint-rim sprocket, H-112. B = 29.93 in. pd, 36t, plate-center, flint-rim sprocket, RO-882.

7 Return idler shafts, plain, $1^{11}/_{16}$″ dia × about 2′ 6″ long
14 Babbitted bearing pillow blocks, $1^{11}/_{16}$″
28 Solid malleable-iron safety collars, $1^{11}/_{16}$″
14 Single-flange rollers, flint-rim, 10″ dia, $2^{1}/_{4}$″ tread
1 Trough consisting of 15 sections hard white (cast) iron, 16″ wide inside × 6″ high sides, each section 5′; fitted together in concrete trough and grouted in (lengths of sections may vary according to manufacturer; aprx weight of each section is 320 lb)

6.9 SCRAPER FLIGHT CONVEYORS IN A CANNERY

In northern locations, most canneries operate only during the growing season, say, for about 4–5 months. Then, all canning machinery shuts down until the next season, so conveyor supports and troughs are made of wood, easily dismantled, to be set up in another location, if necessary. Scraper flight conveyors are operated at a medium speed, and the flights usually are also of wood. Malleable chains of the detachable type are used. If, after a few years, they corrode from juice or water, they are disposed of and new chain is installed at little cost.

6.10 LOADING BALES OF SCRAP INTO RAILROAD CARS DIRECTLY FROM BALING MACHINE

Large dealers in scrap iron and steel compress miscellaneous items into compact bales and load them directly into railroad cars for shipment to steel mills. This scrap consists of bed springs, steel barrels, whole automobile bodies, car fenders, wire fences, bedsteads, bicycle frames, wheels, etc. A baling press ejects bales, some of them weighing as much as 700 lb, and measuring up to 20 × 20 × 18 in. This conveyor is 30 in. wide and operates at a speed of 28 fpm on an incline of 38 degrees, with a center length of 31 ft. It is mounted on a structural steel frame resting on concrete piers at the head end, and in a shallow concrete pit at the foot end. The conveyor consists of two strands of C-$102^{1}/_{2}$ combination steel and malleable iron chain, with four special K2 attachments, to which are attached 6 in. × 18 lb ship channels, to push bales of scrap up incline and discharge them down hinged chute to railroad cars. A 3-hp motor operates this unit. See figure 6.12 and paragraph 14.32.

6.11 DRAG AND SCRAPER FLIGHT CONVEYORS IN A PULP, PAPER, AND SAW MILL

Many pulp and paper mills use scraper conveyors to dispose of sawdust and wood refuse, and it is common practice to use a single strand of malleable chain, often with wood flights spaced at intervals, traveling in a flared trough. Operations in a pulp and paper mill continue all year around, and refuse must be disposed of mechanically. Sawmill trimmings usually are long and stringy, requiring two or three strands of H-class chain to operate as one conveyor. These chains are made of heat-treated malleable iron or welded steel with a combination of malleable iron and steel for the sidebars. The

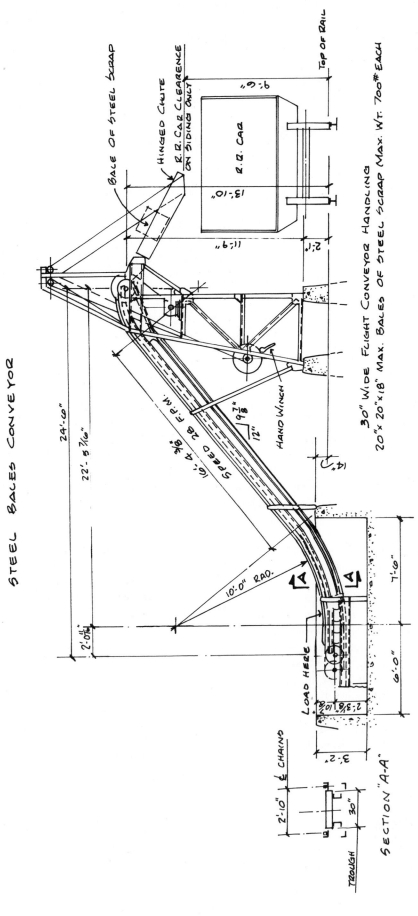

Figure 6.12 A 30-in. wide flight conveyor handling 20 in. × 20 in. × 18 in. (maximum) bales of steel scrap, weighing a maximum of 700 lb each.

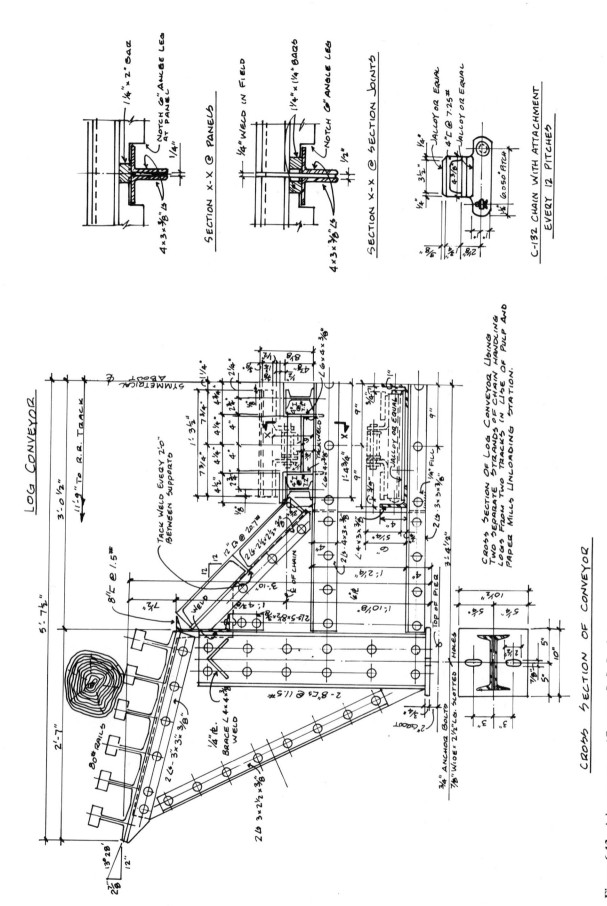

Figure 6.13 A log conveyor. *(Courtesy S. S. Snyder)*

block links and sidebars of Esco Corporation's drag chains are made of weldable or cast alloys with a Brinell hardness of 477 or greater. They will withstand abrasion better, and can be used for temperatures up to 1000°F. They are used frequently in log-handling conveyors.

6.12 STANDARD METHOD OF HANDLING LOGS BY CHAIN (SCRAPER) CONVEYORS

Figure 6.13 shows one method of handling logs from railroad cars or ships to mill storage, using two strands of drag chain. This detail shows the heavy construction necessary to withstand the abuse of logs rolling down to the conveyor chain. Two strands of chain operating as one conveyor are used to take care of a very large required capacity. The steel trough is of very heavy construction. For chipper feeding, some have designed their equipment around belt conveyors. Chains in this service usually last about eight years with normal wear, and the mills in this country continue to use chain conveyors (refer to paragraph 14.20).

6.13 CABLE DRAG SCRAPER (POWER HOE)

Where coal, stone, or similar material is stored in piles and moved to different storage, a cable drag scraper can be used to advantage.

The scraper or hoe is moved to and fro by a steel

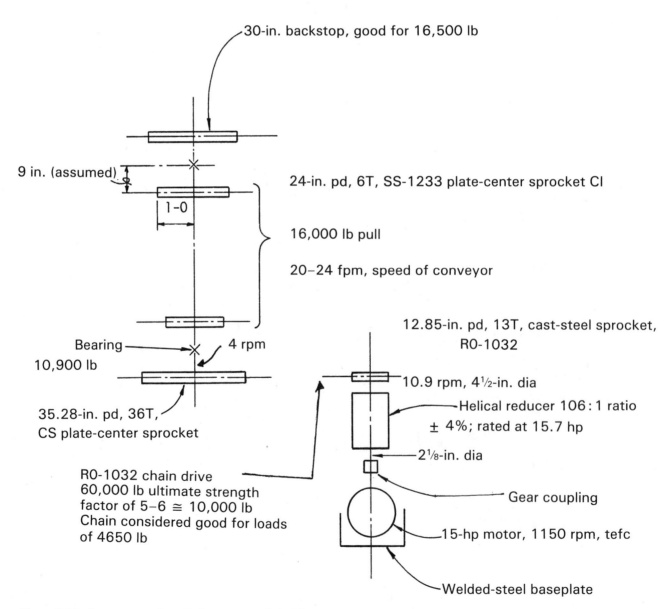

Figure 6.14 Arrangement of a cylinder conveyor chain drive.

cable, the ends of which are fastened to driving drums in the operator's house, which is placed to give a clear view of the storage area. The cable passes around sheaves located to serve the storage area. The scraper (hoe) is first pulled behind or on the side of the initial pile. When its direction of travel is reversed, it is filled automatically and dragged to storage, where it discharges its contents as it begins the return trip. The empty hoe rides on top of the coal until it returns to the initial pile, when the cycle of operation is repeated. The whole storage area is covered by occasionally moving the guide sheaves from point-to-point between the back posts or other form of anchorage located around the yard. When reclaiming from storage, the scraper's attachment to the cable is reversed, thus moving the coal in the opposite direction to elevator, then by conveyor or other means to boiler room.

This system can be used for any shape of pile. It is efficient and simple to operate. With the introduction of the bulldozer, these units have been replaced because of the flexibility of the bulldozer and its lower cost, but they still have a place in the storage of bulk materials where space is not a factor. These scrapers as manufactured by Beaumont Birch Co. and Sauerman Bros. Co.

6.14 DESIGN OF CYLINDER CONVEYOR (figure 6.14)

6.14.1 Data

Required: a double-strand conveyor for handling cylinders, each weighing an assumed 4000 lb, five on the incline conveyor side and four on the horizontal side at one time (see sketch); stop-start intermittent service. It can be fully loaded before starting operation. The speed should be about 20–24 fpm.

6.14.2 Rolling Friction (rf) of Chain

Where d = the diameter of the pin or bushing, D = the diameter of the roller, and 0.25 is used for a steel pin on a bored roller (greased),

$$\text{rf} = 0.25 \times \frac{d}{D}$$

$$= 0.25 \times \frac{1.25}{4}$$

$$= 0.0775$$

Use 0.10 to allow for the chain not being greased regularly.

6.14.3 Load on Run A

5 cylinders (4000 lb each)	= 20,000 lb
5 chain cradles (assumed 250 lb each) =	1250 lb
2 strands chain, each 18 ft long	
(20 lb per strand-ft)	= 720 lb
Total	= 21,970 lb

Use 24,000 lb because of variation in weight.

$$\text{Pull} = \dot{P} \times [\sin + (\text{rf} \times \cos)]$$

$$= 24,000 \times [0.425 + (0.10 \times 0.905)]$$

$$= 24,000 \times 0.516$$

$$= 12,384 \text{ lb}$$

6.14.4 Load on Run B

4 cylinders (4000 lb each)	= 16,000 lb
4 chain cradles (assumed 250 lb each) =	1000 lb
2 strands chain, each 13 ft long	
(20 lb per strand-ft)	= 520 lb
Total	= 17,520 lb

Use 18,000 lb.

$$18,000 \times 0.10 = 1800 \text{ lb}$$

Nat sin of angle (25°8') = 0.425

Nat cos of angle (25°8') = 0.905

6.14.5   Load on Run C

13 ft chain =   520 lb
4 cradles   = 1000 lb
Total   = 1520 lb

$$1520 \times 0.10 = 152 \text{ lb}$$

6.14.6   Load on Run D

2 strands chain, each 15 ft long =   600 lb
5 cradles (250 lb each)   = 1250 lb
Total   = 1850 lb

Use 1900 lb.

$$\text{Pull} = P \times [-\sin + (rf \times \cos)]$$
$$= 1900 \times [-0.425 + (0.10 \times 0.905)]$$
$$= 1900 \times (-0.425 + 0.091)$$
$$= -635 \text{ lb}$$

These calculations indicate that run D would roll down backwards by itself under perfect conditions, and that it would take 635 lb to stop it.

6.14.7   Total Pull at Head Shaft

Pull at A = 12,384 lb
    at B = 1,800
    at C =   152
    at D =   −635
  Total = 13,701 lb

Call it 16,000 lb pull because of stop-start intermittent operation.

6.14.8   Rechecking Conveyor Chain Strength

SS-1233 chain is good for a working load of 9200 lb per strand; two strands of chain = 18,400 lb. The total pull on the conveyor is 16,000 lb. Therefore, the two strands of chain are safe with an allowable pull per strand of 9200 lb.

6.14.9   Horsepower

$$\text{Theoretical hp} = \frac{16,000 \text{ lb} \times 24 \text{ fpm}}{33,000}$$
$$= 11.6 \text{ at head shaft}$$

Use a 15-hp motor to drive the conveyor.

6.14.10   Size of Head Shaft

The total bending moment is

$$\frac{16,000 \text{ lb} \times 9 \text{ in.}}{2} = 72,000 \text{ in.-lb}$$

and the total torsional moment is

$$16,000 \text{ lb} \times 12 \text{ in.} = 192,000 \text{ in.-lb}$$

Consequently, a 6½-in. shaft (from figure 9.3) is required.

6.15   SPECIFICATION FOR CHAIN DRAG CONVEYOR

A.   Required

Design, fabricate, and deliver the chain drag conveyor specified below to _____.
One 24″ × 8″ flight inclined chain drag conveyor 39 ft ± long, c to c sprockets, for unloading wood chips.

B.   Equipment to be Supplied

Furnish all items required to make the chain drag conveyor a completely self-contained mechanical and structural unit. It shall include a steel plate casing with chain carrying tracks, chain, chain sprockets with shafts and bearings, adjustable takeup tail shaft, steel plate flights, drive unit, drive chain and sprockets, machinery guards, and miscellaneous parts for a self-contained unit.

C.   Equipment not in Contract

Owner shall furnish components and facilities such as wiring, disconnect switches, electric power supply, concrete pit, receiving hopper, chutes, and foundation for drive unit.

D.   Design Criteria

Service: 8 to 16 hours per day, 7 days per week.
Capacity: 15 to 30 tph.
Material: Wood chips—damp.
Density: 20 lb/ft³.
Temperature: Ambient.
Casing: Structural frame with ¼″ bottomplate, covered with ¼″ 0.4% carbon steel liner plates fastened with countersunk bolts, 3/16″ side plates, no. 10 gauge steel cover plate. Chain carrying tracks shall be angles with

3″ × 3⁄8″ high-carbon steel wearing bar fastened with countersunk bolts.

Carrying chain: no. 678 HT (heat treated); 85,000 lb ultimate strength.

Flights: 24″ × 8″, made of 1⁄4″ plate bent for attachment to extended pins of carrying chain.

Drive unit: The drive unit shall be an assembled unit. It shall include a heavy duty enclosed dust- and oil-tight speed reducer meeting the requirements of AGMA specifications. The motor shall be mounted on the base of the speed reducer with a flexible coupling to the reducer. A secondary speed reducer shall be by chain-sprocket drive.

Drive chain and sprockets: The drive chain and sprockets shall be 2.609″ average pitch, 25,000 lb average ultimate strength, 2,750 lb allowable chain pull.

Bearings: Antifriction bearings shall be used for grease lubrication.

Lubrication: Grease using Alemite hydraulic type 1610 fittings.

Power: 440 V, 3 ph, 60 Hz.

Control: 120 V, 1 ph, 60 Hz.

Motor: 440 V, 3 ph, 60 Hz, tefc, ball bearing, 1800 rpm synchronous speed.

E. Painting

The exterior of the drag conveyor casing shall be painted with manufacturer's standard finish.

F. Inspection Test and Service

No special conditions shall be required except that the unit shall be a completely self-contained manufacturer's unit, completely assembled, that will require minimum field erection work.

G. Print Requirements

Submit two sets of drawings showing the arrangement and clearance dimensions of the inclined drag conveyor. Upon approval and completion of the drawings, furnish one sepia reproducible of all arrangement and detail drawings and five copies of lubrication instructions, operating information, and instructions for all parts of the equipment supplied with the units.

H. Information to Vendors

All questions pertaining to purchasing terms and procedures shall be directed to: _____
_____ .

I. Drawing Attached

Arrangement of Wood Chip Unloading Drag Conveyor.

Feeders and Vibrating Conveyors

7.1 GENERAL—FEEDERS

Feeders are short conveyors or other devices used to take material from the bottoms of bins, silos, or hoppers, and to transfer them to other types of conveyors. They are used to regulate the flow of material to crushers, screens, conveyors, or other equipment. Single reciprocating feeders give pulsating flow. Twin units that discharge alternately are an improvement. Apron feeders give a more continuous flow and distribute the lumps in the flow. Electromechanical and electromagnetic vibrating feeders can be equipped to screen the fines ahead of the crusher, and all have easily adjusted feed rate. Feeders must be wired to stop promptly when the crusher stops.

7.2 RECIPROCATING FEEDERS

Reciprocating feeders are not very popular at the present time; however, many of them are in use and information about them may become scarce.

The reciprocating feeder consists of a reciprocating steel plate, supported on flanged rollers and driven forward and backward by an eccentric unit, mounted on a driven shaft. On the forward stroke of the reciprocating plate, which forms the hopper bottom, material is carried from the hopper. On the return stroke, the plate slides back under the material, thereby unloading a certain amount of material at the end of the plate. The backward movement of the material on the plate is re-strained by the new material which filled the space created on the forward stroke. These feeders are one of the most economical methods of feeding bulk materials from a track hopper to a belt conveyor or bucket elevator. They will handle large lumps of bulk material, such as coal, ore, stone, and frozen lumps of sand from a pit operation. They are not self-cleaning. The accumulation of material can be reduced somewhat, however, by declining the feeder plate, not to exceed 10°. Normally, the slope is held to 3–5° for damp material.

Figure 7.1 shows a typical arrangement of track hopper and reciprocating feeder discharging to an inclined belt conveyor. Many coal-handling installations are arranged so that the feeder discharges directly to a bucket elevator. Most of these feeders are equipped with a variable-movement eccentric drive so capacity of feeders can be increased or decreased as desired. They run normally at 25–50 fpm, depending on material handled. This type of feeder is commonly used for chemicals and coal, and is limited to material weighing under 150 lb/ft^3.

Figure 7.2 shows the arrangement of a reciprocating feeder receiving material from a car hopper. This sketch shows the drive and a regulating gate to control the feed.

7.2.1 Space for Reciprocating Feeder

Available space for the installation of any feeder enters into the selection of such a unit. If the space is small

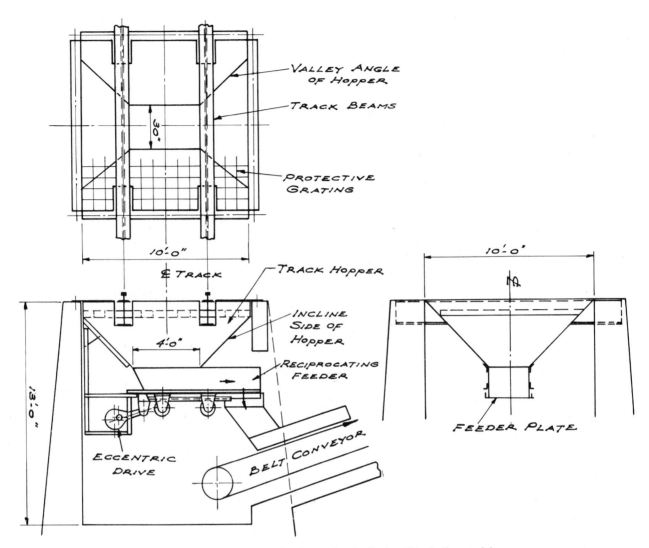

Figure 7.1 Typical arrangement of a track hopper and reciprocating feeder handling bulk material.

between unloading track and building, a reciprocating feeder may fit in well, and it may be impossible to install any other type of feeder. The reciprocating feeder has few working parts, which means low operating costs and low maintenance costs, resulting in long service life with minimum downtime. Feeder plates can be lined, and the lining can be easily renewed when worn.

7.2.2 Capacity of Reciprocating Feeder

Required capacity: 60 tph.
Material: sand and gravel weighing 100 lb/ft³.

The distance between feeder stationary side plates is 15 in., with an unrestricted distance (at the discharge point)

of about 12 in. The cross-section of the area at this point is $1 \times 1.25 = 1.25$ ft².

Use an adjustable throw eccentric set to give a 6-in. stroke (total movement forward and backward). For each 6-in. stroke, $\frac{1}{2} \times 1.25 \times 100 = 62.5$ lb of material will drop off the feeder plate.

If the drive shaft is turning at 40 rpm, there would be 40 strokes, and the capacity would be

$$62.5 \times 40 \times \frac{60}{2000} = 75 \text{ tph}$$

Because of jams and outdoor operation, it is best to assume 60 tph.

Based on a previous test, a 3-hp motor was selected.

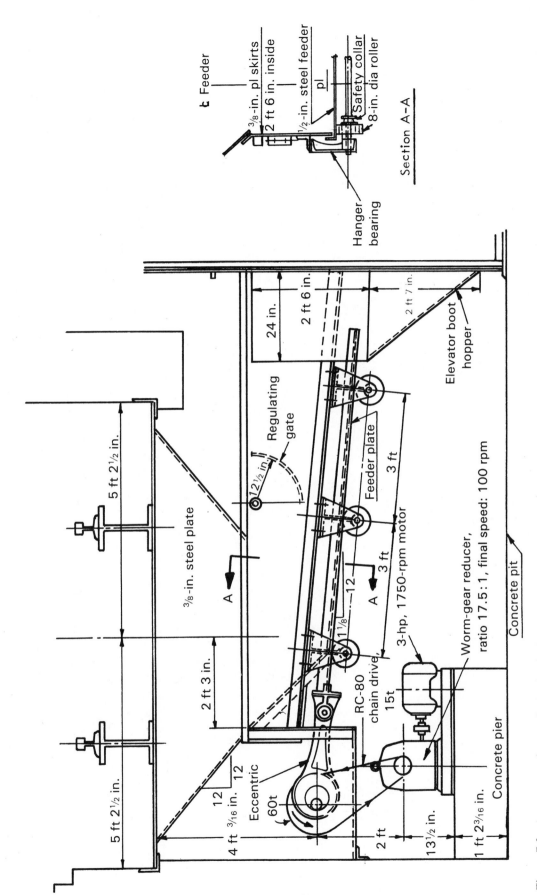

Figure 7.2 Arrangement for a reciprocating feeder.

213

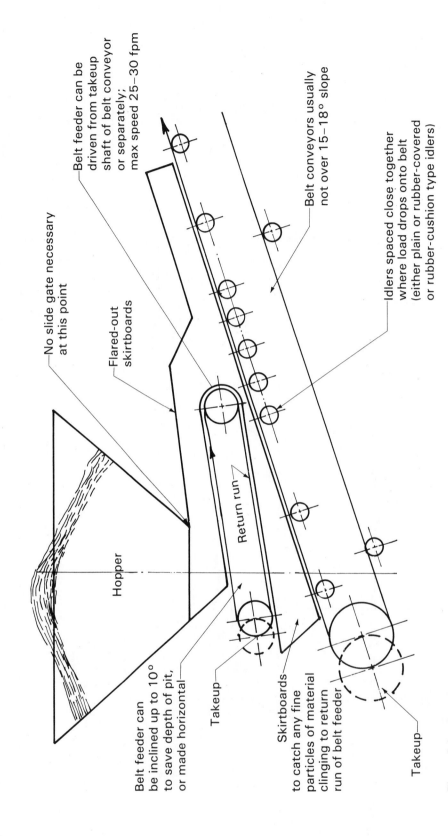

Belt feeder can be driven from takeup shaft of belt conveyor or separately; max speed 25–30 fpm

No slide gate necessary at this point

Flared-out skirtboards

Belt conveyors usually not over 15–18° slope

Idlers spaced close together where load drops onto belt (either plain or rubber-covered or rubber-cushion type idlers)

Hopper

Return run

Belt feeder can be inclined up to 10° to save depth of pit, or made horizontal

Takeup

Skirtboards to catch any fine particles of material clinging to return run of belt feeder

Takeup

Figure 7.3 Belt feeder under a dump hopper.

214

7.3 BELT FEEDERS

A belt feeder can be used under certain conditions, if the speed of the belt does not exceed 25–30 fpm. At 40 fpm, the belt may pull out under the load. If greater capacity is required, it is best to widen the feeder and not to increase the speed.

7.3.1 Material Handled

The feeder will handle material weighing up to 100 lb/ft^3 and is usually used under bins of 15–20-ton capacity. Belt feeders should be used for damp chemicals that would affect steel aprons. The rubber at the bottom of skirt plates wears and has to be replaced from time to time.

7.3.2 Belt Feeders Under Dump Hoppers (see figure 7.3)

These feeders are generally used in sand and gravel plants under dump hoppers to feed material uniformly onto a belt conveyor. Usually, due to high production requirements, a dump hopper is kept fairly full of run-of-bank material which could be wet, damp, or dry—depending on the weather. No cleats of any kind are used, nor are they necessary at this speed. Some people do place metal cleats about 6 in. long and about 2 in. high in the center of the belt, but these cleats do not stay put very long; they either tear away from the belt or bend over and become a nuisance. Troughing idlers are placed under the top side of the belt, as close as possible, to keep the belt from sagging. These belt feeders are effective, the maintenance is low, and the replacement belt is not too costly for the tonnage handled. Under the dump hopper, the idlers are spaced about one foot apart and are rubber cushioned to soften the load coming down on the belt. Where there is continuous flow into the hopper, the belt takes the load right away.

7.4 APRON FEEDERS

An apron feeder is essentially a leakproof, or non-leak, type feeder. The standard design allows 3/16-in. clearance between the overlapping pans. These pans are attached to two strands of roller chain, and the apron supporting frame is designed to have stationary side plates to form the feeder sides. Apron feeders are quite costly. Chain joint wear and maintenance are high. These feed-

ers are usually 6 in., 9 in., or 12 in. pitch, and each joint in the chain has a roller turning on it. These feeders can be inclined normally to 25°, depending on the material handled and size of material. The speed of an apron feeder should be 10–15 fpm to prevent excessive maintenance and to have smooth action over the rollers (see figure 7.4).

7.4.1 Handling Considerations

When handling heavy ores or large lumps, the openings in the bottom of the hopper and the supports of the conveyor should be checked. To prevent shock from loading, the hoppers should be kept partly filled when stopped.

Hot materials should use deep pan buckets to keep chains away from the buckets.

7.4.2 Open-Top Deep-Bucket Carrier

This unit is heavily built and is intended to carry material weighing close to 100 lb/ft^3 up a steep incline, due to limited space. The chain consists of 2½ in. wide steel flats. The 4 in.-diameter roller has a single flange, Alemite lubrication through 1-in.-diameter pins, and 1¼-in. OD steel bushings. The deep buckets are made not less than ¼ in. thick and are continuously welded on the inside.

On a unit of this kind at or about 40°, the weight of the empty bucket or pan with two strands of chain (all rolling down hill on tee-rail track) is heavy enough to move by itself, and the force exerted can help the carrying run. For this reason, the amount of horsepower necessary to drive the head shaft is small compared with a conveyor of this type if at a lesser angle. This does not decrease the bending or torsional pulls on the head shaft. There still are heavy loads (carrier and material) to lift.

The operator must maintain complete control of the apron feeder so it delivers rock to the crusher uniformly, so balance of equipment is not overloaded. The apron feeder must always be stopped before the hopper is completely empty to assure a bed on the apron to cushion the impact of the next load.

From experience, it is much better to have the apron at rest when dumping into the hopper, so as to allow the large pieces to find their place and settle; otherwise, large pieces could get into such a position (if the apron is moving) and wedge together, causing possible jamming against the feeder skirt sides. If this happens, there is much downtime during production runs, until the rock

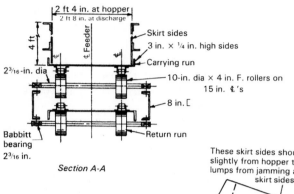

Section A-A

Figure 7.4A A chrome ore apron feeder. *Data:* 36-in. wide heavy-duty apron feeder handling chrome ore at 140 lb/ft³; capacity = 50 tph at 5 fpm or 150 tph at 15 fpm; size of ore = 15 in. and under; apron pans are 1/2-in. thick on a double strand of RS-944+ steel chain.

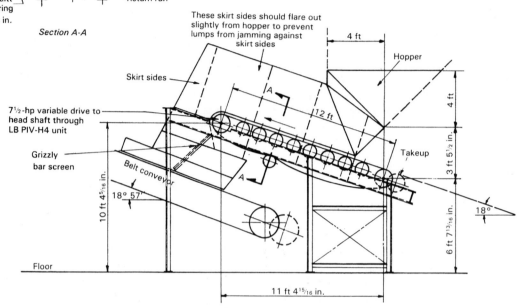

Figure 7.4B A long, deep-pan apron feeder, 36-in. wide, for handling 6-in. and under chrome ore.

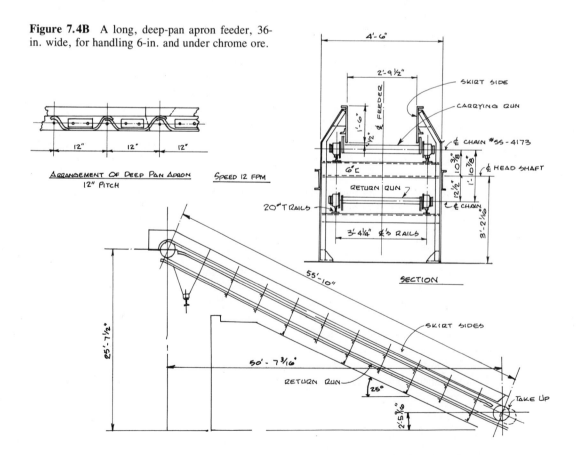

is released by steel handpoles or other mechanical means provided overhead, such as electrically operated tongs for an emergency.

7.5 DESIGN OF APRON FEEDER
(see figure 7.5)

7.5.1 Horsepower

Weight of apron (½-in. pans, ¼-in. sides)
plus 2 chains   = 130 lb/ft

Weight of ore   = 390 lb

Total weight   = 520 lb/ft

Carrying run A (refer to paragraph 5.6.6 for formula)

$$12 \text{ ft} \times 520 \text{ lb}$$
$$= 6240 \times [0.309 + (0.11 \times 0.951)]$$
$$= 2581 \text{ lb}$$

The 0.11 factor is used here instead of the 0.10 recommended in the manufacturers' catalogs as it results in only about 2% error on the side of safety and eliminates a multiplicity of coefficients.

Return run B

$$12 \text{ ft} \times 130 \text{ lb} \times 0.11 = 172 \text{ lb}$$

Because of returning on one roller with some sag between, there would be considerable dead weight of apron (from field experience, add 200 lb). If more than one return roller is used, add 200 lb per roller assembly. The return roller incline formula (paragraph 5.6.6) is neglected in this type of short, heavy conveyor. Total pull on the head shaft = 2581 + 172 + 200 = 2953 lb, say, 3000 lb.

Theoretical hp on maximum speed of 15 fpm equals

$$\frac{3000 \times 15}{33,000} = 1.36$$

A 7½ hp, 900-rpm motor was used, because of the very low speed of the apron and friction developed by chain drives, reducer, and variable-speed reducer connected to motor by V-belt drive. If there is a jam of any kind, the V-belt would have a chance to slip and save the other parts of the drive from harm.

7.5.2 Size of Head Shaft
(see figure 7.6)

$$\text{Bending M} = \frac{3000 \text{ lb}}{2} \times 9 \text{ in.}$$
$$= 13,500 \text{ in.-lb}$$
$$\text{Torsional M} = 3000 \text{ lb} \times 7 \text{ in.}$$
$$= 21,000 \text{ in.-lb}$$

Using factors of 3.0 for bending and torsional moments, and referring to figure 9.3, a $3^{15}/_{16}$ in. shaft is required, and was used (for suddenly applied loads).

7.5.3 Capacity
(see figure 7.4)

2.33 (width of feeder) × 4 (hopper opening)
$$\times 1.0 \text{ (length of feeder)}$$
$$= 9.32 \text{ ft}^3/\text{ft of feeder}$$

Because of odd shape of pieces (70% voids), use 2.80 ft^3/ft.

$$2.80 \text{ ft}^3/\text{ft} \times 140 \text{ lb} = 391 \text{ lb/ft (say, 390 lb)}$$
$$390 \text{ lb} \times 5 \text{ fpm} \times 60 \text{ min} = 117,000 \text{ lb}$$
$$= 58 \text{ tph at 5 fpm}$$
$$= 175 \text{ tph at 15 fpm}$$

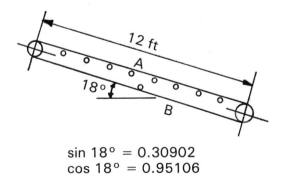

sin 18° = 0.30902
cos 18° = 0.95106

Figure 7.5 Diagram of an apron feeder.

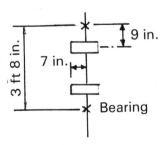

Figure 7.6 Head shaft diagram.

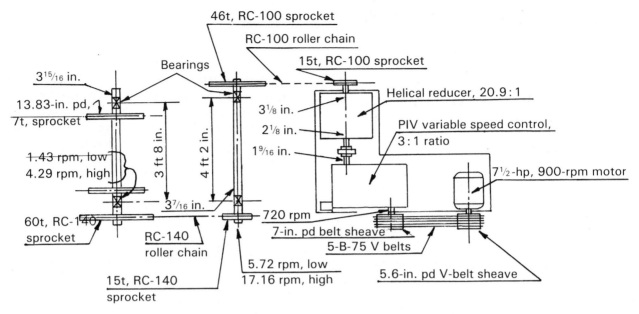

Figure 7.7   Apron feeder drive diagram.

RF (rolling friction) of the supporting rollers (assuming that they will be well greased so that every roller turns) is $0.25 \times (2.1875/10) = 0.05$, where f = 0.25, the diameter of the bushing = 2.1875, and the diameter of the roller = 10.0 in.

Field tests, however, indicated a value of 0.10 (because of dust and poor lubrication). A value of 0.11 was used (refer to paragraph 7.5.1).

7.5.4   Details

The chain and apron are supported on 10-in.-diameter steel rollers, spaced 15 in. on centers to absorb any shock load. The speed should be kept low to have smooth action over rollers. This apron discharges to an inclined belt conveyor with much of the ore first coming in contact with the fixed grizzly bar screen. Many fine pieces (down to dust) go through the bar screen and fall on the belt to form a cushion for the larger lumps falling directly on the belt. The belt conveyor could not properly handle the required capacity or the large lumps without a feeder.

A no-leak apron, with tee rails welded underneath each pan, and arranged to travel on a ¾ in. steel wear plate on top of supporting beams, as shown on figure 5.9, could have been used instead of this arrangement, but if the hopper can always be kept partially filled to absorb the shock of the next load of ore being dumped, this design is good. The apron of this design is less expensive than a no-leak type.

The maximum speed should not exceed 15 fpm, although it could operate at 20 fpm or even 25 fpm, but maintanence would be much higher, and the operation would not be smooth.

7.6   BILL OF MATERIAL*
(see figure 7.7)

1 Head shaft, $3^{15}/_{16}$" dia $\times$ 5'6" $\pm$, ks in line 5" long
2 Series 2-1500 pillowblock bearings, $3^{15}/_{16}$", with Alemite fittings
1 Solid malleable-iron safety collar, $3^{15}/_{16}$"

2 Flint-rim (chilled tooth) sprockets, 13.83" pd, 7t, RS-944 plate-center, with standard hubs, bore and ks in line, each with 2 setscrews at 90°
1 RC-140 steel sprocket, 33.438" pd, 60t, with standard hubs, bore and ks-ss
1 RC-140 roller chain with loose offset coupler, 20 lin ft
1 Countershaft, $3^{7}/_{16}$" $\times$ 5'6" $\pm$, ks with keys 5" long

---

*Part numbers are from Link-Belt catalog.

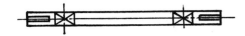

2 series 2-1500 pillowblock bearings with Alemite fittings

2 Solid malleable-iron safety collars

1 RC-140 steel sprocket, 8.417 pd, 15t, standard hubs, bore, and ks-ss

1 RC-100 steel sprocket, 18.317″ pd, 46t, with standard hubs, bore and ks-ss

1 RC-100 roller chain with loose offset coupler, 20 lin ft

1 RC-100 steel sprocket, 6.012″ pd, 15t, standard hubs, bore to suit reducer, ks and ss

1 Helical reducer, ratio 20.9 : 1

1 Gear-type coupling to connect reducer and PIV unit

1 PIV variable-speed drive, 3 : 1 ratio, with electrical remote-control unit and pushbutton arrangement

1 5-B-75 V-belt sheave on PIV unit, 7″ pd

1 5-B-75 V-belt sheave on motor, 5.6″ pd

1 Lot of 5-B-75 V-belts

1 Motor, 7½ hp, 900 rpm, tefc, for 440 V, 3 ph, 60 Hz

1 Welded-steel baseplate for reducer, PIV, and motor

1 Plain foot shaft, 3⁷⁄₁₆″ × 5′0″, no ks

2 Babbitted takeups, 3⁷⁄₁₆″ × 12″ type, Alemite fittings

4 Solid malleable-iron safety collars, 3⁷⁄₁₆″

2 Flint-rim sprockets, 13.83″ pd, 7t, RS-944 plate-center, with standard hubs large enough to rebore to 3¹⁵⁄₁₆″ to suit head shaft in an emergency; no ss, sprockets to run loose*

9 Idler shafts (8 on carrying run, 1 on return run), 2³⁄₁₆″ dia × 4′4″ ±

18 Babbitted pillow-block bearings, 2³⁄₁₆″, Alemite fittings

18 Solid malleable iron safety collars, 2³⁄₁₆″

18 Cast-iron rollers, 10″ × 4″ straight face, setscrew only

1 Style B chain tightener B-1 (for chain drives)

1 RC-140 steel sprocket, 7.313″ pd, 13t

1 RC-100 steel sprocket, 6.012″ pd, 15t

1 Heavy-duty apron, 30 lin ft, 36″ wide, overlapping pans, close clearance, ½″ steel, with 3″ × ¼″ high sides welded to pans, and equipped with a double strand of RS-944, K1 steel strap chain on 2′2″ centers.

---

*This feeder will always be started from a stand-still position, with the hopper partly filled with ore to reduce shock on apron, when dumping next load. The chain pins under such severe conditions will, in time, develop a litle play. The loose sprockets will compensate for this condition should it appear.

7.7 DESIGN OF NO-LEAK TYPE APRON FEEDER (see figure 7.8)

7.7.1 Data

Material: 2 in. (maximum) lumps to dust, weighing 100 lb/ft³
Required capacity: 70 tph
Centers: 25 ft, inclined 15°
Cross rods: every 12 in. through chain links and not chain joints
Speed: 15 fpm

7.7.2 Actual Capacity

$$1.79 \times 1.25 \times 1 \text{ ft } 0 \text{ in.} \times 0.70 \ (30\% \text{ voids})$$

$$= 1.57 \text{ ft}^3$$

$$1.57 \times 100 \text{ lb} \times 60 \text{ min.} \times 15 \text{ ft (speed)}$$

$$= 141,300 \text{ lb (say, 70 tons)}$$

7.7.3 Horsepower at Head Shaft (see figure 7.9)

RF (rolling friction) of a 4-in., single-flanged roller operating on a 1 in. OD bushing, where $\chi = 0.25$ (for steel pin or bushing on a greased, bored roller).

$$\text{RF} = \chi \times \frac{\text{diameter of bushing}}{\text{diameter of roller}}$$

$$= 0.25 \times \tfrac{1}{4} \text{ in.}$$

$$= 0.0625 \ (\text{say, } 0.10)$$

Weight per foot of 24 in. apron, including two strands of chain and cross rods with rollers	= 115 lb
Weight of material on feeder	= 157 lb/ft
Total	= 272 lb/ft

Total load P, carrying run A equals

$$wL(\sin\theta + f\cos\theta)$$

$$= 25 \times 272 \left[0.258 + (0.10 \times 0.966) \right]$$

$$= 2411 \text{ lb}$$

Return run B equals

$$wL(-\sin\theta + f\cos\theta)$$

$$= 25 \times 115 \ (-0.258 + 0.0966)$$

$$= -464 \text{ lb}$$

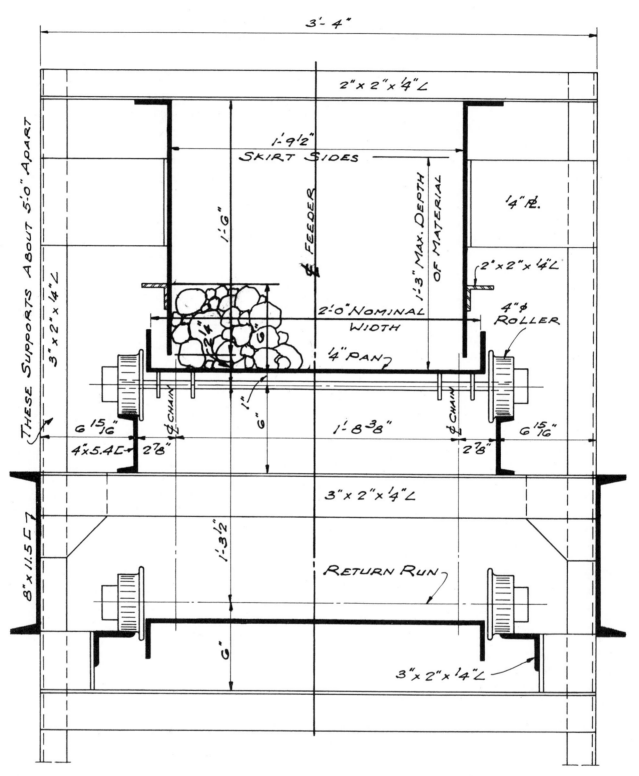

Figure 7.8 Typical cross-section of a 24-in. no-leak apron feeder with a double strand of RS-953 chain.

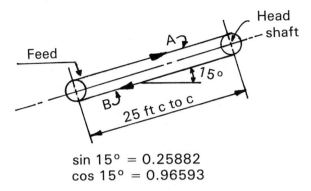

$$\sin 15° = 0.25882$$
$$\cos 15° = 0.96593$$

Figure 7.9 Diagram of a no-leak apron feeder.

Thus we have

$$2411 \text{ lb} - 464 \text{ lb} = 1947 \text{ lb (say, 2100 lb)}$$

to compensate for any spill on the tracks. Therefore,

$$\text{hp} = \frac{2100 \times 15}{33,000} = 0.95$$

Because of friction on the sides, the sharp edges of

the material, and (primarily) field experience with the product, a 3-hp motor was used (see figure 7.10).

7.8 SELECTION OF TYPE

Since a reciprocating feeder is not made much over 10 ft long, it cannot be used if the distance from the track hopper to a bucket elevator or crusher is much over that.

A belt feeder might work but, if a fairly large capacity is required, the belt would have to be wide due to operating at a low speed. This unit would have a number of objections under these conditions.

An apron feeder would be the best to use, with a sizable maintenance department. An apron feeder, which is generally self-cleaning, might be preferred over a reciprocating feeder or even a belt feeder.

7.9 ROTARY TABLE FEEDER

The rotary table feeder consists of a power-driven circular plate, rotating directly below a bin opening. An adjustable feed collar, located above the rotating table,

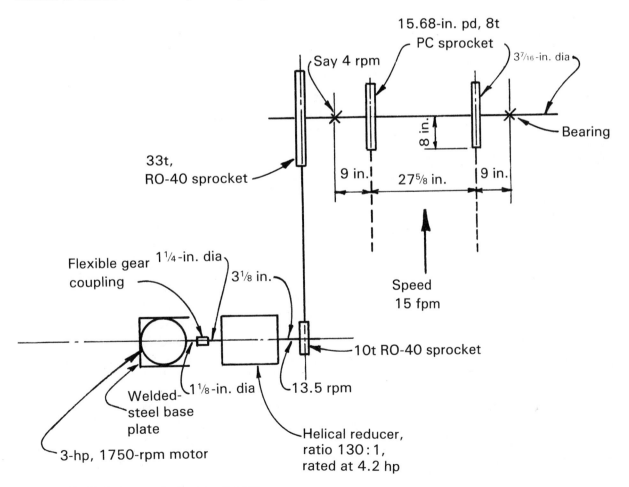

Figure 7.10 No-leak apron feeder drive diagram.

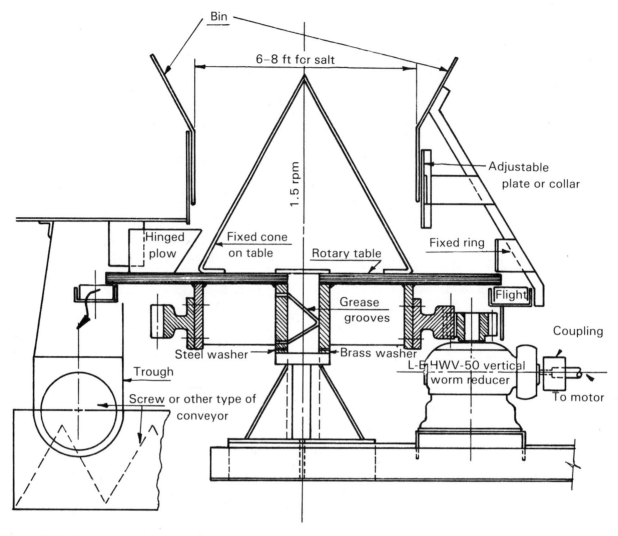

Figure 7.11 Cross-section of a rotary table feeder. *Note:* The opening in the bottom of the bin is smaller than the diameter of the bin. This can be done with materials that do not tend to cling and have a low moisture content.

determines the volume of material delivered. These feeders are used generally with round, vertical bins, and for handling materials which have a tendency to arch; such as damp sand, wood chips, and fine salt. The feed is fairly uniform for most materials. The adjustable collar determines the feed to the table, and the hinged plow is arranged to divert the material to a trough where flights attached to the bottom of the table move the material to an opening in the trough to another conveyor. The opening in the bottom of the round, vertical bin is large, so material cannot arch over it. Some designs are worked out with mechanical arch breakers, operated through the vertical feeder shaft. These work better on a fine material that does not pack too hard (see figures 7.11 and 7.12).

7.10 SCREW FEEDERS

A close clearance between flights and inside of trough is usually maintained throughout the length for free-flowing materials; otherwise, 1/2-in. clearance is used (see figures 7.13 and 7.14). It is advisable to use one size larger than required, and to move material at a slower speed, when using a short-pitch tapered screw. They are good for granular material and lower capacity.

7.11 MECHANICAL VIBRATING FEEDERS

Mechanical vibrating feeders are compact, low-headroom machines, for feeding a wide range of bulk ma-

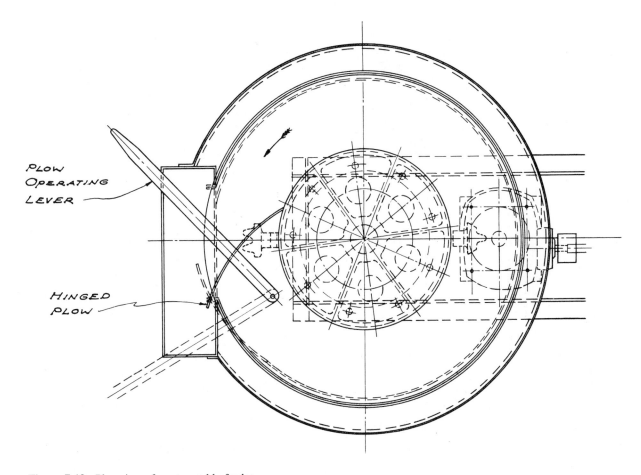

PLOW
OPERATING
LEVER

HINGED
PLOW

Figure 7.12 Plan view of a rotary table feeder.

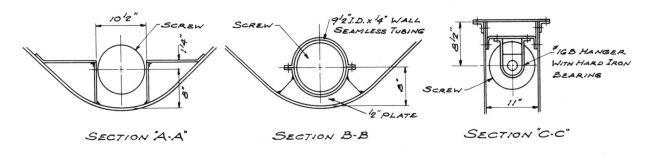

SECTION "A-A"

SECTION B-B

SECTION "C-C"

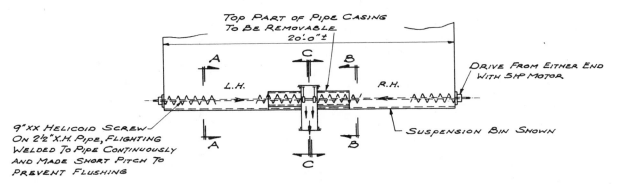

Figure 7.13 A reclaim screw feeder with 9-in. screw feeders to handle 25 tph of stucco (gypsum plaster), −140 mesh at 55 lb/ft³, 68 rpm from the bottom of the suspension bin.

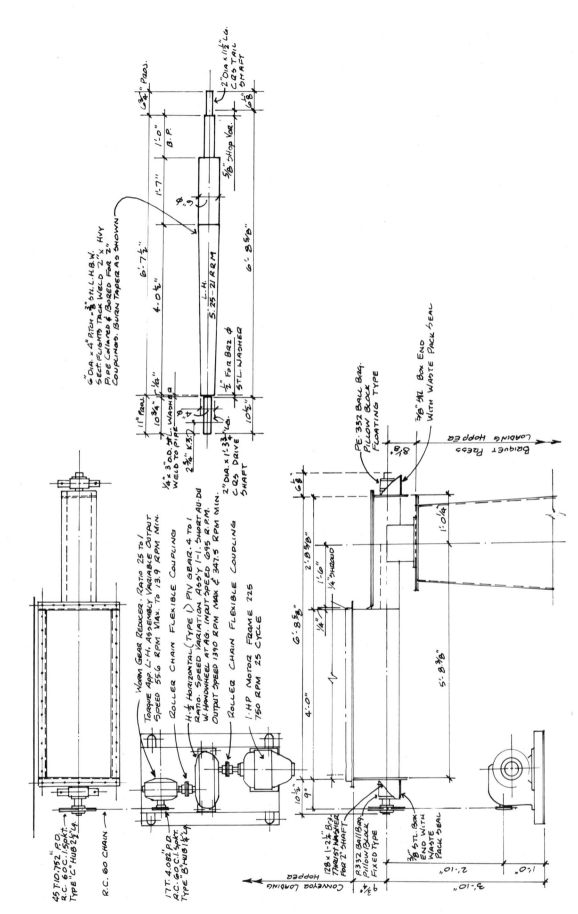

Figure 7.14 A short screw feeder.

224

terials at a uniform rate from bins, hoppers, storage piles, or conveyors. A heavy-duty trough and motorized vibrator drive are mounted directly to a rugged channel frame. These feeders require only a single standard electrical connection, and no adjustments have to be made once the feeder is in operation.

Mechanical vibrating feeders can be floor-mounted or suspended by cables. Stephens-Adamson make a solid-stroke mechanical variable-rate vibrating feeder, heavy duty and extra-heavy duty. The solid-stroke feeder is a sturdy mechanical feeder. It is built to mount directly under the hopper load, without a special chute or shelf to distribute material. Varying head loads in bins and hoppers have no effect on the feed rate; the discharge rate remains constant regardless of material depth. The feed rate can be controlled while in operation, to feed at any point in a fully operating range from zero to maximum. FMC (Link-Belt) uses this type feeder for the most severe duty, such as at a truck dump.

7.12 VIBRATING FEEDERS— ELECTROMAGNETIC AND ELECTROMECHANICAL

7.12.1 General

Vibrating feeders (figure 7.15) are used to control the movement of bulk materials from a bin to a conveyor or process machinery. They are good for uniform feeding of many materials, and are made in capacities varying from a few ounces to 2000 tons per hour. Moisture can range from dripping wet to bone dry. Ambient temperatures can vary from $-30°F$ ($-34°C$) to about $130°F$ ($54°C$) and, with special design, can operate at high temperatures, up to $300°F$ ($149°C$).

The feeders can operate either indoors, outdoors, and in clean to extremely dusty, corrosive atmospheres. They work well under bins or in tunnels. Variable feed rate is standard, controls can respond to various inputs. The particle sizes may range from 40 mesh to large lumps. Finer materials often will travel only with a low bed depth which limits capacity. Connections to vibrating feeders must be through flexible connections.

The two main types are electromagnetic and electromechanical. The electromagnetic type is based on the drive being powered by alternating current energizing a coil and creating magnetic forces to move an armature which is connected to a pan. There are no sliding or rotating parts (see figure 7.16a). The pan weight is tuned to a specific drive size.

The electromechanical types have a number of designs. Some are based on the principle of natural frequency vibration, comprising three basic elements: the trough, a resonant heavy-duty natural-frequency spring system, and a squirrel-cage alternating-current motor (see figure 7.16b).

Some of the manufacturers of the above types are Eriez Magnetics, Stephens-Adamson, Syntron (FMC Corp.), Carrier, and Jeffrey.

7.12.2 Design Options

The trays (pans) come in a wide variety, from a standard steel pan to special trays with screens, grizzlies, dust covers, abrasive liners, and heated liners.

The vibrating feeders, with their variable feed rates, can be controlled by ac-operated variable transformers, SCR (semiconductor-controlled rectifier) electric circuits, and/or a combination of electric motor, rotating weights, heavy-duty coil springs, pneumatic air bag (spring) and variable-voltage controllers.

7.12.3 Bulk Materials

Each manufacturer of vibrating feeders has done some testing in handling bulk materials and should be consulted for their advice on the feeding characteristics of the material on their equipment. In general, bulk material may range in density from 3 lb/ft^3 to 5 lb/ft^3 (cereals) and from 80 lb/ft^3 to 250 lb/ft^3 (ferro-alloys).

7.12.4 Typical Designs

Paragraph 7.12.5 gives the general formula for capacity. Paragraphs 7.12.5.1 and 7.12.5.2 show selections of feeders for limestone and sand and gravel, respectively.

7.12.5 Capacity Calculations

C = Capacity (tph)
d = Burden depth (in.)
T = Throat opening (in.)
H = Gate opening (in.)
W = Width of trough (in.)
FR = Flow rate (fpm)
D = Density (lb/ft³)

The flow rate for electromagnetic feeders is generally about 30–35 fpm, and for electromechanical feeders is about 50 fpm. It is best to consult with manufacturers or people familiar with the product or actually to perform tests.

The width of opening for random-size material should

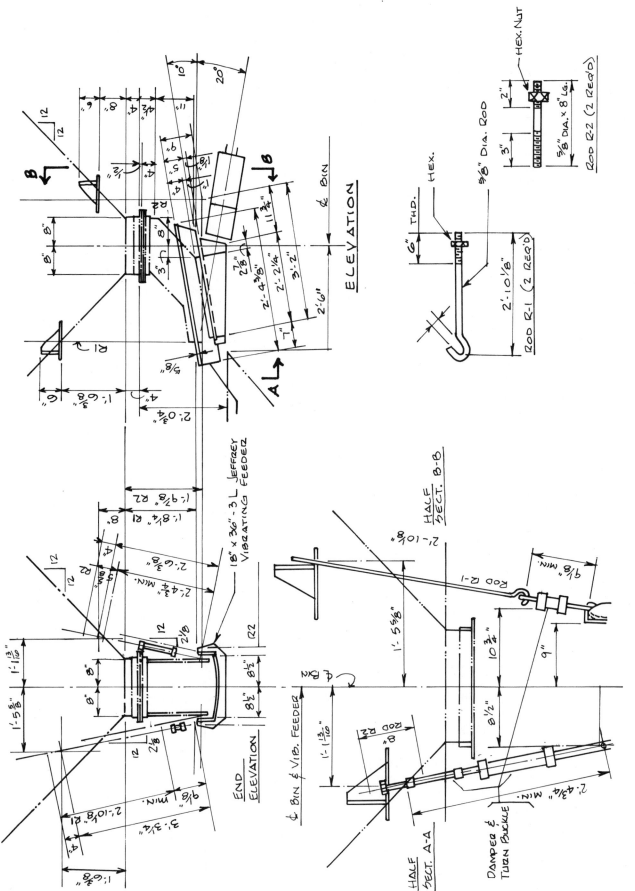

Figure 7.15 Details of a vibrating feeder.

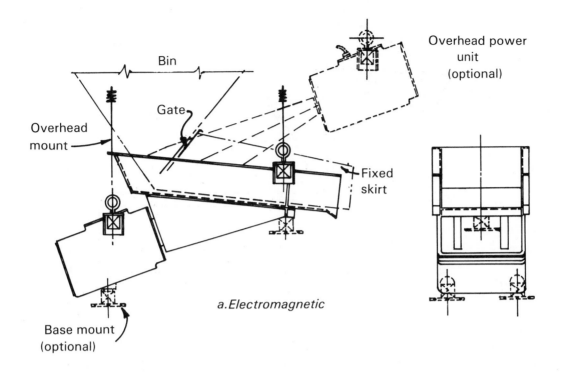

a.Electromagnetic

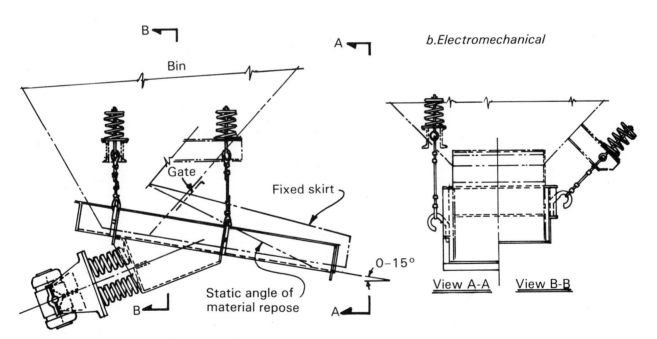

b.Electromechanical

Figure 7.16 Electromagnetic (top) and electromechanical (bottom) vibrating feeders.

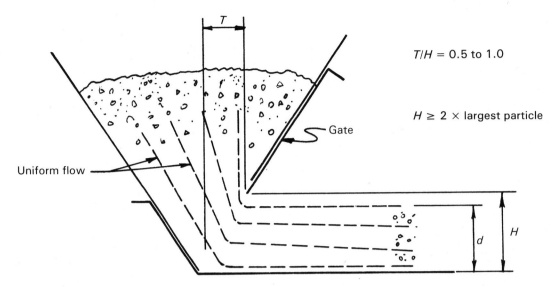

$$T/H = 0.5 \text{ to } 1.0$$

$$H \geq 2 \times \text{largest particle}$$

Figure 7.17   Minimum gate openings.

be 2½ times the largest particle size; for nearly equal-size particles, the width should be 5 times the largest particle (see figure 7.17).

H should be between 1.2 and 1.5 times *d*, where

$$d = \frac{C \times 4800}{W \times FR \times D}$$

7.12.5.1   Feeder for Limestone (figure 7.18)

$D = 100 \text{ lb/ft}^3$
$W = 2.5 \times 10 = 25$ in. effective
  or 30 in. gross
$FR = 30\text{--}35 \text{ fpm}$
$C = 350 \text{ tph}$

Consider an electromagnetic feeder with the bottom and sides having liners. The size of the material to be moved (with a 10% moisture content) is 10 in. $\times D$, 6 in. on the average. Because the stone is mostly of the larger size, but not enough to consider it uniform, a 36-in. wide (33 in. effective) trough is used.

$$d = \frac{350 \times 4800}{33 \times 35 \times 100} = 14.5 \text{ in. (say, 15 in.)}$$

or

$$\frac{35}{30} \times 15 = 17.5 \text{ in.}$$

for an *FR* equalling 30 fpm.

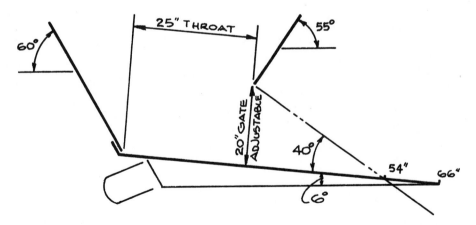

Figure 7.18   Limestone feeder gate opening with a 40° assumed angle of repose, which intersects at 54 in. + 6 in. (to assure no free flow) = 60 in. Use a 66-in. long trough, which is standard for F-450.

The gate height of opening is

15 in. (depth) × 1.3 (gate height factor)

= 19.5 in. (use 20 in.)

The throat opening is

20 in. (gate) × 0.75 (throat opening factor)

= 15 in.

but because of the size of the stone (10 in.), for the throat opening use

2.5 × 10 = 25 in.

Syntron Model F-440, with the trough sloped 6° downgrade and measuring 36 in. × 54 in., is rated at a capacity of 350 tph. Because of 10 in. maximum lump size, however, a 36 in. × 66 in. trough is required, and F-450 is needed to drive the heavier trough with ¼ in. stainless-steel bottom and side liners; below deck suspension mounting; remote control. Approximate power consumption is 11,000 W, 48 A, 230 V, 60 Hz. The feeder discharges into a 36 in. belt, in line.

7.12.5.2 Feeder for Sand and Gravel (figure 7.19)

$D = 100$ lb/ft^3
$W = 5 × 4 = 20$ in. effective
  or 24 in. gross
$FR = 30–35$ fpm
$C = 150$ tph

Consider an electromagnetic feeder to be used to move material of an unknown moisture content and being a uniform 4 in. × D in size.

$$d = \frac{150 × 4800}{21 × 35 × 100} = 9.8 \text{ in. (say, 10 in.)}$$

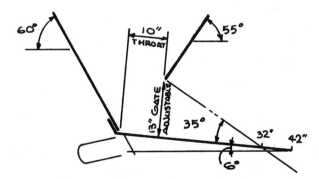

Figure 7.19 Sand and gravel feeder gate opening with a 35° angle of repose, which intersects at 32 in. + 6 in. (to assure no free flow) = 38 in. Use a 42-in. long trough, which is standard for F-330.

or

$$\frac{35}{30} × 10 = 12 \text{ in.}$$

for an *FR* equalling 30 fpm.

The gate height of the opening is

10 in. (depth) × 1.3 (gate height factor)

= 13 in.

The throat opening is

13 in. (gate) × 0.75 (throat opening factor)

= 9.75 in. (say, 10 in.)

Use Syntron F-330, 24 in. wide × 42 in. long trough with ³/₁₆ in. stainless steel bottom and side liners, below deck, suspended mounting. Approximate power consumption is about 5300 W, 23 A, 230 V, 60 Hz. The feeder discharges into a 30 in. belt, in line.

7.13 BAR SCRAPER FEEDERS

Bar feeders are used mostly for handling slack bituminous coal ($-¾$ in. and under), where soil conditions are poor (see figures 7.20 and 7.21).

An apron feeder has to carry the load on the top side, and would greatly reduce the slope of the hopper sides if the unloading pit bottom has to be kept high because of poor soil or high ground water. Besides, the cost of an apron feeder would be considerably higher for both equipment and installation. An 18-in.-wide bar feeder consists of two strands of either C-188 or C-131 combination malleable-iron and steel chain and 1½ in. steel flights welded to chain attachments every 10 in. Operating at a speed of 10 fpm, it will move about 12 tph of coal (50 lb/ft^3). It should stay under 30 fpm to prevent considerable maintenance. Where the weather is fairly dry, the coal flows freely, and a 50-ton car can be unloaded in about 5 hours, barring any unforeseen delays. In winter, coal that gets wet enroute from the mines (even small sizes) tends to bunch into about 3 in. lumps, but easily falls apart from rough handling. Run-of-mine coal (10 in. lumps or under) should not be handled on this type of feeder, because of small clearances. The capacity of the bar feeders is low. When necessary, the feeders can operate on a 30° slope, and can thus be made to feed another conveyor or elevator.

7.14 WEIGH FEEDERS

The Tecweigh feeder can be connected to the bottom of the hopper to read the belt speed, weight of material on the belt, and display instantaneous reading of the ma-

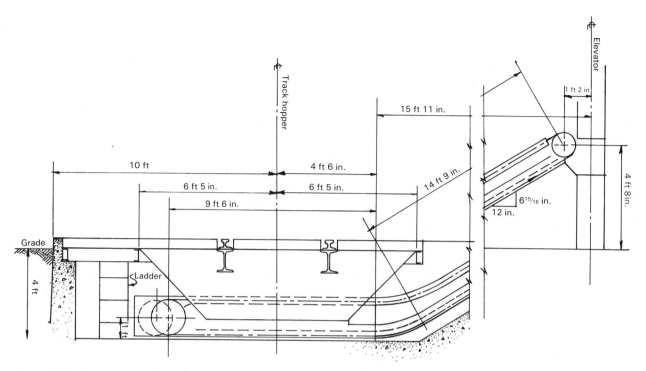

Figure 7.20 Arrangement of a shallow track hopper with bar feeder.

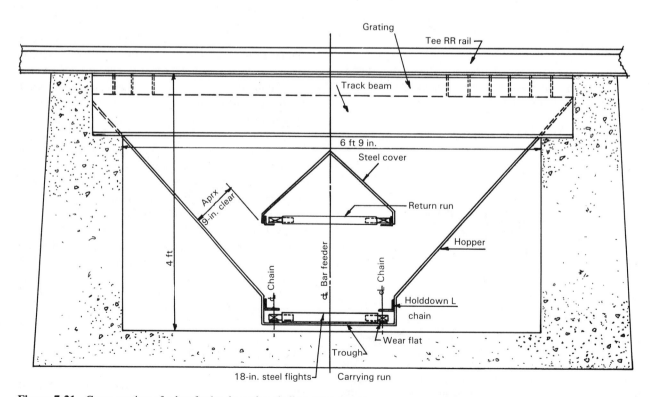

Figure 7.21 Cross-section of a bar feeder through a shallow track hopper.

terial in tons per hour. It can be used on free-flowing powder or granular material, such as sand or grain. Maximum size of particle is ¾ in. Minimum density of material is 20 lb/ft^3. Power required is 115 V or 230 V, 60 Hz, single phase.

7.15 SIZING APRON FEEDER UNDER TRACK HOPPER
(see figure 7.22)

7.15.1 Data

Material: coke, minus 2 in. to dust, weighing 52 lb/ft^3
Speed: 20 fpm recommended, but up to 50 fpm possible, so use 30 fpm
Capacity: 75 tph
Hopper opening between rails: 4 ft 8½ in., say 4 ft.

7.15.2 Design

Try a 24-in. wide apron feeder with 18 in. skirt sides. Available cross-section of feeder equals

$$1.75 \times 1.25 \times 1 \text{ ft } 0 \text{ in.} = 2 \text{ ft}^3/\text{ft of feeder.}$$

Due to voids in lumpy coke, use 80% factor.

$$2 \text{ ft}^3 \times 0.80 \times 52 \text{ lb/ft}^3 = 83.2 \text{ lb/lin ft}$$

$$83.2 \times 30 = 2496 \text{ lb coke per minute}$$

or

$$2496 \times 60 = 149,760 \text{ lb coke per hr.}$$

Where ice or snow are involved, a 30 in. apron feeder (27½ in. inside) would be better.

$$2.29 \times 1.25 \times 1.0 = 2.86 \text{ ft}^3 \text{ of space/ft of feeder}$$

$$2.86 \times 0.8 = 2.29 \text{ ft}^3;$$

$$2.29 \times 52 = 119 \text{ lb/ft. of feeder}$$

$$\frac{150,000}{119 \times 60} = 21.0 \text{ fpm.}$$

7.16 ROTARY FEED VALVES

Rotary valves are placed under the hoppers or bins to arrange for a uniform feed of material to conveying equipment. The valves will handle powdered, pelletized, and granular products and help reduce hazards from explosive dust and volatile material. The tightness of the valve can be varied to suit the product handled. The motor is mounted on the body and assembled as a unit. Air leakage may be important where toxicity is involved. The feeder should be checked for size of pockets and size of material to be handled. It should not be used for hard or abrasive material. For mildly abrasive material, it should be oversize and travel at slow speed. In pressure systems, outboard bearings must be used.

In addition to regular rotary feed valves, Sprout-Waldron, a Division of Koppers makes a "Blow-Through" airlock feeder that is used in connection with pneumatic loading and unloading of material. It becomes an integral part of the pipeline in either positive

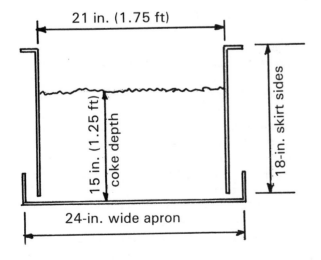

Apron Feeder Cross-section

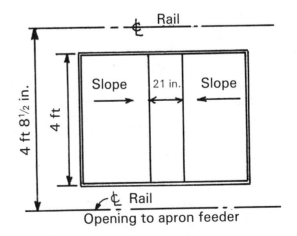

Plan of Track Hopper

Figure 7.22   Cross-section and plan of a track hopper.

or negative pressure systems, and will accurately feed against pressure of 25 psi.

The Fuller Company's airlock feeders are made for delivering mildly abrasive material into pneumatic conveying systems, under pressure of vacuum. The feeder must be vented. A gear motor can be mounted directly on the airlock.

All standard feeders (airlocks) are designed to handle material having a maximum temperature of 170°F. Above 170°F requires additional peripheral and end clearances. From 700°F to 900°F requires additional clearance and a water-cooled shaft. Above 900°F, and not in excess of 1200°F, requires ductile iron construction, special bearings, and a water-cooled rotor.

When feeding into pressure or vacuum systems, the valves must be supplied with airlocks.

Rotor vanes may be coated or made of special steels. For sticky materials, teflon coating should be used, but some materials coat the vanes and harden.

The application of a feeder can usually be broken down into two categories: (1) rotary feeder and (2) rotary airlock. Although the actual feeder is the same in both cases, the efficiency of each application is different, making the drive speed different. A rotary feeder meters material. There is no pressure differential across the feeder; however, a rotary feeder can be used under a head of material, if

a. it is located immediately beneath an Airslide bottom bin,
b. it is located beneath a small batch hopper having aeration pads,
c. if the head of material above the feeder is held to a minimum by the use of a vented expansion hopper, or
d. if located beneath a bag dump hopper where the head is limited.

The volumetric efficiency of a rotary feeder is approximately 85%.

The rotary airlock application operates with either a pressure or vacuum differential, but does not necessarily meter material. Airlocks are normally used either as a charging device for low-pressure pneumatic conveying systems, or as a suction lock for dust collectors. The volumetric efficiency of a rotary airlock is approximately 67%.*

7.17   CAR LOADING DEVICES

See also Package Conveyors (section 17) and Pneumatic Conveyors (section 18).

There are several special devices for loading cars.

---

*J. H. Clauser of The Fuller Company.

Stephens-Adamson, makes a centrifugal thrower unit for loading grain, sand, and other similar bulk materials. In the operating position, the thrower is centered on the car door, and a pivoted chute is placed in the center of the car where a pivoted thrower unit distributes the material through the car. The unit has a capacity of 30 tph for material weighing 25 lb/ft^3, and 50 tph for material weighing 50 lb/ft^3 and over. The size of lumps should not exceed 1½ in. The device cannot be used for very abrasive material and is handicapped by dust problems.

Superior Systems Company makes a collapsible, clear-view spout for dust-free loading of fine dry bulk materials into cars or trucks. They also make heavy-duty abrasive-resistant spouts for handling coal or limestone products. Pushbutton control for loading cars is available. They also make a spout positioner, which is remote-controlled. Open-bed trucks can be loaded with minimum dusting.

7.18   GENERAL—VIBRATING (OSCILLATING) CONVEYORS

Oscillating conveyors move material in a uniform, continuous flow by the upward and forward oscillating motion of a continuous metal trough, mounted on sturdy, inclined reactor legs and supported by a spring system, or leaf springs. A constant-stroke eccentric drive provides a powerful surge-proof conveying action. Trough covers can be provided for dust-tight construction.

7.19   LIGHT-DUTY OSCILLATING CONVEYORS (UP TO 25 TONS PER HOUR)

In general, these units are used to handle granular, free-flowing materials ⅛ in. mesh and larger, but +100 mesh may work if moisture content is low. They can operate as one unit up to 100 ft long (figure 7.23c). Materials that are moist or sticky, such as prepared foundry sand, should be carefully examined and, in most cases, laboratory tests should be made to determine the speed of flow and capacity to be expected. As an aid in determining the conveyability of various materials on oscillating conveyors, refer to table 7.1.

An 8 in. × 4 in. light-duty unit was installed to handle coke dust from dust collectors (200 mesh) with an expected capacity of 10 tph. The coke dust, when discharged from the dust collectors, dropped onto the conveyor a distance of 18 in., and the coke dust seemed to settle densely with little air in the dust. What little air remained tried to work its way up through the dense mass and bubble out at the top. It appeared that the dust

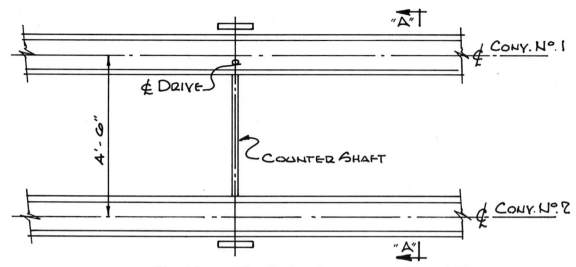

a. Plan View of Oscillating Conveyors on Top of Silo

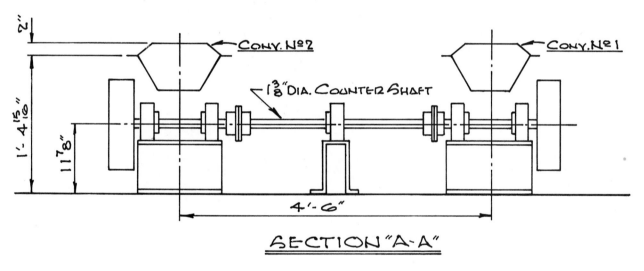

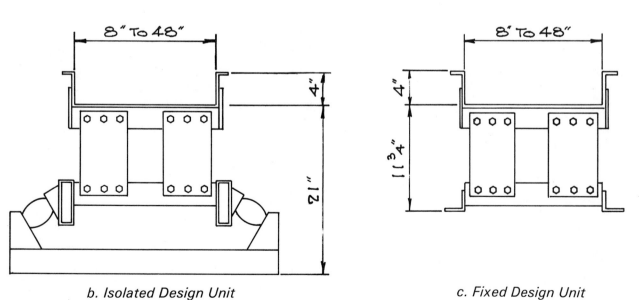

b. Isolated Design Unit   *c. Fixed Design Unit*

Figure 7.23 Oscillating conveyors.

Table 7.1   Conveyability of Materials

Material Characteristics	*Recommendation*
Size	
Very fine: minus 100 mesh	Lab test or test unit may be required. Consult manufacturer.
Fine: 100 mesh to 1/8 in.	Good if low moisture content. For materials that tend to stick or ball up, consult manufacturer.
Granular: 1/8–1/2 in.	Excellent
Lumpy: containing lumps 1/2 in. and over	Excellent
Irregular: fibrous, stringy, or the like	Good. For stringy material such as long turnings, consult manufacturer.
Flowability	
Very free-flowing; angle of repose to 30°	Excellent
Free-flowing; angle of repose 30–45°	Excellent
Sluggish; angle of repose 45° and up	Fair. Depends on moisture content. Lab test may be required. Consult manufacturer.
Abrasiveness	
Nonabrasive	Excellent
Mildly abrasive	Excellent
Very abrasive	Good for fine or granular particle sizes. Can be rubber lined or special heavy construction for large particles. Consult manufacturer.
Other	
Contaminative, affecting use or salability	May require special material or lining. Consult manufacturer.
Hygroscopic	Excellent if dust-tight cover provided.
Highly corrosive	Consult manufacturer. May require stainless steel or galvanizing.
Mildly corrosive	Consult manufacturer. May require stainless steel or galvanizing.
Gives off dust or fumes harmful to life	Excellent if provided with dust-tight cover.
Contains explosive dust	Good if provided with cover. Unit may need additional ground strap.
Degradable, affecting use or salability	Excellent
Very light and fluffy	Excellent
Interlocks or mats to resist digging	Excellent
Aerates and becomes fluid	Poor. Consult engineer for test or recommendations.
Packs under pressure	Poor. Consult engineer for test or recommendations.

was a dead mass and would only travel very slowly, giving only about 2 tph. Electric vibrating units acted the same way. Most very fine materials, −100 mesh, do not handle well in any light-duty oscillating conveyors. Light-duty units have been successful in handling scrap from milling machines, small metal parts, metal powders, food products, and chemicals.

7.20 MEDIUM- AND HEAVY-DUTY OSCILLATING CONVEYORS

Medium units handle about 80 tph, depending on weight per ft³ and are more durable than the light-duty units. They handle material up to 1 in. with fines in it. Their legs are made of malleable iron and they are backed up in oscillating motion by coil springs in each unit. Open troughs can be covered to be made dust-tight.

In medium units, the rpm of the eccentric shaft is 450. Their motors are slightly larger than light duty. They can be put on a floor made of concrete, or on bins, if the base is isolated or balanced. They can be lined with stainless steel or enameled for handling food prod-

ucts, and they are easy to clean. Antifriction roller bearings are on each shaft end. These are equipped with flared troughs. Natural frequency plus positive action help maintain capacity despite load variations.

High-capacity units are of the same general construction as medium-duty units, except components are built for heavier duty. The throw is up to ¾ in. The heavy units are used for handling hot materials. They are ideal for hopping foundry sand to get heat out and cool down. Sometimes they are used as picking conveyors, if placed on a floor where there are men in asbestos suits to pick off good castings. They can be used for drying materials.

Figure 7.23a shows the general arrangement of a series of 10 in. × 6 in. oscillating conveyors on top of storage bins. This is an interesting installation, but usually isolated units are preferred.

An arrangement of two conveyors, operating side-by-side on 4 ft-6 in. centers, each driven by a 5-hp motor, but connected together by a 1⅜ in. diameter countershaft with gear-type couplings was used. These conveyors were arranged to operate in phase; that is, while one conveyor was moving forward, the other conveyor

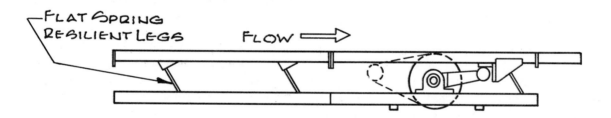

Link-Belt Flexmount Conveyor

for Light- and Medium-Duty Service

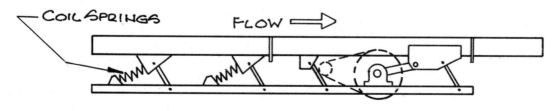

Link-Belt Coilmount Conveyor

for Medium- and Heavy-Duty Service

Figure 7.24 Link-Belt vibrating conveyor.

was moving backward. This so-called balance motion practically eliminates any vibration to the supporting steel (which was designed to take vibration) on top of bins about 30 ft above the floor. Air-operated gates, located in the bottom of the trough, were remotely controlled. Each conveyor was equipped with a 14-gauge steel cover with rubber gasket, to keep dust in conveyor. Stephens-Adamson and FMC also make a rugged natural-frequency coil-mounted unit with troughs as above, that can be supported on the floor or hung by cables. Unit is on an isolated base. It uses a spring-supported trough with the springs spaced close together, and depends on natural frequency plus positive action for smooth operation. This type of conveyor, with covers, has given good results in handling graphite scrap, coke, sand, and mixtures on top of storage bins, and air-operated gates located in bottom of trough. These gates are remotely controlled.

Jeffrey makes a rugged heavy-duty unit, in which the trough is supported by a series of spring steel slots spaced close together.

All vibrating conveyors should be placed on the level (see figure 7.24 for illustration of types).

7.21 NATURAL FREQUENCY

Natural frequency is a function of weight of trough, weight of material, and drive speed. There is just one combination of weight and speed wherein the natural frequency is theoretically reached, without changing the number of spring reactors. Adding positive action means that the drive is directly connected to the carrying trough, providing the additional energy required for loads beyond the natural-frequency range. Uneven loads with surges throw the conveyor out of natural-frequency balance, and require a supplement of positive action.

7.22 SUPPORTS

Vibrating conveyors located on top of bins must have heavy rigid supports, and be well tied into the bin structure. Dynamically balanced or isolated-base drive units will eliminate a large part of the transmitted vibrating forces.

7.23 ISOLATING CONVEYOR

The FMC Corporation is producing vibrating conveyors as isolated-design units. This computer-designed system is located between the conveyor base and the supporting structure (figure 7.23b).

7.24 MAINTENANCE

Even with a well designed conveyor, a bearing can malfunction and cause serious problems. Maintenance on vibrating conveyors is high when handling products that weigh over 100 lb/ft^3. The heat problem also is a factor in many cases.

7.25 VIBRATING DRUM PACKERS (VIBRATING TABLES)

7.25.1 General

It is best to send samples of the material to the manufacturers for their recommendation. A part of a sample specification for drums storing carbide follows. It was based on performed tests and is not to indicate that the manufacturer mentioned is necessarily preferred at this time, nor that it performed well for other products at the time the test was made.

7.25.2 Partial Specification

The vibrating drum packer shall be a heavy-duty electromagnetically operated unit, operating at a frequency of 3600 vibrations per minute, with a variable straight-line stroke on packer table, adjustable while the packer is in operation. This stroke shall be at least $1/16$-in. at maximum setting, when gross load of 750 lb or less is mounted on the packer table. The stroke shall be capable of adjustment downward from this maximum by remote electrical control, and without mechanical adjustment on the packer itself. The packer table shall be sufficiently rigid to ensure uniform vibration of all points during maximum vibration. The packer table shall be complete with support frame and shock absorbers for floor mounting.

Manufacturer's standard electromagnetic vibrating table, Model VP181C1, manufactured by FMC Corporation, Material Handling Equipment Division (Syntron), shall be furnished as described above. The vibrator table shall operate on half-wave rectified current from a power supply of 460 V, single-phase, 60 Hz.

A totally enclosed (NEMA 4 type), separate, wall-mounted control shall be furnished. The control shall be suitable for operation on 460-V, single-phase, 60-Hz alternating current, and for continuous operation at rated power output. The control shall include an on-off toggle switch, fusing, rectifier, and a rheostat for control of packer stroke.

The equipment shall be completely assembled, operated, and tested at the manufacturer to ensure immediate and successful operation when delivered.

WIRE MESH CONVEYORS

8.1 GENERAL

Wire cloth conveyors operate slowly. They are used in canneries and food-processing plants moving products through washing, sorting, blanching, scalding, cooking, and draining processes. They can be washed clean and made sanitary. Quick freezing of vegetables or fruits can be done on these conveyors while they pass through freezing chambers. Meat packing, including seafood, also is handled in this way. The wire cloth does not retain any small scraps, and is not affected by changes in temperature. Where food is involved, the cloth is made of stainless steel.

The glass and ceramic industry uses many wire cloth conveyors, handling glass bottles and electric light bulbs through annealing furnaces with a minimum of breakage; as well as milk bottles for dairies, eyeglass lenses, safety glass, Christmas tree ornaments, decorated china through a kiln, pottery of all kinds, glazing of many pieces, and sintering operations for powder metallurgy parts.

8.2 HEAT-RESISTING ALLOY BELTS

Many wire cloth manufacturers can furnish heat-resisting alloys. Their properties have been calculated on quick-pull tensile tests; creep-strength values based on maximum temperatures to produce a certain percentage of elongation in a 10,000-hr operation scaling, temperatures, and oxidizing atmospheric conditions, in degrees Fahrenheit.

8.3 TYPES OF WIRE BELTS

Wire belts are obtainable with any size opening in the mesh from $\frac{1}{64}$ to $2\frac{1}{2}$ in., and can be woven of any thickness of wire from 3 to 32 W&M gauge (Washburn and Moen gauge is considered a standard for wire cloth in this country).

Many wire cloth conveyors are furnished with double strands of various standard chains operating over specially designed sprockets to form an apron conveyor. Cross rods, depending on conditions, are furnished to stiffen the cloth. Table 8.1 serves as a guide for selecting the proper chain.

These flexible conveyor belts have greatly changed methods of production and are widely used in all types of industry.

Calculation parameters are given on figure 8.1.

8.4 ANSI CHAIN NUMBERS

Tables 8.5, 8.6, 8.7, 8.8, and 8.9 are from Cambridge Wire Cloth Co. data. RC chain numbers correspond to ANSI numbers given in Cambridge data, but are used here to correspond with descriptions elsewhere in the text.

Table 8.1   Drag Chain Selection

Chain Type	Recommended Speed	Recommended Maximum Temperature	Recommended Service
Drag			
Detachable	30 ft/min	Malleable iron—600°F Promal—1000°F	Light duty with nonabrasive materials
Pintle	50 ft/min	Malleable iron—600°F Promal—1000°F High-temp alloys—1600°F	Medium to heavy duty with moderately abrasive material (Promal can be used in more abrasive environment)
Combination	50 ft/min	Malleable iron—600°F Promal—1000°F	Same as pintle
Welded	50 ft/min	Steel—1000°F	Heavy duty, abrasive environment
Roller			
Cast roller	100 ft/min	Malleable iron—600°F Promal—1000°F	Same as pintle and combination
Drag or Roller			
Bush (semi-precision)	Drag: 50 ft/min Roller: 200 ft/min	Carbon steel—600°F	Heavy duty, dirty, and abrasive environment
Precision	Drag: 50 ft/min Roller: 300 ft/min	500–700°F, depending on conditions	Light to heavy duty; good for long life and extra smooth operation

Source: Courtesy Cambridge Wire Cloth Co.

Note: High-temperature applications should be referred to the manufacturer.

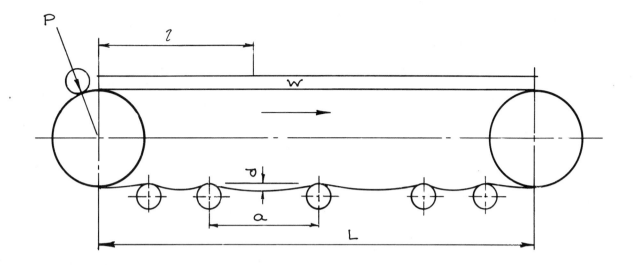

L	CENTER DISTANCE OF END PULLEYS IN FEET
W	LOAD ON BELT PER LIN. FOOT - UNIFORMLY DISTR.
w	WEIGHT OF BELT PER LINEAR FOOT
f	COEFF. OF FRICTION - (see table 8.2)
K	CONSTANT - (see table 8.3)
a	SPACING OF SUPPORTING ROLLS IN INCHES
d	FESTOONING BETWEEN ROLLS IN INCHES
P	PRESSURE ROLL PRESSURE
T	MAXIMUM BELT TENSION TO DRIVE
T_1	BELT TENSION PRODUCED BY BELT & LOAD
l	BELT TENSION DIST. FROM CHARGE-OR DISCHARGE PULLEY
t	BELT TENSION AT l - PLAIN FRICTION DRIVE
t_1	" " " - PRESSURE ROLL DRIVE - DISCHARGE END
t_2	" " " " " " - CHARGE END
t_3	BELT TENSION AT ROLL SUPPORTS (see table 8.4)
T	T_1 FOR PRESSURE ROLL DRIVE
T	$T_1\left(1+\frac{1}{K}\right)$ FOR PLAIN FRICTION DRIVE
T_1	$[(W+w) \times L \times f] + [w \times L \times f]$
t	$T - [(W+w)(L-l) \times f]$
t_1	$[(W+w) \times l \times f] + [w \times L \times f]$
t_2	$(W+w) \times l \times f$

TOTAL NORMAL PRESSURE OF PRESSURE ROLL TO PRODUCE "O" SLACK TENSION

FOR PLAIN PULLEY $\quad P = \dfrac{T_1}{(K+1) \times .3}$

FOR LAG PULLEY $\quad P = \dfrac{T_1}{(K+1) \times .6}$

BELT TENSION AT ROLL SUPPORTS $\quad t_3 = \dfrac{(W+w) \times a}{24}\sqrt{1+\dfrac{a^2}{16d^2}}$

OR WITH KNOWN TENSION t_3 TO FIND FESTOON "d" $= \dfrac{a}{4} \times \dfrac{1}{\sqrt{\left(\dfrac{24\,t_3}{(W+w) \times a}\right)^2 - 1}}$

HORSEPOWER TO DRIVE

$$HP = \frac{W \times S \times f}{33000}$$

W = TOTAL WEIGHT OF BELT & LOAD IN POUNDS
S = SPEED OF BELT IN FEET PER MINUTE
f = COEFF. OF FRICTION
THE ABOVE HP FORMULA IS THEORETICAL AND IS CALCULATED AT HEAD OR DRIVING PULLEY ONLY. IT DOES NOT CONSIDER FRICTIONAL LOSSES OR EFFICIENCY OF SPEED REDUCER.

Figure 8.1 Design calculation parameters. Note: check your figures with the manufacturers of wire cloth.

Table 8.2 Coefficient of Friction

Temperature (°F)	Metal Hearth	Brick Hearth
Up to 1000	0.35	0.50
1200	0.37	0.52
1400	0.40	0.54
1600	0.44	0.57
1800	0.49	0.60
2000	0.55	0.65

Table 8.3 Constant *K*

Degree Wrap on Drive Pulley	Plain Pulley	Lagged Pulley
150°	1.19	3.81
165	1.37	4.63
180	1.57	5.56
195	1.78	6.71
210	2.00	8.02
225	2.25	9.55
240	2.51	11.35
255	2.80	13.45
270	3.11	15.90
285	3.45	18.78
300	3.81	22.14

Table 8.4 Roll Support

	Start	Motion
Lub. bearing at room temp	0.12	0.08
Roller chain	0.20	0.15
Steel or mall. drag chain on stl	0.35	0.25
Steel on hardwood, dry	0.50	0.40

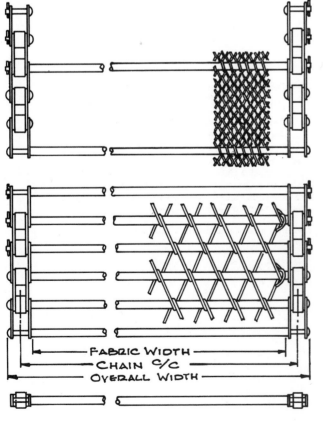

Table 8.5 Metal Mesh Belt with Standard Roller Chain and Journaled Rods

RC Chain No.		Chain Pitch (in.)	Rod Diameters		Allowable Chain Pull (lb/strand)
Standard Roller	Large Roller		Major Size (in.)	Minor Size (in.)	
35	—	0.375	0.156 to 0.250	0.135	250
40	—	0.500	0.192 to 0.375	0.156	450
50	—	0.625	0.207 to 0.500	0.192	750
60	—	0.750	0.250 to 0.500	0.234	1050
80	—	1.000	0.375 to 0.625	0.312	1800
C-2050	C-2052	1.250	0.192 to 0.250	0.192	750
C-2060	C-2062	1.500	0.250 to 0.375	0.234	1050
C-2080	C-2082	2.000	0.375 to 0.500	0.312	1800
C-2120	C-2122	3.000	0.500 to 0.625	0.437	4250

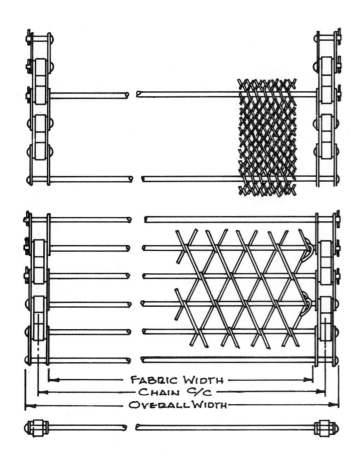

Table 8.6 Metal Mesh Belt with SBS or RS Bushed Roller Chain

Chain No. and Type	Chain Pitch (in.)	Rod Size (in.)	Allowable Chain Pull (lb/strand)
SBS-188	2.604	0.500	2750
RS-3013	3.000	0.438	2100
RS-4013	4.000	0.438	2100
RS-4113	4.000	0.438	2150
RS-4019	4.000	0.500	2450
RS-1114	6.000	0.625	3800

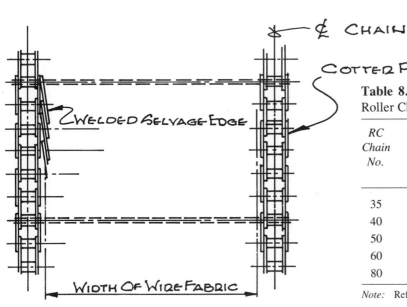

Table 8.7 Metal Mesh Belt with Standard Precision Roller Chain

RC Chain No.	Pitch of Chain (in.)	Size of Transverse Rods (in.)	Allowable Chain Pull (lb/strand)
35	3/8	0.135	250
40	1/2	0.156	450
50	5/8	0.200	750
60	3/4	0.234	1050
80	1	0.312	1800

Note: Refer table 9.6.

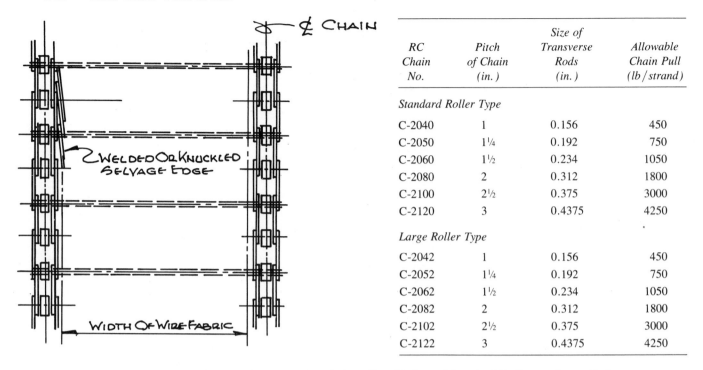

RC Chain No.	Pitch of Chain (in.)	Size of Transverse Rods (in.)	Allowable Chain Pull (lb/strand)
Standard Roller Type			
C-2040	1	0.156	450
C-2050	1¼	0.192	750
C-2060	1½	0.234	1050
C-2080	2	0.312	1800
C-2100	2½	0.375	3000
C-2120	3	0.4375	4250
Large Roller Type			
C-2042	1	0.156	450
C-2052	1¼	0.192	750
C-2062	1½	0.234	1050
C-2082	2	0.312	1800
C-2102	2½	0.375	3000
C-2122	3	0.4375	4250

Table 8.8 Metal Mesh Belt with Extended-Pitch Roller Chain and Standard Rollers or Large Rollers

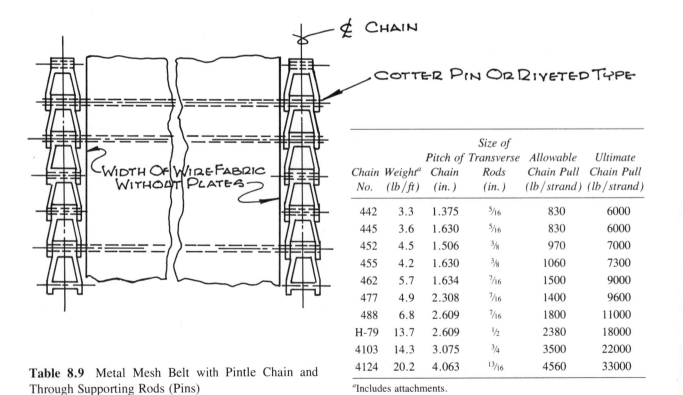

Chain No.	Weight[a] (lb/ft)	Pitch of Chain (in.)	Size of Transverse Rods (in.)	Allowable Chain Pull (lb/strand)	Ultimate Chain Pull (lb/strand)
442	3.3	1.375	5/16	830	6000
445	3.6	1.630	5/16	830	6000
452	4.5	1.506	3/8	970	7000
455	4.2	1.630	3/8	1060	7300
462	5.7	1.634	7/16	1500	9000
477	4.9	2.308	7/16	1400	9600
488	6.8	2.609	7/16	1800	11000
H-79	13.7	2.609	1/2	2380	18000
4103	14.3	3.075	3/4	3500	22000
4124	20.2	4.063	13/16	4560	33000

Table 8.9 Metal Mesh Belt with Pintle Chain and Through Supporting Rods (Pins)

[a]Includes attachments.

Drives

9.1 SPEED REDUCERS

Machinery generally operates at speeds slower than common fixed-speed power sources; e.g., electric motors, turbines, engines. It is more economical and practical to provide a speed reducer for converting power to the proper combination of speed, torque, and direction than to provide a special slow-speed power source.

Reducers are manufactured in many standard configurations and ratios in any given size, and are selected on the basis of hp (or torque), speed, and application with special considerations for mounting, space limitations, efficiency, overhung loads, weight, angle of transmission, special features, etc.

Backstops prevent counter-rotation and generally are available as an integral part of a reducer, if required.

If variable speed is needed, this can be accomplished by the use of variable-pitch sheaves, mechanical variators, fluid drives, dc motors, eddy-current couplings, or variable-frequency ac motors. Variable-speed reducers often are used in connection with dust collectors, dryers, or any other equipment having variable load requirements.

A final reduction using roller chain often is used to provide more versatility in speeds and mounting (see figure 9.1, arrangement 1). For screw conveyors with 1200 (1150 under load) or 1800 (1750) rpm, 10-hp or less motors, a 4.5-in. sheave is the minimum recommended for the drive shaft. The diameter of the sprockets or sheaves should not be less than the values in table 9.1, multiplied by 1.25 for machined pinion gear, 1.5 for V-belt drive, and 2.5 for flat-belt drive.

9.2 SHAFT COUPLINGS

Couplings are generally used to connect shaft to shaft and are available either rigid or mechanically (axially) flexible, torsionally stiff or flexible, lubricated or nonlubricated, and combinations of these; also, for special purposes such as brake wheels, slip, spacer, etc., a coupling that is both mechanically and torsionally flexible is recommended.

Selection is made by application, horsepower, rpm, bore capacity, and maximum speed. Refer to manufacturers' catalogues for procedure.

9.3 DRIVES

Figure 9.1 shows three arrangements of drives for conveyors or bucket elevators. Arrangement 2 is generally maintenance-free, more expensive; arrangement 1 is most practical; arrangement 3 is practical and inexpensive.

To select the proper reducer, the motor hp (or brake hp) is multiplied by a service factor. A partial listing of service factors by application is shown in table 9.2 and applies only to the category of speed reducers.

Gear motors, motor reducers, shaft-mounted, and screw-conveyor drives are selected from Class I, II, or III (uniform, moderate shock, or heavy shock). For material handling, it is well to use Class II, and for crushers, Class III.

Worm-gear drive load classifications used by some are U, M, and H (uniform, moderate shock, and heavy

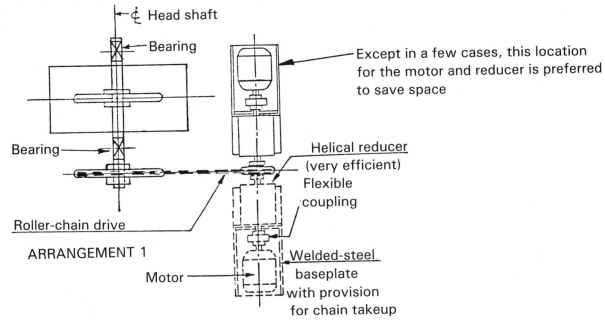

ARRANGEMENT 1

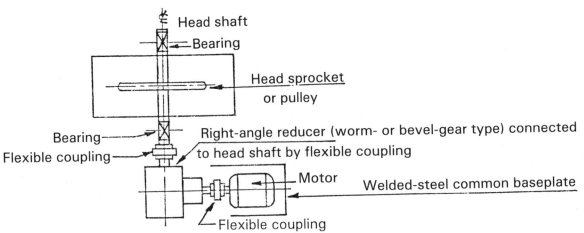

ARRANGEMENT 2

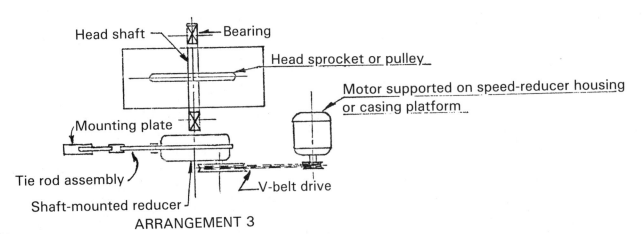

ARRANGEMENT 3

Figure 9.1 Typical drive details. Note that arrangement 2 does not have the flexibility of arrangement 1. In arrangement 1, using a drive chain between the head shaft and reducer, it is possible, if necessary, to increase or decrease the speed of the head shaft a few rpm. In arrangement 2, speed reducers mounted on the head shaft, with a V-belt drive connecting the motor, save space and cost, and allow versatility.

Table 9.1 Recommended Minimum Sheave or Sprocket Diameters for Electric Motors (in.)

Motor Horsepower	*Motor rpm*	
	1150	*1750*
2	2.5	2.5
3	3.0	2.5
5	3.0	3.0
7½	3.75	3.0
10	4.5	3.75
15	4.5	4.5
20	5.25	4.5
25	6.0	4.5
30	6.75	5.25

Source: Courtesy NEMA

shock), which are converted to a numerical value based on prime mover and hours of usage per day. Others use numerical values.

For each of these categories, it is necessary to refer to the manufacturer's catalog that has the complete AGMA listings and selection tables taking into account other factors, such as input speed, ratio, thermal hp, and so on.

Mechanical efficiencies are covered in table 9.3. As an example of the application of the overall drive efficiency—the result of combining equipment unit efficiencies—consider a belt conveyor drive consisting of a double-helical-gear speed reducer and an open-guarded roller chain on cut sprockets. The approximate overall efficiency, according to table 9.3, is $(0.94)(0.93) = 0.874$. If the calculated minimum horsepower at the drive shaft is 13.92 hp, then the required motor horsepower is $13.92/0.874 = 15.9$ hp. Therefore, it is necessary to use at least a 20-hp motor.

Shaft coupling manufacturers each have their own method of determining service factors.

Much of the equipment manufacturers' design data is figured on the basis of using babbitt bearings.

Conveyor head-shaft sizes can be taken from manufacturers' catalogs or calculated as shown in paragraph 9.6. Theoretically, the torsional moment is caused by the unbalanced load on each side of the shaft, but to allow for the effect of starting, teeth in sprocket, mis-

Table 9.2 Service Factors Listed by Application for Electric-Motor Drives

Application	*Service Hours*		*Application*	*Service Hours*	
	10	*24*		*10*	*24*
Agitators			Cement kilns	—	1.50
Pure liquids	1.00[a]	1.25	Dryers and coolers	—	1.50
Liquids and solids	1.25	1.50	Skip hoist	1.25	1.50
Liquids, variable density	1.25	1.50	Stokers	1.00	1.25
Blowers			Elevators		
Centrifugal, vane	1.00	1.25	Bucket, UL	1.00	1.25
Car dumpers	1.75	2.00	Bucket, HD	1.25	1.50
Car pullers	1.25	1.50	Fans		
Conveyors (UL)[b]			Centrifugal	1.00	1.25
Apron, belt, bucket, chain,			Forced draft	—	1.25
flight, screw	1.00	1.25	Induced draft	1.25	1.50
Conveyors (HD)[c]			Feeders		
Apron, belt, bucket, chain,			Apron, belt, screw	1.25	1.50
flight, screw	1.25	1.50	Disc	1.00	1.25
Conveyors, vibrating	1.75	2.00	Reciprocating	1.75	2.00
Crushers			Pebble mills	—	1.50
Ore or stone	1.75	2.00	Screens		
Ball and rod mills			Air washing	1.00	1.25
with spur ring gear	—	2.00	Rotary, for sand or gravel	1.25	1.50
with helical ring gear	—	1.50			
direct connected		2.00			

Source: Courtesy AGMA

Notes: Recommendations are minimum and normal conditions are assumed.
See Table 9.4 for V-belt drive recommendations.

[a]Lower service factors may be used when gear manufacturer or its customer has control of design of entire system, consisting of prime mover, gear, and driven equipment, *and* has verified service factors by recorded load measurements or by actual operating experience of successful installations of same design.
[b]UL = uniform load or feed.
[c]HD = heavy duty (not uniformly loaded).

Table 9.3 Mechanical Efficiencies of Speed-Reduction Mechanisms

Type of Speed Reduction Mechanism	Approximate Mechanical Efficiency
V-belts and sheaves	0.94
Roller chain and cut sprockets, open guard	0.93
Roller chain and cut sprockets, oil-tight enclosure	0.95
Single-reduction helical or herringbone gear speed reducer or gearmotor	0.95
Double-reduction helical or herringbone gear speed reducer or gearmotor	0.94
Triple-reduction helical or herringbone gear speed reducer or gearmotor	0.93
Double-reduction helical gear, shaft-mounted speed reducers	0.94
Low-ratio (up to 20:1 range) worm-gear speed reducers	0.90
Medium-ratio (20:1 to 60:1 range) worm-gear speed reducers	0.70
High-ratio (60:1 to 100:1 range) worm-gear speed reducers	0.50
Cut spur gears	0.90
Cast spur gears	0.85

Source: Courtesy CEMA

alignment, and so on, many engineers use the total load figured in bending as the torsion load.

Figure 9.2 shows a standard screw conveyor drive package offered by most speed reducer manufacturers, providing maximum economy and convenience plus versatility. It is available in sizes from ½ hp to 100 hp (for most manufacturers, only to 75 hp for screw conveyors).

9.4 TORQUE LIMITERS

Torque limiters are control devices that slip under excessive load. As soon as torque requirements are reduced below a set limit, the device will again transmit the load. The limiters can be adjusted to slip at any desired overload. No resetting is required after the torque limiter has slipped. These torque limiters are simple and positive, and prevent machine damage, product damage, costly downtime caused by shock loads, overloads, or machine jams.

9.5 DRIVE SELECTION

Example: Application: continuous bucket elevator with 25 in. pd (pitch diameter), head sprocket operating at 125 fpm, requires 13.8 hp. Chain and sprocket combination, 40t/14t (teeth) to a speed reducer, coupling-connected to an electric motor (see figure 9.1, arrangement 1).

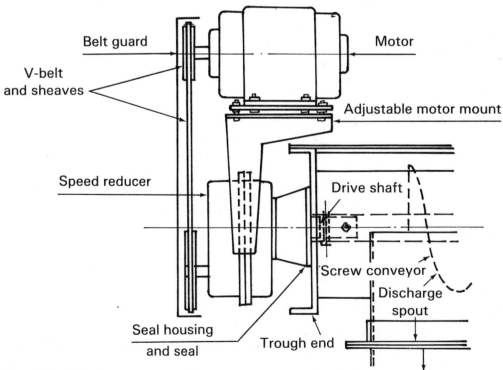

Figure 9.2 Typical screw-conveyor drive.

Table 9.4 Service Factors for V-Belt Drives

| Driven Machine Types[a] | Driver Types, Electric Motors | | | | | |
| | AC Normal Torque AC Squirrel-Cage | | | AC Hi-Torque AC Hi-Slip | | |
	Intermittent Service[b]	Normal Service	Continuous Service	Intermittent Service	Normal Service	Continuous Service
Blowers and exhausters Fans up to 10 hp Light-duty conveyors	1.0	1.1	1.2	1.1	1.2	1.3
Belt conveyors for sand, grain, etc. Fans over 10 hp Revolving and vibrating screens	1.1	1.2	1.3	1.2	1.3	1.4
Bucket elevators Conveyors (drag, pan, screw) Hammermills Positive-displacement blowers Pulverizers	1.2	1.3	1.4	1.4	1.5	1.6
Crushers (gyratory, jaw, roll) Mills (ball, rod, tube) Hoists	1.3	1.4	1.5	1.5	1.6	1.8

Source: Courtesy Goodyear Tire & Rubber Co.

[a]Driven machine types are representative samples. Select a category approximating your application. If idlers are used, add the following to the service factor:

idler on slack side (inside)   none
idler on slack side (outside)   0.1
idler on tight side (inside)   0.1
idler on tight side (outside)   0.2

[b]The correct service factor is determined by (1) the extent and frequency of peak loads; (2) the number of operating hours per year, broken down into average hours per day of continuous service; and (3) the proper service category (intermittent, normal, or continuous). Select the service category that most closely approximates your application conditions.

Intermittent service, service factor 1.0–1.5: light duty (not more than 6 hr/day and never exceeding rated load.

Normal service, service factor 1.1–1.6: daily service 6–16 hr/day and occasional starting or peak load does not exceed 200% of full load.

Continuous service, service factor 1.2–1.8: continuous service 16–24 hr/day and starting or peak load exceeds 200% of full load or where starting or peak loads and overloads occur frequently.

Determine head shaft rpm:

rpm = speed ÷ circumference (ft)

$$= 125 \div \frac{\pi \times 25}{12}$$

$$= 19$$

Determine reducer out rpm:

rpm = head shaft rpm × sheave ratio

$$= 19 \times \frac{40}{14}$$

$$= 54$$

Reducer ratio = 1750/54 = 32.4 : 1.

Select the closest nominal ratio from the gear man-ufacturers' catalog, with an output speed of 56 rpm for a 15-hp, 1750-rpm motor having a service factor of 1.0.

Because the small gear has only 14 teeth to engage the 40 teeth of the larger gear, it will have to make 40/14 revolutions while the large gear makes only one revolution.

9.6 SELECTION OF SHAFTS SUBJECT TO TORSION AND BENDING

The ASME formula for shafting is:

$$d = \sqrt[3]{\frac{16}{\pi S} \sqrt{(K_b M_b)^2 + (K_t M_t)^2}}$$

Table 9.5 Stock Sprockets for Roller Chain

RC-40		RC-50		RC-60		RC-80		RC-100		RC-120		RC-140		RC-160	
No. of Teeth	Pitch Dia. (in.)	No. of Teeth	Pitch Dia. (in.)	No. of Teeth	Pitch Dia. (in.)	No. of Teeth	Pitch Dia. (in.)	No. of Teeth	Pitch Dia. (in.)	No. of Teeth	Pitch Dia. (in.)	No. of Teeth	Pitch Dia. (in.)	No. of Teeth	Pitch Dia. (in.)
9	1.462	9	1.827	9	2.193	9	2.924	11	4.437	13	6.268	12	6.762	11	7.099
10	1.618	10	2.023	10	2.427	10	3.236	12	4.830	14	6.741	13	7.313	12	7.727
11	1.775	11	2.219	11	2.662	11	3.549	13	5.223	15	7.215	14	7.864	13	8.357
12	1.932	12	2.415	12	2.898	12	3.864	14	5.617	16	7.689	15	8.417	14	8.988
13	2.089	13	2.612	13	3.134	13	4.179	15	6.012	17	8.163	16	8.970	15	9.620
14	2.247	14	2.809	14	3.371	14	4.494	16	6.407	18	8.638	17	9.524	16	10.252
15	2.405	15	3.006	15	3.607	15	4.810	17	6.803	19	9.113	18	10.078	17	10.885
16	2.563	16	3.204	16	3.844	16	5.126	18	7.198	21	10.064	19	10.632	18	11.518
17	2.721	17	3.401	17	4.082	17	5.442	19	7.595	26	12.444	21	11.742	19	12.151
18	2.879	18	3.599	18	4.319	18	5.759	20	7.991	35	16.734	26	14.518	21	13.419
19	3.038	19	3.797					21	8.387	45	21.503	35	19.523	26	16.593
20	3.196	20	3.995					22	8.783	60	28.661	45	25.087	35	22.312
21	3.355							24	9.577	70	33.434	60	33.438	45	28.671
22	3.513							26	10.370	80	38.207	70	39.006	60	38.215
23	3.672							30	11.958			80	44.575	70	44.578
24	3.831							32	12.753					80	50.943
								35	13.945						
								40	15.932						
								45	17.920						
								48	19.112						
								54	21.498						
								60	23.884						
								70	27.862						
								80	31.839						

Notes: RC-40, RC-50, etc. refer to FMC nomenclature. For ANSI nomenclature, omit the letters RC.

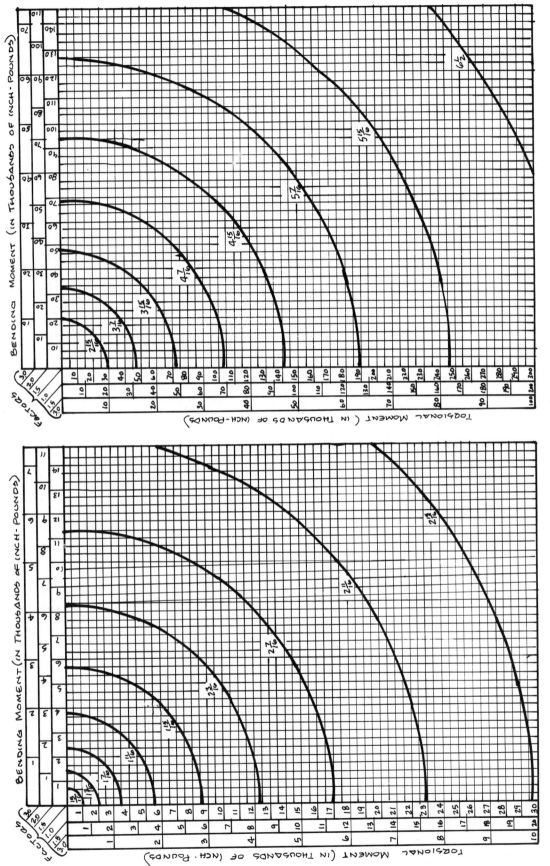

Figure 9.3 Selection of shafts subject to torsion and bending. Charts are based on S = 6,000 psi.

249

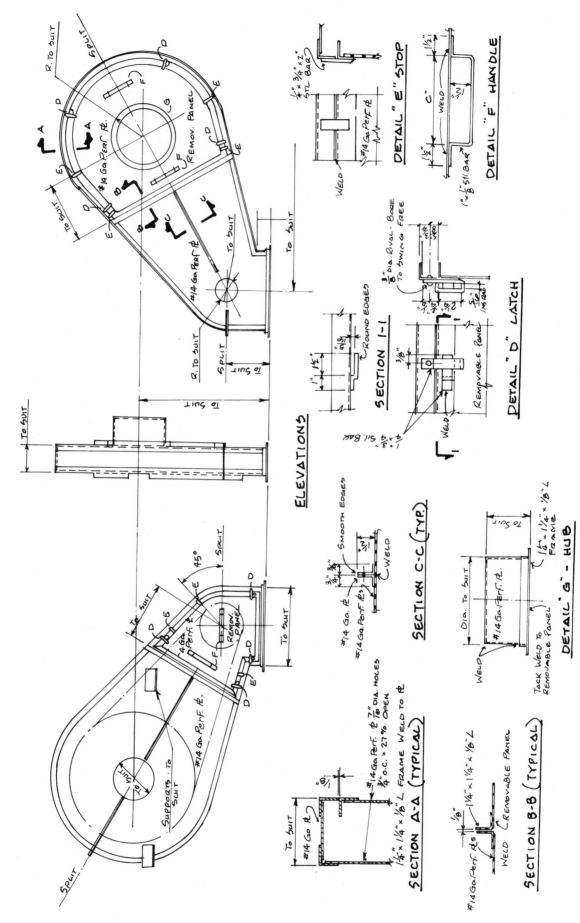

Figure 9.4 Chain drive guards.

Table 9.6 Roller Chain—Ultimate Strength

Chain No.[a]	Pitch (in.)	Average Strength (lb)[b]	Weight lb/ft
RC-40	½	3700	0.41
RC-50	⅝	6100	0.68
RC-60	¾	8500	0.99
RC-80	1	14,500	1.73
RC-100	1¼	24,000	2.51
RC-120	1½	34,000	3.69
RC-140	1¾	46,000	5.00
RC-160	2	58,000	6.53

[a]ANSI numbers are the same, except the letters RC are omitted.
[b]Ultimate strength of single chain.

where

M_b = bending moment in in-lb.
K_b = shock and fatigue factor for M_b
M_t = torsion moment in in-lb.
K_t = shock and fatigue factor for M_t
S = safe shearing stress in psi

K_b is generally assumed as 1.5 for gradual loading, 2.0 for minor shock loading, and 3.0 for heavy shock loading. K_t is generally assumed as 1.0 for gradual loading, 1.5 for minor shock loading, and 3.0 for heavy shock loading.

Figure 9.3 gives the size of shafts subject to torsion and bending.

S is taken as 6,000 psi for shafts with keyseats, and 8,000 psi for shafts without keyseats.

To illustrate the use of the formula, the diameter of the shaft selected in paragraph 7.5 will be computed. Where K_b = 3.0; K_t = 1.5; M_b = 13,500 in.-lb; M_t = 21,000 in.-lb; and S = 6000 psi,

$$d = \sqrt[3]{\frac{16}{\pi S}} \sqrt{(3.0 \times 13,500)^2 + (1.5 \times 20,000)^2}$$

$$= 3.5 \text{ in.}$$

Therefore, use $3\,^{15}/_{16}$-in. diameter. This value may also be obtained from the chart on figure 9.3. For different

Table 9.7 Multiple Strands

No. of Strands	Multiple Strand Factor
2	1.7
3	2.5
4	3.3

Table 9.8 Chain Working Factor

Chain Speed (rpm)	Working Factor (W_f)[a]
Up to 50	7
50–100	8
100–150	10
150–200	12
200–250	14
250–300	16

[a]$W_f \times P$ = minimum ultimate strength, where P is allowable tension in lb/strand.

values of shear S', d' (diameter of shaft)

$$d = \sqrt[3]{\frac{6000}{S'}}$$

and is thus equal to $d \times 0.908$ for S' = 8000 psi, and $d' = d \times 0.843$ for S' = 10,000 psi and $d' = d \times 0.794$ for S' = 12,000 psi.

9.7 DRIVE GUARDS

A typical guard for chain drives is shown on figure 9.4. These are required by safety-oriented government agencies, and safety departments of most industrial plants.

9.8 SELECTION OF TYPE OF V-BELT

For 900-rpm motors (870 under load), use type 3 V-belts up to 20 hp and type 5 V-belts up to 100 hp.

For 1200 rpm motors (1150 under load), use type 3 V-belts up to 25 hp and type 5 V-belts up to 100 hp.

For 1800-rpm motors (1750 under load) use type 3 V-belts up to 40 hp and type 5 V-belts above 40 hp.

Service factors for V-belt drives are given in table 9.4.

Table 9.9 Minimum Size of Driver (Small) Sprockets

Speed	No. of Teeth
Slow	17
Moderate	21
High	25

Note: The size is a function of chain tension, pitch, and overhung load.

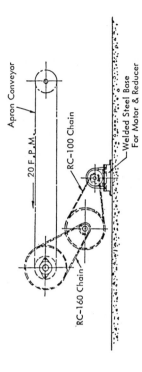

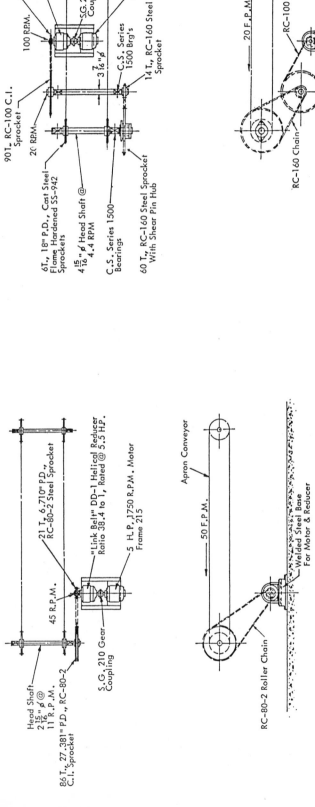

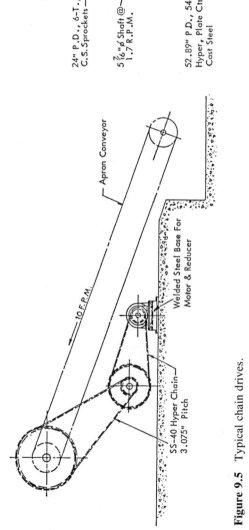

Figure 9.5 Typical chain drives.

9.9 CHAIN FOR DRIVES

Riveted-type roller chain is considered standard in the smaller sizes, up to and including ¾-in. pitch, and will be supplied unless detachable type is specified.

Riveted-type chains are recommended for high-speed drives as a greater rigidity of the pins and sideplates is secured from this construction.

Most roller chain manufacturers produce shear-pin hubs and sprockets that depend on a shear pin of proper diameter to do the safety work intended. These, while they look good on paper, at times are faulty, due to a careless mechanic installing a bolt where a necked shear pin should be.

9.10 SPROCKETS FOR ROLLER CHAIN DRIVES

Table 9.5 lists only standard stock sprockets. Similar items can be furnished by the Rexnord and Stephens-Adamson Manufacturing companies. Many other sprocket sizes are available, as shown in the catalogs. Ultimate strength of roller chain is given in table 9.6. Multiple strands are covered in table 9.7, and chain working factors in table 9.8.

As an example of sprocket selection, assume RC-100 chain, a 5-hp motor, and that a speed reduction of 55 rpm to 23.4 rpm, or 2.35:1, is desired. The service factor for a belt conveyor working fewer than 10 hr/day is 1.0 (table 9.2). Design the sprocket for 5 hp × 1.0 = 5.0 hp. Table 9.9 calls for a minimum of 17 teeth for the driver sprocket. Use 20 teeth. For the driven sprocket, 47 teeth would give the exact ratio required: 47/20 = 2.35:1, but 47-tooth sprockets are not stock items. Use 48 teeth, giving a ratio of 48/20 = 2.4:1.

From manufacturers' catalogs or American Chain Association's "Chains for Power Transmission and Material Handling", select 20- and 48-tooth sprockets for RC-100 chain and a 5-hp motor.

Some typical chain drives are shown on figure 9.5.

Crushers and Screens

10.1 INTRODUCTION TO CRUSHERS

Crushing and pulverizing reduces the material to a uniform size, or to a graduation of sizes for immediate use or further processing. The various manufacturers make specialized crushers for various processes, such as reduction of long coils of metals into short lengths, reduction in size of run-of-mine coal down to one inch and under in one operation, and crushing of rock or ores, grinding garbage, bone shredding, etc. In addition, the equipment manufacturers have, during the years, assembled a vast amount of data on all products, and their advice should be solicited.

Crushers must be clear of material before crushing starts, even though some single-roll crushers can start under load, especially with friable material. Gyratory crushers with hydraulic mainshaft support can be started under load by lowering the mainshaft.

10.2 SELECTION OF CRUSHERS

Most, if not all, crusher manufacturers maintain laboratories where samples of the materials to be crushed are analyzed and actually processed. They are in a position to make recommendations once they are given the information required. Along with the samples, the following information on the properties of the material to be handled should be given to the manufacturer of the crushers or grinders.

Toxic or explosive hazards.

Size and gradation of feed, as well as desired finished product.

Required maximum and minimum capacity. Surge or peak loads must be known.

How material is delivered to equipment, and how the product is handled.

Available power.

Reducing the size of the feed by the use of secondary crushers will tend to reduce the amount of fines.

10.3 HAMMERMILLS (Swing Hammermills)

The hammermill is used for the crushing or pulverizing of semiabrasive to nonabrasive materials, or the shredding or reducing in size of practically every kind of fibrous or pulpy material. Swing hammer grinders used in combination with bar screen have been used successfully for disposal of sewage plant screenings. Except for nonfriable materials, the hammermill is replacing the jaw and gyratory crushers in many instances.

Hammermills with ring-type hammers generally are operated at considerably lower speeds than those with T-head or bar-type hammers (see figure 10.1). For example, a hammermill with a T-head hammer generally operates at a peripheral speed of 10,000–12,000 fpm, while one with a ring-type hammer would generally operate between 4,000 fpm and 8,000 fpm. These hammers crush, grind, and shatter lumps through a semicircular grid spanning the discharge opening. Spacing of the grid bars determines the maximum product size. The

BAR HAMMER

T-HEAD HAMMER

**RING-TYPE HAMMER
(PLAIN)**

**RING-TYPE HAMMER
(TOOTHED)**

SHREDDER HAMMERS

Figure 10.1 Basic hammer types. (*Courtesy of Pennsylvania Crusher Corp.*)

hammers may be circular, triangular, or serrated circular, and for shredding, tooth or bar shaped. The grinding plate, similar to the breaker plate, is usually adjustable. Foreign matter too hard to crush is thrown by centrifugal action into a trap. Some hammermills are reversible, permitting symmetrical sharpening of hammers, avoiding shutdowns for turning hammers and helping to maintain the best crushing surface. The lumps will generally be from 1¼ in. to ⅛ in. The hammermills can be made effective in handling materials with a high moisture content. They are efficient in their use of power, and generally are lower in installation cost. In a one-stage operation, reduction can be 10:1. The feed for crushers should be with the direction or rotation of crusher. Feed to hammermills should be even and spread across the rotor width. Crushers should have some relief to prevent being air bound.

The flexibility of application and operation of the swing hammermill lies in its design and operating principle. The feed enters the machine from above and falls into the path of rapidly revolving hammers. As it passes over a screen bar or perforated plate cage, that which is sufficiently reduced is discharged, while the remainder is carried around the unit for subsequent reduction. The size reduction obtained is to a large extent determined by the intensity of the hammer blow. Often different degrees of reduction can be obtained by varying the speed of the rotor. Generally, for a given bar setting, more fines are produced at a high hammer-tip speed than at a lower hammer-tip speed. The hammermill is particularly suited to the reduction of materials when product requirements are 3 in. and under, down to and including 90% minus 10 mesh. Different materials and different conditions of the same material can and will narrow or widen this range.

Slow-speed hammermills (tip speed of 2,000–7,000 fpm) were developed for the purpose of obtaining a swing hammermill-type unit that would reduce the amount of fines. Due to the interior changes made in order to obtain the correct force on the hammers, these basic slow-speed units are also ideal for application in both the slow and high operating speed ranges as a primary size reduction unit. Special modifications of hammermills now permit their application to materials of extremely high moisture content and extraordinary cohesive properties, as well as to materials when feeds are 10–12 in. for reduction in one stage or more to products of 3 in. and under, down to and including 90% minus 10 mesh. If the feed cannot be controlled, it may be advisable to add a flywheel.

A general-purpose machine is designed to take care of average requirements and moderate capacities. It can be furnished with or without a metal catcher. The metal catcher or tramp iron pocket is furnished as a means of protecting the crusher from damage when metal is inadvertently contained with the crusher feed. The presence of a tramp iron pocket is not to separate metals, and while it does perform this function, such metals should be removed prior to the crusher by use of magnets, metal detectors, etc. Any large piece of metal entering the crusher may cause severe equipment damage. A similar unit is designed for larger feed size and heavier work as well as finer reduction. It can handle limestone, shale, slate, clay, chalk, marl, gypsum, phosphate rock, asbestos rock, garbage, and tankage.

A similar small unit pulverizer is made for laboratory work, reducing materials to a fine, uniform product, or in a production line where heavy duty or large capacities are not required.

Materials containing silica, compounds of silica, or those of a similar nature will cause more than ordinary wear on any exposed parts of swing hammermills. The greatest wear will be on the hammers and to a lesser degree on the screen bars and liners. All of these parts are classed as wearing parts and are replaceable. They can also be specified to be made from a high abrasion-resisting material if this characteristic of the material is known at the time of selection of the unit. The use of swing hammermills for materials with a silica content of 5% or less is permissible, for those with 10–15% is doubtful, and for those with 20% or more is dangerous.

10.3.1 High-Speed Hammermills

The advantages of using high-speed (over 7,000 fpm) hammermills are:

Wide range of application

Low installation cost in comparison to other types of size-reduction equipment

High ratio of size-reduction for one-stage operation

The disadvantages are:

Minimum control of fines

High maintenance cost on abrasive materials

High power consumption per ton crushed due to high reduction ratio

10.3.2 Slow-Speed Hammermills

Slow-speed hammermills are used on the same materials as high-speed hammermills when a maximum control of fines for hammermill operation is required. Their particular application is in the preparation of products requiring a maximum percentage passing a top size of 1¼–¾ in. with a minimum percentage passing ¼ in. to 10 mesh in a one-stage operation. The advantages of slow-speed hammermills are:

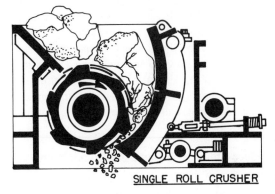

Figure 10.2 Roll crushers. (*Courtesy of Pennsylvania Crusher Corp.*)

Maximum control of fines for hammermill operation

Minimum headroom

High capacity

Low power consumption

The disadvantages are:

Small upper crushing chamber

Higher maintenance on abrasive materials than roll-type size-reduction units.

10.4 INTRODUCTION TO ROLL CRUSHERS

Roll crushers (figure 10.2) are used essentially for the size-reduction of friable to medium-hard materials when products are required containing a minimum amount of fines. Each of these crushers has its own particular field and one is not a substitute for the other, although their application does coincide at various points.

10.5 SINGLE-ROLL CRUSHERS

In the single-roll crusher, a toothed roll crushes the material against a breaker plate, reducing the lumps to desired size and discharging the product from the lower part of the roll. Shifting the breaker plate determines the size of the product. Relief springs protect the roll from damage by foreign matter. The plate swings away from the roll when harder particles are encountered. Sometimes a shear pin is used on the driving pulley, which breaks when hard materials resist turning. The drive is generally by V-belt. The product size usually produced by single-roll crushers is 10–2 in., since the single-roll crusher is used as a primary crusher. Product sizes less than 2 in. are more normally produced by a double-roll crusher.

The single-roll crusher has its primary application in the size-reduction of bituminous coal. It is used on other medium-hard materials, such as salt, alum, petroleum coke, bones, gypsum, limestone, oil shale, etc. The single-roll crusher products range from approximately 100% minus 8 in. down to and including approximately 90% minus 1¼ in. The ratio of size reduction is high for a roll crusher: from 2 to 8 : 1, depending on the roll diameter when materials as above are being reduced. Its advantages are:

Reduces in one stage large or medium lumps
Minimum of fines
Readily adjustable to a variety of product sizes
Minimum overall height

Its disadvantages are:

Unsuitable for abrasive materials
These machines are not normally used to produce product size under 2 in. While 90% minus 1¼ in. may be possible if feed conditions and reduction ratio are ideal, normally a double-roll crusher would be used to produce this product size.
Produces slabby product on materials of slabby nature or stratified in one plane.
Single-roll crushers are not designed to be started under load. It is a matter of probability how the feed material will orient itself between the roll and breaker plate, and certain orientation of the material may prevent starting under load. Loaded starts are possible most of the time. It would be a mistake, however, to count on a loaded start unless special drive equipment is used to ensure this.

10.6 DOUBLE-ROLL CRUSHERS

In the double-roll crusher, two rolls revolve and crush the material between them. Relief springs and floating bearings permit one roll to yield, so hard materials can be passed. Size adjustment during operation is possible with well designed crushers. Size produced can be reduced to 1¼ in. V-belt drives are commonly used. When fitted with small teeth or flutes on the rolls, they will produce ½ in. lumps. Double-roll crushers have a wide range of applications from that of a heavy-duty primary unit to that of the final unit to make a finished product. The wide range of application of the double-roll crusher is due to the larger number of various types of segments that can be applied to the rotating rolls. The double-roll crusher, however, does have the disadvantage of a very small size-reduction ratio between feed and product. This ratio will vary to a large extent with the type of segment used.

10.6.1 Maximum Feed Size

For smooth-faced segments, a safe guide to use is opening between roll faces +¼ in. Thus, if the maximum opening between roll faces for a 30-in. diameter roll is 1.20 in., the maximum feed size will be 1.20 + 0.25 = 1.45 in.

For corrugated-faced segments, add four times the depth of corrugation. Thus, for a 30-in. roll diameter with a maximum opening between the faces of 1.20 in., and corrugation ½ in. deep, the maximum size of feed would be 1.20 + 4(½) = 3.20 in.

For tooth-faced segments, the feed size is generally given in three dimensions: length × width × thickness. Teeth are generally ¼ in. to ½ in. less in height than the desired product. Length should be a maximum of 80% of roll width. Thus, for a 30-in. diameter roll, the maximum feed length would be 0.8 × 30 = 24 in. The maximum thickness of the feed for a 30 in. roll, with 8 in. desired product, would be 1.20 in. (the same as for a smooth roll) + 8 in. (desired product) + 1½ × 7½ (tooth height = 8 in. minus ½ in.) = 1.20 in. + 8.0 in. + 11.25 = 20.45 in.

Width should be average of length and thickness. Thus, for 30-in. diameter rolls, 30 in. wide and 8-in. product:

$$\frac{24 + 20.45}{2} = 22.225 \text{ in. maximum width}$$

Note: the above dimensions are maximum. Small sizes are suitable as feed.

10.6.2 Maximum Feed Summary

The maximum feed sizes stated are a safe guide. These limits can be exceeded, depending upon the friability of the material involved, the capacity, and the feed-material abrasiveness. The opening between the roll faces can be obtained from the manufacturers.

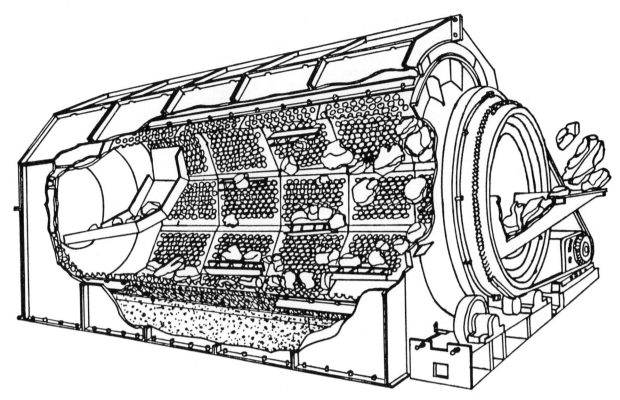

Figure 10.3 Bradford breakers. (*Courtesy of Pennsylvania Crusher Corp.*)

10.7 BRADFORD BREAKERS

In the Bradford Breakers (figure 10.3), the material is continuously charged at the loading end. Passing sizes are immediately screened out through rotating perforated plates. Larger lumps are raised by radial lifting shelves and then dropped. Gravity impact breaks the material as it falls on the heavy perforated screen plates. The units are chain driven at a speed of 12–18 rpm. Debris that cannot pass screen openings travels the full length of the breaker and is automatically discharged along with tramp iron, timber, or other refuse. Some large-capacity breakers have a modified hammermill fitted inside the main cylinder, substantially increasing capacity and breaking lumps that might not break up merely by tumbling. Most Bradford Breakers are used to produce product sizes of 3–6 in. It is extremely unusual for product sizes less than 2 in. to be produced in this type of equipment any more. A number of factors have influenced this, the most important of which is the mining techniques currently being employed. Continuous miners and similar equipment produce a large percentage of fines at the mine face and the dust suppressing sprays combined with these fines tend to make a fairly sticky material. Hole sizes smaller than 2 in. are very susceptible to plugging, and this has virtually eliminated the Bradford Breaker for use on these smaller product sizes. Additionally, the higher capacities prevalent in most modern coal mines would require multiple units for such small product sizes.

10.8 JAW AND GYRATORY CRUSHERS

In a jaw crusher, the material is fed between one movable and one fixed surface set at a small angle between them. They are used primarily as prime crushers for medium to hard material, and for capacities up to about 800 tph. They operate at about 100–400 rpm. They can be set to accept material up to 60 in. In the gyratory crusher, the material is fed into the top of the unit between an inverted stationary cone in the center of the opening, and a movable truncated cone surface. The crushers operate at 300–500 rpm, and can handle material up to 12 in. The gyratory crusher generally will:

Cost more for housing, but slightly less for foundation

Costs less to operate per unit

Lowest in maintenance cost

Most efficient on hard ore

Better control of size

Table 10.1 KVS Primary Gyratory Crushers Capacity Chart

Crusher Size	Feed Opening (in.)	Max Feed Size (in.)	Rglr Feed Size (in.)	Min Set Open Side (in.)	Intmd Set Open Side (in.)	Max Set Open Side (in.)	Cap tph	hp	rpm	Opening (in.)[a]	Weight (lb)
No. 17	16	16	80%	2½			180				
			minus		3		170	75	150	17 × 58	74,700
			10½			3½	210				
No. 20	20	20	80%	3			210				
			minus		3½		270	100	200	20 × 72	105,000
			14			4	300				
No. 22	22	22	80%	3			220				
			minus		3½		290	100	200	22 × 75	114,000
			15			4	340				
No. 26	26	26	80%	3			260				
			minus		3½		380	150	180	26 × 100	170,000
			16			4½	475				
No. 30	30	30	80%	4½			475				
			minus		5		515	150	180	30 × 115	213,000
			20			5½	625				
No. 36	36	36	80%	4½			500				
			minus		5½		725	250	170	36 × 132	232,000
			24			6½	900				
No. 42	42	42	80%	5½			520				
			minus		6½		715	300	145	42 × 162	405,000
			28			7	790				
No. 48	48	48	80%	6½			600				
			minus		7½		855	450	125	48 × 180	450,000
			32			8	1000				

Source: Courtesy Kennedy Van Saun Corporation

Note: Capacity in closed circuit, based on 100 lb/ft^3 limestone, and short ton. Crusher capacity will vary with differing field conditions of hardness, toughness, fracture pattern, moisture, gradation, etc., of the material fed and method of feeding.

[a] Two openings, each of approximate size given in table.

The capacity of crushers depends on:

Feed of material
Type of crusher selected
Reduction ratio
Speed of crushing elements
Toughness of material handled
Percentage of various sizes required

There are formulas for figuring capacity, but actual production does not seem to have any relation to calculated values, and it is suggested that crusher manufacturers be consulted for recommendation, if data is not available for a given product. KVS quotes this rule of thumb: "80% of the feed material should be smaller than two-thirds of the crusher opening." In other words, a 30-in. crusher opening should have a feed with 80% passing a $\frac{2}{3} \times 30$ in., or 20 in. opening.

Table 10.1 is a reproduction of a KVS table giving the maximum feed size, capacity, hp, rpm, size of opening and weight of the primary gyratory crushers they manufacture (figure 10.4). The screen analysis of the product for various settings of discharge openings of crushers is given in Tables 10.2 and 10.3. Tables 10.4 and 10.5 give similar data for KVS jaw crushers.

Table 10.2 Screen Analysis, KVS Gyratory Crusher—Closed Circuit

Product Size (in.)	Closed Side Setting of Crusher					
	$1^{1}/_2$ in. %	$1^{1}/_4$ in. %	1 in. %	$^{3}/_4$ in. %	$^{5}/_8$ in. %	$^{1}/_2$ in. %
$+1^{1}/_4$	10					
$-1^{1}/_4 + 1$	23	10				
$-1 + ^{3}/_4$	22	27	10			
$-^{3}/_4 + ^{1}/_2$	15	23	27	16	6	
$-^{1}/_2 + ^{3}/_8$	7	10	17	27	23	10
$-^{3}/_8 + ^{1}/_4$	8	10	17	26	36	40
$-^{1}/_4$	15	20	29	31	35	50
	100	100	100	100	100	100

Source: Courtesy Kennedy Van Saun Corporation
Note: Product analysis will vary depending upon type of material, feeding method, moisture content, cleanliness of material, and fracture pattern.

Table 10.3 Screen Analysis, KVS Gyratory Crusher—Open Circuit

Product Size (in.)	Closed Side Setting of Crusher							
	2 in. %	$1^{1}/_2$ in. %	$1^{1}/_4$ in. %	1 in. %	$^{3}/_4$ in. %	$^{5}/_8$ in. %	$^{1}/_2$ in. %	$^{1}/_4$ in. %
$-3 + 2^{1}/_2$	15							
$-2^{1}/_2 + 2$	40	10	5					
$-2 + 1^{3}/_4$	5	20	5	5				
$-1^{3}/_4 + 1^{1}/_2$	5	20	20	5	10			
$-1^{1}/_2 + 1^{1}/_4$	5	8	20	20	13			
$-1^{1}/_4 + 1$	5	9	10	20	14	10	5	
$-1 + ^{3}/_4$	11	8	10	13	13	27	19	3
$-^{3}/_4 + ^{1}/_2$	9	9	10	12	17	23	26	14
$-^{1}/_2 + ^{3}/_8$	5	3	5	7	8	10	12	11
$-^{3}/_8 + ^{1}/_4$		4	5	6	8	10	13	12
$-^{1}/_4$		9	10	12	17	20	25	60
	100	100	100	100	100	100	100	100

Source: Courtesy Kennedy Van Saun Corporation
Note: Product analysis will vary depending upon type of material, feeding method, moisture content, cleanliness of material, and fracture pattern.

10.9 RAYMOND® MILL

Raymond® roller mills, by C-E Raymond, are efficient, sturdy units for pulverizing ½ in. to ¾ in. lumps to fines. The High Side Mill (figure 10.5) can reduce the materials to 95% passing the 200 mesh screen for the Single

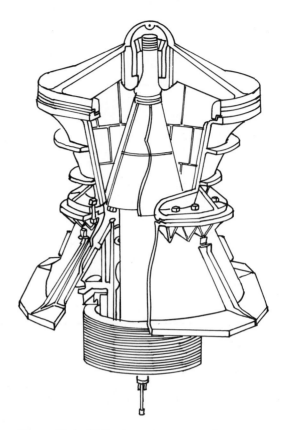

Figure 10.4 KVS primary gyratory crushers.

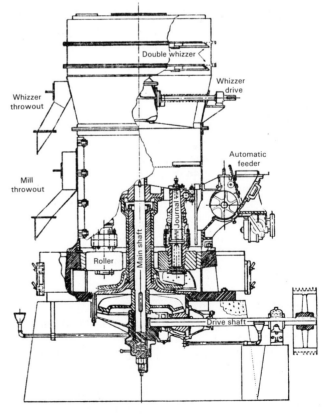

Figure 10.5 Section of a high-side Raymond mill.

Table 10.4 KVS Swing Jaw Crushers

Feed Opening (in.)	Flywheel (rpm)	hp	Discharge Opening—Closed (in.)													Crusher Weight (lb)
			7/8	1	1½	2	2½	3	3½	4	4½	5	6	7	8	
7 × 10	275	7	3	5	8	10										6,000
10 × 24	275	15		16	22	25	33									14,000
14 × 24	275	25			26	33	41	50								20,000
18 × 36	250	40				60	76	93	108	123						45,000
24 × 36	200	75					95	113	132	150	168					71,500
30 × 42	200	100						148	175	198	225	249	298			113,500
36 × 48	200	150							239	270	298	330	389	445		163,500
48 × 60	175	200								350	384	425	500	615	685	310,000

Source: Courtesy Kennedy Van Saun Corporation
Notes: Crusher capacity: short tons/hr.
Crusher capacities will vary with differing field conditions of hardness, fracture pattern, roughness, moisture, gradation, etc., of the material fed and method of feeding.

Whizzer type, and to 99.9% passing 325 mesh for the Double Whizzer type. The large ducts cannot be supported on the mill, but must have additional supports at roof level, and flexible connections below.

When the pulverizer is handling explosive dust, an inert gas must be introduced to reduce O_2 content. The amount of inert gas equals the leakage into the system. Figure 10.6 shows the piping for such a Raymond mill installation. In this case, a waste-heat boiler was used to produce CO_2, and the design for it is given below. Other types of inert gas generators are available.

10.9.1 CO_2 System

The Raymond mill requires 1000 cfm at the start and 600–700 cfm during operation. Since 359 ft³ of any gas weighs its molecular weight in lb, 359 ft³ of CO_2 weighs $12 + (2 \times 16) = 44$ lb. The mill will therefore require 12 lb carbon. The mixture fed the mill should be about 12% CO_2, that is, 0.12(1000) or 120 ft³ CO_2 is required per min.

$$\frac{120}{359} \times 12 = 4.01 \text{ lb/min coke}$$

$$4.01 \times 60 = 240 \text{ lb/hr coke}$$

Assume a pound of coke will yield 12,000 Btu or, at 60% efficiency, 7,200 Btu. With latent heat of steam equalling 970 Btu and 34.5 lb of steam per hp,

$$\frac{7200}{34.5 \times 970} = 0.215 \text{ hp/lb coke}$$

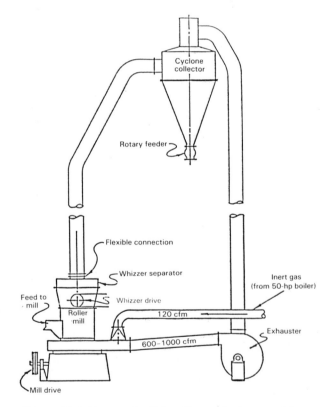

Figure 10.6 Raymond mill handling explosive material.

Required hp for the boiler is

$$240(0.215) = 51.5 \text{ hp}$$

Therefore, use a 50-hp boiler.

Table 10.5 Product Analysis for KVS Overhead Eccentric and Swing Jaw Crushers

Percentage Passing (left axis, 100 → 0) and *Percentage Retained* (right axis, 0 → 100) versus *Crusher Discharge Opening (in.)*

¼	½	¾	1	1¼	1½	1¾	2	2½	3	3½	4	5	6	7	8	9	10	11
¼		¾	1	1¼	1½	1¾	2	2½	3	3½	4	5	6	7	8	9	10	11
										3¼								
							1¾	2¼	2¾	3½	4½		6½			8½		10½
		⅝			1¼	1½		2½	3			5		6½		8		
3/16	⅜	¾	1			1½	2	2¼	3	4			5½			6½		8½
					1¼	1¾		2½		3½			6		7	8		
	5/16	½	⅝		1		1¼	2	2¼	2½								
				¾	1	1½	1¾	2		2½	3½	4½			5½		6½	
⅛	¼	⅜	½	⅝	¾		1		1¾		3		4		5		6	
									1½							4½		
		3/16	5/16	⅜	½	⅝	⅝	¾	1	1¼	1¼	1½	2					
		¼		⅜	½		⅝	¾	1	1½			2¾		3		4	
1/16	⅛	3/16	¼		⅜	½	½		¾	1	1							
				¼		⅜	⅜	½	½	¾	1	1½			2		2¼	
	1/16	⅛	⅛		¼	¼	¼	⅜	½	½	½							
		1/16			⅛	⅛	⅛	⅛	¼	¼	¼		½		1½		1½	2
			1/16		1/16	1/16	1/16		⅛									

Source: Courtesy Kennedy Van Saun Corporation.

Note: Product analysis will vary with type of material, moisture, crushing cavity, and cleanliness of material.

Note that different size mills will require different amounts of inert gas.

10.9.2 Flue Gas

Inlet temperature: 450°F
Outlet temperature: 95°F
Specific heat: 0.24 Btu/lb
Specific gravity: 0.081

A 50-hp boiler will yield 115,000 ft³/hr of flue gas at 450°F and 60 psig steam pressure. The output will be

$$50 \times 34\tfrac{1}{2} \ \text{lb/hr} \times 1177.6 \ \text{Btu/lb (above 32°F)}$$
$$= 2,031,360 \ \text{Btu/hr}$$

The input of fuel to the boiler will be 240 lb/hr coke at 12,000 Btu/lb = 2,880,000 Btu/hr. Consequently, the boiler efficiency is

$$\frac{2,031,360}{2,880,000} = 70.5\%$$

(If the heating value of coke byproduct is 12,700 Btu/lb, the efficiency will be 66.6%.)

10.9.3 Other Raymond Machines

Raymond roller mills can process a variety of other non-metallic minerals and can easily be built to incorporate Raymond Flash Dryers. C-E Raymond also makes the Imp Mill, a swing-hammer impact mill ideally suited to fine and medium-fine grinding of softer nonmetallic minerals and coal. The Vertical Mill fits highly uniform, superfine grinding applications, and the Raymond Bowl Mill is intended primarily for pulverizing coal. Raymond VR Mills (high capacity) are integrated systems that incorporate pulverizing, drying, and classifying.

10.10 COMPUTATION OF HORSEPOWER REQUIRED FOR GYRATORY CRUSHERS*

10.10.1 General

The A-C (Allis-Chalmers) primary gyratory crusher is shown on figure 10.7. Initial power estimates can be made by comparing data for the proposed project with operating conditions at similar installations. Horsepower requirements can also be approximated with a formula based on the Work Index (Wi).

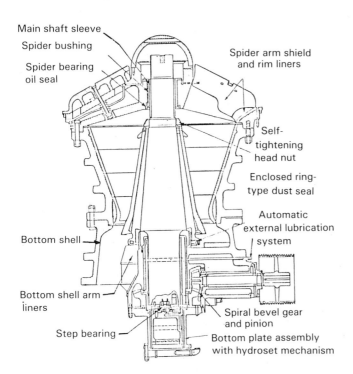

Main shaft sleeve
Spider bushing
Spider bearing oil seal
Spider arm shield and rim liners
Self-tightening head nut
Enclosed ring-type dust seal
Automatic external lubrication system
Bottom shell
Bottom shell arm liners
Spiral bevel gear and pinion
Step bearing
Bottom plate assembly with hydroset mechanism

Figure 10.7 Allis-Chalmers primary gyratory crusher.

---

*From Allis-Chalmers data.

Bond Work Index (Wi) (see paragraph 10.10.3)

$$= 2.59 \, \frac{\text{average impact value}}{\text{specific gravity}}$$

Feed size: estimate 80% of feed size passing a square opening equal to two thirds of the crusher receiver opening.

Product size: product analysis curves can be used to estimate product size with sufficient accuracy.

10.10.2 Design Formula

In calculating feed (F) and product (P) sizes, 1 in. = 25,400 μm (micrometers).

$F = 80\%$ passes _____ in., or _____ μm.
$\sqrt{F \; \mu m} =$ _____ .
$P = 80\%$ passes _____ in., or _____ μm.
$\sqrt{P \; \mu m} =$ _____ .

$$\sqrt{F} - \sqrt{P} = \underline{\qquad}.$$

$$\sqrt{F} \times \sqrt{P} = \underline{\qquad}.$$

10.10.3 Design Example

The design capacity for a proposed mining facility is 1550 short tph. The A-C Test Center reports the following characteristics:

Average impact values:	21.4 ft-lb/in.
Specific gravity:	2.85
Work index:	19.4

A 54-74 gyratory crusher has been selected to operate at 6-in. open side setting. Feeding will be directly into the crusher (refer to table 10.6).

Feed size (F): 80% passes 36 in. (66% of feed opening) (0.66 × 54 = 35.6, say 36 in.).

$$F = 914{,}400 \; \mu m \; (36 \times 25{,}400)$$

$$\sqrt{F} = 956 \; (\text{paragraph } 10.6.1))$$

Product size (P): 80% passes 5½ in. (per product curve).

$$P = 139{,}700 \; \mu m \; (5.5 \times 25{,}400)$$

$$\sqrt{P} = 374$$

$$956 - 374 = 582$$

$$956 \times 374 = 357{,}544$$

$$\text{Hp/short ton} = \frac{19.4 \times 13.4 \times 582}{357{,}544} = 0.423$$

Total hp = 1550 × 0.423 × 0.75 = 492

The recommended motor size is 500 hp.

Table 10.6A Primary A-C Gyratory Crushers—Performance Table

Crusher Size	Aprx Feed Opening	Gyr per Min	Pinion rpm	Max hp	Ecc Throw (in.)	Open Side Setting of Discharge Opening (in.)																			
						2½	3	3½	4	4½	5	5½	6	6½	7	7½	8	8½	9	9½	10	10½	11	11½	12
30-55	30 × 78	175	585	150	⅝	150	205	270	335	390	450	510													
30-55	30 × 78	175	585	180	¾		240	320	400	480	550	620													
30-55	30 × 78	175	585	240	1			425	485	540	600	660													
30-55	30 × 78	175	585	300	1¼				605	675	735	800													
36-55	36 × 90	175	585	180	¾				270	310	350	380													
36-55	36 × 90	175	585	240	1					380	440	500													
36-55	36 × 90	175	585	300	1¼						515	585													
42-65	42 × 108	150	497	265	1					540	660	790	920	1040	1170										
42-65	42 × 108	150	497	330	1¼						700	850	1000	1140	1300										
42-65	42 × 108	150	497	400	1½								1040	1260	1490										
48-74	48 × 120	135	497	300	1						930	1000	1080	1150	1230	1300	1380								
48-74	48 × 120	135	497	385	1¼							1280	1390	1480	1580	1680	1780								
48-74	48 × 120	135	497	425	1⅜								1530	1640	1780	1860	1980								
48-74	48 × 120	135	497	500	1⅝									1920	2060	2180	2340								
54-74	54 × 132	135	497	300	1							960	1040	1100	1160	1240	1330								
54-74	54 × 132	135	497	385	1¼								1340	1410	1500	1590	1700								
54-74	54 × 132	135	497	425	1⅜									1560	1650	1750	1870								
54-74	54 × 132	135	497	500	1⅝										1950	2070	2210								
60-89	60 × 145	110	435	330	1								1130	1170	1240	1310	1400	1480	1610						
60-89	60 × 145	110	435	410	1¼									1450	1540	1630	1740	1850	2000						
60-89	60 × 145	110	435	450	1⅜										1680	1780	1900	2020	2200						
60-89	60 × 145	110	435	495	1½											1960	2100	2230	2360						
60-89	60 × 145	110	435	600	1¹³⁄₁₆												2540	2600	2700						
60-109	60 × 150	100	400	1000	1½													3250	3500	3750	4000	4250	4500	4750	5000
72-109	72 × 174	100	400																						

265

Table 10.6B   Secondary A-C Gyratory Crushers—Performance Table

Crusher Size	Aprx Feed Opening	Gyr per Min	Pinion rpm	Max hp	Ecc Throw (in.)	Open Side Setting of Discharge Opening (in.)									
						1	1½	2	2½	3	3½	4	4½	5	5½
16-50	16 × 55	225	764	100	¾	150	190	250	310	360					
16-50	16 × 55	225	764	125	1		250	308	350	390					
16-50	16 × 55	225	764	150	1¼			350	376	415					
24-60	24 × 66	175	585	175	¾	195	230	268	305	340	380	415			
24-60	24 × 66	175	585	250	1		325	360	400	440	475	510			
24-60	24 × 66	175	585	300	1¼			415	475	525	575	620			
30-70	30 × 84	150	497	200	¾			300	350	400	460	515	570	620	680
30-70	30 × 84	150	497	265	1				450	500	550	610	665	720	770
30-70	30 × 84	150	497	330	1¼					550	650	750	850	915	1000
30-70	30 × 84	150	497	400	1½						810	940	1060	1180	1310

Source: Courtesy Allis-Chalmers

Notes: Stepped lines indicate recommended minimum and maximum setting for each throw. Capacities in short tons/hr based on 100 lb/ft³ crushed material. Capacities given here are based on field data under average quarry conditions when crushing dry friable material equivalent to limestone. Because conditions of stone and methods of operation vary, capacities given are only approximate.

Where no capacity data is given, the crusher is under development.

Figures under maximum hp are correct only for throw and pinion rpm given above. When speed is reduced, maximum hp must also be reduced proportionally. For example, a 42–65 crusher having a 1-in. throw with 265 hp at 497 rpm, when reduced to 400 rpm will give maximum hp of (265 × 400)/497 = 213.

10.10.4 Impact Testing

Impact value and work indexes are determined by destructive tests on at least 10 material samples, each passing a 3-in. square opening, but not passing a two-inch square. Twin 30-lb pendulum hammers are arranged to strike both sides of selected samples simultaneously. Hammers are dropped in successive steps, with a greater angle of fall at each drop until the sample breaks.

Impact strength is the average number of ft-lb of energy represented by the breaking fall divided by sample thickness in inches. Average impact strength is the average number of ft-lb required to break 10 or more pieces.

10.10.5 Product Analysis of A-C Gyratory Crusher

Product analysis curves approximate size gradation, or screen analysis, of the A-C gyratory crusher product. Actual gradation will vary, depending on the material, quarry conditions and the amount of fines or product size in the feed (figure 10.8). Tables 10.7 and 10.8 list the percentage of product, for various classes of material, that will pass a square screen opening equal to the crusher open side setting. Product percentage changes with the type of feed, and represents nominal conditions. These values are the basis for using the product analysis curves (figure 10.8).

Table 10.7   Product Curve Selection, A-C Gyratory Crusher

Material	Weight (lb/ft³)		
	Run of Quarry	Scalped	Scalped and Recombined with Fines
Limestone	90	85	88
Granite	82	75	80
Trap rock	75	70	75
Ores	90	85	85

Table 10.8   Cumulative Percentages as Read from Figure 10.8

Square Opening in Sleeves (in.)	Approximate % Passing
7	97
6	91
5	80
4	67
3	52
2	37
1	23

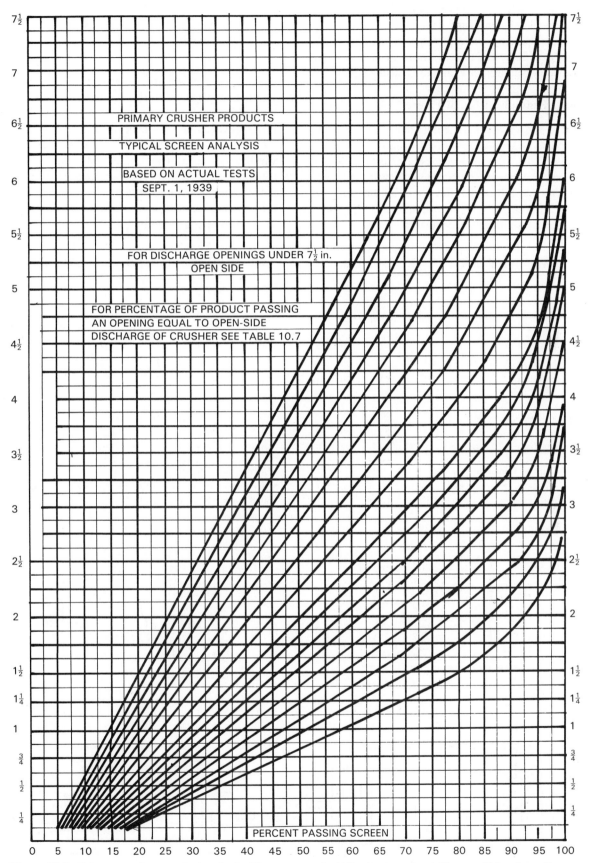

Figure 10.8 Primary crusher—product analysis. The screen analysis from any crusher will vary widely, depending on the character of the material, quarry conditions, and the amount of fines or product size in the initial feed at the time. (*Courtesy Allis-Chalmers*)

Horizontal axes on figure 10.8 represent cumulative percentages passing corresponding screen openings. Vertical axes represent material sizes in inches, based on a square opening. See example for using tables.

10.10.6 Example of Use of Figure 10.8

Feed is a run of quarry, medium-hard limestone, including fines. 42-65 A-C gyratory crusher is set at 6 in., open side. Table 10.8 shows 90% of product should pass a 6-in. square opening. Select the 90% vertical line on figure 10.8 and follow it to the horizontal line for 6-in. opening. Follow the curve nearest the intersecting point to get this approximate screen analysis.

10.11 GUNDLACH CRUSHERS (Rexnord)

Gundlach makes single- and two-stage crushers (table 10.9). The two-stage crusher has been used successfully in crushing coal and graphite electrodes. The upper and lower rolls can be independently adjusted to produce a minimum of fines. They also control the top size. By varying the speed of the crusher, the particle size distribution can be varied. The two-stage crusher is essentially a double-roll crusher with the upper rolls acting as primary breakers and the lower rolls as final crushers. They operate at a speed of 150–300 rpm.

Table 10.9A Gundlach Two-Stage Crusher Data

Model	Approx. Weight (lb.)	A Length	B Width	C Height	D	E
18DA	7800	72	73	52	19	39
27DA	8600	72	83	52	29	44
36DA	9400	72	93	52	39	49
45DA	10400	72	103	52	49	54
56DA	11200	72	113	52	59	59
70DA	15200	79	138	53	69	73
80DA	16100	79	148	53	79	78

Source: Courtesy Gundlach

*On drawing below, 8 in. on 18DA through 56DA; 5 in. on 70DA and 80DA. Dimensions are correct to the nearest inch.

Table 10.9B Gundlach Single-Stage Crusher Data

Model	Approx. Weight (lb.)	A Length	B Width	C Height	D	E
18SS	4000	72	64	32	19	29
27SS	4400	72	74	32	29	34
36SS	5000	72	84	32	39	39
45SS	5500	72	94	32	49	44
56SS	6000	72	104	32	59	49
70SS	7900	80	125	33	69	61
80SS	8400	80	135	33	79	66

Source: Courtesy Gundlach

*On drawing below, 8 in. on 18SS through 56SS; 5 in. on 70SS and 80SS. Dimensions are correct to the nearest inch.

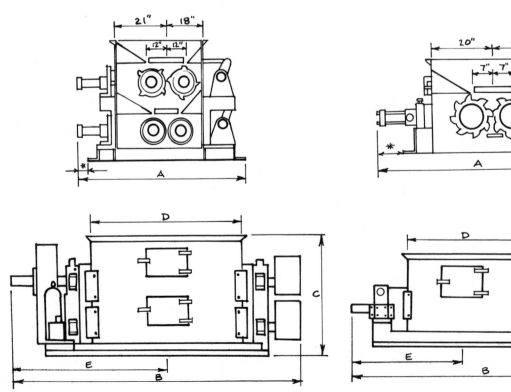

10.12 INTRODUCTION TO SCREENS

Vibrating screens are made in one- to three-deck size and are used primarily for grading of material by size. They can also be used for washing material, dewatering, descaling, or any combination of these operations. Electrically heated screens can be furnished to solve many screening problems involving high moisture and fine materials that tend to adhere to the screening surface. For special problems, it is best to conduct a test with the material to be screened. Material with angle of repose of 25° or less will screen well.

Dust enclosures are made semidusttight or fully enclosed. Dusty, corrosive, or toxic materials are completely confined by the enclosure, and dust retainers seal the openings through which the vibrator assembly projects.

Automatic spray cleaning was developed to eliminate blinding (reduction of open area) of the screen cloth on liquid dewatering screens. This is of help in many different screening problems commonly associated with sticky or greasy materials, which have a tendency to blind the fine screen cloth. The lighter the wire, the less blinding.

Linings should be used for abrasive material. Rubber linings such as Armabond can be used for temperatures up to 150°. Above that, carbon steel or stainless steel can be used, depending on the product. Armabond can also be obtained in flame-resistant construction.

10.13 FINE MATERIALS

If there are many fines in the material to be screened, especially where such treatment as drying is required, it may be advisable to place a dust collector ahead of the screening. This will not only improve the operation, but should result in cost reduction. Sifters (sifter screens) will work better with fines (48 mesh and under).* Since the sifters are used where fine materials are involved, the mesh and the wires are small. In handling abrasive material, especially where the capacity is, say five tons per hour or over, it is best to introduce an auxiliary screen of heavier wire to remove the larger particles and protect the finer mesh screens below. Sifter screens should not be over five feet wide.

10.14 VIBRATING SCREENS

Two types of vibration mounting are available. Figure 10.9 shows the support assembly for a suspension-type

---

*Mesh is defined as the number of openings per inch of length.

mounting, while figure 10.10 shows a floor-mounted screen. The angle at which the screen is set varies with the material and size of particles. It can be anywhere from 20° to 35°. For metallurgical coke, the angle should be 30°, for sand it should be 33°. Horizontal vibrating screens can be used where headroom is limited.

10.15 TYLER SCREENS

C. E. Tyler Company manufactures the V-50 high speed, electrically vibrated screen known as the Hummer screen (figure 10.11), in which the material comes in contact with a vigorously vibrated surface. The coarser particles are forced up and away from the surface, and the finer particles work their way down to the separation surface. This method of separation is known as stratification.

The V-85 has a rotary electric vibrator and is recommended for separation of flaky, sticky, hard-to-screen materials. It is a low-energy consumption unit, and requires little maintenance. It is available in either fixed- or variable-speed drive up to 3600 rpm, delivering 9 Gs of vibratory force. It is used in the chemical, fertilizer, food, and coal industries. Field tests show a noise level of 85 dBa or less. Rubber dust covers further dampen noise and control dust. The V-85 retrofit unit can be fitted to existing Hummers.

The Ty-Level X is a horizontal vibrating screen used for heavy-duty, high-capacity service. Its noise level is 80 dBa. It is mechanically vibrated with a double eccentric shaft that delivers a straight-line stroke at 45° to horizontal. It is designed for selective replacement of parts. It can be cable- or base-mounted. It can be used for sizing, dewatering, washing, and conveying.

The Ty-Speed is designed for difficult wet or dry separation of wet or dry chemicals, silica, fertilizer, sand, gravel, or limestone. It operates up to 3600 rpm with variable stroke combinations.

They also make Testing Sieves and a full line of laboratory standard screen scale testing sieves.

10.16 EXOLON SCREENS

The Exolon Company developed a multiform grader, consisting of a series of vibrating screens which provide a wave-like motion to the material being screened, resulting in a close contact with the screening surface and, in turn, giving high grading efficiency. In operation, the oversized material is carried off through chutes at the end of each screen; the undersized passing through the screen into a pan and then through a wide chute onto the next-finer screen. Because of the compact construction of the grader and the care taken to keep the screens

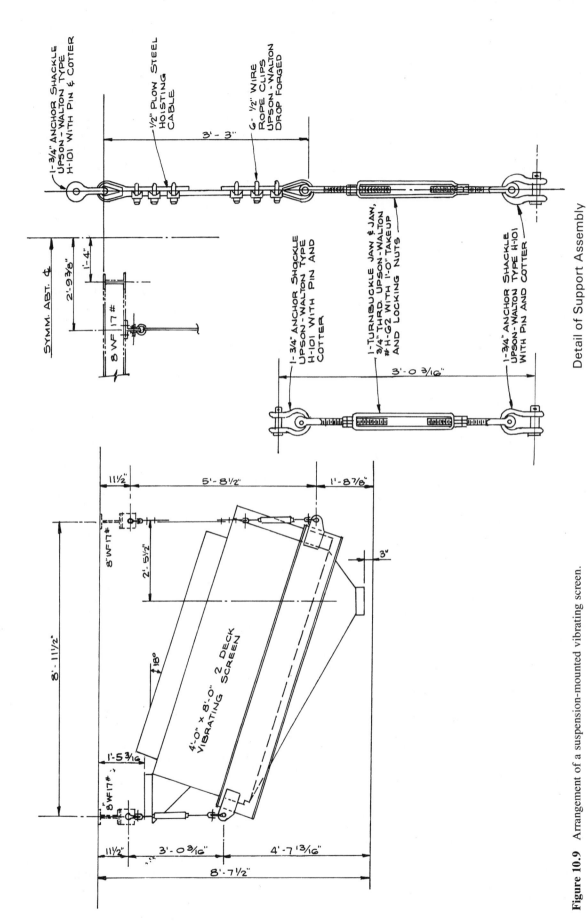

1-3/4" ANCHOR SHACKLE UPSON-WALTON TYPE H-101 WITH PIN & COTTER

1/2" PLOW STEEL HOISTING CABLE

6- 1/2" WIRE ROPE CLIPS UPSON-WALTON DROP FORGED

3'- 3"

SYMM. ABT. ₵

2'-9⅝"

1'-4"

8 WF 17 #

1-3/4" ANCHOR SHACKLE UPSON-WALTON TYPE H-101 WITH PIN AND COTTER

1-TURNBUCKLE JAW & JAW, 3/4" THRD. UPSON-WALTON #H-G2 WITH 1'-0" TAKEUP AND LOCKING NUTS

1-3/4" ANCHOR SHACKLE UPSON-WALTON TYPE H-101 WITH PIN AND COTTER

3'- 0 3/16"

Detail of Support Assembly

11½"

5'- 8½"

1'- 8⅞"

8" WF 17 #

2'- 5½"

3"

18°

4'-0" x 8'-0" 2 DECK VIBRATING SCREEN

8'- 11½"

1'-5 3/16"

8 WF 17 #

11½"

3'- 0 3/16"

4'-7 13/16"

8'- 7½"

Figure 10.9 Arrangement of a suspension-mounted vibrating screen.

270

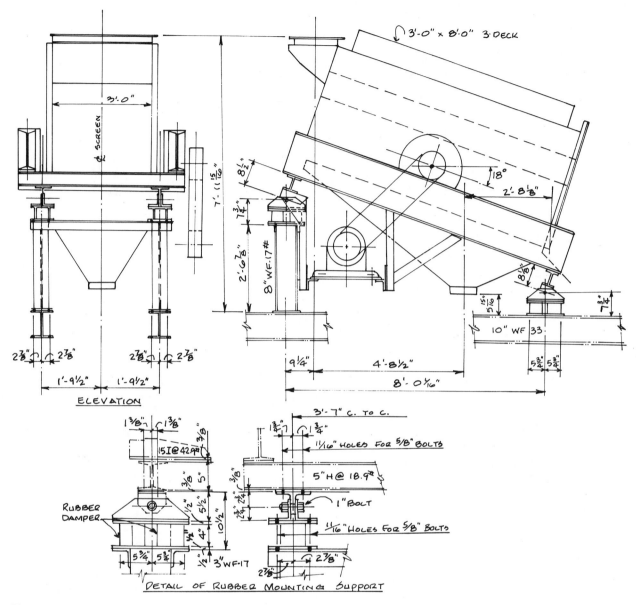

Figure 10.10 Arrangement of a floor-mounted vibrating screen.

true, a uniformly even flow of material is assured, and every square inch of screen surface is used. Since all the screens are vibrated as a unit, the screening action of each sieve is identical with that of the others.

10.17 ALLIS-CHALMERS VIBRATING SCREENS

Allis-Chalmers makes SH Ripl-Flo screens (refer to table 10.10). They are made in sizes from 4 × 8 ft to 10 × 20 ft and are two-bearing, circle throw, inclined 20° screens. They can be used for light scalping. The maximum aperture is 5 in. The XH is a heavy-duty model used for scalping and coarse sizing, wet or dry. The

maximum aperture for the XH screen is 10 in., and the size is up to 8 × 20 ft.

The Levl-Flo screens are horizontal, floor-mounted, low-headroom screens and can be single-, double-, and three-deck. The Levl-Flo screens are designed for long-throw applications. They are made in sizes from 4 × 8 ft to 6 × 16 ft. The maximum aperture is 3 in. The vibrator action is provided by two geared eccentric shafts designed to produce a force at a 45° angle of motion to the deck.

The Low-Head screens are for limited headroom locations. They are floor-mounted and come in only one- or two-deck models. They come in sizes from 3 × 12 ft to 10 × 24 ft. The maximum feed size is 8 in.

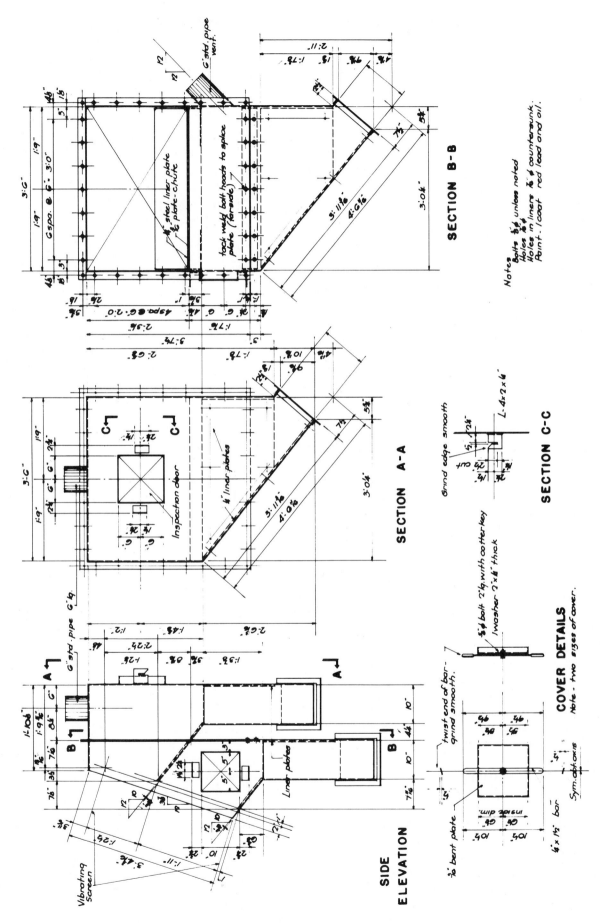

Figure 10.11 Discharge chute for a two-surface vibrating screen.

Table 10.10 Recommendations for A-C Types of Screen

Application	Maximum Feed Size (in.)	Aperture Range	Screen
Scalping before primary crusher	36	3 to 10 in.	XH Ripl-Flo
Scalping after primary crusher	24	1 to 10 in.	XH Ripl-Flo
Scalping after secondary crusher	10	¼ to 5½ in.	SH Ripl-Flo
Dry sizing	10	20 mesh to 5½ in.	SH Ripl-Flo
Dry sizing	8	⅛ to 2½ in.	Low-Head
Dry sizing	10	⅛ to 5½ in.	ST Ripl-Flo
Dry sizing	10	⅛ to 4 in.	Levl-Flo
Wet sizing	10	⅛ to 4 in.	Levl-Flo
Wet sizing	10	20 mesh to 5½ in.	SH Ripl-Flo
Wet sizing	8	¼ to 2½ mm	Low-Head
Media recovery	8	¼ to 2 mm	Low-Head
Dewatering	6	⅛ to 1 mm	Low-Head

Source: Courtesy Allis-Chalmers
Note: Refer also to paragraph 10.17.

10.18 SIMPLICITY VIBRATING SCREENS

Simplicity screens have crowned decks, single crowns for all widths 6 ft and under, double crowns for 7 ft and 8 ft widths. Simplicity screens provide a full screening area in a given size. For example, a 4 × 8-ft model provides a full 48 in. of screening area between side rails and 96 in. of clear screening length. The standard setting of Simplicity screens is 20° down-slope. Each subsequent deck on a multideck screen is set ahead to allow full use of screening length.

The four-bearing screen uses a counterbalanced eccentric shaft that provides an exact counterbalance for the frame and deck. Simplicity four-bearing screens can be mounted by supporting the main frame from below or by suspending it by rods or cables attached to plant superstructure. With either type of mounting, positive action of the screen's counterbalanced shaft will maintain full effectiveness and minimize transfer of vibration to the supporting structure. The main frame is rigid.

The two-bearing screen was designed to allow installation of a full-size screen where lack of space is a problem. Minimum width is obtained by omitting main frame and outboard bearings required on four-bearing screens. The screen can be mounted by supporting from below on coil springs provided with pedestal mounts, or it can be suspended from above.

Horizontal vibrating screens operate in stationary and in portable plants where headroom is limited. They are used for sizing and dewatering materials such as coal, gravel, and ores. They can be built for floor or suspension support. The rugged shaft assemblies with two counter-rotating shafts properly geared together provide the extra power needed to operate large areas on an efficient basis.

10.19 ROTEX SCREENS

Rotex screens are used widely for separating dry, lightweight material in sizes ³⁄₁₆ in. to 200 mesh and finer (figure 10.12). They are dust-tight, and screens can be changed without special tools. The high screening efficiency is obtained because of its gyratory action.

The Rotex gyratory action is induced by the eccentric drive, which is circular. At the feed end, the material is spread and stratified, with smaller particles going through the screen.

In the middle of the screen, there is a horizontal elliptical action which imparts a conveying action to the finer particles in contact with the screens, so they will pass through.

At the discharge end of the screen, the gentler motion allows the remainder of the near size particles to pass through the screen. The machine motion causes the rubber balls in the screen pockets to be deflected against bevel strips and bounce continuously on the underside of the screen mesh, keeping the screen clean and preventing screen blinding. The ball action also agitates the material.

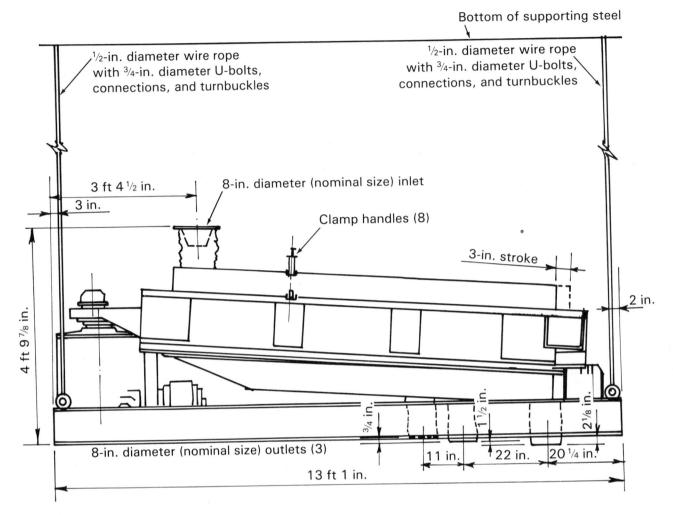

Figure 10.12 Rotex screen.

10.20 MAGNETIC SEPARATORS

Magnetic separators are used to separate dry, granular magnetic particles from nonmagnetic particles. In the Exolon separator, magnetic particles cling to a rotor longer than nonmagnetic. The units can separate different materials with different magnetic properties. The capacity of their laboratory model is 20–30 lb per hour in one pass. Two production models are made, in capacities of 1–2 tons per hour and 2–4 tons per hour.

10.21 SIZE OF SCREEN

Many plant engineers know the square feet required per ton of material, the best shape and angle to use for their material. In addition, the screen manufacturers have accumulated, through the years, information on screening various materials.

If no data is available it is always best to send a one

ft^3 sample of the material to be screened to those manufacturers from whom you are soliciting bids, and ask them to recommend the size of the screen and the angle at which to place it. A close check on resulting sizes can be obtained by using the Tyler Laboratory screens. If the materials result from a crushing operation, refer to paragraphs 10.1–10.11 for sizes to be expected.

The relation of length to width will depend on screen intake and the velocity of material on the screen.

10.22 SPECIFICATION FOR SIFTER SCREEN

Material: metallurgical coke, very abrasive
   will arrive in trucks with ⅝ in. and over, 7–18%
   retained on 4-mesh screen, 16–25%
   retained on 50-mesh screen, 7%
   pass through 50-mesh screen, 7%

Capacity: 10 tph

Screening required: top deck: ⅝ in. clear opening, 0.120 in. diameter, 304 ss. (This screen is interposed to protect the required two lower screens. Furnish third deck on screen to protect 50-mesh screen.)

middle deck: 4 mesh, 0.035-in. diameter wire, 304 ss.

bottom deck: 50 mesh × 0.0045-in. diameter wire, 304 ss.

Include motors for 460 V, 3 hp, 60 Hz power, and belt guards.

Also include: 8-in. diameter ring and cone for inlet, 8-in. diameter neoprene sleeve, 8-in. diameter ring, and screen tension tool.

Edging must be tight and abrasion resistant.

Edging for ⅝ in. deck, ss.

Hopper section: to be lined with material to withstand abrasion and temperature of 225°F.

10.23 TYPICAL INDUSTRIAL SCREEN REQUISITION (refer to table 10.11)

Furnish one 4-ft wide by 10-ft long two-surface vibrating screen similar to Tyler Type 38 tandem Hummer, equipped with two vibrators. Include dust covers and dustproofing throughout, with rubber gaskets. The screen's angle of inclination shall be 33°. The bottom shall be a steel collecting hopper with flanged round opening for bolting to the screen body.

Power: 230 V, 60 Hz

Screens: two 4 × 5 ft sections of 5 mesh, 0.041 wire cloth with reinforced hooked edges, and two 4 × 5 ft sections of 66 mesh, 0.004-in. diameter wire, with 8 mesh 0.025-in. rolled-tinned backing cloth forming a half sandwich hooked section.

Material: metallurgical coke

Samples: one ft³ of material will be sent on request.

It is expected that 90% of the material will be retained on 3-mesh screen.

10.24 BLINDING AND PLUGGING

Blinding and plugging reduce effective screening area, and thus cause a reduction in separation accuracy. Plugging is caused by near-size particles (relative to the hole size) getting wedged in the holes; a correction of amplitude can help prevent blinding. Heated decks and ball-tray decks can also be used if material is damp and sticky.

Table 10.11 Net Effective Screening Area (ft²)

Screen Size (ft)	Top Deck	2nd Deck	3rd Deck
1 × 2.5	1.9	1.7	—
1 × 4	3.0	2.7	—
1.5 × 3	4.0	3.6	3.3
1.5 × 4	5.3	4.8	4.5
2 × 4	6.0	5.4	4.9
2 × 6	9.0	8.1	7.3
3 × 4	10.0	9.0	8.1
3 × 6	15.0	13.5	12.1
3 × 8	20.0	18.0	16.2
3 × 10	25.0	22.5	20.0
3 × 12	30.0	27.0	24.3
3 × 14	35.0	31.5	28.4
3 × 16	40.0	36.0	32.4
4 × 6	21.0	18.9	17.0
4 × 8	28.0	25.2	22.7
4 × 10	35.0	31.5	28.4
4 × 12	42.0	37.8	34.0
4 × 14	49.0	44.1	39.7
4 × 16	56.0	50.4	45.4
5 × 6	27.0	24.3	21.9
5 × 8	36.0	32.4	29.2
5 × 10	45.0	40.5	36.5
5 × 12	54.0	48.6	43.7
5 × 14	63.0	56.7	51.0
5 × 16	72.0	64.8	58.3
5 × 20	90.0	81.0	72.9
6 × 6	33.0	29.7	26.7
6 × 8	44.0	39.6	35.6
6 × 10	55.0	49.5	44.6
6 × 12	66.0	59.4	53.5
6 × 14	77.0	69.3	62.4
6 × 16	88.0	79.2	71.3
6 × 20	110.0	99.0	89.1
7 × 12	78.0	70.2	63.2
7 × 14	91.0	81.9	73.7
7 × 16	104.0	93.6	84.2
7 × 20	130.0	117.0	105.3
8 × 12	90.0	81.0	72.9
8 × 14	115.0	103.5	93.2
8 × 16	120.0	108.0	97.2
8 × 20	150.0	135.0	121.5

Source: Courtesy Allis-Chalmers

10.25 WATER SCREENS

10.25.1 General

The rapid accumulation of debris and fish on stationary racks in front of intake screens, and on the screens, tends to clog both the racks and the screens. A velocity of about 2 ft/s at intake will lessen the fish problem. Any increase in speed of water seems to drive them away. Most of the fish trouble occurs when the temperature becomes extremely low. It seems that the fish cannot stand this low temperature, and migrate to points of refuge along the shore that might be somewhat warmer.

10.25.2 Protection of Intake

Intakes can be equipped with a curtain wall extending approximately 8 ft below the water surface to prevent wreckage or flotsam from entering the flume leading to the screens. A wall should be placed about 12 in. from the screen, with water circulating in front of the screen, at not more than 1½ fps. Wind velocities of about 25 mph or over will create problems of debris on the racks and screens. The debris can be controlled by travelling screens and stationary bar screens in front of the travelling screens. Travelling intake water screens are in general use. Most of them operate on a time cycle automatically: rest for 30 minutes and operate for 30 minutes. Most standard screens are designed to take a differential head of water of 5–6 ft.

If, for any reason, refuse or a school of small fish should block the openings in the screen cloth and cause a loss of head water which may seriously affect the flow through screens, there is an arrangement available using

an automatic float system, which flashes a red light in the power house to let operators know something is wrong at the intake.

10.25.3 Flow Through Water Screen (refer to table 10.12)

Screen opening: 0.375×0.375 in. = 0.1406 in.^{2}
Diameter of 10-gauge wire = 0.135 in.
Total area of opening (incl wire) = $(0.375 + 0.135)^2$ = 0.2601 in.^{2}

$$\frac{\text{clear opening}}{\text{c to c of wire}} = \frac{0.1406}{0.2601} = 54\% \text{ eff open area}$$

For an 11-ft width of well, screen is 9 ft 6 in. wide (see figures 10.13 and 10.14).
Net area of screen = $1.67 \times 9.5 \times 0.54 = 8.567$ ft^2.
Assume a water velocity of 2 ft/sec.
Screen to be made up of 5 panels.
One ft^3 of water = 7.54 gal.
Flow = $5 \times 8.567 \times 2 \times 7.54 \times 60 = 38,575$ gal/min.

10.25.4 Loss of Head

The following formula has been used for approximating the drop in feet of water in front of a screen compared to that on the opposite side (differential head) of the screen panel (tray):

$$H = \frac{[Q^2/K \times A]}{64.4^*}$$

---

*This figure determined from tests.

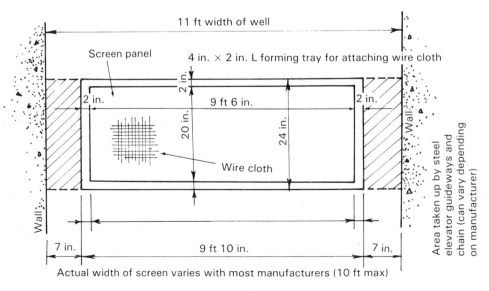

Figure 10.13 Vertical section through a screen well based on using 24-in. pitch elevator chain.

Table 10.12 Wire Cloth Standards for Square Openings (Dry Screening Only)

Clear Square Opening (in.)	A Dia (in.)	A Open Area (%)	B Dia (in.)	B Open Area (%)	C Dia (in.)	C Open Area (%)	D Dia (in.)	D Open Area (%)	Feed Size (in.) I[a]	Feed Size (in.) II	Feed Size (in.) III
1/8	0.041[b]	57	0.054[b]	48	0.072[b]	40	0.080[b]	37	5/8	3/4	1
3/16	0.047	64	0.080	49	0.092	45	0.105	41	3/4	1	1 1/4
1/4	0.063	64	0.105	49	0.120	46	0.135	42	1	1 1/2	2
5/16	0.072	66	0.120	52	0.135	49	0.148	46	1 1/2	2	2 1/2
3/8	0.080	68	0.135	54	0.148	51	0.162	49	1 1/2	2	2 1/2
7/16	0.092	71	0.148	56	0.162	53	0.177	51	2	2 1/2	3
1/2	0.105	68	0.162	57	0.177	55	0.192	52	2	2 1/2	3
9/16	0.105	70	0.177	58	0.192	56	0.207	54	2 1/2	3 1/4	3 3/4
5/8	0.120	70	0.177	61	0.192	58	0.225	54	2 1/2	3 1/4	3 3/4
3/4	0.135	72	0.192	63	0.207	61	0.250	56	3	3 3/4	4 1/2
7/8	0.148	73	0.207	65	0.225	63	0.250	61	3	3 3/4	4 1/2
1	0.162	74	0.225	67	0.250	64	5/16	58	3 1/2	4 1/2	5 1/4
1 1/8	0.177	75	0.225	69	0.250	67	5/16	61	3 1/2	4 1/2	5 1/4
1 1/4	0.192	75	0.250	69	5/16	64	3/8	59	4	5	6
1 3/8	0.207	76	0.250	72	5/16	66	3/8	62	4	5	6
1 1/2	0.225	76	0.250	73	5/16	69	3/8	64	4	5	6
1 5/8	1/4	75	5/16	70	3/8	66	7/16	62	4 1/2	5 1/2	7
1 3/4	1/4	77	5/16	72	3/8	68	7/16	64	4 1/2	5 1/2	7
1 7/8	1/4	78	5/16	73	3/8	69	7/16	66	4 1/2	5 1/2	7
2	5/16	75	3/8	71	7/16	67	1/2	64	5	6 1/2	8
2 1/4	5/16	77	3/8	73	7/16	70	1/2	67	5	6 1/2	8
2 1/2	5/16	79	3/8	76	7/16	72	1/2	69	5	6 1/2	8
2 3/4	5/16	81	3/8	77	7/16	74	1/2	72	5	6 1/2	8
3	3/8	79	7/16	76	1/2	74	5/8	69	6	7 1/2	9
3 1/2	3/8	82	7/16	79	1/2	77	5/8	72	6	7 1/2	9
4	7/16	81	1/2	79	5/8	75	3/4	71	7	8 1/2	10

Source: Courtesy Allis-Chalmers

Notes: A = light, 50–75 lb/ft³, coal, nonabrasive; B = standard, 75–100 lb/ft³, limestone, sand, and gravel; C = medium heavy, 100–120 lb/ft³, average ores, moderate abrasives; D = heavy, 120–140 lb/ft³, heavy ores, high abrasives.
Selections below the solid lines require wedge-type clamping bars. Perforated plate is recommended for openings below the dotted lines.
[a] The wire diameters listed are suitable for feed size not exceeding that listed in column I. When feed size exceeds column I but not column II, use the next larger wire diameter. When it exceeds column II but not column III, increase the wire diameter two sizes. When it exceeds column III, a relief deck is recommended to increase the life of the wire.
[b] Rectangular openings are perferred to obtain heavier wire or more open area.

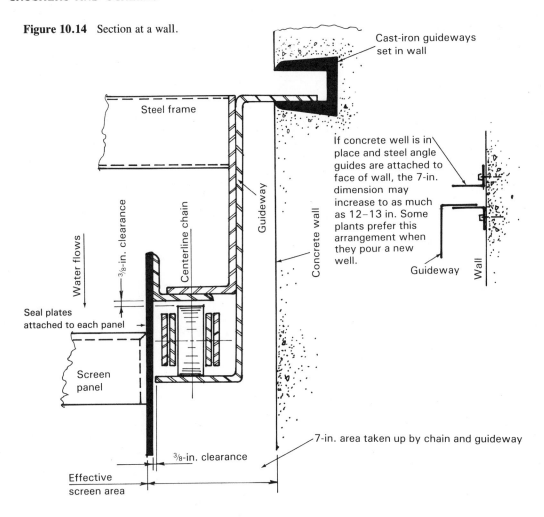

Figure 10.14 Section at a wall.

where

H = drop in feet

Q = cubic feet of water per second

K = 0.95 for 100% clear screen panel, 0.90 for 90% and 0.80 for 60%

A = net clear area of screen below water line, or

$$\frac{\text{capacity in ft}^3/\text{sec}}{\text{velocity through screen in ft/sec}}$$

10.25.5 Spray Nozzles

Spray nozzles form a very important part of a water screen installation. A series of them, depending on the width of the screen, are located just below the head shaft and arranged to spread the force of water over the surface of the screen panel (tray) to clean the debris from the wire cloth and lips of the panel. They are nonclogging and easily assembled on a pipe. One type of nozzle is arranged with a shank threaded to fit a threaded hole in a pipe. The clamp type fits over a plain hole drilled in a pipe.

10.25.6 Accumulated Sand

If a large amount of fine sand accumulates around the boot, lifting shelves and digging lips must be added. Provision must be made to move the screen at low speed, or set it in motion, when a heavy load (fish, leaves, and refuse) suddenly appears.

10.26 SCREEN IN LABORATORIES

To protect the environment from dangerous bacteria, a travelling screen can be placed in the wall of a laboratory, with the bottom quarter of the screen immersed in a germicide solution. The speed of travel should be such as to provide the proper time for the solution to act on the bacteria. The screen is made up of removable sections, so that any one can be replaced on the clean side. Controls can be provided that stop the screen in case of problems.

Skip Hoists

11.1 GENERAL

The skip hoist consists of a bucket mounted on four rollers traveling on a tee rail or angle track structural runway, with or without counterweights or two buckets balancing each other, a winding machine, wire rope, equipment for loading and unloading the buckets, and a system of electrical apparatus and wire-rope sheaves.

This is an effective method of unloading bulk materials, and is especially applicable to high lifts. It is suitable for various combinations of vertical and inclined paths of travel. It has a few moving parts. A skip hoist can handle abrasive or corrosive material and large lumps as well as fine material; but very fine, light, fluffy, and contaminable materials should be handled by other means.

11.2 TYPES

Skip hoist buckets are made as standard units in capacities of 20 through 200 ft^3 and operate at speeds from 40 to 450 fpm.

Slow speed is considered to be 40–100 fpm.
Medium speed is considered to be 100–150 fpm.
High speed is considered to be 150–450 fpm.

The single-bucket, uncounterweighted type is commonly used for small capacities, especially when intermittent peak loads are not detrimental to the maximum demand on the power system. It travels at about 100 fpm. The single-bucket, counterweighted type is well suited for medium capacities and lifts where a small-size motor and a smooth electric load curve are desirable.

The two-bucket, balanced type is preferable for large capacities and high lifts. It travels up to 150 fpm. It is the most economical. The capacity with a given size bucket can be increased by decreasing the length of time required for the bucket to travel between the loading and emptying zones, or by decreasing the loading and dumping time. Single-speed bucket travel is preferable for low lifts, up to about 75 ft.

Two- or three-speed travel are used for high lifts and large capacities, to allow proper slow speed of bucket for operating the loader and for emptying the bucket at the discharge point. The slower the speed, the lighter the stress on the loader and the motor.

Automatic loading should be used for fully automatic operation, and for maximum capacity with minimum labor and attention. The balanced skip travels 150–250 fpm. It has antifriction rollers.

11.3 LOADING METHODS

Skip buckets usually receive the material at their lowest point of travel. The method of loading the bucket depends mainly upon the rate at which the material is to be handled, as well as on local conditions. Where a small quantity of material is to be handled, and when it is brought to the loading point in small push cars, the contents are usually dumped or shoveled direct into the bucket. This loading method is designated as manual loading and requires an attendant to place the material in the bucket. When any material is received in large quantities, as by truck or railroad car, a receiving hopper is ordinarily provided to serve as a reservoir from which the material is withdrawn a skipful at a time to be hoisted to the emptying point. The receiving hopper can be provided with a mechanical loader for manual or automatic delivery, loading, and cutting off of the flow of material to the bucket, thereby eliminating the services of an attendant and, at the same time, ensuring continuous loading as long as there is material in the receiving hopper. Modern practice favors automatic loading wherever possible, but there are often local conditions or special requirements which necessitate manual loading of either the direct or indirect type.

11.4 MANUAL LOADING— DIRECT TYPE

The direct method of manual loading is used where material is dumped or shoveled direct into the bucket. A

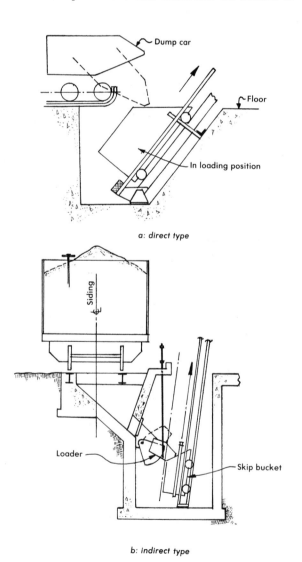

a: direct type

b: indirect type

Figure 11.1 Manual loading.

common method of loading by the use of a small car is shown on figure 11.1a. One man can easily push and dump a car of 18 ft^3 capacity of this type. Larger cars may be used, but they require more effort in handling. The skip bucket should have a capacity slightly greater than that of the push car from which the material is dumped. This will avoid spillage. Buckets of 20 ft^3 and 30 ft^3 capacity are commonly used for cars of 18 ft^3 and 27 ft^3 capacity.

11.5 MANUAL LOADING— INDIRECT TYPE

The indirect method of manual loading consists of a hand-operated, undercut gate and chute, attached to a dump hopper. This hopper permits storage of material ready to be discharged to the bucket as soon as it reaches the loading position, as shown on figure 11.1b.

The damming type of loading serves the purpose of automatically controlling the delivery of material into the bucket. With this type of loader, the bucket continues to be hoisted and lowered whether or not there is material in the hopper.

11.6 FULL BUCKET CONTROL SYSTEM

When the bucket reaches the loading position, the full-bucket-control-type loader permits the material to flow from the hopper until the bucket is full, at which time the mechanism instantly causes the bucket to be hoisted, thereby moving the loader into its closed position. It remains there as the bucket continues to travel to the emptying point, and until bucket returns to the loading position, where it is again loaded automatically. The bucket continues to load and empty automatically as long as there is sufficient material to fill it. The full-bucket-control-type accomplishes the additional important functions of causing the bucket and its associated moving elements to remain idle in the loading position ready for loading when there is insufficient material to fill the bucket, thereby eliminating lost motion. Instead of permitting the empty bucket to be hoisted and lowered, which is the condition existing with all automatic loaders, except the more complicated and expensive weighing type, which is not used very often. Also, this type eliminates the need for an attendant to close and open the electric circuit that causes the bucket to be hoisted when there is material in the hopper, and causes it to stop when the hopper is empty, except to open the main switch as a safety precaution when the equipment is not in use.

11.7 BALANCED SKIP HOIST

The two-bucket installation, operating on separate tracks, although connected to one hoisting machine, is a specially designed unit. While one bucket is in loading position, the other bucket is dumping. A counterweight arrangement also is included, to take care of the starting and dumping phases of the operating cycle. The motor horsepower is low on a unit of this kind, considering the large tonnage delivered. These units are used principally by steel mills.

11.8 MATERIALS HANDLED

Most loaders and buckets are similar in construction, but the hoisting machines vary. One type is equipped with herringbone-gear reduction; another type is either horizontal or vertical with enclosed worm gear and internal spur gear reductions. Such material as large lumps of ferro alloys, foundry coke, which often is made with 8-in. lumps, and charcoal, where the pieces are 6 to 8-in. long, handle nicely in a skip bucket. Clay, clinkers, cullet (broken glass), lime, gravel, stone, and steel or cast iron turnings, also are handled satisfactorily.

11.9 SELECTION OF A SKIP HOIST

The selection of a skip hoist depends on the following criteria:

1. Material handled: Weight per cubic foot of material handled and nature of material, whether abrasive or corrosive, and size of lumps. These factors govern the size and shape of bucket, whether lined or not, and the loading and discharge arrangement recommended.
2. Height of lift: Determined by method of loading and net distance bucket will travel. Lifts up to 200 feet are practicable.
3. Capacity in tph: Also the average number of hours per day the skip hoist will be operated. From this data, the type of hoist machine can be determined.
4. Path of bucket travel: Whether bucket will be hoisted vertically or at an angle from which the supporting rope pull can be obtained.

11.10 SPECIFICATION FOR VERTICAL-LIFT DRUM CONVEYOR

The equipment covered by this specification consists of a Swinging Carriage for installation in _____ .

The equipment is to convey empty steel drums from first level to second level and also return them from second level to first level. The drums are of one size only: _____ outside diameter × _____ long, weighing _____ each, and are of the all-steel type with three chimes, one at each end and one in the center. The drums are handled on their side with the axis across the face of the carriages, so they may be rolled to and from the loading and unloading stations.

The conveyor is to be set in a pit at the first level, with the first-level loading station at floor level so the drums may be conveniently rolled into the loading station. Purchaser may elect to use runways or skids from cars to station which, if so elected, are to be furnished and installed complete by purchaser, with any manual assistance provided, if and as required, at the loading station, to ensure that the drums will be delivered at the proper time to be properly picked up by the ascending carriages. Once in the station, the drum will be picked up by the next ascending carriage, conveyed to the top of the machine, around the head sprocket, and down the descending side, where it is delivered to the second-level discharge station, being automatically combed from the carriage as the fingers pass through the station fingers, and rolling off of its own weight onto the floor.

In returning the drums from second to first level, the same procedure is used, but the second-level discharge station is manually thrown out of position by means of the control levers provided. The second-level loading station also is equipped with the same type of lever arrangement, so it may be thrown out of position when service from first to second level is required. It will be impossible, therefore, to provide the two types of service simultaneously. The capacity of the conveyor is to be a maximum of seven drums per minute.

Work To Be Performed By Others

First: Preparation and finishing of all necessary floor openings and pits, as required, together with any protection that may be necessary for same, such as guardrails, doors, etc.

Second: Clearing of the site, including removal of any obstructions such as piping, conduits, machinery, etc, which might interfere with the work of installation.

Third: Performing of all field wiring, including mounting of starters and pushbutton control stations. The Contractor, however, is to mount the motor unit and the limit switches that comprise the electric safety devices, and to provide a wiring diagram to show the hookup of these limit switches in the control circuit.

Equipment To Be Provided By Contractor

One Swinging-Carriage Vertical-Lift Continuous-Chain Conveyor, with one fixed and one hinged loading station with manual feed, and one fixed and one hinged automatic discharge station, and eight carriages.

Speed: Approximately 45 ft per minute.

Capacity: Approximately seven drums per minute.

Drive Mechanism: To consist of a 1½ hp, 1200 rpm, 440 V, 60 Hz, 3 ph motor, direct-connected to an enclosed worm-gear reducer by means of a flexible coupling. The motor unit is to be furnished by Purchaser, but the reducer and couplings by the Contractor. The slow-speed shaft of the reducer unit drives the conveyor head shaft through a 4-dp spur-gear reduction with 5-in. pitch diameter, 20-tooth 2½-in. face-cut steel pinion, and 38-in. pitch diameter, 152-tooth, 2½ in. face-cut cast-iron gear. The motor and reducer unit shall be mounted on a structural steel stand built integral with the conveyor head framing. The motor which Purchaser is to furnish is to be provided with an extended shaft to permit the applying of a solenoid brake, which also is to be furnished by the Contractor.

Control: Automatic starter with necessary Stop and Start pushbutton control stations, to be furnished complete by Purchaser.

Framework: Structural steel throughout, approximately 25 ft 8½ in. high overall, of the 4-post type, with main members of 5 in. × 3½ in. × ⁵⁄₁₆ in. angles and auxiliary members of 1½ in. × 1½ in. × ³⁄₁₆ in. carriage-guide angles, rigidly braced and crossbraced as required. (Auxiliary guides to be provided around the head sprockets that engage with offset guidewheels mounted on carriage arms and which ensure that the carriages remain in a true horizontal position in passing from the ascending to the descending side around the head sprocket.)

Chain: Two endless strands of No. C-188 malleable combination type 2.069 in. pitch of 14,000-lb ultimate strength per strand, with each strand approximately 48 ft 9 in. long.

Carriage: To be of the fingered type with three V-shaped, or cradle fingers, of heavy structural steel and designed so as to intermesh with the loading and unloading station fingers spaced on approximately 6 ft 1 in. centers with 3 in. channel hangers 24 in. long, guide wheels, arms, etc., complete as required.

Stations: To be of the double-leaf fingered type, of structural steel construction, with two sets of fingers in each leaf. The loading and unloading stations on the second level are to be provided with a hinge arrangement and hand-lever control, permitting them to be thrown out of position when not in use.

Head and Tail Sprockets and Shafts: Head shaft to be 2³⁄₁₆ in. dia with babbitt pillow-block bearings, and with 23 gap-tooth (normal 46-tooth), 38.23 in. pd cast-iron sprockets mounted on the shaft. Tail shafts to be of the stub type, 1⁵⁄₁₆ in dia; four are required, on each of which is mounted a 7 gap-tooth (normal 14-tooth), 11.72 in. pd cast-iron sprocket, with the cross angles on which the four stub shafts are mounted provided with a hand-screw takeup mechanism, and the necessary framework to permit vertical adjustment of the carrying chain, as required to keep chain at proper operating tension.

Safety Device: Both unloading stations are to be provided with electric-type automatic safety devices consisting of two limit switches for each station, with arms and mounting so connected with the control circuit that, in the event a drum fails to discharge properly or is held on the station for any reason, the conveyor will automatically come to an instantaneous stop, and cannot be started again until the offending drum has been cleared.

Alternative No. 1: If, in lieu of the drive mechanism specified, with the motor furnished by Purchaser and the balance of the drive by Contractor, a geared-head motor, complete with brake and controls, of 440 V (nominal), 60 Hz, 3 ph power description is provided, to be furnished complete by Contractor, including an automatic enclosed starter with two pushbutton control stations—the balance of the equipment is to remain as specified.

Shipment: Approximately _____ weeks after receipt of order, with the work of installation estimated to require approximately _____ days after arrival of material.

Drawing _____ attached.

Hoists and Cranes

12.1 GENERAL

This section deals with hoists; jib, gantry, and overhead cranes; and monorail systems.

12.2 HOISTS

The electrically operated hoist is most common. The motors are high torque, high slip, with the starting torque 2½ to 3 times the rating of the motor; they are flange-mounted and designed for direct application to hoists and trolleys. These hoists provide many different mountings, each with its own advantages for a particular type of hoisting service. These hoists operate parallel to rails or perpendicular to the rails. They are available in hook mounting. A motor-driven trolley (figure 12.1) is wired through the hoist, using one pushbutton pendant to operate both trolley and hoist. They have a high factor of safety, with the cable capable of lifting five times the weight of the hoist's rated capacity. Some hoist units are particularly adapted to low headroom conditions, where placement of the hoist drum at right angles to the mounting beam permits the load block to be drawn up along side the drum for highest possible lift. A two-speed hoist provides high speed for quick handling and low speed for careful handling and precise spotting.

Most electric hoists have two brakes; a ratchet roller-type mechanical-load brake and a spring-set electric-release solenoid-type brake. The latter consists of a series of discs, with alternate discs anchored. When power is cut off, a spring sets the brake. Lighter capacity electric hoists may be designed so that the motor can be removed without lowering the load. The hoists normally are designed with the brakes on one side and the motor on the other side of the drum.

Hand chain hoists are limited not so much by capacity as by the length of chain overhauled to make the lift. Because of gearing, the length of chain overhauled is many times the distance that the load is lifted. Their use is limited to intermittent duty and low capacity; say 0.5–5 tons. Table 12.1 gives dimensions and details for P & H (Harnischfeger) Hevi-Lift hoists.

Harnischfeger uses a dual braking system, which includes a safety-torque system that eliminates the mechanical brake. Thus, under a simultaneous power failure and electric brake failure, the load will lower itself to the ground at a controlled speed.

Air-operated hoists are used primarily in foundries, where extreme heat and dust conditions exist. They lift loads of ¼ ton to 10 tons at speeds of 10–100 fpm, depending on the throttling control valve and air supply.

The Hoist Manufacturers' Institute publishes "Standard Specifications for Electric Rope Hoists." These should be adapted to the user's requirements. There is a useful appendix attached to the specifications.

12.2.1 Hoist Sheaves and Cables

Sheaves should have minimum diameter 18 times the diameter of the cable.

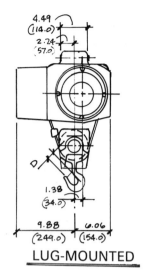

LUG-MOUNTED

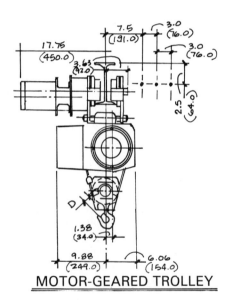

MOTOR-GEARED TROLLEY

Figure 12.1   P&H Hevi-Lift hoists.

12.2.2   Hoists on Traveling Bridges

Hoists may be deck-mounted on a trolley on a traveling bridge. The bridge runs on tracks and may be a double-girder bridge for extremely heavy loads (10 tons and up). The hoist is mounted on a trolley deck between the trolley wheels. Lighter units (4–5 tons) can be mounted on a swivel type of mounting.

12.2.3   Hoists—Electrical

Pushbutton controls have a chain anchored to a control box on the hoist supporting the pushbutton station, so

there is no strain on the electric cable. The pushbutton station is offset from the hoist cables so there will be no mechanical interference between the free-swinging devices. Hoists are available in the twin-hook type.

With a short run of single cable, a typical hoist comes with a 15-ft flexible cord or cable (National Electrical Code Article 400). They are used for intermittent duty.

12.3   SWINGING JIB CRANES

Jib cranes (figure 12.2) are cranes cantilevered from either the building frame or independent columns. Most jib cranes have booms, trolleys, and hoisting mechanisms. It is an economical method where limited coverage is desired. If supported from independent columns, they will require more expensive foundations to prevent overturning.

12.4   GANTRY CRANES

Normally, gantry cranes (figure 12.3) are of the dual-legged type, travelling on two horizontal girders, or on ground rails, with each leg supported on wheels. Sometimes a gantry travels on one leg supported at a desirable elevation on a building frame, with the other leg travelling on rails at ground level. It can have a main hoist and an auxiliary hoist. Gantry cranes have been built with spans of a hundred feet and more.

12.5   CRANE CLASSIFICATION

CMAA has adopted industry-wide classifications for cranes, depending on their use. The classification below is taken from CMAA "Specifications for Electrical Overhead Traveling Cranes":

Service classes have been established to specify the most economical crane for the installation. Specific requirements are shown for these components where design is influenced by classifications. All classes of cranes are affected by the operating conditions so, for the purpose of these definitions, it is assumed that the crane will be operating in normal ambient temperatures (0–100°F) and normal atmospheric conditions (free from excessive dust, moisture, and corrosive fumes).

CLASS A

This class is further divided into two subclasses, because of the nature of the loads to be handled.

Class A1 (Standby Service)

This service class covers cranes used in installations such as: power houses, public utilities, turbine rooms, nuclear reactor buildings, motor rooms, nuclear fuel handling, and trans-

TABLE 12.1   *285*

Table 12.1 P & H (Harnischfeger) Beta Hevi-Lift Hoists

Capacity	Power (V-ph-Hz)	Lift (ft-in.)	Speed (fpm)	Motor (hp)	Lug Weight (lb)
½ ton 500 kg 1100 lb	200-3-60 230-3-60 460-3-60 575-3-60	23-7	28	1	210
½ ton 500 kg 1100 lb	200-3-60 230-3-60 460-3-60 575-3-60	23-7	37	1½	210
1 ton 1000 kg 2200 lb	200-3-60 230-3-60 460-3-60 575-3-60	23-7	21	1½	210
1 ton 1000 kg 2200 lb	200-3-60 230-3-60 460-3-60 575-3-60	23-7	28	2	210
2 ton 2000 kg 4400 lb	200-3-60 230-3-60 460-3-60 575-3-60	16-5	14	2	240
2 ton 2000 kg 4400 lb	200-3-60 230-3-60 460-3-60 575-3-60	16-5	21	3	240
3 ton 3000 kg 6600 lb	200-3-60 230-3-60 460-3-60 575-3-60	19-1	9	2	290
3 ton 3000 kg 6600 lb	200-3-60 230-3-60 460-3-60 575-3-60	19-1	14	3	290
5 ton 5000 kg 11,000 lb	200-3-60 230-3-60 460-3-60 575-3-60	16-5	11	4	335

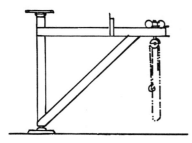

Figure 12.2 Jib crane. *(Courtesy Whiting Corp.)*

former stations, where precise handling of valuable machinery at slow speeds with long idle periods between lifts is required. Capacity loads may be handled for initial installation of machinery and for infrequent maintenance.

Class A2 (Infrequent Use)

These cranes will be used in installations such as: small maintenance shops, pump rooms, testing laboratories, and similar operations, where the loads are relatively light, the speeds are slow, and a low degree of control accuracy is required. The loads may very anywhere from no load to full capacity, with a frequency of a few lifts per day or month.

CLASS B (Light Service)

This service covers cranes such as those used in repair shops, light assembly operations, service buildings, light warehousing, etc, where service requirements are light and the speed is slow. Loads may vary from no load to full rated load, with an average load of 50% of capacity, with 2 to 5 lifts per hour, averaging 15 feet, not over 50% of the lifts at rated capacity.

CLASS C (Moderate Service)

This service covers cranes such as those used in machine shops, papermill machine rooms, etc, where the service requirements are medium.

In this type of service, the crane will handle loads which average 50% of the rated capacity, with 5 to 10 lifts per hour, averaging 15 feet, not over 50% of the lifts at rated capacity.

CLASS D (Heavy Duty)

This service covers cranes, usually cab-operated, such as are used in heavy machine shops, foundries, fabricating plants, steel warehouses, lumber mills, etc, and standard-duty bucket and magnet operation, where heavy-duty production is required, but with no specific cycle of operation. Loads approaching 50% of the rated capacity will be handled constantly during the working period. High speeds are desirable for this type of service, with 10 to 20 lifts per hour, averaging 15 feet, not over 65% of the lifts at rated capacity.

CLASS E (Severe Duty-Cycle Service)

This type of service requires a heavy-duty crane capable of handling the rated load continuously, at high speed, in repetition throughout a stated period per day, in a predetermined cycle of operation. Applications include magnet, bucket, magnet-bucket combinations of cranes for scrap yards, cement mills, lumber mills, fertilizer plants, etc, with 20 or more lifts per hour, all at rated capacity. The complete cycle of operation should be specified.

CLASS F (Steel Mill AISE Specification)

Cranes in this class are covered by the current issue of The Association of Iron and Steel Engineers Standard No. 6, for ''Electric Overhead Traveling Cranes for Steel Mill Service.''

Note: Most installations involved in handling bulk material will be Class D, or Class C on occasion.

12.6 CONVENTIONAL OVERHEAD CRANES: TWO TYPES

Underhung cranes (figure 12.4) have runways that are suspended from the structural frame. The points of suspension should be as close to the columns as possible.

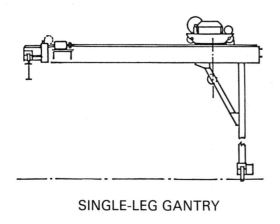

SINGLE-LEG GANTRY

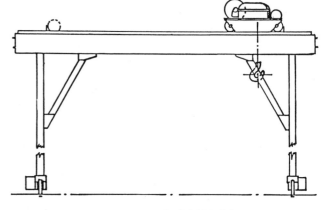

DUAL-LEG GANTRY

Figure 12.3 Gantry cranes. *(Courtesy Whiting Corp.)*

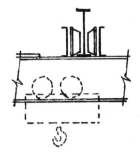

Figure 12.4 Underhung crane.

The capacity of these cranes should be limited to 10 tons. The runways can have spurs, and thus extend into other bays, or can be interlocked at one end into a monorail system. The trolley can be hand-pushed or electrified. The switches can be manual or electric, or automatic. All switches should have automatic locking devices.

Top riding, heavier, overhead cranes ride on track beams that are supported by columns. This type of crane may be as small as 5-ton capacity, and may extend to

very heavy ones. Fifty- and 100-ton cranes are not rare, and they are built to much heavier capacity. Cranes 10 tons or under can rest on brackets on the building columns. Heavier crane runways should rest on stub columns tied to the building columns, or on building columns with one flange cut away (bayonet).

12.6.1 Dimensions

It is advisable to buy the cranes early, so that clearance dimensions can be determined before structural design is started. The crane manufacturer's diagrams will determine dimensions of building frame and size of structural members. Further, crane deliveries are slow. It is a great advantage in time and cost to have the crane available during construction, and to erect it when the structural steel is being erected.

12.6.2 Diagram Notes

The diagram on figure 12.5 is for Class D cranes, and shows required dimensions for pendant-operated (con-

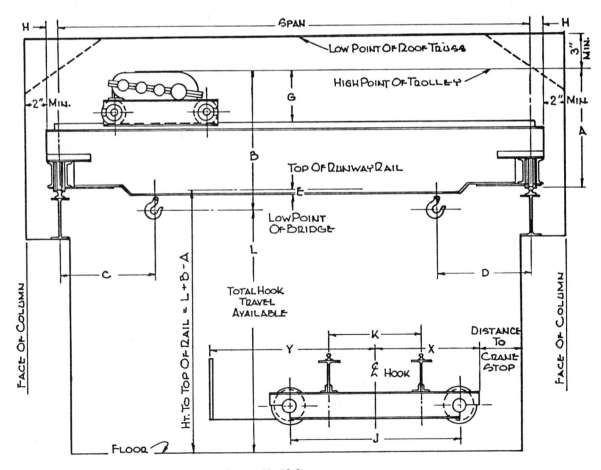

Figure 12.5 A pendant-operated crane (refer to table 12.2).

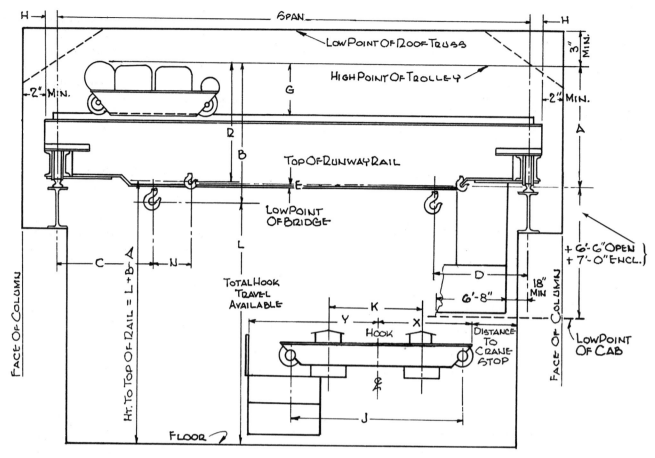

Figure 12.6   A cab-operated crane (refer to table 12.3).

trolled) cranes. Figures 12.6 and 12.7 give dimensions for cab-operated (controlled) cranes, and are for Class C and D cranes (refer to paragraph 12.6.4). In these diagrams:

Dimensions indicated by letter are furnished by the crane manufacturer.

Height from top of runway rail to bottom chord of truss equals $A + 3$ in. minimum.

Available hook travel is shown by L and elevation to top of rail equals $L + B - A$.

Dimensions to center line of crane runways from nearest positions of hook are shown by C and D.

Dimensions from end of runway to extreme positions of crane equal $X +$ crane stop allowance and $Y +$ crane stop allowance.

Clearance from center line of crane runway to inside face of columns equals $H + 2$ in. minimum.

(*Text continues on page 299.*)

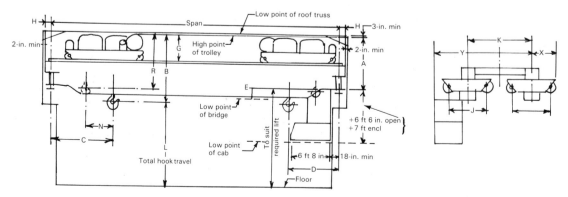

Figure 12.7   Cab-controlled overhead cranes with two trolleys (refer to table 12.4).

Table 12.2 Pendant-Operated Overhead Cranes (see figure 12.5)

Capacity (tons)	Span (ft)	A	B	C	D	E	G	H	J	K	L	X	Y	Max. Wheel Load (lb)	Runway Rail Wt. (lb)	Trolley Wt. (lb)	Total Crane Wt. (lb)	Type Girder
5	20	4'4¾"†	4'10"	2'6"	2'9"	5"	2'2"	5½"	8'8"	6'0"	56'7"	5'6½"	8'8¾"	8500	30	3900	10269	Beam
	25	4'4¾"†	4'10"	2'6"	2'9"	5"	2'2"	5½"	8'8"	6'0"	56'7"	5'6½"	8'8¾"	8980	30	3900	11018	Beam
	30	4'4¾"†	4'10"	2'6"	2'9"	5"	2'2"	5½"	8'8"	6'0"	56'7"	5'6½"	8'8¾"	9430	30	3900	11862	Beam
	35	4'4¾"†	4'10⅛"	2'6"	2'9"	5"	2'2"	5½"	8'8"	6'0"	56'7"	5'6½"	8'8¾"	10050	30	3900	13500	Beam
	40	4'4¾"†	4'7"	2'6"	2'9"	5"	2'2"	5½"	8'8"	6'0"	56'7"	5'6½"	8'8¾"	10500	30	3900	14569	Beam
	45	4'5¾"	4'5"	2'6"	2'9"	5"	2'2"	5½"	8'8"	6'0"	56'7"	5'6½"	8'8¾"	11240	30	3900	16770	Beam
	50	4'5¾"	4'5"	2'6"	2'9"	5"	2'2"	5½"	8'8"	6'0"	56'7"	5'6½"	8'8¾"	12270	30	3900	20100	Beam
	55	4'5¾"	4'5"	2'6"	2'9"	5"	2'2"	5½"	8'8"	6'0"	56'7"	5'6½"	8'8¾"	12780	30	3900	21523	Beam
	60	4'6"	4'5"	2'6"	2'9"	5"	2'2"	5½"	8'8"	6'0"	56'7"	5'6½"	8'8¾"	14180	30	3900	26300	Beam
	65	4'3½"	4'5"	3'0"	3'6"	8"	2'2"	5⅝"	12'0"	6'0"	56'7"	7'1"	8'11¾"	13600	30	3900	23600	Box
	70	4'3½"	4'5"	3'0"	3'6"	11"	2'2"	5⅝"	12'0"	6'0"	56'7"	7'1"	8'11¾"	14300	30	3900	25900	Box
	75	4'3½"	4'5"	3'0"	3'6"	13"	2'2"	5⅝"	12'0"	6'0"	56'7"	7'1"	8'11¾"	15100	30	3900	28200	Box
	80	4'3½"	4'5"	3'0"	3'6"	16"	2'2"	5⅝"	12'0"	6'0"	56'7"	7'1"	8'11¾"	15900	30	3900	31000	Box
10	20	4'4¾"†	5'0"	2'6"	2'9"	5"	2'2"	5½"	8'8"	6'0"**	37'9"**	5'6½"	8'8¾"	13410	30	4600	11500	Beam
	25	4'4¾"†	5'0"	2'6"	2'9"	5"	2'2"	5½"	8'8"	6'0"**	37'9"**	5'6½"	8'8¾"	14020	30	4600	12248	Beam
	30	4'4¾"†	5'0¼"	2'6"	2'9"	5"	2'2"	5½"	8'8"	6'0"**	37'9"**	5'6½"	8'8¾"	14730	30	4600	13743	Beam
	35	4'5¾"	4'7"	2'6"	2'9"	5"	2'2"	5½"	8'8"	6'0"**	37'9"**	5'6½"	8'8¾"	15460	30	4600	15579	Beam
	40	4'5¾"	4'7"	2'6"	2'9"	5"	2'2"	5½"	8'8"	6'0"**	37'9"**	5'6½"	8'8¾"	16050	30	4600	16982	Beam
	45	4'5¾"	4'7"	2'6"	2'9"	5"	2'2"	5½"	8'8"	6'0"**	37'9"**	5'6½"	8'8¾"	17010	30	4600	19868	Beam
	50	4'5¾"	4'7"	2'6"	2'9"	5"	2'2"	5½"	8'8"	6'0"**	37'9"**	5'6½"	8'8¾"	17560	30	4600	21292	Beam
	55	4'6"	4'7"	2'6"	2'9"	5"	2'2"	5½"	8'8"	6'0"**	37'9"**	5'6½"	8'8¾"	18960	30	4600	25974	Beam
	60	4'8¾"	4'7"	2'6"	2'9"	5"	2'2"	5½"	8'8"	6'0"**	37'9"**	5'6½"	8'8¾"	20040	30	4600	29446	Beam
	65	4'9½"	4'7"	3'0"	3'6"	7"	2'2"	5⅝"	12'0"	6'0"**	37'9"**	7'1"	8'11¾"	19300	30	4600	26400	Box
	70	4'9½"	4'7"	3'0"	3'6"	10"	2'2"	5⅝"	12'0"	6'0"**	37'9"**	7'1"	9'0¼"	20100	30	4600	28800	Box
	75	4'9½"	4'7"	3'0"	3'6"	12"	2'2"	5⅝"	12'0"	6'0"**	37'9"**	7'1"	9'0¾"	20900	30	4600	31400	Box
	80	4'9½"	4'7"	3'0"	3'6"	15"	2'2"	5⅝"	12'0"	6'0"**	37'9"**	7'1"	9'1¼"	21900	30	4600	34400	Box

†High point of crane is top of control panel.

*For "K" = 7'0", available lift "L" = 49'4".

289

Table 12.2 (*continued*)

Capacity (tons)	*Span (ft)*	*A*	*B*	*C*	*D*	*E*	*G*	*H*	*J*	*K*	*L*	*X*	*Y*	*Max. Wheel Load (lb)*	*Runway Rail Wt. (lb)*	*Trolley Wt. (lb)*	*Total Crane Wt. (lb)*	*Type Girder*
15	20	4'4¾"†	5'7"	2'6"	2'9"	5"	2'2"	5½"	8'8"	6'0"‡	22'8"	5'6½"	8'8¾"	17900	30	4900	11750	Beam
	25	4'4¾"†	5'4"	2'6"	2'9"	5"	2'2"	5½"	8'8"	6'0"‡	22'8"	5'6½"	8'8¾"	18800	30	4900	13127	Beam
	30	4'5⅝"	5'2"	2'6"	2'9"	5"	2'2"	5½"	8'8"	6'0"‡	22'8"	5'6½"	8'8¾"	19680	30	4900	14907	Beam
	35	4'5¾"	5'2"	2'6"	2'9"	5"	2'2"	5½"	8'8"	6'0"‡	22'8"	5'6½"	8'8¾"	20650	30	4900	17407	Beam
	40	4'5¾"	5'2"	2'6"	2'9"	5"	2'2"	5½"	8'8"	6'0"‡	22'8"	5'6½"	8'8¾"	21470	30	4900	19520	Beam
	45	4'6"	5'2"	2'6"	2'9"	5"	2'2"	5½"	8'8"	6'0"‡	22'8"	5'6½"	8'8¾"	22720	40	4900	23384	Beam
	50	4'8⅝"	5'2"	2'6"	2'9"	5"	2'2"	5½"	8'8"	6'0"‡	22'8"	5'6½"	8'8¾"	23760	40	4900	26510	Beam
	55	4'8⅝"	5'2"	2'6"	2'9"	5"	2'2"	5½"	8'8"	6'0"‡	22'8"	5'6½"	8'8¾"	24450	40	4900	28438	Beam
	60	4'9"	5'2"	3'0"	3'6"	8"	2'2"	5⅝"	8'8"	6'0"	22'8"	5'6½"	8'8¾"	25600	40	4900	25700	Beam
	65	4'9½"	5'2"	3'0"	3'6"	11"	2'2"	5⅝"	12'0"	7'0"	29'9"	7'1"	9'5¾"	24650	40	4900	28300	Box
	70	4'9½"	5'2"	3'0"	3'6"	14"	2'2"	5⅝"	12'0"	7'0"	29'9"	7'1"	9'6¼"	25450	60	4900	30800	Box
	75	4'9½"	5'2"	3'0"	3'6"	16"	2'2"	5⅝"	12'0"	7'0"	29'9"	7'1"	9'7¼"	26500	60	4900	34000	Box
	80	4'9½"	5'2"	3'0"	3'6"	19"	2'2"	5⅝"	12'0"	7'0"	29'9"	7'1"	9'7¾"	27500	60	4900	37360	Box

†High point of crane is top of control panel.

‡For "K" = 7'0", available lift "L" = 29'9".

Capacity (tons)	*Span (ft)*	*A*	*B*	*C*	*D*	*E*	*G*	*H*	*J*	*K*	*L*	*X*	*Y*	*Max. Wheel Load (lb)*	*Runway Rail Wt. (lb)*	*Trolley Wt. (lb)*	*Total Crane Wt. (lb)*	*Type Girder*
20	20	4'9"	4'10"	3'0"	3'6"	5"	2'5"	5⅝"	10'0"	7'0"	39'0"	6'5¼"	8'11"	23630	60	7440	17800	Beam
	25	4'9"	4'10"	3'0"	3'6"	5"	2'5"	5⅝"	10'0"	7'0"	39'0"	6'5¼"	8'11"	24780	60	7440	19030	Beam
	30	5'0"	4'10"	3'0"	3'6"	5"	2'5"	5⅝"	10'0"	7'0"	39'0"	6'5¼"	8'11"	25680	60	7440	20260	Beam
	35	5'0"	4'10"	3'0"	3'6"	5"	2'5"	5⅝"	10'0"	7'0"	39'0"	6'5¼"	8'11"	26530	60	7440	21790	Beam
	40	5'3"	4'10"	3'0"	3'6"	5"	2'5"	5⅝"	10'0"	7'0"	39'0"	6'5¼"	8'11"	27520	60	7440	24140	Beam
	45	5'3½"	4'10"	3'0"	3'6"	5"	2'5"	5⅝"	10'0"	7'0"	39'0"	6'5¼"	8'11"	28750	60	7440	27690	Beam
	50	5'6¼"	4'10"	3'0"	3'6"	5"	2'5"	5⅝"	10'0"	7'0"	39'0"	6'5¼"	8'11"	29480	60	7440	29400	Beam
	55	5'6¾"	4'10"	3'0"	3'6"	5"	2'5"	5⅝"	10'0"	7'0"	39'0"	6'5¼"	8'11"	30780	60	7440	33430	Beam
	60	5'9¼"	4'10"	3'0"	3'6"	5"	2'5"	5⅝"	10'0"	7'0"	39'0"	6'6¼"	8'11"	32050	60	7440	37420	Beam
	65	5'5"	4'10"	3'0"	3'6"	13"	2'5"	7"	12'0"	7'0"	39'0"	7'8¾"	9'1¾"	32150	60	7440	36950	Box
	70	5'5"	4'10"	3'0"	3'6"	14"	2'5"	7"	12'0"	7'0"	39'0"	7'8¾"	9'3¾"	33250	80	7440	40314	Box
	75	5'5"	4'10"	3'0"	3'6"	19"	2'5"	7"	12'0"	7'0"	39'0"	7'8¾"	9'2¾"	34350	80	7440	43714	Box
	80	5'5"	4'10"	3'0"	3'6"	22"	2'5"	7"	12'0"	7'0"	39'0"	7'8¾"	9'3¼"	35450	80	7440	47334	Box

Capacity (tons)	Span (ft)	A	B	C	D	E	G	H	J	K	L	X	Y	Max. Wheel Load (lb)	Runway Rail Wt. (lb)	Trolley Wt. (lb)	Total Crane Wt. (lb)	Type Girder
25	20	4′9″	4′10″	3′0″	3′6″	5″	2′5″	5¾″	10′0″	7′0″	31′2″	6′6¼″	8′11″	28320	60	7650	19320	Beam
	25	4′9″	4′10″	3′0″	3′6″	5″	2′5″	5¾″	10′0″	7′0″	31′2″	6′6¼″	8′11″	29700	60	7650	20870	Beam
	30	5′0″	4′10″	3′0″	3′6″	5″	2′5″	5¾″	10′0″	7′0″	31′2″	6′6¼″	8′11″	30790	60	7650	22410	Beam
	35	5′3″	4′10″	3′0″	3′6″	5″	2′5″	5¾″	10′0″	7′0″	31′2″	6′6¼″	8′11″	31520	60	7650	23200	Beam
	40	5′3½″	4′10″	3′0″	3′6″	5″	2′5″	5¾″	10′0″	7′0″	31′2″	6′6¼″	8′11″	32820	60	7650	26560	Beam
	45	5′6¼″	4′10″	3′0″	3′6″	5″	2′5″	7″	10′0″	7′0″	31′2″	6′10¾″	8′11″	34700	80	7650	32330	Beam
	50	5′6¾″	4′10″	3′0″	3′6″	5″	2′5″	7″	10′0″	7′0″	31′2″	6′10¾″	8′11″	35480	80	7650	34130	Beam
	55	5′9¼″	4′10″	3′0″	3′6″	5″	2′5″	7″	10′0″	7′0″	31′2″	6′10¾″	8′11″	36750	80	7650	37950	Beam
	60	5′9¾″	4′10″	3′0″	3′6″	5″	2′5″	7″	10′0″	7′0″	31′2″	6′10¾″	8′11″	38290	80	7650	42880	Beam
	65	5′5″	4′10″	3′0″	3′6″	16¼″	2′5″	7″	11′6″	7′0″	31′2″	7′7¾″	9′1¾″	37500	80	7650	38630	Box
	70	5′5″	4′10″	3′0″	3′6″	19¼″	2′5″	7″	11′6″	7′0″	31′2″	7′7¾″	9′2¼″	34500	80	7650	41830	Box
	75	5′5″	4′10″	3′0″	3′6″	22¼″	2′5″	7″	11′6″	7′0″	31′2″	7′7¾″	9′2¾″	39600	80	7650	45270	Box
	80	5′5″	4′10″	3′0″	3′6″	24¼″	2′5″	7″	11′6″	7′0″	31′2″	7′7¾″	9′3¼″	40800	80	7650	48890	Box
30	20	5′0″	5′0″	3′0″	3′6″	5″	2′5″	7″	11′0″	8′0″	29′2″	7′3¼″	9′5″	33390	80	8950	21610	Beam
	25	5′0″	5′0″	3′0″	3′6″	5″	2′5″	7″	11′0″	8′0″	29′2″	7′3¼″	9′5″	34860	80	8950	22880	Beam
	30	5′3½″	5′0″	3′0″	3′6″	5″	2′5″	7″	11′0″	8′0″	29′2″	7′3¼″	9′5″	36230	80	8950	25050	Beam
	35	5′3½″	5′0″	3′0″	3′6″	5″	2′5″	7″	11′0″	8′0″	29′2″	7′3¼″	9′5″	37750	80	8950	28500	Beam
	40	5′3½″	5′0″	3′0″	3′6″	5″	2′5″	7″	11′0″	8′0″	29′2″	7′3¼″	9′5″	39160	80	8950	32030	Beam
	45	5′6¾″	5′0″	3′0″	3′6″	5″	2′5″	7″	11′0″	8′0″	29′2″	7′3¼″	9′5″	40050	80	8950	33920	Beam
	50	5′9¼″	5′0″	3′0″	3′6″	5″	2′5″	7″	11′0″	8′0″	29′2″	7′4¾″	9′5″	41350	80	8950	37570	Beam
	55	5′9¾″	5′0″	3′0″	3′6″	5″	2′5″	7″	11′0″	8′0″	29′2″	7′4¾″	9′5″	42900	80	8950	42310	Beam
	60	5′9″	5′0″	3′0″	3′6″	5″	2′5″	7″	11′0″	8′0″	29′2″	7′4¾″	9′5″	44940	80	8950	48850	Beam
	65	5′5″	5′0″	3′0″	3′6″	5″	2′5″	7″	12′6″	8′0″	29′2″	8′1¾″	9′7¾″	43100	80	8950	40990	Box
	70	5′5″	5′0″	3′0″	3′6″	5″	2′5″	7″	12′6″	8′0″	29′2″	8′1¾″	9′8¼″	44200	80	8950	44350	Box
	75	5′5″	5′0″	3′0″	3′6″	5″	2′5″	7″	12′6″	8′0″	29′2″	8′1¾″	9′8¾″	45400	80	8950	47895	Box
	80	5′5″	5′0″	3′0″	3′6″	5″	2′5″	7″	12′6″	8′0″	29′2″	8′1¾″	9′9¼″	47000	80	8950	53120	Box

Source: Courtesy Whiting Corp.

Table 12.3 Design Data for Cab-Operated Cranes (see figure 12.6)

Capacity (tons)	Span (ft)	A	B	C	D	E	G	H	J	K	L	N	R	X	Y	Max. Wheel Load	Run-way Rail	Trolley Wt.	Total Crane Wt.	Type of Girder
5	30	4'4½"	5'4"	2'7"	3'4"	2"	2'6½"	7½"	8'0"	5'0"	38'6"	—	—	4'10"	6'10"	13800	40	5600	21900	Beam
	40	4'10½"	5'4"	2'7"	3'4"	2"	2'6½"	7½"	8'0"	5'0"	38'6"	—	—	4'10"	6'10"	15400	40	5600	27500	Beam
	50	4'11"	5'4"	2'7"	3'4"	2"	2'6½"	7½"	8'0"	5'0"	38'6"	—	—	4'10"	6'10"	17300	40	5600	33100	Beam
	60	5'0"	5'4"	2'5"	3'4"	1'3"	2'6½"	8"	9'0"	5'0"	38'6"	—	—	5'9"	6'11"	19400	40	5600	35600	Box
	70	5'2"	5'4"	2'8"	3'4"	1'5"	2'6½"	8"	10'0"	5'0"	38'6"	—	—	6'8"	7'0"	22600	40	5600	42600	Box
	80	5'5"	5'4"	2'9"	3'2"	1'8"	2'6½"	8"	11'6"	5'6"	47'1"	—	—	7'3"	7'4"	24200	40	5790	49900	Box
	90	5'10½"	5'4"	2'9"	3'2"	1'5"	2'6½"	8"	13'0"	8'0"	88'9"	—	—	7'4"	8'9"	25400	40	6465	59900	Box
	100	6'1½"	5'4"	2'9"	3'2"	1'8"	2'6½"	8¼"	14'6"	7'6"	80'0"	—	—	9'3"	8'6"	31100	40	6330	70150	Box

Add or deduct 7'6" lift and 200 lb trolley wt for each 6" change of "K"

Capacity (tons)	Span (ft)	A	B	C	D	E	G	H	J	K	L	N	R	X	Y	Max. Wheel Load	Run-way Rail	Trolley Wt.	Total Crane Wt.	Type of Girder
10	30	5'3½"	6'0"	2'5"	3'6"	4"	2'11"	7½"	8'0"	5'0"	31'4"	—	—	5'4"	7'3"	20500	40	7880	28000	Beam
	40	5'3½"	6'0"	2'5"	3'6"	4"	2'11"	7½"	8'0"	5'0"	31'4"	—	—	5'4"	7'3"	21600	40	7880	33700	Beam
	50	5'6½"	6'0"	2'5"	3'6"	4"	2'11"	7½"	8'0"	5'0"	31'4"	—	—	5'4"	7'3"	23000	40	7880	36000	Beam
	60	5'6"	6'0"	2'5"	3'6"	1'5"	2'11"	8"	9'0"	5'0"	31'4"	—	—	5'2"	6'11"	24200	40	7880	39200	Box
	70	5'11"	6'0"	2'5"	3'6"	1'3"	2'11"	8¼"	10'0"	5'0"	31'4"	—	—	6'3"	7'0"	28700	40	7880	47200	Box
	80	6'3"	6'0"	2'5"	3'3"	1'8"	2'11"	8¼"	11'6"	5'6"	40'2"	—	—	7'4"	7'4"	31200	40	8100	55400	Box
	90	6'4"	6'0"	2'5"	3'3"	1'10"	2'11"	8¼"	13'0"	6'0"	49'0"	—	—	8'6"	8'0"	35200	40	8330	64800	Box
	100	6'6½"	6'0"	2'5"	3'3"	1'8"	2'11"	8¼"	14'6"	7'6"	74'7"	—	—	9'3"	8'6"	39800	40	9000	77900	Box

Add or deduct 8'0" lift and 250 lb trolley wt for each 6" change of "K"

Capacity (tons)	Span (ft)	A	B	C	D	E	G	H	J	K	L	N	R	X	Y	Max. Wheel Load	Run-way Rail	Trolley Wt.	Total Crane Wt.	Type of Girder
15	30	5'3½"	6'8"	2'10"	3'6"	2"	2'11"	8¼"	9'6"	5'6"	22'7"	—	—	5'9"	7'3"	25300	40	8960	31000	Beam
	40	5'4"	6'8"	2'10"	3'6"	2"	2'11"	8¼"	9'6"	5'6"	22'7"	—	—	5'9"	7'3"	26800	40	8960	37000	Beam
	50	5'11"	6'8"	2'10"	3'6"	10"	2'11"	8¼"	9'6"	5'6"	22'7"	—	—	5'8"	7'1"	29500	40	8960	39300	Box
	60	5'11"	6'8"	2'10"	3'6"	1'3"	2'11"	8¼"	9'6"	5'6"	22'7"	—	—	5'7"	7'2"	31200	40	8960	43400	Box
	70	5'11"	6'8"	2'10"	3'6"	1'8"	2'11"	8¼"	10'0"	5'6"	22'7"	—	—	6'0"	7'3"	33800	60	8960	49100	Box
	80	6'4"	6'8"	2'10"	3'4"	1'10"	2'11"	8¼"	11'6"	5'6"	22'7"	—	—	7'4"	7'4"	39600	60	8960	57500	Box
	90	6'4"	6'8"	2'10"	3'4"	1'10"	2'11"	8¼"	13'0"	7'6"	41'3"	—	—	7'9"	8'6"	41000	60	9860	68700	Box
	100	6'8"	6'8"	2'10"	3'4"	2'4"	2'11"	8¾"	14'6"	8'6"	51'1"	—	—	8'2"	9'0"	44000	60	10300	86700	Box

Add or deduct 4'9" lift and 250 lb trolley wt for each 6" change of "K"

Capacity (tons)	Span (ft)	A	B	C	D	E	G	H	J	K	L	N	R	X	Y	Max. Wheel Load	Run-way Rail	Trolley Wt.	Total Crane Wt.	Type of Girder
15 5 Aux	30	5'5½"	6'9"	3'8"	7'11"	2"	3'1"	8¼"	9'6"	5'6"	22'7"	4'10"	5'10"	5'9"	7'3"	27100	40	12900	35000	Beam
	40	5'9"	6'9"	3'8"	7'11"	2"	3'1"	8¼"	9'6"	5'6"	22'7"	4'10"	5'10"	5'9"	7'3"	28600	40	12900	41000	Beam
	50	6'1"	6'9"	3'8"	7'11"	10"	3'1"	8¼"	9'6"	5'6"	22'7"	4'10"	5'10"	5'8"	7'1"	31250	40	12900	43300	Box
	60	6'1"	6'9"	3'5"	7'11"	1'3"	3'1"	8¼"	9'6"	5'6"	22'7"	4'10"	5'10"	5'1"	7'2"	33000	40	12900	47400	Box
	70	6'1"	6'9"	3'5"	7'11"	1'8"	3'1"	8¼"	10'0"	5'6"	22'7"	4'10"	5'10"	6'0"	7'2"	35500	60	12900	53200	Box
	80	6'6"	6'9"	3'5"	7'11"	1'10"	3'1"	8¼"	11'6"	5'6"	22'7"	4'10"	5'10"	7'4"	7'4"	41350	60	12900	61500	Box
	90	6'6"	6'9"	3'5"	7'11"	1'10"	3'1"	8¼"	13'0"	7'6"	41'6"	4'10"	5'10"	7'9"	8'6"	43200	60	14300	72700	Box
	100	6'10"	6'9"	3'5"	7'11"	2'4"	3'1"	8¾"	14'6"	8'6"	51'1"	4'10"	5'10"	8'2"	9'0"	46750	60	15000	90700	Box

Add or deduct 4'9" lift and 400 lb trolley wt for each 6" change of "K"

Capacity (tons)	Span (ft)	A	B	C	D	E	G	H	J	K	L	N	R	X	Y	Max. Wheel Load	Run-way Rail	Trolley Wt.	Total Crane Wt.	Type of Girder
20	30	5'4"	6'9"	2'10"	3'6"	2"	2'11"	8¼"	9'6"	5'6"	20'1"	—	—	5'10½"	7'3"	33500	60	9500	37100	Beam
	40	5'7"	6'9"	2'10"	3'6"	2"	2'11"	8¼"	9'6"	5'6"	20'1"	—	—	5'10½"	7'3"	35000	60	9500	42300	Beam
Add or deduct 4'0" lift and 300 lb trol-ley wt for each 6" change of "K"	50	5'11"	6'9"	2'10"	3'6"	1'2"	2'11"	8¼"	9'6"	5'6"	20'1"	—	—	5'10½"	7'3"	36000	60	9500	43500	Box
	60	5'11"	6'9"	2'10"	3'6"	1'8"	2'11"	8¼"	9'6"	5'6"	20'1"	—	—	5'10½"	7'3"	36900	60	9500	45800	Box
	70	6'3"	6'9"	2'10"	3'6"	1'8"	2'11"	8¾"	10'0"	5'6"	20'1"	—	—	6'4½"	7'3"	41600	60	9500	55600	Box
	80	6'6"	6'9"	2'10"	3'4"	1'8"	2'11"	8¾"	11'6"	5'6"	20'1"	—	—	7'4"	7'4"	48200	60	9500	63600	Box
	90	6'6"	6'9"	2'10"	3'4"	1'9"	2'11"	8¾"	13'0"	7'6"	37'3"	—	—	7'9"	8'6"	46600	60	10560	80300	Box
	100	6'8"	6'9"	2'10"	3'4"	2'7"	2'11"	8¾"	14'6"	8'6"	46'2"	—	—	8'2"	9'0"	51600	80	11100	94000	Box
20 5 Aux	30	5'6"	6'10"	3'5"	7'11"	2"	3'1"	8¼"	9'6"	5'6"	20'1"	4'10"	6'0"	5'10½"	7'3"	35900	60	14700	42300	Beam
	40	5'9"	6'10"	3'5"	7'11"	2"	3'1"	8¼"	9'6"	5'6"	20'1"	4'10"	6'0"	5'10½"	7'3"	37600	60	14700	47500	Beam
Add or deduct 4'0" lift and 500 lb trol-ley wt for each 6" change of "K"	50	6'1½"	6'10"	3'5"	7'11"	1'2"	3'1"	8¼"	9'6"	5'6"	20'1"	4'10"	6'0"	5'10½"	7'3"	38600	60	14700	48700	Box
	60	6'1½"	6'10"	3'5"	7'11"	1'8"	3'1"	8¼"	9'6"	5'6"	20'1"	4'10"	6'0"	5'10½"	7'3"	39500	60	14700	54300	Box
	70	6'5½"	6'10"	3'5"	7'11"	1'8"	3'1"	8¾"	10'0"	5'6"	20'1"	4'10"	6'0"	6'4½"	7'3"	44200	60	14700	60300	Box
	80	6'8½"	6'10"	3'5"	7'11"	1'8"	3'1"	8¾"	11'6"	5'6"	20'1"	4'10"	6'0"	7'4"	7'4"	50800	60	14700	68800	Box
	90	6'8½"	6'10"	3'5"	7'11"	1'9"	3'1"	8¾"	13'0"	7'6"	34'6"	4'10"	6'0"	7'9"	8'6"	50600	60	16520	86500	Box
	100	6'10½"	6'10"	3'5"	7'11"	2'7"	3'1"	8¾"	14'6"	8'6"	42'6"	4'10"	6'0"	8'2"	9'0"	54800	80	17420	100200	Box
25	30	5'6"	7'1"	3'1"	3'7"	2"	3'1"	8¼"	12'0"	8'0"	27'9"	—	—	7'0"	9'0"	38200	60	11500	39000	Beam
	40	6'0"	7'1"	3'1"	3'7"	3"	3'1"	8¾"	12'0"	8'0"	27'9"	—	—	7'0"	9'0"	40600	60	11500	47000	Beam
Add or deduct 2'7" lift and 300 lb trol-ley wt for each 6" change of "K"	50	6'4"	7'1"	3'1"	3'7"	9½"	3'1"	8¾"	12'0"	8'0"	27'9"	—	—	7'0"	9'0"	40000	60	11500	45000	Box
	60	6'4"	7'1"	3'1"	3'4"	1'3½"	3'1"	8¾"	12'0"	8'0"	27'9"	—	—	7'0"	9'0"	43800	60	11500	51000	Box
	70	6'5"	7'1"	3'1"	3'4"	1'9"	3'1"	8¾"	12'6"	8'0"	27'9"	—	—	7'4½"	9'0"	48400	60	11500	61000	Box
	80	6'8"	7'1"	3'1"	3'4"	1'9"	3'1"	8¾"	12'6"	8'0"	27'9"	—	—	7'1½"	9'0"	54000	60	11500	70500	Box
	90	6'8¼"	7'1"	3'1"	3'4"	1'9"	3'1"	8¾"	13'0"	8'0"	27'9"	—	—	7'7"	9'0"	54300	80	11500	89900	Box
	100	7'3"	7'1"	3'1"	3'4"	2'1"	3'1"	8¾"	14'6"	8'0"	27'9"	—	—	7'10"	9'3"	53100	80	11500	99300	Box
25 5 Aux	30	5'9"	7'1"	3'7"	7'7"	2"	3'3"	8¼"	12'0"	8'0"	27'9"	4'10"	5'11"	7'0"	9'0"	41600	60	18300	45800	Beam
	40	6'3"	7'1"	3'7"	7'7"	3"	3'3"	8¾"	12'0"	8'0"	27'9"	4'10"	5'11"	7'0"	9'0"	44000	60	18300	53800	Beam
Add or deduct 2'7" lift and 500 lb trol-ley wt for each 6" change of "K"	50	6'7"	7'1"	3'7"	7'7"	9½"	3'3"	8¾"	12'0"	8'0"	27'9"	4'10"	5'11"	7'0"	9'0"	43400	60	18300	51800	Box
	60	6'7"	7'1"	3'7"	7'10"	1'3½"	3'3"	8¾"	12'0"	8'0"	27'9"	4'10"	5'11"	7'0"	9'0"	47200	60	18300	57800	Box
	70	6'8"	7'1"	3'7"	7'10"	1'9"	3'3"	8¾"	12'6"	8'0"	27'9"	4'10"	5'11"	7'4½"	9'0"	51800	60	18300	67800	Box
	80	6'11"	7'1"	3'7"	7'10"	1'9"	3'3"	8¾"	12'6"	8'0"	27'9"	4'10"	5'11"	7'1½"	9'0"	57400	60	18300	77300	Box
	90	6'11"	7'1"	3'7"	7'10"	1'9"	3'3"	8¾"	13'0"	8'0"	27'9"	4'10"	5'11"	7'7"	9'0"	57700	80	18300	96700	Box
	100	7'6"	7'1"	3'2"	7'10"	2'1"	3'3"	8¾"	14'6"	8'0"	27'9"	4'10"	5'11"	7'10"	9'3"	56500	80	18300	106100	Box

Table 12.3 (continued)

Capacity (tons)	Span (ft)	A	B	C	D	E	G	H	J	K	L	N	R	X	Y	Max. Wheel Load	Run-way Rail	Trolley Wt.	Total Crane Wt.	Type of Girder
30	30	5'6"	7'1"	3'1"	3'7"	3"	3'1"	8¾"	12'0"	8'0"	27'9"	—	—	7'0"	9'0"	44200	60	11600	41600	Beam
	40	6'0"	7'1"	3'1"	3'7"	3"	3'1"	8¾"	12'0"	8'0"	27'9"	—	—	7'0"	9'0"	46000	60	11600	49000	Beam
	50	6'4½"	7'1"	3'1"	3'7"	9"	3'1"	8¾"	12'0"	8'0"	27'9"	—	—	7'0"	9'0"	46300	60	11600	50000	Box
	60	6'4½"	7'1"	3'1"	3'4"	1'3"	3'1"	8¾"	12'0"	8'0"	27'9"	—	—	7'0"	9'0"	48900	60	11600	56900	Box
	70	6'5½"	7'1"	3'1"	3'4"	1'8"	3'1"	8¾"	12'6"	8'0"	27'9"	—	—	7'4½"	9'0"	53200	60	11600	67100	Box
	80	6'11½"	7'1"	3'1"	3'4"	1'5"	3'1"	8¾"	12'6"	8'0"	27'9"	—	—	7'2"	9'0"	56000	80	11600	80700	Box
	90	6'11½"	7'1"	3'1"	3'4"	2'0"	3'1"	8¾"	13'0"	8'0"	27'9"	—	—	7'8"	9'0"	59700	80	11600	93000	Box
	100	7'3½"	7'1"	3'1"	3'4"	2'2"	3'1"	8¾"	14'6"	8'0"	27'9"	—	—	7'10"	9'3"	62000	80	11600	103800	Box

Add or deduct 2'7" lift and 300 lb trolley wt for each 6" change of "K"

Capacity (tons)	Span (ft)	A	B	C	D	E	G	H	J	K	L	N	R	X	Y	Max. Wheel Load	Run-way Rail	Trolley Wt.	Total Crane Wt.	Type of Girder
30 5 Aux	30	5'8"	7'1"	3'7"	7'7"	3"	3'3"	8¾"	12'0"	8'0"	27'9"	4'10"	5'11"	7'0"	9'0"	47400	60	18300	48200	Beam
	40	6'2"	7'1"	3'7"	7'7"	3"	3'3"	8¾"	12'0"	8'0"	27'9"	4'10"	5'11"	7'0"	9'0"	49200	60	18300	55700	Beam
	50	6'7"	7'1"	3'7"	7'7"	9"	3'3"	8¾"	12'0"	8'0"	27'9"	4'10"	5'11"	7'0"	9'0"	49500	60	18300	56700	Box
	60	6'7"	7'1"	3'7"	7'10"	1'3"	3'3"	8¾"	12'0"	8'0"	27'9"	4'10"	5'11"	7'0"	9'0"	52100	60	18300	63900	Box
	70	6'8"	7'1"	3'7"	7'10"	1'8"	3'3"	8¾"	12'6"	8'0"	27'9"	4'10"	5'11"	7'4½"	9'0"	56400	60	18300	74800	Box
	80	7'2"	7'1"	3'7"	7'10"	1'5"	3'3"	8¾"	12'6"	8'0"	27'9"	4'10"	5'11"	7'2"	9'0"	59200	80	18300	87700	Box
	90	7'2"	7'1"	3'2"	7'10"	2'0"	3'3"	8¾"	13'0"	8'0"	27'9"	4'10"	5'11"	7'8"	9'0"	62900	80	18300	99700	Box
	100	7'6"	7'1"	3'2"	7'10"	2'2"	3'3"	9½"	14'6"	8'0"	27'9"	4'10"	5'11"	7'10"	9'3"	65200	80	18300	110000	Box

Add or deduct 2'7" lift and 500 lb trolley wt for each 6" change of "K"

Capacity (tons)	Span (ft)	A	B	C	D	E	G	H	J	K	L	N	R	X	Y	Max. Wheel Load	Run-way Rail	Trolley Wt.	Total Crane Wt.	Type of Girder
40 10 Aux	30	7'5½"	6'10"	4'9"	8'0"	3"	4'0"	8¾"	12'0"	8'0"	40'8"	4'11"	7'6"	7'3"	9'0"	57000	80	31700	64500	Beam
	40	7'4"	6'10"	4'9"	8'0"	7"	4'0"	9½"	12'6"	8'0"	40'8"	4'11"	7'6"	7'6"	9'0"	65000	80	31700	72000	Box
	50	7'6"	6'10"	4'9"	8'0"	11"	4'0"	9½"	12'6"	8'0"	40'8"	4'11"	7'6"	7'6"	9'0"	66100	80	31700	78400	Box
	60	7'11"	6'10"	4'9"	8'0"	6"	4'0"	9½"	12'6"	8'0"	40'8"	4'11"	7'6"	7'6"	9'0"	71000	80	31700	91000	Box
	70	7'11"	6'10"	4'9"	8'0"	1'5"	4'0"	9½"	12'6"	8'0"	40'8"	4'11"	7'6"	7'6"	9'0"	76000	100	31700	99700	Box
	80	8'0"	6'10"	4'9"	8'0"	1'11"	4'0"	9½"	12'6"	8'0"	40'8"	4'11"	7'6"	7'6"	9'0"	80000	100	31700	118500	Box
	90	8'2"	6'10"	4'9"	8'0"	2'1"	4'0"	10¼"	13'0"	8'0"	40'8"	4'11"	7'6"	7'10"	9'3"	84900	100	31700	132000	Box
	100	8'6"	6'10"	4'9"	8'0"	2'7"	4'0"	10¼"	14'6"	8'0"	40'8"	4'11"	7'6"	7'10"	9'3"	88000	100	31700	145000	Box

Add or deduct 4'2" lift and 700 lb trolley wt for each 6" change of "K"

Capacity (tons)	Span (ft)	A	B	C	D	E	G	H	J	K	L	N	R	X	Y	Max. Wheel Load	Run-way Rail	Trolley Wt.	Total Crane Wt.	Type of Girder
50 10 Aux	30	7'6"	6'10"	4'9"	8'0"	3"	4'0"	9½"	12'0"	8'0"	35'10"	4'11"	7'6"	7'3"	9'0"	70000	80	31900	65600	Beam
	40	7'6"	6'10"	4'9"	8'0"	7"	4'0"	9½"	12'6"	8'0"	35'10"	4'11"	7'6"	7'6"	9'0"	74000	100	31900	74000	Box
	50	7'9"	6'10"	4'9"	8'0"	1'0"	4'0"	9½"	12'6"	8'0"	35'10"	4'11"	7'6"	7'6"	9'0"	78000	100	31900	81800	Box
	60	8'0"	6'10"	4'9"	8'0"	1'1"	4'0"	9½"	12'6"	8'0"	35'10"	4'11"	7'6"	7'6"	9'0"	82200	100	31900	91600	Box
	70	8'3"	6'10"	4'9"	8'0"	1'4"	4'0"	10¼"	12'6"	8'0"	35'10"	4'11"	7'6"	7'6"	9'3"	86300	100	31900	105000	Box
	80	8'3"	6'10"	4'9"	8'0"	2'1"	4'0"	10¼"	12'6"	8'0"	35'10"	4'11"	7'6"	7'6"	9'3"	90000	100	31900	119500	Box
	90	8'3"	6'10"	4'9"	8'0"	2'6"	4'0"	10¼"	13'0"	8'0"	35'10"	4'11"	7'6"	7'10"	9'3"	94700	135	31900	132500	Box
	100	8'7"	6'10"	4'9"	8'0"	2'10"	4'0"	10¼"	14'6"	8'0"	35'10"	4'11"	7'6"	7'10"	9'3"	98200	175	31900	146000	Box

Add or deduct 4'2" lift and 700 lb trolley wt for each 6" change of "K"

Capacity (tons)	Span (ft)	A	B	C	D	E	G	H	J	K	L	N	R	X	Y	Max. Wheel Load	Run-way Rail	Trolley Wt.	Total Crane Wt.	Type of Girder
60	30	7′10″	9′2″	4′5″	8′1″	5″	4′4″	9½″	13′0″	9′0″	40′0″	5′0″	7′7″	7′10″	9′6″	81800	100	38050	76500	Beam
10 Aux	40	8′4″	9′2″	4′5″	8′1″	7″	4′4″	10¼″	13′6″	9′0″	40′0″	5′0″	7′7″	8′1″	9′6″	87300	100	38050	85000	Box
Add or deduct 3′6″	50	8′6″	9′2″	4′5″	8′1″	7″	4′4″	10¼″	13′6″	9′0″	40′0″	5′0″	7′7″	8′1″	9′6″	91000	135	38050	95000	Box
lift and 900 lb trol-	60	8′8″	9′2″	4′5″	8′1″	9″	4′4″	10¼″	14′0″	9′0″	40′0″	5′0″	7′7″	8′6″	9′6″	95600	135	38050	104800	Box
ley wt for each 6″	70	8′8″	9′2″	4′5″	8′1″	1′10″	4′4″	10¼″	14′0″	9′0″	40′0″	5′0″	7′7″	8′6″	9′9″	99600	175	38050	116900	Box
change of "K"	80	8′8″	9′2″	4′5″	8′1″	2′4″	4′4″	10¼″	14′0″	9′0″	40′0″	5′0″	7′7″	8′6″	9′9″	104300	175	38050	130700	Box
	90	9′0″	9′2″	4′5″	8′1″	2′10″	4′4″	10¼″	14′0″	9′0″	40′0″	5′0″	7′7″	8′6″	9′9″	107800	175	38050	143400	Box
	100	9′0″	9′2″	4′5″	8′1″	2′10″	4′4″	10¼″	14′6″	9′0″	40′0″	5′0″	7′7″	8′9″	10′0″	113000	175	38050	164000	Box
75	40	9′5″	7′2″	4′3″	9′6″	6″	4′11″	10¼″	14′6″	10′0″	38′0″	5′9″	7′8″	8′7½″	10′3″	104600	175	45400	96000	Box
15 Aux	50	9′9″	7′2″	4′3″	9′6″	10″	4′11″	10¼″	14′6″	10′0″	38′0″	5′9″	7′8″	8′7½″	10′3″	109600	175	45400	105800	Box
Add or deduct 3′6″	60	9′9″	7′2″	4′3″	9′6″	11″	4′11″	10¼″	14′6″	10′0″	38′0″	5′9″	7′8″	8′7½″	10′3″	114200	175	45400	117500	Box
lift and 1250 lb	70	9′9″	7′2″	4′3″	9′6″	1′8″	4′11″	10¼″	14′6″	10′0″	38′0″	5′9″	7′8″	8′7½″	10′3″	120000	175	45400	132000	Box
trolley wt for each	80	10′3″	7′2″	4′3″	9′6″	1′10″	4′11″	10¼″	14′6″	10′0″	38′0″	5′9″	7′8″	8′7½″	10′3″	124700	175	45400	145800	Box
6″ change of "K"	90	10′4″	7′2″	4′3″	9′6″	2′1″	4′11″	11″	14′6″	10′0″	38′0″	5′9″	7′8″	8′9″	10′5″	130000	175	45400	168000	Box
	100	10′4″	7′2″	4′3″	9′6″	3′0″	4′11″	11″	14′6″	10′0″	38′0″	5′9″	7′8″	8′9″	10′5″	137500	175	45400	182000	Box
	110	10′6″	7′2″	4′3″	9′6″	3′0″	4′11″	11″	16′0″	10′6″	41′6″	5′9″	7′8″	9′6″	10′11″	140300	175	46500	201000	Box
100	40	9′10″	7′4″	4′6″	8′2″	4″	5′0½″	11″	16′0″	12′0″	46′6″	4′1″	7′10″	9′6″	11′0″	134300	175	61050	117200	Box
15 Aux	50	10′2″	7′4″	4′6″	8′2″	7″	5′0½″	11″	16′0″	12′0″	46′6″	4′1″	7′10″	9′6″	11′0″	141300	175	61050	131500	Box
Add or deduct 2′9″	60	10′4″	7′4″	4′6″	8′2″	1″6″	5′0½″	11″	16′6″	12′0″	46′6″	4′1″	7′10″	9′9″	11′3″	147000	175	61050	143700	Box
lift	70	10′5″	7′4″	4′6″	8′2″	1′9″	5′0½″	11½″	16′6″	12′0″	46′6″	4′1″	7′10″	10′0″	11′3″	153700	175	61050	162000	Box
60	90	9′0″	9′2″	4′5″	8′1″	2′10″	4′4″	8¾″	5′0″	9′6″	43′6″	5′0″	7′7″	8′6″	9′10″	54700	80	39000	147000	Box
10 Aux.	100	9′0″	9′2″	4′5″	8′1″	3′0″	4′4″	8¾″	5′0″	9′6″	43′6″	5′0″	7′7″	8′6″	9′11″	57300	80	39000	165400	Box
75	50	9′9″	7′2″	4′3″	9′6″	10″	4′11″	8¾″	4′6″	10′0″	38′0″	5′9″	7′8″	8′3″	10′1″	55600	80	45400	108000	Box
15 Aux	60	9′9″	7′2″	4′3″	9′6″	11″	4′11″	8¾″	4′6″	10′0″	38′0″	5′9″	7′8″	8′3″	10′1″	57500	80	45400	120500	Box
Add or deduct 3′6″	70	9′9″	7′2″	4′3″	9′6″	1′8″	4′11″	8¾″	5′0″	10′0″	38′0″	5′9″	7′8″	8′6″	10′1″	60600	80	45400	137000	Box
lift and 1250 lb	80	10′3″	7′2″	4′3″	9′6″	1′10″	4′11″	8¾″	5′0″	10′0″	38′0″	5′9″	7′8″	8′6″	10′1″	63500	80	45400	150900	Box
trolley wt for each	90	10′4″	7′2″	4′3″	9′6″	2′1″	4′11″	9½″	5′0″	10′0″	38′0″	5′9″	7′8″	8′9″	10′4″	66400	80	45400	169000	Box
6″ change of "K"	100	10′4″	7′2″	4′3″	9′6″	3′0″	4′11″	9½″	5′0″	10′0″	38′0″	5′9″	7′8″	8′10″	10′4″	69200	80	45400	185000	Box
	110	10′6″	7′2″	4′3″	9′6″	3′0″	4′11″	9½″	5′6″	10′6″	41′6″	5′9″	7′8″	9′0″	10′7″	72100	100	46500	202000	Box
100	40	9′10″	7′4″	4′6″	8′2″	4″	5′0½″	9½″	4′6″	12′0″	46′6″	4′1″	7′10″	9′6″	11′1″	67500	100	61050	117200	Box
15 Aux	50	10′2″	7′4″	4′6″	8′2″	7″	5′0½″	9½″	4′6″	12′0″	46′6″	4′1″	7′10″	9′6″	11′1″	71000	100	61050	134000	Box
Add or deduct 2′9″	60	10′4″	7′4″	4′6″	8′2″	1′6″	5′0½″	9½″	4′6″	12′0″	46′6″	4′1″	7′10″	9′6″	11′1″	73600	100	61050	144700	Box
lift and 1250 lb	70	10′5″	7′4″	4′6″	8′2″	1′9″	5′0½″	9½″	5′0″	12′0″	46′6″	4′1″	7′10″	9′9″	11′4″	77000	100	61050	160000	Box
trolley wt for each	80	10′6″	7′4″	4′6″	8′2″	2′0″	5′0½″	9½″	5′0″	12′0″	46′6″	4′1″	7′10″	9′9″	11′4″	79000	100	61050	173100	Box
6″ change of "K"	90	10′7″	7′4″	4′6″	8′2″	2′7″	5′0½″	10¼″	5′6″	12′0″	46′6″	4′1″	7′10″	10′1″	11′7″	83000	100	61050	196700	Box
	100	10′8″	7′4″	4′6″	8′2″	3′0″	5′0½″	10¼″	5′6″	12′0″	46′6″	4′1″	7′10″	10′1″	11′7″	85700	100	61050	216100	Box
	110	11′0″	7′4″	4′6″	8′2″	3′0″	5′0½″	10¼″	5′6″	12′0″	46′6″	4′1″	7′10″	10′1″	11′7″	88100	135	61050	235400	Box
	120	11′9″	7′4″	4′6″	8′2″	3′0″	5′0½″	10¼″	5′6″	12′0″	46′6″	4′1″	7′10″	10′1″	11′7″	91300	135	61050	255200	Box

Table 12.3 (continued)

Capacity (tons)	Span (ft)	A	B	C	D	E	G	H	J	K	L	N	R	X	Y	Max. Wheel Load	Run-way Rail	Trolley Wt.	Total Crane Wt.	Type of Girder
125	50	10'0"	7'10"	5'3"	8'4"	1'7"	5'2"	10¼"	4'6"	13'0"	43'0"	4'2½"	8'0"	10'1"	11'7"	84800	100	63400	144500	Box
	60	10'6"	7'10"	5'3"	8'4"	1'7"	5'2"	10¼"	5'0"	13'0"	43'0"	4'2½"	8'0"	10'4"	11'10"	88000	135	63400	161500	Box
	70	11'0"	7'10"	5'3"	8'4"	2'1"	5'2"	10¼"	5'0"	13'0"	43'0"	4'2½"	8'0"	10'4"	11'10"	91000	135	63400	175700	Box
20 Aux	80	11'1"	7'10"	5'3"	8'4"	2'5"	5'2"	10¼"	5'0"	13'0"	43'0"	4'2½"	8'0"	10'4"	11'10"	94000	135	63400	194200	Box
Add or deduct 3'6"	90	11'2"	7'10"	5'3"	8'4"	2'8"	5'2"	10¼"	5'6"	13'0"	43'0"	4'2½"	8'0"	10'4"	11'10"	97500	175	63400	216000	Box
lift and 1450 lb	100	11'3"	7'10"	5'3"	8'4"	2'8"	5'2"	10¼"	5'6"	13'0"	43'0"	4'2½"	8'0"	10'4"	11'10"	100500	175	63400	235400	Box
trolley wt for each	110	11'8"	7'10"	5'3"	8'4"	3'2"	5'2"	10¼"	5'6"	13'0"	43'0"	4'2½"	8'0"	10'4"	11'10"	104000	175	63400	257600	Box
6" change of "K"	120	12'1"	7'10"	5'3"	8'4"	3'2"	5'2"	10¼"	5'6"	13'0"	43'0"	4'2½"	8'0"	10'4"	11'10"	108000	175	63400	282000	Box
150	50	11'0"	8'6"	6'6"	8'0"	1'5"	5'8"	10¼"	4'6"	15'0"	51'9"	4'6"	8'11"	11'1"	12'7"	97200	175	79000	168000	Box
	60	11'6"	8'6"	6'6"	8'0"	1'6"	5'8"	10¼"	5'0"	15'0"	51'9"	4'6"	8'11"	11'4"	12'10"	102800	175	79000	183000	Box
	70	11'9"	8'6"	6'6"	8'0"	1'11"	5'8"	10¼"	5'0"	15'0"	51'9"	4'6"	8'11"	11'4"	12'10"	105700	175	79000	201000	Box
25 Aux	80	11'9"	8'6"	6'6"	8'0"	2'5"	5'8"	10¼"	5'6"	15'0"	51'9"	4'6"	8'11"	11'7"	13'1"	109100	175	79000	219500	Box
Add or deduct 2'7"	90	12'0"	8'6"	6'6"	8'0"	2'7"	5'8"	10¼"	5'6"	15'0"	51'9"	4'6"	8'11"	11'7"	13'1"	113300	175	79000	242000	Box
lift and 1750 lb	100	12'3"	8'6"	6'6"	8'0"	2'8"	5'8"	10¼"	5'6"	15'0"	51'9"	4'6"	8'11"	11'7"	13'1"	116200	175	79000	261500	Box
trolley wt for each	110	12'7"	8'6"	6'6"	8'0"	3'2"	5'8"	10¼"	5'6"	15'0"	51'9"	4'6"	8'11"	11'7"	13'1"	120000	175	79000	287500	Box
6" change of "K"	120	12'10"	8'6"	6'6"	8'0"	3'0"	5'8"	10¼"	5'6"	15'0"	51'9"	4'6"	8'11"	11'7"	13'1"	123700	175	79000	310000	Box
200	50	13'6"	14'8"	7'0"	9'0"	10"	6'8"	11"	5'6"	17'0"	59'0"	4'3"	10'6"	13'1"	14'3"	133000	175	128200	240000	Box
	60	13'9"	14'8"	7'0"	9'0"	1'7"	6'8"	11"	5'6"	17'0"	59'0"	4'3"	10'6"	13'1"	14'3"	138500	175	128200	258500	Box
	70	14'0"	14'8"	7'0"	9'0"	1'4"	6'8"	11"	5'6"	17'0"	59'0"	4'3"	10'6"	13'1"	14'3"	143400	175	128200	281500	Box
25 Aux	80	14'4"	14'8"	7'0"	9'0"	1'7"	6'8"	11"	5'6"	17'0"	59'0"	4'3"	10'6"	13'1"	14'3"	147000	175	128200	301000	Box
Add or deduct 3'0"	90	14'9"	14'8"	7'0"	9'0"	1'11"	6'8"	11½"	6'0"	17'0"	59'0"	4'3"	10'6"	13'4"	14'6"	153000	175	128200	334000	Box
lift and 2400 lb	100	15'0"	14'8"	7'0"	9'0"	1'7"	6'8"	11½"	6'0"	17'0"	59'0"	4'3"	10'6"	13'4"	14'6"	158000	175	128200	361000	Box
trolley wt for each	110	15'3"	14'8"	7'0"	9'0"	1'5"	6'8"	11½"	6'0"	17'0"	59'0"	4'3"	10'6"	13'4"	14'6"	162500	175	128200	389500	Box
6" change of "K"	120	15'3"	14'8"	7'0"	9'0"	1'5"	6'8"	11½"	6'0"	17'0"	59'0"	4'3"	10'6"	13'4"	14'6"	167800	175	128200	423000	Box
250	50	15'6"	15'0"	8'3"	8'9"	1'1"	8'0"	11½"	6'0"	18'0"	90'0"	3'5"	13'0"	13'10"	15'0"	154500	175	144000	274100	Box
	60	15'9"	15'0"	8'3"	8'9"	1'5"	8'0"	11½"	6'0"	18'0"	90'0"	3'5"	13'0"	13'10"	15'0"	162000	175	144000	295000	Box
	70	16'9"	15'0"	8'3"	8'9"	1'4"	8'0"	12"	*	18'0"	90'0"	3'5"	13'0"	16'6"	16'6"	86000†	175	144000	351000	Box
25 Aux	80	17'0"	15'0"	8'3"	8'9"	1'3"	8'0"	12"	*	18'0"	90'0"	3'5"	13'0"	16'6"	16'6"	89000†	175	144000	375000	Box
Add or deduct 3'6"	90	17'0"	15'0"	8'3"	8'9"	1'3"	8'0"	12"	*	18'0"	90'0"	3'5"	13'0"	16'6"	16'6"	92000†	175	144000	408300	Box
lift and 2800 lb	100	17'3"	15'0"	8'3"	8'9"	1'0"	8'0"	12"	*	18'0"	90'0"	3'5"	13'0"	16'6"	16'6"	96400†	175	144000	459000	Box
trolley wt for each	110	17'6"	15'0"	8'3"	8'9"	9"	8'0"	12"	*	18'0"	90'0"	3'5"	13'0"	16'6"	16'6"	100000†	175	144000	504000	Box
6" change of "K"	120	17'6"	15'0"	8'3"	8'9"	9"	8'0"	12"	*	18'0"	90'0"	3'5"	13'0"	16'6"	16'6"	105000†	175	144000	547000	Box

Source: Courtesy Whiting Corp.

*Wheel spacing = 4'6"–3'0"–4'6"–6'0"–4'6"–3'0"–4'6"

†16 wheels per crane

Table 12.4 Cab-Operated Overhead Cranes, Two Trolleys

Capacity (tons)	Span (ft)	A	B	C	D	E	G	H	J	K	L	N	R	X	Y	Max. Wheel Load* (lb)	Runway Rail Wt. (lb)	Trolley Wt. (lb)	Total Crane Wt. (lb)	Type of Girder
100	40	8'10"	6'10"	8'0"	8'0"	4"	4'0"	10¼"	12'6"	8'0"	35'10"	4'11"	7'6"	7'8"	9'3"	108000[4]	175	31900	114300	Box
2 – 50/10 trolleys 7-Motor	50	9'2"	6'10"	8'0"	8'0"	7"	4'0"	10¼"	12'6"	8'0"	35'10"	4'11"	7'6"	7'8"	9'3"	120300[4]	175	31900	127000	Box
Add or deduct	60	9'4"	6'10"	8'0"	8'0"	1'5"	4'0"	11"	12'6"	8'0"	35'10"	4'11"	7'6"	7'10"	9'3"	130600[4]	175	31900	143500	Box
4'2" lift and	70	9'5"	6'10"	8'0"	8'0"	1'8"	4'0"	11"	12'6"	8'0"	35'10"	4'11"	7'6"	7'10"	9'3"	138600[4]	175	31900	158500	Box
1400 lb trolleys	80	9'6"	6'10"	8'0"	8'0"	1'10"	4'0"	9½"	5'0"	10'0"	52'0"	4'11"	7'6"	8'10"	10'6"	74000[8]	100	35000	177000	Box
wt for each	90	9'7"	6'10"	8'0"	8'0"	2'6"	4'0"	9½"	5'6"	10'0"	52'0"	4'11"	7'6"	9'1"	10'9"	77600[8]	100	35000	197200	Box
6" change of "K"	100	9'8"	6'10"	8'0"	8'0"	2'11"	4'0"	10¼"	5'6"	10'0"	52'0"	4'11"	7'6"	9'3"	10'9"	81100[8]	100	35000	221000	Box
	110	10'0"	6'10"	8'0"	8'0"	3'0"	4'0"	10¼"	5'6"	10'0"	52'0"	4'11"	7'6"	9'3"	10'9"	84400[8]	100	35000	242000	Box
	120	10'9"	6'10"	8'0"	8'0"	3'0"	4'0"	10¼"	5'6"	12'0"	69'0"	4'11"	7'6"	10'3"	11'9"	90200[8]	100	38000	268000	Box

Minimum distance between main hooks = 10'3"

Capacity (tons)	Span (ft)	A	B	C	D	E	G	H	J	K	L	N	R	X	Y	Max. Wheel Load* (lb)	Runway Rail Wt. (lb)	Trolley Wt. (lb)	Total Crane Wt. (lb)	Type of Girder
150	50	10'3"	7'2"	9'6"	9'6"	1'4"	4'11"	9½"	4'6"	10'0"	38'0"	5'9"	7'8"	8'7"	10'2"	82000[8]	100	45400	169000	Box
2 – 75/15 trolleys 7-Motor	60	10'9"	7'2"	9'6"	9'6"	1'6"	4'11"	10¼"	5'0"	10'0"	38'0"	5'9"	7'8"	8'11"	10'6"	89400[8]	100	45400	189600	Box
Add or deduct	70	11'0"	7'2"	9'6"	9'6"	1'11"	4'11"	10¼"	5'0"	10'0"	38'0"	5'9"	7'8"	8'11"	10'6"	95000[8]	135	45400	206400	Box
3'6" lift and	80	11'0"	7'2"	9'6"	9'6"	2'5"	4'11"	10¼"	5'0"	10'0"	38'0"	5'9"	7'8"	8'11"	10'6"	100300[8]	175	45400	225200	Box
2500 lb trolleys	90	11'3"	7'2"	9'6"	9'6"	2'7"	4'11"	10¼"	5'0"	10'0"	38'0"	5'9"	7'8"	8'11"	10'6"	104700[8]	175	45400	245000	Box
wt for each	100	11'6"	7'2"	9'6"	9'6"	2'8"	4'11"	10¼"	5'0"	10'0"	38'0"	5'9"	7'8"	8'11"	10'6"	109300[8]	175	45400	264800	Box
6" change of "K"	110	11'10"	7'2"	9'6"	9'6"	2'10"	4'11"	10¼"	5'6"	10'0"	38'0"	5'9"	7'8"	9'2"	10'9"	114100[8]	175	45400	290700	Box
	120	12'1"	7'2"	9'6"	9'6"	3'0"	4'11"	10¼"	5'6"	12'0"	52'0"	5'9"	7'8"	10'2"	11'9"	120800[8]	175	49800	326000	Box

Minimum distance between main hooks = 10'0"

Capacity (tons)	Span (ft)	A	B	C	D	E	G	H	J	K	L	N	R	X	Y	Max. Wheel Load* (lb)	Runway Rail Wt. (lb)	Trolley Wt. (lb)	Total Crane Wt. (lb)	Type of Girder
200	50	11'6"	7'4"	8'2"	8'2"	1'2"	5'0½"	10¼"	5'6"	12'0"	46'6"	4'1"	7'10"	10'1"	11'9"	112700[8]	175	61050	229700	Box
2 – 100/15 trolleys 7-Motor	60	11'9"	7'4"	8'2"	8'2"	1'11"	5'0½"	10¼"	5'6"	12'0"	46'6"	4'1"	7'10"	10'1"	11'9"	121300[8]	175	61050	248700	Box
Add or deduct	70	11'9"	7'4"	8'2"	8'2"	1'11"	5'0½"	10¼"	5'6"	12'0"	46'6"	4'1"	7'10"	10'1"	11'9"	128400[8]	175	61050	265200	Box
2'9" lift and	80	12'3"	7'4"	8'2"	8'2"	2'0"	5'0½"	11"	5'6"	12'0"	46'6"	4'1"	7'10"	10'3"	11'9"	135100[8]	175	61050	289000	Box
2500 lb trolleys	90	12'6"	7'4"	8'2"	8'2"	2'7"	5'0½"	11"	6'0"	12'0"	46'6"	4'1"	7'10"	10'6"	12'0"	140000[8]	175	61050	315300	Box
wt for each	100	12'9"	7'4"	8'2"	8'2"	2'2"	5'0½"	11"	6'0"	12'0"	46'6"	4'1"	7'10"	10'6"	12'0"	146100[8]	175	61050	341600	Box
6" change of "K"	110	13'3"	7'4"	8'2"	8'2"	1'9"	5'0½"	11"	6'0"	12'0"	46'6"	4'1"	7'10"	10'9"	12'0"	152000[8]	175	61050	376900	Box
	120	13'3"	7'4"	8'2"	8'2"	1'9"	5'0½"	11"	6'0"	12'0"	46'6"	4'1"	7'10"	10'9"	12'0"	158400[8]	175	61050	414200	Box

Minimum distance between main hooks = 10'4"

Table 12.4 (continued)

Capacity (tons)	Span (ft)	A	B	C	D	E	G	H	J	K	L	N	R	X	Y	Max. Wheel Load* (lb)	Runway Rail Wt. (lb)	Trolley Wt. (lb)	Total Crane Wt. (lb)	Type of Girder
250	50	12'3"	7'10"	8'4"	8'4"	8"	5'2"	11"	5'6"	13'0"	43'0"	4'2½"	8'0"	10'10"	12'3"	132500[8]	175	63400	243300	Box
	60	12'6"	7'10"	8'4"	8'4"	11"	5'2"	11"	5'6"	13'0"	43'0"	4'2½"	8'0"	10'10"	12'3"	142000[8]	175	63400	261100	Box
2 – 125/20 trol-	70	12'9"	7'10"	8'4"	8'4"	1'1"	5'2"	11"	5'6"	13'0"	43'0"	4'2½"	8'0"	10'10"	12'3"	151200[8]	175	63400	285500	Box
leys 7-Motor	80	13'3"	7'10"	8'4"	8'4"	2'3"	5'2"	11½"	6'0"	13'0"	43'0"	4'2½"	8'0"	11'3"	12'6"	158800[8]	175	63400	316200	Box
Add or deduct	90	13'6"	7'10"	8'4"	8'4"	2'0"	5'2"	11½"	6'0"	13'0"	43'0"	4'2½"	8'0"	11'3"	12'6"	165200[8]	175	63400	346500	Box
3'6" lift and	100	14'0"	7'10"	8'4"	8'4"	1'4"	5'2"	12"	4'6"†	16'6"	67'0"	4'2½"	8'0"	15'8"	15'8"	92000[16]	175	73400	450300	Box
2900 lb trolleys	110	14'3"	7'10"	8'4"	8'4"	1'4"	5'2"	12"	4'6"†	16'6"	67'0"	4'2½"	8'0"	15'8"	15'8"	95700[16]	175	73400	496600	Box
wt for each 6" change of "K"	120	14'6"	7'10"	8'4"	8'4"	1'1"	5'2"	12"	4'6"†	16'6"	67'0"	4'2½"	8'0"	15'8"	15'8"	99200[16]	175	73400	539000	Box

Minimum distance between main hooks = 11'4"

Capacity (tons)	Span (ft)	A	B	C	D	E	G	H	J	K	L	N	R	X	Y	Max. Wheel Load* (lb)	Runway Rail Wt. (lb)	Trolley Wt. (lb)	Total Crane Wt. (lb)	Type of Girder
300	60	13'6"	8'6"	8'0"	8'0"	6"	5'8"	12"	4'6"†	16'6"	59'0"	4'6"	8'11"	15'8"	15'8"	88000[16]	175	85000	364000	Box
	70	13'9"	8'6"	8'0"	8'0"	6"	5'8"	12"	4'6"†	16'6"	59'0"	4'6"	8'11"	15'8"	15'8"	93000[16]	175	85000	390000	Box
2 – 150/25 trol-	80	14'3"	8'6"	8'0"	8'0"	9"	5'8"	12"	4'6"†	16'6"	59'0"	4'6"	8'11"	15'8"	15'8"	98000[16]	175	85000	418000	Box
leys 7-Motor	90	14'9"	8'6"	8'0"	8'0"	9"	5'8"	12"	4'6"†	16'6"	59'0"	4'6"	8'11"	15'8"	15'8"	102000[16]	175	85000	488100	Box
Add or deduct	100	15'0"	8'6"	8'0"	8'0"	12"	5'8"	12"	4'6"†	16'6"	59'0"	4'6"	8'11"	15'8"	15'8"	106300[16]	175	85000	538000	Box
2'7" lift and	110	15'3"	8'6"	8'0"	8'0"	9"	5'8"	12"	4'6"†	16'6"	59'0"	4'6"	8'11"	15'8"	15'8"	111400[16]	175	85000	588000	Box
3500 lb trolleys wt for each 6" change of "K"	120	15'3"	8'6"	8'0"	8'0"	9"	5'8"	12"	4'6"†	16'6"	59'0"	4'6"	8'11"	15'8"	15'8"	115000[16]	175	85000	638800	Box

Minimum distance between main hooks = 13'6"

Source: Courtesy Whiting Corp.

*Superscripts represent the number of wheels per crane.

†Wheel spacing = 4'6"–3'0"–4'6"–4'6"–4'6"–3'0"–4'6"

298

Location and elevation of crane walkway cab floor and repair platforms.

Where column brackets are used, check trolley approach.

High point of crane from floor equals $L + B$.

Where more than one crane operates on a runway, the nearest approach of the second crane can be limited by placing separators (extended bumpers) on the crane.

The wheel loads can be halved by doubling the number of wheels, which will increase the depth of the crane by about 12 in.

12.6.3 Bridge, Trolley, and Hoist Speeds for Table 12.2

Wheel loads in table 12.2 do not include cab and walkway. Add 1000–2500 lb for walkway. The tabulated data are based on:

Bridge speeds (all): 150 fpm
Trolley speeds (all): 75 fpm
Hoist speeds:
  5 tons: 30 fpm (std); 55 fpm (optional)
  10 tons: 15 fpm (std); 30 fpm (optional)
  15 tons: 9 fpm (std); 18 fpm (optional)
  20 tons: 14 fpm (std); 18 and 21 fpm (optional)
  25 tons: 11 fpm (std); 14 and 17 fpm (optional)
  30 tons: 9 fpm (std); 12 and 14 fpm (optional)

The data in tables 12.2, 12.3, and 12.4 can be used for preliminary design. They should be corrected with the data for the actual crane purchased.

12.6.4 Notes for Tables 12.3 and 12.4

Maximum wheel load is figured with the trolley and rated load at the end of the bridge. Maximum wheel loads do not include impact, but do include cab and walkway.

B dimensions are determined with paddle-type limit switch. Add 9 in. for other types. C and D dimensions are based on cab and runway conductors located at the right-hand end with no allowance for knee braces. C and D dimensions also make no allowance for cable reels or auxiliary equipment on crane hooks.

These tables are based on lifts shown by dimension L. Additional lift may be obtained by increasing K and all related dimensions. See note for each capacity.

Dimension X does not include idler girder walk or service platform. For spring bumpers, add 12 in. to dimension X. For wood bumpers, add 4 in. to dimension X. The Y dimension is based on an open cab and Whiting controls. Add 2 ft 6 in. to dimension Y for cranes with enclosed cabs.

Weights shown in the tables are based on plain magnetic controls, ac motors and brakes, wire conductors, open cabs, no bumpers, class C and D service 5–30 tons capacity, and class C service above 30 tons capacity.

12.7 AUXILIARY HOOKS

Auxiliary hooks are of great advantage in handling buckets that have to be turned in dumping or scooping up material, in handling baskets or pellets. They are generally of small capacity in relation to the main hook. Special hooks are available for special purposes. If bridge motion is used to turn over an item, a special trolley is required.

12.8 SPEED OF TRAVEL

Table 12.5 (from CMAA specifications) gives the suggested operating speeds in fpm for hoist, trolley, and bridge of overhead cranes. Other speeds, if required, can be obtained by arrangement with the crane manufacturer.

12.9 SELECTION OF CRANE RAIL

Table 12.6 (from CMAA specifications) recommends rail sections for various types of loadings. Bethlehem section 171CR has a flat head and finds many uses where a flat-head rail is advantageous.

12.10 EXPLOSION-PROOF AREAS

Where the system has to be installed in an explosion-proof area, use a festoon cable where the run is 60 ft or less, and explosion-proof cable reels for longer distances. Explosion-proof means that the housing is capable of withstanding an explosion, should it occur inside the housing. Crane and hoist systems are made of metals that cannot spark by contact. Wheels usually are made of phosphor bronze. Hoists have stainless steel cable. Load block and hook are bronze. Hoist motor is in a special explosion-proof enclosure. Trolley wheels, crane runway wheels, etc, are of phosphor bronze.

12.11 SPECIFICATIONS

Except for steel mill use, the "Specifications for Electric Overhead Traveling Cranes" prepared by CMAA are very satisfactory.

For steel mill service, an excellent specification (actually a textbook) that includes design data and is pub-

Table 12.5 Suggested Operating Speeds (fpm)

Capacity (tons)	Hoist			Trolley*			Bridge*		
	Slow	Medium	Fast	Slow	Medium	Fast	Slow	Medium	Fast
3	20	35	70	125	150	200	200	300	400
5	20	35	70	125	150	200	200	300	400
7½	20	35	70	125	150	200	200	300	400
10	20	30	60	125	150	200	200	300	400
15	15	30	50	125	150	200	200	300	400
20	15	25	40	125	150	200	200	300	400
25	15	25	30	100	150	175	200	300	400
30	15	25	30	100	125	175	150	250	350
35	10	15	25	100	125	150	150	250	350
40	8	15	25	100	125	150	150	250	350
50	5	10	20	75	125	150	100	200	300
60	5	10	20	75	100	150	100	200	300
75	5	10	18	50	100	125	75	150	200
100	5	8	12	50	100	125	50	100	150
150	5	8	12	30	50	100	50	75	100

*For floor-controlled cranes, it is recommended that trolley and bridge speeds not exceed those given in the "slow" columns.

lished as AISE Standard No. 6, should be used. The requirements do, however, make the crane expensive, and difficult to get competitive bids. Special conditions, if required, can be added to or deducted from the standards.

12.12 MONORAIL SYSTEMS

Monorail systems are made with various types of hoists and controls. Track may be straight or have switches so the load may be transferred to a different track. The switches must operate easily and align accurately, with positive latch mechanisms, and must be provided with adequate safety baffles for all open tracks, to prevent accidental trolley derailment.

The system is very flexible, as any number of turns, horizontal or vertical, are possible. The system can be automated to pick up material at any one of several points, and discharge at another. As an illustration, by putting different marks on different products, mixed carloads of material can be processed from a central warehouse control station.

It is important to note that a short wheelbase trolley will result in heavier concentrated loads and a heavier track. It is advisable to use a wide wheelbase, say 4 feet, as a minimum.

ANSI Standard MH27.1, "Specifications for Un-

derhung Cranes and Monorail Systems" sponsored by Monorail Manufacturers Association, Inc, can be adapted for any special requirements of the particular project. The specifications require that the loads on the lower load-carrying flange be assumed to act at a point central with the wheel tread, and the stress shall not exceed one fifth of the ultimate strength. Hand-power or automatic track openers or lift-out links should be used to allow for the closing of doors, either manually or as a result of the parting of a fuse, in case of fire, where the track runs through the door.

12.13 OVERHEAD TROLLEY CONVEYORS

Overhead trolley conveyors are used for handling a variety of things, from extremely heavy loads down to light loads, such as clothing in dry-cleaning plants, and food products through processing plants. Automobile manufacturers use miles of overhead trolley conveyors hung from overhead steel hangers to keep production moving, without interfering with work at floor level. Using standard designs developed by the manufacturers of these conveyors, the path of travel is extremely flexible. They can dip down to an operator of a machine, go up again to avoid interference, make all kinds of curves and, at times, cover a complete operation from the beginning

Table 12.6 Guide for Maximum Bridge and Trolley Wheel Loadings (lb) (P)

Rail Section

CMAA* Service Class	Wheel Dia (D) (in.)	ASCE 20 lb	ASCE 25 lb	ASCE 30 lb	ASCE 40 lb	ARA-A 90 lb	ASCE 60 & 70 lb / ARA-B 100 lb	ASCE 80 & 85 lb / ARA-A 100 lb / Beth 104 lb / USX 105 lb	ASCE 100 lb	BETH & USX 135 lb	BETH & USX 175 lb	BETH 171 lb
Class A1 & A2 Power house and infrequent service	8	10800	12800	13610	16000	23900	25200	36000	40800			
	9	12150	14400	15310	18000	26600	28000	45000	51000			
	10	13500	16000	17010	20000	31900	33600	54000	61200			
	12		19200	20410	24000	39800	42000	63000	71400			
Class B Light service	15			25510	30000	47800	50400	72000	81600			
	18			30610	36000	55800	58800	81000	91800			
Class C Moderate service P = 1600 WD	21				42000	63800	67200	90000	102000	75600	105000	117600
	24								125500	86400	120000	134400
	27									97200	135000	151200
	30									108000	135000	168000
	36									130000	180000	202000
Class D Heavy-duty service P = 1400 WD	8	9450	11200	11900	14000	20900	22050	31500	35700			
	9	10630	12600	13390	15750	23200	24500	39380	44630			
	10	11820	14000	14880	17500	27900	29400	47250	53550			
	12		16800	17860	21000	34900	36750	55130	62480			
	15			22320	26250	41800	44100	63000	71400			
	18			26790	31500	48800	51450	70880	80330			
	21				36750	55800	58800	78750	89250	66150	91880	102900
	24								107200	75600	105000	117600
	27									85050	118130	132300
	30									94500	131250	147000
	36									113600	157800	176500

Table 12.6 (continued)

		Rail Section										
CMAA* Service Class	Wheel Dia (D) (in.)	ASCE 20 lb	ASCE 25 lb	ASCE 30 lb	ASCE 40 lb	ARA-A 90 lb	ASCE 60 & 70 lb / ARA-B 100 lb	ASCE 80 & 85 lb / ARA-A 100 lb / Beth 104 lb / USX 105 lb	ASCE 100 lb	BETH & USX 135 lb	BETH & USX 175 lb	BETH 171 lb
Class E†	8	8100	9600	10200	12000	17900	18900					
Severe duty-cycle service	9	9120	10800	11480	13500	19900	21000					
	10	10130	12000	12760	15000	23900	25200					
P = 1200 WD	12		14400	15310	18000	29900	31500	27000	30600			
	15			19130	22500	35900	37800	33750	38250			
	18			22960	27000	41800	44100	40500	45900			
	21				31500	47800	50400	47250	53550	56700	78750	88200
	24							54000	61200	64800	90000	100800
	27							60750	68850	72900	101250	113400
	30							67500	76500	81000	112500	126000
	36								92000	97300	135000	151000
Effective Width of Rail Head (W) (in.) (Top of head minus corner radii)		0.844	1.000	1.063	1.250	1.656	1.750	1.875	2.125	2.250	3.125	3.500

Note: Bethlehem and USX sections are crane rail sections rolled to meet ASTM A-759. See Bethlehem and USX catalogs for properties of CR section.

*CMAA specification no. 70.

†The loading limits for Class E are also recommended wherever travel speeds exceed 400 fpm.

302

of the manufacture of an article to its final destination in the shipping room. Very long conveyors may reach a mile or more in length and, depending on the load to be handled and path of conveyor, the unit may require several booster drives, tied in together and located along the path of travel.

12.13.1 Uses

As examples, transformers are carried through washing, finishing, and drying operations, a tremendous saving over handling each operation separately on the floor. Rubber tire manufacturers are large users of these trolley conveyors, handling green rubber tires to machines for shaping and vulcanizing operations. Coil handling

in a rod mill is an important part of production, with these trolley conveyors automatically picking up 600-lb coils of wire heated to 800°F, and transporting them to the wire mill (see figure 12.8). Speed of travel is 25–100 fpm.

An unusual use of an overhead trolley conveyor is the assistance it gives to miners entering and leaving a mine. This "man-tow" is designed to take approximately 60% of a man's weight off his feet and permit him to travel at a constant speed.

12.13.2 Chain and Trolley

These trolley conveyors consist of a single strand of rivetless chain. The standard sizes are known as 458 and

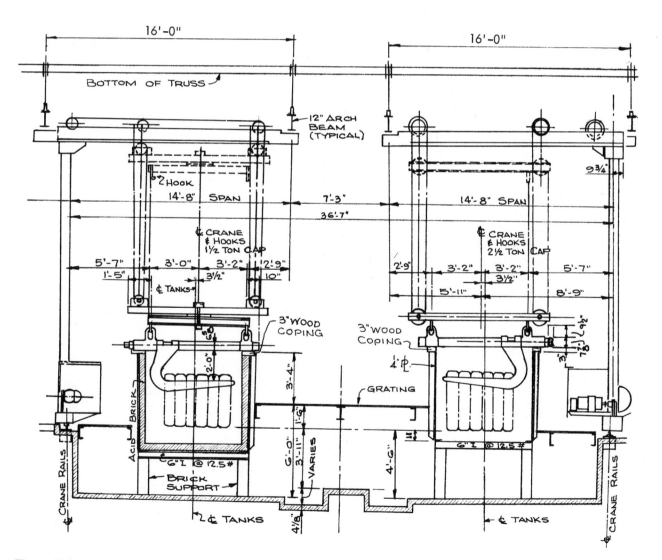

Figure 12.8 A coil-handling overhead trolley for wire cleaning and coating tanks.

678, and are made of forged steel in varying pitches and strength to suit conditions. Attached to these chains would be ball-bearing trolleys with grease fittings riding on the lower flange of an I-beam track when horizontal. Various shapes of baskets, hoppers, or frames hold production material.

12.13.3 Power and Free Trolley Conveyors

These conveyors consist of a combination of powered trolley conveyors and unpowered monorail type of free conveyors.

Two sets of tracks are used, usually suspended one above the other, although a side-by-side track arrangement can also be supplied. The upper track is a structural I-beam, and carries the powered trolley conveyor. The lower, or free track, usually consists of two structural channels or angles and carries four-wheel load trolleys from which carriers are suspended.

Load trolleys have two pivoted dogs that engage pushers attached to the power conveyor trolleys or chain. They are switched to and from either powered or unpowered free tracks on either or both sides of the main power conveyor and free conveyor while the power conveyor is moving. Load trolleys are hand- or gravity-propelled on free track runs.

12.13.4 Floor Conveyors

Modern warehouses are equipped with floor conveyors for the movement of trucks or conveyors to any location, loaded with merchandise, generally speeding up the delivery to outgoing trucks. These are specialized conveyor units, and it would be well to consult with the manufacturers.

Platforms that can be moved manually, mechanically, or electrically anywhere, and can be raised or lowered, carrying a person who can reach into storage bins to remove or replace tools or small equipment, are made by several manufacturers.

The Webb-Triax Company makes a sophisticated movable carriage operated on ground-floor track, and located between two multilevel platforms storing basket containers. The carriage can be moved to any location where any desired basket can be retrieved and lowered on the elevator, or moved out to the proper location and raised for deposition in its proper niche. The whole operation is electromechanically controlled from the ground floor or any desired place.

12.13.5 Design Data for Overhead Trolley Conveyors

In calculating chain pull, use 0.03 as coefficient of friction, and assume 4.6 lb/ft as the weight of No. 458 chain, and 9.6 lb/ft. for No. 678 chain (see figure 12.9). Speed of travel generally is 25–100 fpm.

12.14 NUMBER OF CRANES

The number of cranes required in any operation depends on the time needed for the crane to move to the required location, lifting time (there may be quite a delay when item has to be handled very gently, when gulpers or vacuum tanks are attached to the crane, or similar conditions), time to unload the crane and move it to a new location.

Allowance must be made for downtime, servicing of crane, just idling, or delays by other operations. A service platform with stairs leading to it must be provided for each crane. If only two cranes are involved, a ser-

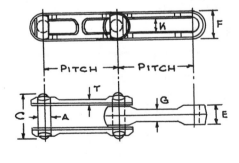

CHAIN NUMBER		458
PITCH		4 ¹/₃₂
A		⁵/₈
C		2 ³/₁₆
E	INCHES	1
F		1 ³/₈
G		⁵/₈
K		¹¹/₁₆
T		⁵/₁₆

CHAIN NUMBER		678
PITCH		6 ¹/₃₂
A		⁷/₈
C		3 ¹/₈
E	INCHES	1 ⁹/₃₂
F		2
G		¹³/₁₆
K		1
T		¹/₂

Figure 12.9 Chains 458 and 678.

vice platform is provided at each end of the building. If more cranes are involved, additional platforms are required.

The additional cranes can be serviced from platforms that have no access stairs, since entrance to platform can be provided by another crane, and extra stairways are sometimes in the way.

12.15 AERIAL CABLEWAYS

Double-cable system, in which twin fixed cables serve as the runway or track from which spaced buckets are suspended and towed by an endless cable.

Shuttle system, in which buckets or cars travel to a discharge point after the manner of a balanced skip hoist. It is well adapted to the disposal of waste from a plant to a dump pile, if the distance and capacity are not too great. The reverse of the function is the transfer of material from the top of a slope to a discharge point at the bottom. Automatic loading is easily provided.

Primarily, the field of the cableway is for transport of materials over long distances across rough country and heavily wooded routes, preferably along a straight line, though curves are possible. It is free from interference with surface traffic, requires no bridges, cuts or fills and, with suitable carriers, it can transport material

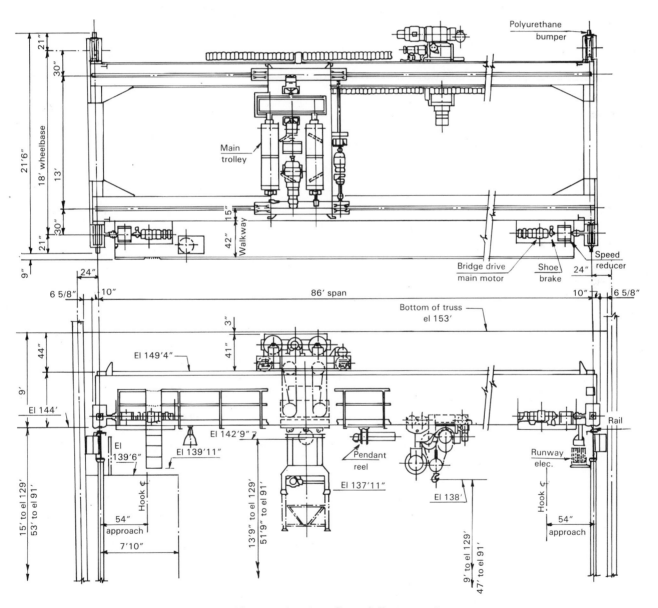

Figure 12.10 A 35/15-ton radwaste crane. *(Courtesy American Crane & Equipment Corp.)*

that a belt conveyor cannot handle, such as logs, lumber, and sacks. The usual speed is 500–600 fpm. Most installations are of moderate length, but some have been built 10 miles long. The spans may be as long as 5000 ft, but long spans involve costly supporting structures. If each bucket carries 40 ft^3 of material weighing 50 lb/ft^3, and the speed is 600 fpm, with buckets spaced 300 ft apart, the capacity per hour would be:

$$\frac{40 \times 50}{2000} \times \frac{600}{300} \times 60 = 120 \text{ tph}$$

There must be an interval for loading the buckets while they are stationary. Loading two buckets per minute is fast work.

Ski tows are actually modifications of aerial cableways.

12.16 35/15-TON RADWASTE HANDLING CRANE

This crane system offers a unique automated approach to remote handling of hazardous contaminated radiation waste. The crane system utilizes an independent main and auxiliary hoist, which are operable from a remote operator's console through both manual and automated computer controls (figure 12.10). The crane system not only operates in the automated mode, but also maintains a complete computerized record of all inventory handled, including radiation isotopic data. In addition to the computer equipment, a complete closed-circuit TV system also is on the crane; it is used to aid the operator in picking up loads with the remote grappling devices.

Gates, Chutes, and Spouts; Bins and Hoppers

13.1 INTRODUCTION TO GATES, CHUTES, AND SPOUTS

Gates are normally used on the bottom or sides of bins, tanks, or hoppers to feed material onto conveyors for further processing, or into trucks or railroad cars for shipment to other points. Numerous styles of gates are available. Many of these are general-purpose gates suitable for handling coal, sand, crushed stone, etc.

Gates are built in a wide range of sizes and materials, such as carbon steel, stainless steel, cast iron, or plastics to meet all normal requirements.

13.1.1 Types of Gates

All gates shown on figure 13.1 are of the manual-operation type but, with a slight design change, can be made to be operated by an electric motor or air cylinder.

Figure 13.1*a* shows a gate designed for installation on the bottom of bins or hoppers. It is made of steel plate with welded construction.

Figure 13.1*b* shows a gate designed for installation on the bottom of bins or hoppers. It is made of gray iron.

Figure 13.1*c* shows an undercut gate designed for installation on the bottom of bins or hoppers. It is made of gray iron.

Figure 13.1*d* shows an undercut gate designed for installation at the end of a chute. Complete details are given.

Figure 13.1*e* shows a duplex-type gate designed

for installation on the bottom of bins or hoppers. It is made of steel plate with welded construction. The linkage system of this gate results in quick-opening and smooth-acting operation.

Figure 13.1*f* shows a rack-and-pinion gate designed for installation on the bottom of bins or hoppers. It is of welded steel construction. The slide plate is operated by a rack-and-pinion gear arrangement.

13.1.2 Linings

Chutes and spouts are often lined with rubber lining, especially at corners, to reduce noise or wear. Stainless steel linings are used to reduce wear or corrosion.

13.2 FEEDING MORE THAN ONE LOCATION

Feeding from a single source to two locations can be done by a flop gate. These gates are of welded steel construction and can be made to fit any requirement. Figure 13.2 shows typical flop gates.

Figure 13.2*a* shows a typical arrangement of flop gate and two-way chute with removable cover plate. This is a standard piece of equipment, used mostly from the discharge chute of a bucket elevator. These two-way chutes equipped with a flop gate should always be placed in a vertical position

(Text continues on p. 313)

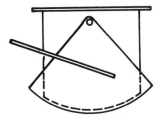

a. For Use on Bottom
of Bin or Hopper

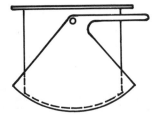

b. For Use on Bottom
of Bin or Hopper

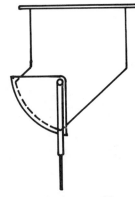

c. Undercut Type

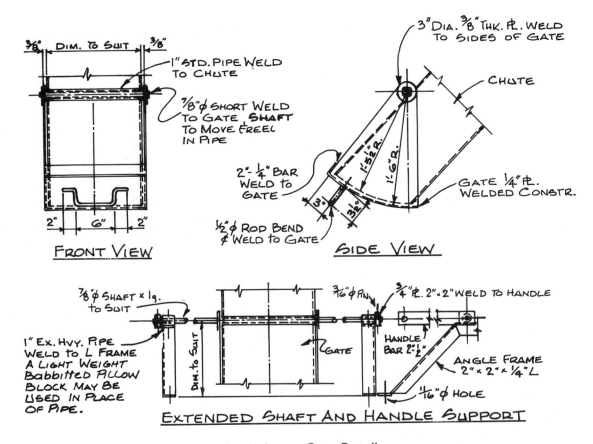

3" DIA. ⅜" THK. ℔. WELD
TO SIDES OF GATE

CHUTE

⅜" DIM. TO SUIT ⅜"

1" STD. PIPE WELD
TO CHUTE

⅞"⌀ SHORT WELD
TO GATE SHAFT
TO MOVE FREEL
IN PIPE

2"- ¼" BAR
WELD TO
GATE

½"⌀ ROD BEND
& WELD TO GATE

2" 6" 2"

FRONT VIEW

1'-5½" R.

1'-6" R.

3" 3½"

GATE ¼" ℔.
WELDED CONSTR.

SIDE VIEW

⅞"⌀ SHAFT x lg.
to SUIT

1" EX. HVY. PIPE
WELD TO L FRAME
A LIGHT WEIGHT
Babbitted Pillow
BLOCK MAY BE
USED IN PLACE
OF PIPE.

DIM. to SUIT

GATE

⁷⁄₁₆"⌀ PIN

¾" ℔. 2"· 2" WELD TO HANDLE

HANDLE
BAR 2"·½"

ANGLE FRAME
2" x 2" x ¼" L

⁷⁄₁₆"⌀ HOLE

EXTENDED SHAFT AND HANDLE SUPPORT

d. Undercut Gate Details

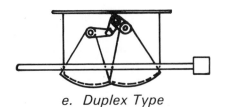

e. Duplex Type

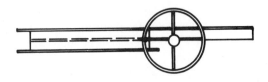

f. Rack-and-Pinion Type

Figure 13.1 Typical gates.

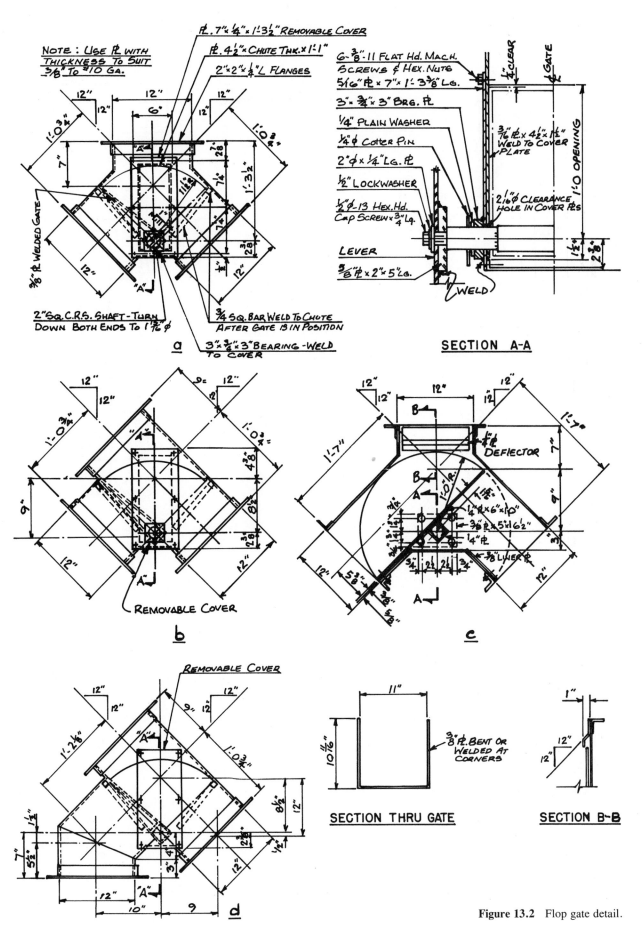

Figure 13.2 Flop gate detail.

309

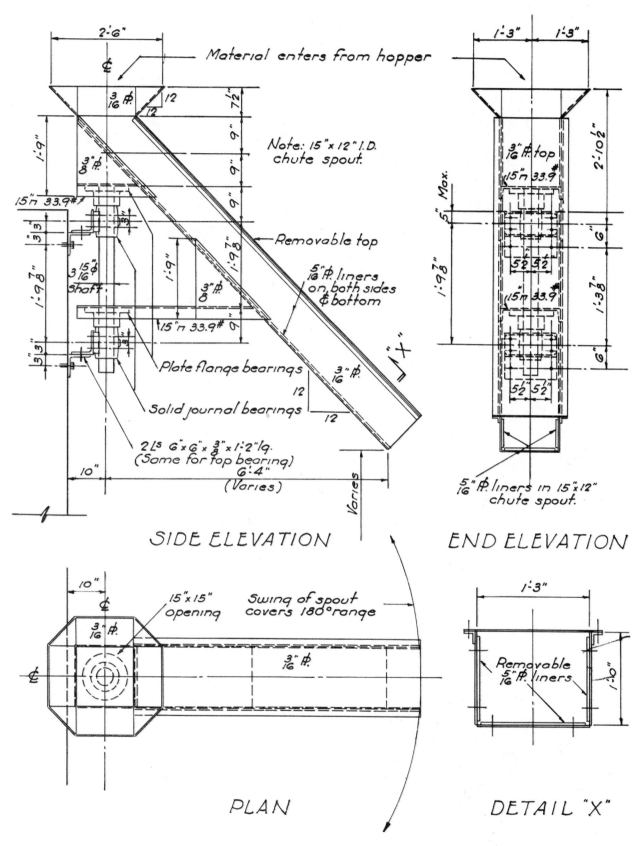

Figure 13.3   Swivel spout with viewing ports to allow the operator to see the load level.

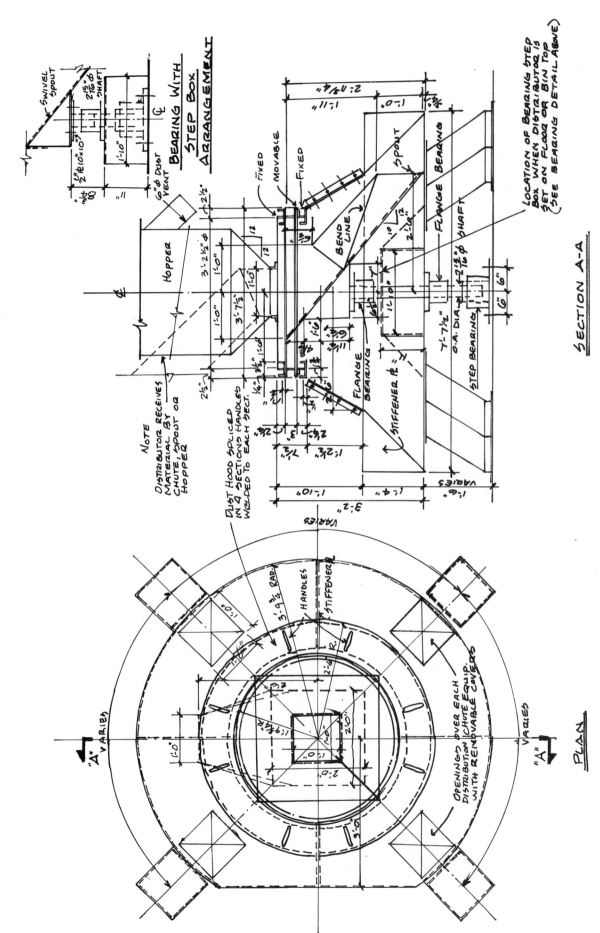

Figure 13.4 Distributor spout assembly.

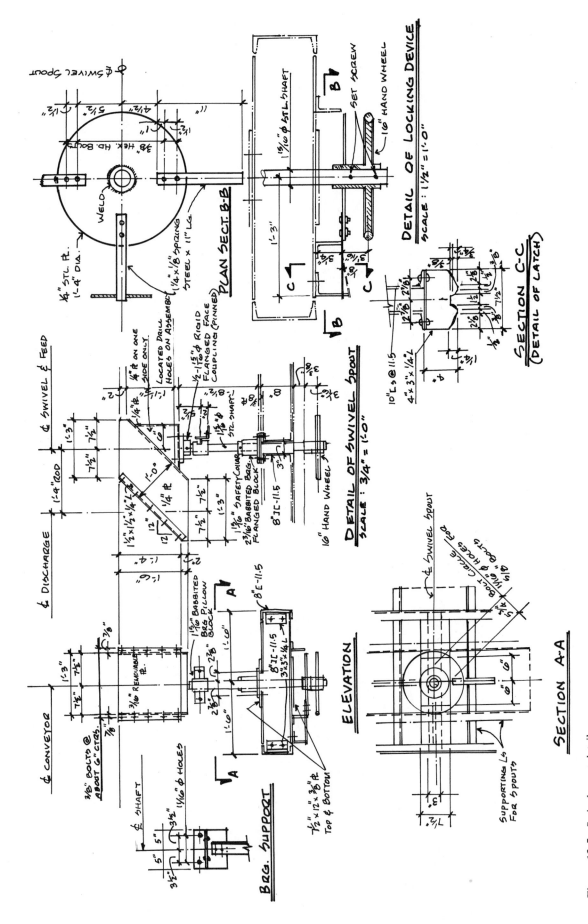

Figure 13.5 Swivel spout details.

because if placed on an angle, the gate plate is likely to bind and is very difficult to operate.

Figure 13.2*b* shows another typical arrangement of chutes with flop gate.

Figure 13.2*c* shows a removable cover, if one is desired.

Figure 13.2*d* shows a different design of flop gate, effective and easily operated.

More than two places can be fed using various spouts and chutes.

Figure 13.3 shows a swivel spout and figure 13.4

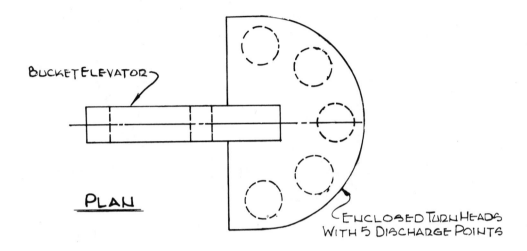

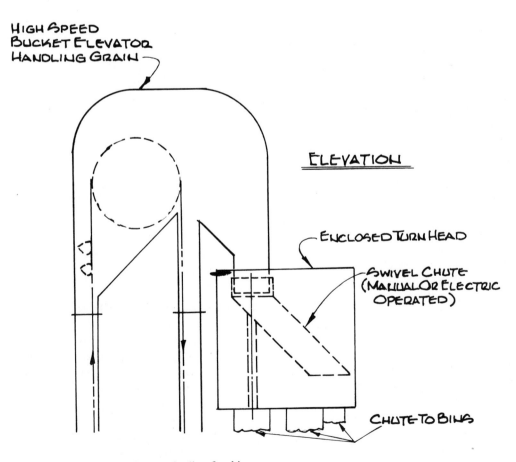

Figure 13.6   A bucket elevator feeding five bins.

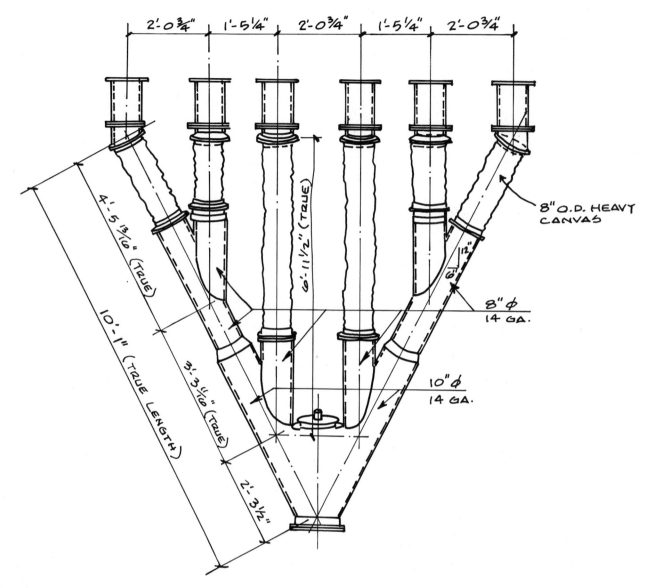

Figure 13.7 Chutes under vibrating screens. Canvas should be riveted outside the flange to the top flange and inside the flange to the bottom flange. Six vibrating screens feed a single location on the floor below. Note the flexible chutes. Stainless steel was used to avoid sparks and corrosion.

shows a distributor spout. These chutes and spouts are of welded steel construction. The swivel chute can be hand-operated if within the reach of an operator, or a sprocket-chain drive can be arranged on a shaft to turn the spout.

Figure 13.5 shows a handwheel-operated swivel spout.

Figure 13.6 shows a typical bucket elevator installation, where five different storage bins are loaded using a turnhead with chutes.

Figure 13.7 shows six vibrating screens feeding a single location below. Note the flexible connections to the bottom section. Stainless steel was used to avoid sparks and corrosion.

Figure 13.8 shows flexible joints for chutes.

Figure 13.9 shows typical spreader detail.

13.3 FLEXIBLE SPOUTS

Sprout, Waldron and Co. has made a specialty of building telescopic, flexible, car-loading spouts called "elephant trunks," for grain-elevator operators. These flexible spouts have been a standard piece of equipment in the grain industry and fit into the production line of loading box cars or trucks. Worn sections can be easily and quickly replaced without taking the entire spout apart. This type of spout is also manufactured by Screw Conveyor Corp. and Ehrsam Co.

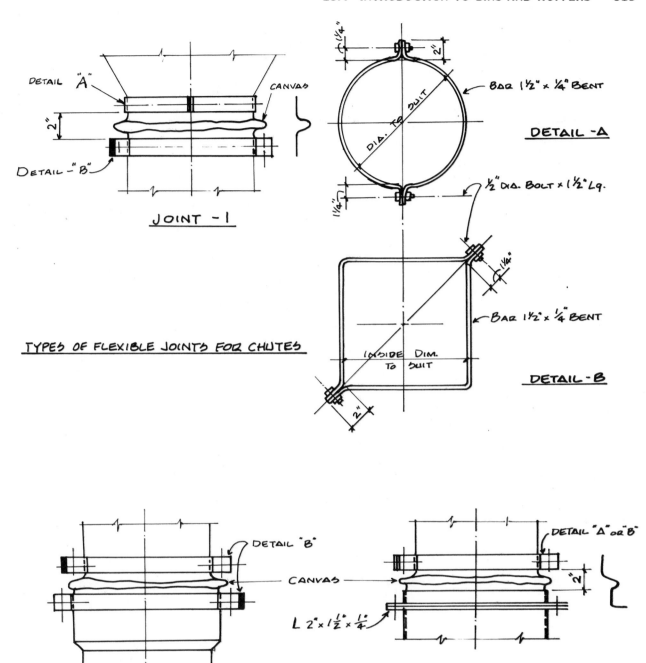

DETAIL "A"

CANVAS

2"

DETAIL - "B"

JOINT - 1

TYPES OF FLEXIBLE JOINTS FOR CHUTES

1/4"

2"

DIA. TO SUIT

BAR 1 1/2" × 1/4" BENT

DETAIL -A

1/4"

1/2" DIA. BOLT × 1 1/2" Lg.

1/4"

BAR 1 1/2" × 1/4" BENT

INSIDE DIM. TO SUIT

DETAIL -B

2"

DETAIL "B"

CANVAS

L 2" × 1 1/2" × 1/4"

DETAIL "A" OR "B"

2"

JOINT - 3

JOINT - 2

Figure 13.8 Types of flexible joints for chutes.

Superior Systems Company manufactures various types of retractable loading spouts. Their E Z View™ Loading Spout is designed for dust-free loading of any dry bulk material into cars, trucks, or barges. In addition to the loading spouts, Superior Systems makes accessories such as spout positioners, turnheads, spreaders, and level sensors.

13.4 INTRODUCTION TO BINS AND HOPPERS

Bulk material is delivered for processing by rail, truck, or ship. It is unloaded through a hopper to be conveyed directly to storage, or through a crusher for transfer to storage. The material may be stored on the ground, on

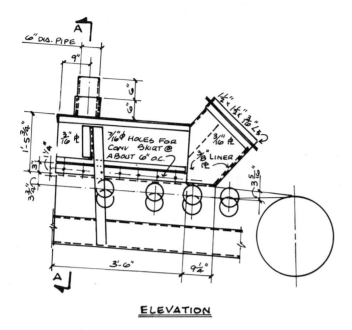

ELEVATION

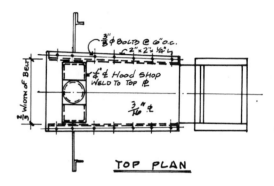

TOP PLAN

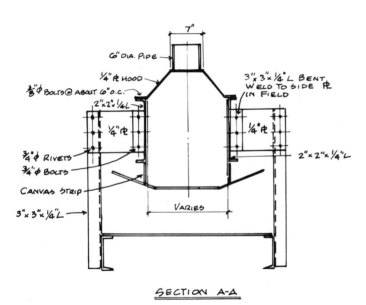

SECTION A-A

Figure 13.9 Details of a typical spreader.

concrete storage pads, or in bins or silos, but materials being processed further should be stored in bins or silos. The capacity of a bin must be more than the amount of each delivery to allow for material that could be in the bin at the time of new delivery, delay in future deliveries, and variation in delivery amounts.

13.5 TYPES OF BINS AND SILOS

The most economical shape for bins or silos is cylindrical. There may, however, be times when square, or even rectangular, configurations are used. Figure 13.10 shows two square bins of different height. Figure 13.11 shows a typical cylindrical installation. A group of hexagonal, square, or circular bins has been used for grain storage. In the case of the circular bins, the interstices also are used for storage. Recently, some old silos have been converted to apartments.

The bottoms of the bins and silos are frustums of cones or truncated pyramids, to aid in the discharge of the material. If the bottoms are flat, the material itself will form a base of proper shape. In an emergency, this base material can be dug out for use.

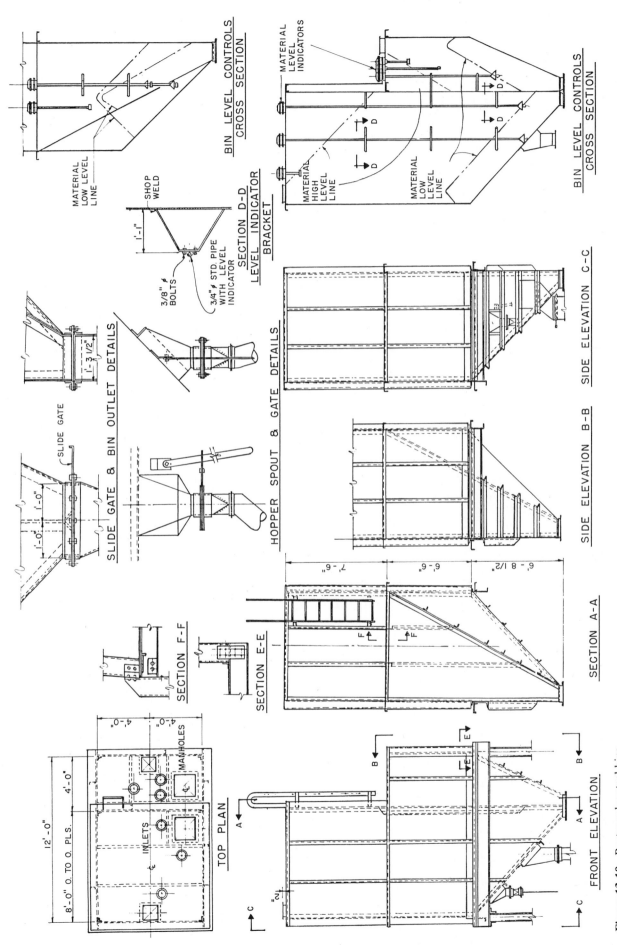

Figure 13.10 Rectangular steel bins.

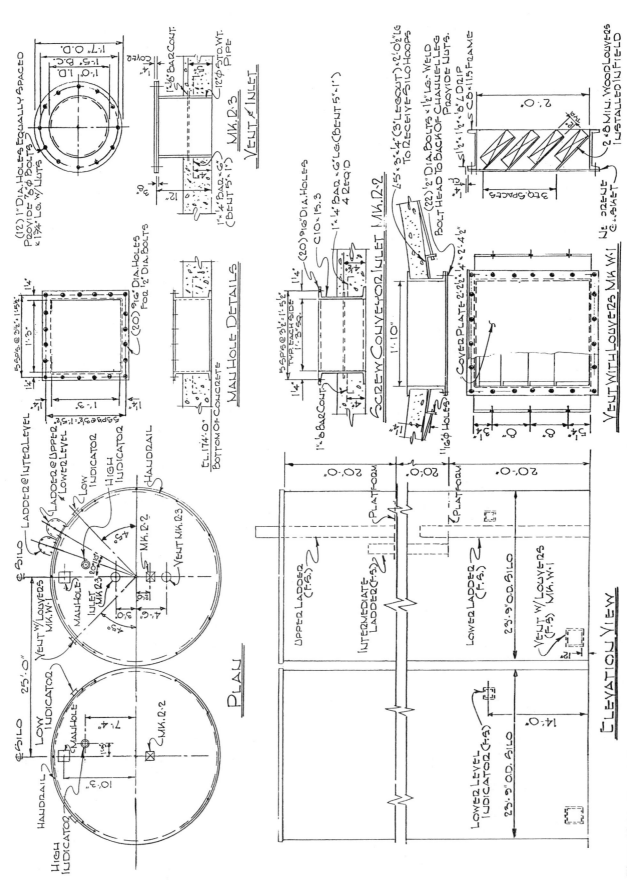

Figure 13.11 Precast silos.

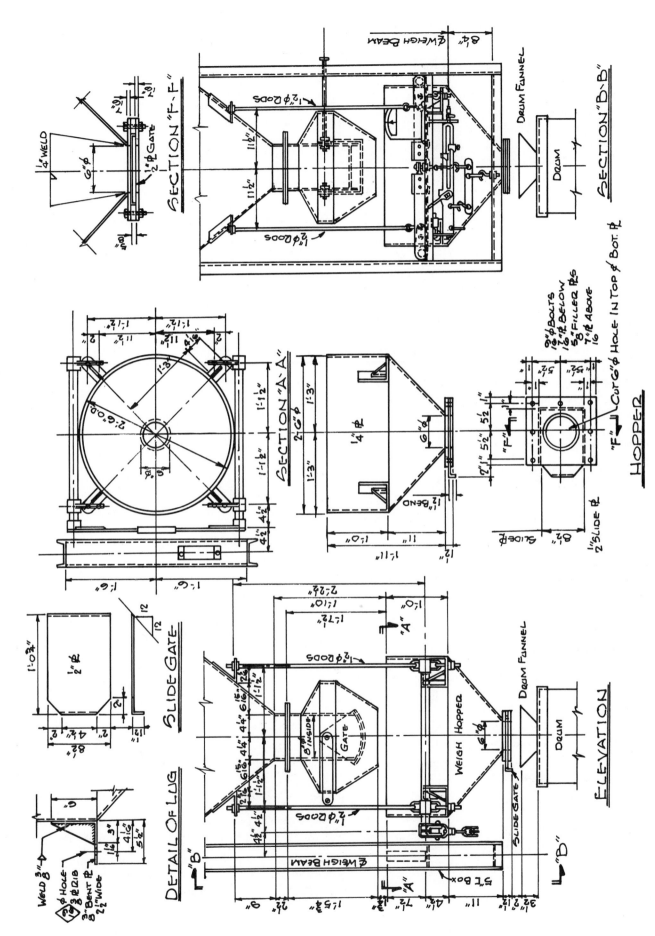

Figure 13.12 A scale hopper.

319

13.6 TRACK OR TRUCK HOPPERS

It is desirable to make the valley angle of a hopper not less than 45°, because of the sluggishness of flow of many materials. This usually makes the inclined sides of the hopper about 50°. If damp material is handled, the valley angle should be about 50°, and sometimes even steeper than that. The track beams should be designed for loading as required by the serving railroad, or loaded trucks.

Protective steel gratings must be provided to prevent people from falling into the hopper. These are required by government regulations and safety departments of the various industrial plants. Grating openings should not be larger than 4 in. × 4 in. This grating may have to be removed when unloading larger pieces, but must be replaced as soon as the car is moved. The grating bars should be at least 4 in. × ¼ in., so that larger lumps can be broken down on the grating to pass through the openings. If they cannot be broken down, they have to be removed by hand. These gratings should provide sections that can be removed temporarily, to allow for passage of material to a crusher, if so desired.

13.7 WEIGH (SCALE) HOPPERS

Figure 13.12 shows a weigh hopper. The slide gate on the weigh hopper is closed, and the bin gate is opened. The hopper is then filled to the desired weight, as shown on the dial scale. The bin gate is then closed, and the weigh hopper gate is opened to fill the drum. This operation generally is automatically controlled. See figure 13.13 for electrical diagram.

13.8 HOPPER SPOUTS

Figure 13.14 shows a hand-operated slide gate and pivoted chute for hoppers in concrete bins.

A hinged hopper spout is shown on Figure 13.15. An air cylinder is used to actuate the spout. Note the counterweight to help pivot the gate.

Figure 13.16 shows a typical hopper gate with a rack-and-pinion side-discharge gate.

13.9 STORING HOT OR COLD COVERED MATERIAL

Often, reclaimed hot material has to be stored in bins. Figure 13.17 shows a bin designed to provide for cool-

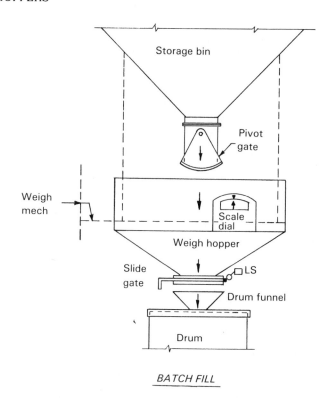

BATCH FILL

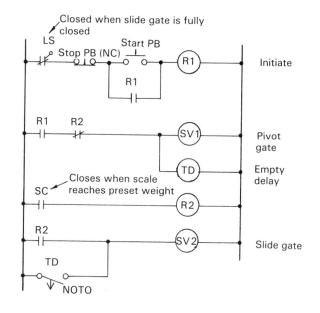

ELECTRICAL DIAGRAM

Figure 13.13 Weigh hopper wiring details. *Sequence:* (1) With the drum in position, START is depressed and the pivot gate opens to fill the weigh hopper to a predetermined weight, as set on the scale dial. (2) When the set weight is reached, the scale contact initiates SV to open the slide gate, which opens for a timed period allowing the hopper to empty. (3) After the timed interval, the slide gate closes and the system is ready to recycle, provided the drum is removed and replaced with an empty drum.

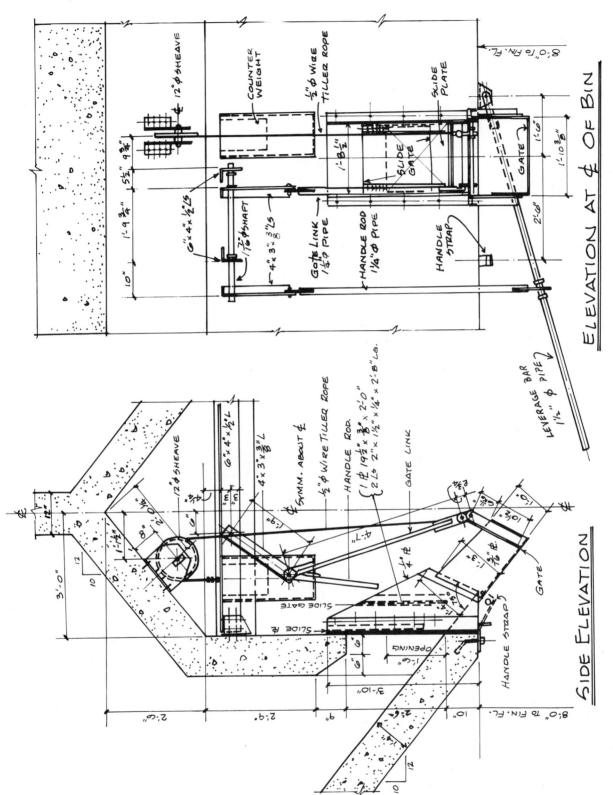

Figure 13.14 Hand-operated slide gates in concrete bins.

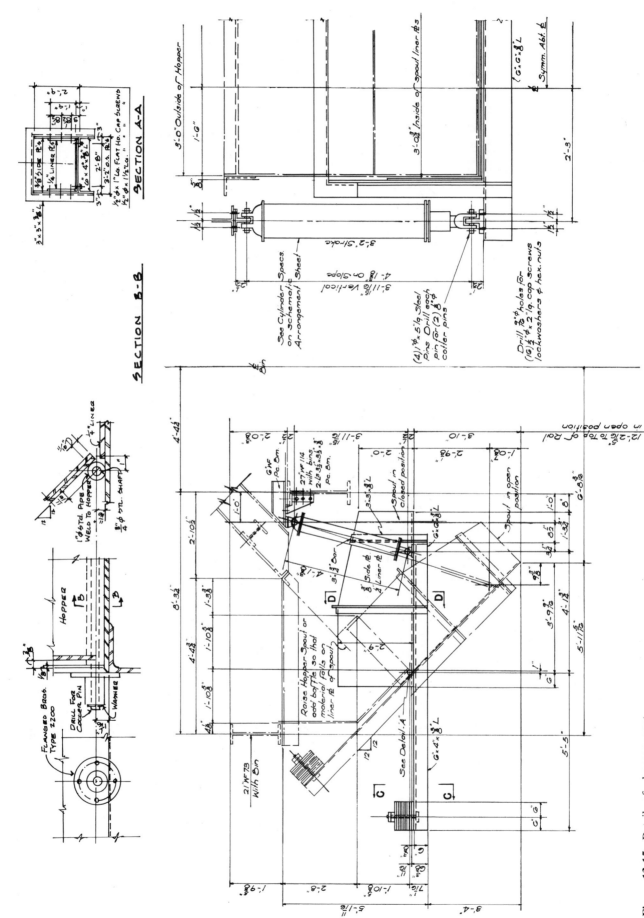

Figure 13.15 Details of a hopper spout.

Figure 13.16 Detail of a typical hopper gate.

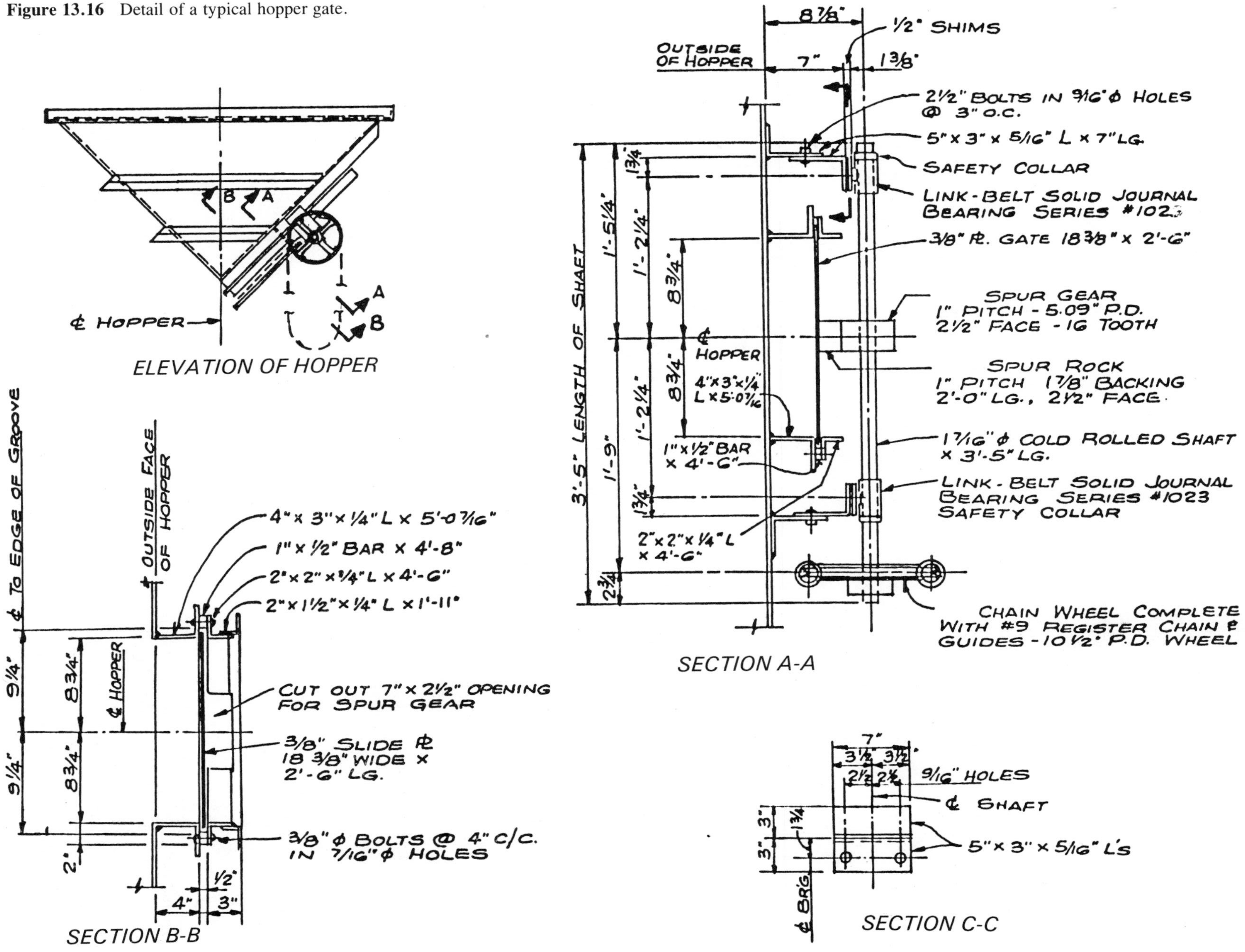

ELEVATION OF HOPPER

₵ HOPPER

SECTION A-A

½" SHIMS

1⅜"

8⅞"

7"

OUTSIDE OF HOPPER

2½" BOLTS IN 9/16"⌀ HOLES @ 3" O.C.

5" x 3" x 5/16" L x 7" LG.

SAFETY COLLAR

LINK-BELT SOLID JOURNAL BEARING SERIES #1023

3/8" ₵ GATE 18⅜" x 2'-6"

SPUR GEAR
1" PITCH - 5.09" P.D.
2½" FACE - 16 TOOTH

SPUR RACK
1" PITCH (1⅞" BACKING
2'-0" LG., 2½" FACE.

1⅝"⌀ COLD ROLLED SHAFT x 3'-5" LG.

LINK-BELT SOLID JOURNAL BEARING SERIES #1023 SAFETY COLLAR

CHAIN WHEEL COMPLETE WITH #9 REGISTER CHAIN & GUIDES - 10½" P.D. WHEEL

₵ HOPPER

8¾"

8¾"

4"x3"x¼ L x 5'-0 1/16"

1"x½" BAR x 4'-6"

2"x2"x¼" L x 4'-6"

1⅜"

1¾"

1'-2¼."

1'-2¼."

1'-9."

1'-5¼."

2¾."

3'-5" LENGTH OF SHAFT

SECTION C-C

7"

3½" 3½"

2½" 2½"

9/16" HOLES

₵ SHAFT

5" x 3" x 5/16" L's

₵ BRG.

1¾"

3" 3"

SECTION B-B

4" x 3" x ¼" L x 5'-0 1/16"

1" x ½" BAR x 4'-8"

2" x 2" x ¼" L x 4'-6"

2" x 1½" x ¼" L x 1'-11"

CUT OUT 7" x 2½" OPENING FOR SPUR GEAR

3/8" SLIDE ₵ 18 3/8" WIDE x 2'-6" LG.

3/8"⌀ BOLTS @ 4" C/C. IN 7/16"⌀ HOLES

½"

3"

4"

2."

OUTSIDE FACE OF HOPPER

₵ HOPPER

8¾."

8¾."

₵ TO EDGE OF GROOVE

9¼."

9¼."

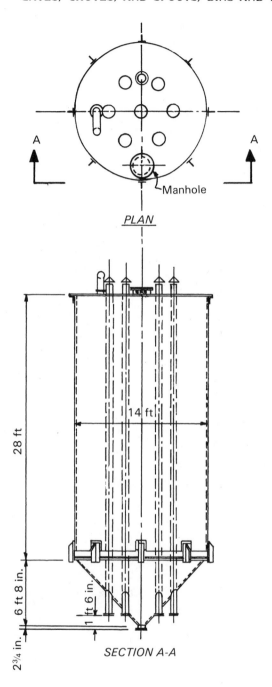

PLAN

14 ft

28 ft

6 ft 8 in.

1 ft 6 in.

2¾ in.

SECTION A-A

—Manhole

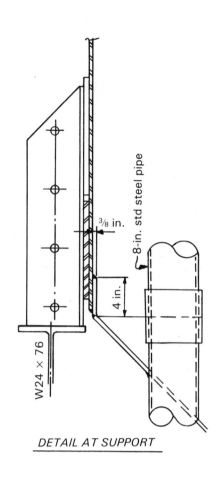

⅜ in.

4 in.

8-in. std steel pipe

W24 × 76

DETAIL AT SUPPORT

Figure 13.17 A circular bin storing hot material.

ing the material. Some cooling will take place while the material is delivered to the bin. More cooling can be provided by the flow of outside air through the vertical pipes placed circumferentially. (The pipes also help prevent arching of the material.) If more cooling is necessary, fans can be introduced.

Refer to paragraph 16.7.1 for a similar piping installation for heating the contents.

13.10 BIN ACCESSORIES

Bins should be provided with high- and low-level indicators and/or controls. These devices can be connected to lights or buzzers located in control rooms, or to operate valves that control the flow of material to and from the bins.

FIGURE 13.18   325

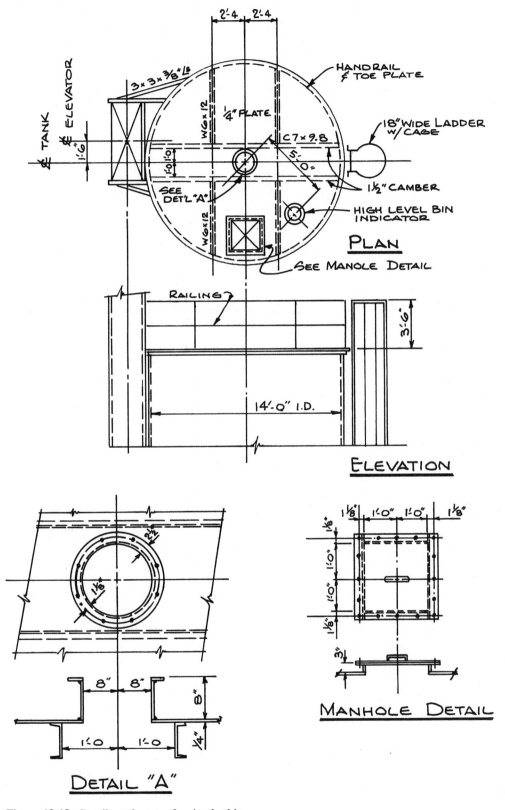

Figure 13.18 Details at the top of a circular bin.

Inlets generally are provided near the center of the top of the bin, with outlets in the conical bottoms. Where the bottom of the bin is flat, manholes must be provided near the bottom for cleaning out the bins. Manholes also must be provided in the top of the bins for servicing and maintenance. Railings are required at the top of the bins for safety.

Figure 13.18 shows typical details at the top of the bin. This bin is a 14-ft diameter bin. It shows the required railing protection, a caged ladder to reach the top of the bin, location of high-level indicator, manhole detail, and hole for feed into bin. A connection for dust ducts is required many times in the top of any storage bin.

Handling Special Materials

14.1 GENERAL

Manufacturers of conveying equipment, through their Conveyor Equipment Manufacturers' Association, and as individuals, issue handbooks that recommend equipment for handling various materials under stated conditions. Tables give the weights of the materials, capacities of equipment, abrasiveness, and other pertinent data. The characteristics of material data are data obtained in the western New York area. They may differ slightly from CEMA values applied nationally. The materials listed below are not intended to form an all-inclusive list, but are intended merely to guard against pitfalls and problems that were encountered by various members of our staff in dealing with the products listed. It is hoped that these will act as a guide for similar products. In general the client, through his engineering department, is very familiar with his raw and finished products. Processes where the consulting engineer makes more of the decisions have been dealt with in greater detail. Some items are repeated under various headings.

This section is intended to cover suggestions for the handling of only some materials. The chemical and physical processes for producing the various materials or using them in process industries is beyond the scope of this book.

Material contaminated with wet clay will coat the bucket of an elevator and eventually fill it with clay. Even a belt conveyor will find the belt coated with clay that has to be scraped, and idlers and pulleys given constant attention. Fine, highly abrasive products cause less wear on a rubber-covered belt than on a metal chain, with its multiplicity of articulations at the chain pins.

When the temperature of the material to be handled is likely to exceed 250°F, chain is to be preferred, and with the selection of quality chain, satisfactory service can be had. There are belts made for higher temperatures as well as those for higher temperatures that may be oil-coated (refer to paragraph 2.4.1). Provision for expansion must be made wherever material is transported at high temperatures.

Consideration should be given to the possibility that the material to be handled may be of lump size with sharp edges that could become lodged between the buckets of an elevator and belt and would damage the belt as the material is picked up in the boot and as the belt passes over the head pulley. For lumps 2 in. and under and materials with sharp edges, it is usually best to use chain. For larger lumps, use of a belt might be best.

Refer to paragraph 14.13 for handling explosive materials.

14.2 ASHES

Weight/ft^3: 40–45 lb
Angle of repose: 42°
Abrasive

Ashes of any size are considered waste material and, at present, offer very little or no return on the money invested to get rid of them. Some stoker ash can be used as fill material, but mostly it is a waste-disposal problem. Keeping the dust down is a consideration, even if the ashes are wetted down. For small plants, a small, one-man front-end loader can be used to move the stoker ash out of the boiler room to the outside removal storage hopper.

Some hand-filled boilers produce large clinkers that are difficult to remove by hand under the grate with a long steel rod hoe. After the ashes are pulled out from under a boiler, they must be disposed of in some way by a wheelbarrow or front-end loader to a pile outside. Modern boilers are equipped with stokers, which produce ash that also must be pulled out beneath the stoker and handled in about the same manner.

Ashes produced by a stoker can be very fine or in lumps. If in lumps, they may have to be broken up to be handled. A centrifugal discharge elevator handles ashes well, if the lumps are not too large, say 2–3 in. A continuous bucket elevator should not be used, as the bottom of the V-shaped buckets soon fills up and, when dry, gets hard, greatly reducing capacity. Ash chutes should be on at least a 60° slope; more is better.

Most industrial plants will install handling equipment to get rid of the ashes with as little dust as possible and with minimum investment. Very often the floor of the boiler plant is 5 ft or 6 ft below the yard level, thus presenting more problems in getting the ashes to the ground or yard level.

Figure 6.9 shows a typical arrangement of a drag conveyor handling ashes below the boiler plant floor level. Ashes can be pulled out of a boiler by hand onto a steel grating, where any large lumps can be broken with a hand shovel. The ashes then fall through the grating to the carrying run of H-112 drag chain, sliding along in hard white (cast) iron (replaceable sections) set in a concrete pit. The ashes can be hot or even glowing, but they cool rapidly, and they can be quenched by water before reaching the conveyor. They do not distort the trough nor the malleable iron chain, which normally can stand temperatures up to 600°F or, if made of "A" metal (Rexnord) or Promal (Link-Belt), a maximum of 1000°F. There is no foolproof conveyor built to fit beneath a boiler to bring the ashes to the drag chain conveyor without hand labor. Apron conveyors below the floor also are used, with the pans not less than ⅜ in.

Depending on the type of ash produced by the stoker and the location of the disposal site, very large steam stations found it economical to use sluicing systems. The material must be fine fly ash or dust, and a considerable amount of water must be available economically. Getting rid of the water in a pond sometimes becomes a problem. A pneumatic system is also available. This re- quires blowers and storage bins. Piping connections must be cast iron and elbows will need replacing from time to time. This method is clean but expensive. The excess amount of air for conveying the ashes must be disposed of, usually in the storage bin.

14.3   BAUXITE ORE

Weight/ft³: 75–85 lb
Angle of repose: 30–44°
Very abrasive

The bauxite deposits are cleared by stripping. The surface is then cleaned by scrapers and hand labor. It is broken up by blasting. The loosened ore is loaded by small shovels into small bottom-dump cars on a narrow-gauge track.

In damp or rainy weather, the rock may have a thin, clayey coating, and bucket elevators will not work well. Belt conveyors give much better results. Any bauxite dust clinging to the rubber cover can usually be scraped off enough on the return run to break down any possible small lumps, and then hosed off further along the return run.

Bauxite abrasives are made by fusing bauxite with carbon in an electric furnace, breaking down the fused product, pulverizing and sizing the grains. If made from pure bauxite, the product is practically an artificial corundum. By varying the grade of bauxite used, the hardness and the toughness of the abrasive may be varied in the finished product.

High-alumina cements, which are essentially calcium aluminates with very low silica, are made by fusing in a small blast furnace a mixture of bauxite, coke, and limestone. The slag, when poured, cooled, and ground, forms the finished cement. This cement contains about 50% lime, 40% alumina, and 10% silica, and so on.

Since bauxite is mined in an open-pit operation, after removing top overburden such as clay, it is subject to all kinds of weather conditions (wet and dry), and should not be handled in a bucket elevator of any kind. In damp or wet weather, with a fine film of clayey material over it, it gives a gummy or plastic feel. It would soon fill a bucket on an elevator, get hard, and be next to impossible to remove in a reasonable length of time. If a belt conveyor system could be installed, it would give much better results. The clay will cling to a rubber belt, but it can be removed by rubber line scrapers on the return run and then hosed off with a water spray. Generally speaking, bauxite handles much like gypsum rock from the quarry.

14.4 CARBORUNDUM

Weight/ft^3: 100 lb
Angle of repose: 20–29°
Very abrasive

Carborundum (silicon carbide) is produced in an electric furnace at about 4000°F. A charge of petroleum coke, sand, sawdust, and salt is fed into these furnaces and, at the end of 36 hours, the crystals are produced. It is cooled for about 48 hours before material is removed. It is crushed and recrushed to obtain the right grain size. The grains are washed of impurities and graded by passing through silk screens with size 6–220 mesh. Finer grains are separated by water flotation. Up to 100 mesh, it is dusty and dry.

Centrifugal discharge elevator with hardened (Brinell 190) malleable iron buckets attached to a belt has proved satisfactory. Chain does not stand up as well as belts of good grade canvas, without rubber cover. Fine material gets into the chain joints. The buckets should be the same size as the belts.

Screw conveyors of heavy construction are used for distribution of crushed or screened material to storage bins. Capacity should be figured on the basis of 15% trough loading. Hard white bearings should be used in the hangers, and coupling shafts should be hardened.

Vibrating conveyors (oscillating type) can be used to convey the coarse grains, but the grains minus 10 mesh should not be handled on this machine, generally speaking, without making a test run to see if the very fine grains will move at the correct speed to give the desired capacity.

The carborundum finished product is very abrasive, and any machinery handling it must be made heavier than standard; also, wherever possible chutes and troughs should have removable lining plates.

14.5 CARBON CHIP REFUSE
(refer to paragraph 4.23)

In the finishing process, graphite chips can be dropped through a grating with one-inch square openings to a trough below the floor, where a screw conveyor in the square concrete trough can take it to its destination. The quantities are generally small, but it is best to use a 9-in. screw conveyor as a minimum to allow for a larger coupling and shaft. The larger lumps can be broken down on the grating. Weight is 45 lb/ft^3.

If floor space is limited and a standard motor and drive cannot be used, a hydraulic pump motor will work well.

14.6 CEMENT

Weight/ft^3: 85 lb warm; 90 lb in bags; 72 lb aerated
Angle of repose: 30–45° when cold
Mildly abrasive

Screw conveyors play a very important part in the conveying system. Portland cement has a fineness of 200 mesh and is very fluid. The design of machinery should be based on 85 lb/ft^3 and capacity on the basis of 72 lb/ft^3.

In figuring the capacity of screw conveyors, the figuring should be on the basis of 30% trough loading. Two-inch clearance is provided at the discharge end of the trough from the end of the screw-pipe section to the inside of the casing (figure 14.1).

To prevent a screw conveyor from jamming when the discharge chute becomes clogged, a hinged or swing baffle plate connected to a roller-type, snap-action limit switch should be used. Flooding will move the swinging baffle, and a roller switch will shut off the motor. When the baffle plate returns to the original position, current will be restored, and the equipment will operate again (figure 14.2).

Screw sections should be equipped with Schedule 80 or heavier pipe, and XX flighting. Hangers should be floating, and expansion should be provided at the discharge end of the trough when carrying cement at 150°F or higher. Some cement companies use square troughs made of ¼-in. steel plate, reinforced at top and bottom with structural angles. The inside-of-trough dimension is made 3 in. wider than standard for sizes 12 in. and over, and 2 in. for sizes 10 in. and under. Square section troughs allow foreign material, such as bolts and nuts, to imbed in the corners. It further allows trough corners to be more readily cleaned. Others prefer to use U-shaped troughs of standard dimensions up to 16 in. Above 16 in., they use square troughs. A little extra clearance has given good results in the industry (figures 14.3 and 14.4).

Continuous bucket elevators are not used for handling cement because of so much carryover down the casing on the return run. Hot or cold cement has a tendency to stick to steel, and in discharging from one continuous bucket to the back of the forward bucket, tends to build up a bead of cement on the back of the forward bucket and prevents the cement from sliding off the back of the forward bucket to the discharge chute, thus carrying a considerable amount of cement down the elevator casing to the boot (figure 14.5).

Cement is handled successfully in centrifugal dis-

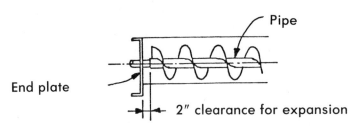

Figure 14.1 An expansion detail.

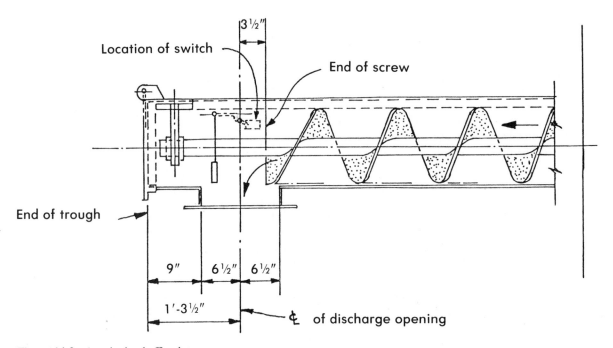

Figure 14.2 A swinging baffle plate.

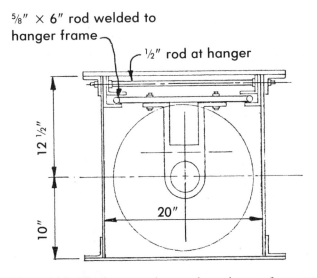

Figure 14.3 Floating expansion-type larger hangers for cement and lime products over 150°F. A modified no. 216 hanger with a hard white-iron bearing for an 18-in. screw in a square trough. The design can be used for 12–20-in. diameter screws.

charge elevators, using a single strand of either C-188, C-102B, C-102½ or C-110 combination malleable iron and steel chain or all steel chain. The malleable iron Style A or AA buckets are spaced close together, as for a continuous bucket, and the chain speed reduced to about 185 fpm. If the cement is hot (above 150°F), it will flow like water, and tests have shown an increase in capacity over manufacturers' tabulated values of up to 50%. The point of discharge for the chute at the top of the elevator should be 12 in. lower than standard.

When figuring capacity of these cement elevators, figure the bucket at water level full, as the cement will fill to this level as it becomes aerated while being handled (figure 14.6).

The elevator should have gravity takeups in the boot to keep the elevator chain tight at all times. It will compensate for stretch in chain and temperature elongation.

Traction wheels should be used on head shafts when the distance between centers of the elevators exceeds 35 ft, because there is enough weight of chain and buckets on the carrying run to hug the traction wheel and not

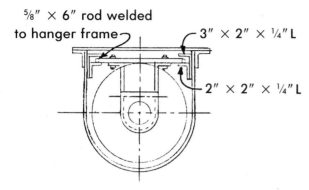

Figure 14.4 Floating expansion-type smaller hangers for cement and lime products over 150°F. A modified no. 216 hanger with a hard white-iron bearing for 9-, 12-, 14-, and 16-in. screws.

⅝" × 6" rod welded to hanger frame

3" × 2" × ¼" L

2" × 2" × ¼" L

slip. (Traction wheels work well on elevators above 35-ft centers, especially for abrasive materials.) Under 35 ft, sprockets must be used. Traction wheels are always used on the foot shaft. These wheels can keep themselves clean and prevent damage to parts. The wheels run loose on the shaft, and are kept in place by safety set collars.

Cement can be stored in overhead bins and reclaimed by the use of a right- and a left-hand screw on the bottom, if the discharge point is at or near the center of the bin. The screw section is exposed in the bottom of the bin, then covered in a pipe. If separate drives are used, each screw can operate by itself. In reclaiming cement from the bottom of a silo, two H-type drag chains, located about 5 ft each side of the silo center line, have

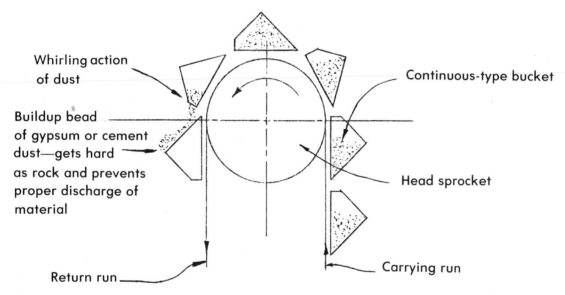

Whirling action of dust

Buildup bead of gypsum or cement dust—gets hard as rock and prevents proper discharge of material

Continuous-type bucket

Head sprocket

Return run

Carrying run

Figure 14.5 Carryover of cement at a head sprocket.

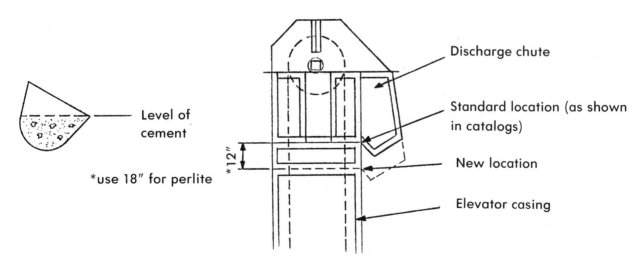

Level of cement

*use 18" for perlite

*12"

Discharge chute

Standard location (as shown in catalogs)

New location

Elevator casing

Figure 14.6 A cement elevator head.

also been used. The speed varies from 36–143 fpm and, with an HC-6110 chain, will give a capacity of 18 to 30 tons per hour of 200-mesh cement. The carrying run of the drag chain is on the bottom of the silo, and hung from it, while the return run is protected through the silo by a continuous V-shaped cover. An adjustable surge plate prevents flooding at the discharge end and regulates the capacity if a variable-speed conveyor is required, and a fixed surge plate at the takeup end prevents leakage at that point (figure 14.7). Some experiments seem to indicate that screw arrangements work better than a drag chain conveyor for discharging out of the bottom of a bin.

In areas that are subject to extreme cold and wet weather, outside storage silos should provide for possible heating of the cement. If possible, storage silos should be enclosed.

A number of cement-producing plants own or lease delivery trucks equipped with a pneumatic system operated by the truck engine, whereby when they deliver a truckload of cement to a ready-mix concrete plant, the truck operator has only to connect his pipe attachment to that of the plant, start his engine, and the cement is discharged to the plant bins. The piping in the plant is simple, and the amount of air with cement is taken care of by a series of baffles in the bins.

This method of cement delivery eliminates clouds of cement dust at the discharge point of the truck. When the cement is freshly made and warm, it tends to flow like water, but is easily handled pneumatically. In bad or rainy weather, these trucks are not likely to be operating, but it takes very unusual weather to keep them from making deliveries. This scheme of delivery is mostly confined within city limits, where the distance of delivery is not great.

For details of an effective head shaft dust seal, see

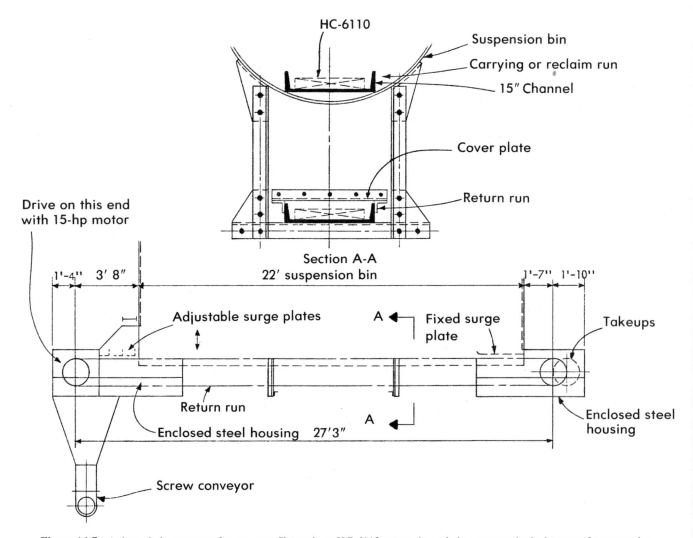

Figure 14.7 A drag-chain conveyor for gypsum. Shown is an HC-6110 stucco drag-chain conveyor in the bottom of a suspension bin. The speed is 150 fpm, with a capacity of 30 tph of cement at 180°F.

figure 3.7. For details of a venting-type inspection door, see figure 3.6. Some type of an inspection door and a dust seal at the head shaft is almost a must.

14.7 CHARCOAL

Weight/ft^3: 18–25 lb
Angle of repose: 35°
Abrasive

Charcoal generally is received in boxcars. When the bottom doors are opened, hand labor often is necessary to get the charcoal to move into the track hopper. An apron feeder works well. Plenty of space should be allowed in the bottom of the hopper to avoid hangups.

Charcoal is shipped in pieces ranging from 4 in. to 8 in. long, in all kinds of shapes, much like the branches of trees. Most chemical plants use it in pieces about ¾ in. to 1½ in. in diameter. Charcoal pieces tangle together and must be handled in a super capacity bucket elevator with flat-bottom centrifugal discharge buckets, even if the capacity required is small, as only a few pieces actually get into a bucket. As a result, there is much void space, and it is well to take the capacity from catalog tables and assume only 40% as the actual load. Charcoal requires wide buckets on two strands of chain. The machinery parts are usually much heavier than required due to the character of the material. A vibrating conveyor will work well. A skip hoist bucket arrangement could also be used if space permits, but these units usually run into considerable cost.

If the charcoal is crushed to a maximum of 2 in. and a 12- or 14-in. screw with a low capacity is used, a screw conveyor will probably work satisfactorily, but no large pieces of charcoal should get into the screw. Charcoal generally breaks up easily, but there will be pieces that just do not break up. Otherwise, screw conveyors should not be used.

14.8 CHROME ORE

Weight/ft^3: 140–150 lb (to allow for voids)
Angle of repose: 30–44°
Extremely abrasive, sharp edges

Chrome ore is shipped in sizes from 15 in. down to dust. Track hopper should be made of not less than ¼-in. plate, with removable liner plates of ¼-in. steel that are replaced when worn. Usually, an inclined apron feeder carries the material from the track hopper to a bucket elevator, belt conveyor, or inclined apron pan conveyor with the pans overlapping and the chain outside. It operates at a speed of about 10–15 fpm. To take the shock of the load when the ore is dumped into the hopper, T-rails are installed under each pan, as shown on figure 5.9. Note the provision for expansion in handling hot material.

The apron pans should be ½-in. thick and mounted on a double strand of steel strap roller chain located underneath the pans, as shown in section A-A, figure 7.4. In the particular project shown, the chain and pans are supported on 10-in. diameter steel rollers, spaced 15 in. on centers, to absorb any shock load. The ore comes in contact with a grizzly, which allows fine pieces of ore down to dust to fall on the belt and act as a cushion. The belt conveyor, operating at 250 fpm, could not properly handle such a capacity or such large lumps without a feeder, because the belt conveyor operating at this speed would simply roll out from under the load, and the ore would not move.

An apron feeder with T-rails welded underneath each pan, and arranged to travel on a ¾-in. steel wear plate on top of supporting beams, as shown on figure 5.9, could have been used instead of the roller-supported apron, as shown on figure 7.4. The roller-type apron can operate at up to 25 fpm, and with the low speed, can have smooth action with little maintenance.

The selection of a bucket elevator or a skip hoist will depend on how much chrome ore is used per week or month. If one carload of ore is used each month, the chances are a bucket elevator of the super capacity type can be used. The size of the ore may be as large as 6-in. lumps and under and, depending where it comes from, some chrome ores weigh as much as 150 lb/ft^3. There is a wide range in the composition of this ore, and the weight per cubic foot and size of the lumps must be checked. If three or four carloads of ore are used per month, it might be well to consider a skip hoist. These units are very efficient, but expensive to install, yet the overall cost balanced against the tonnage of ore handled could result in a decision in favor of the skip hoist. Medium- (140 fpm) or high- (260 fpm) speed skip operation is used when a large tonnage of ore has to be handled.

These hoists do require a deep pit, and foundation work may be expensive, especially if working in water and/or rock. If the depth of the foundation is not a serious item, an automatic loader at the discharge of the apron feeder will work well. When the loader is filled, it can be arranged to shut down the feeder until the operation is repeated. The loader is operated by the skip bucket coming into position to receive its load.

If a skip hoist is selected, it can be obtained with what is called "manual loading," where the skip bucket would travel at a slow speed of 80 fpm. This would be a dangerous unit to operate if the skip bucket had to be filled by an apron feeder. It would mean a person would

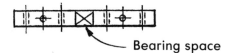

Figure 14.8 Bearing space.

have to start and stop the feeder when the bucket was either empty or full. When the bucket is full, a person would stop the feeder, push a button for the bucket to go up, and the bucket would discharge and come down to rest again until the operation is repeated. This depends on the human element for successful operation and, at times, it does not work.

When chrome ore is crushed to minus 1 in. to dust, it can be handled in a continuous bucket elevator, where the ore is fed to the buckets operating at 125 fpm. A centrifugal discharge elevator will not work well, as the buckets, whether malleable iron or steel, would wear out fast or tear themselves from the chain.

If need be, this crushed ore could be handled by a screw conveyor, with all parts made heavy and designed on the basis of 15% trough loading, to keep the ore low in the trough so it would pass beneath the hangers. Hanger bearings should be chilled and the coupling shafts made of hardened steel (see figure 14.8).

On account of the possible heavy weight per cubic foot, the coupling shafts on conveyors over 50 ft long should have three-bolt connections, with the coupling bolts made of SAE 3140 steel. A three-bolt coupling gives protection against shearing under impact or sudden loads.

14.9   COAL

> For industrial plants with a steam capacity of
> 20,000–200,000 lb steam/hr
> Weight/ft^3: 50–60 lb
> Angle of repose: 35–37°

Most bituminous coal is washed and sized at the mine preparation plant and shipped in the size specified. Run-of-mine coal has about 10-in. lumps and must be crushed at the site. Run-of-mine coal should not be handled on a bar scraper feeder, as the clearances provided are not enough for large lumps. Coal sized to, say, 1¼ in. to dust, can be purchased at a higher cost. For stokers, the coal must be properly sized and delivered to the stoker grate evenly distributed, size varying for each type of stoker. In general, it must be minus 1¼ in. with about 20% fines. In areas subject to severe winter weather, the coal is subject to freezing when shipped in open gon-

dola cars, and there is the danger of entire cars freezing solid. A thawing shed (48 hours) and a car shaker should be provided in those areas. In extremely cold weather, the surface moisture on the coal causes the individual pieces of coal to bunch together hard enough so the coal has to be crushed or squeezed to break this freeze bond. Lumps formed by small sizes bunching together fall apart when handled roughly.

Much of our bituminous coal contains sulfur. When mined, cleaned and screened the sulfur, combined with oxygen in the air, causes corrosion. The lubricating properties of coal will lessen the impact.

Coal can be stored outdoors in piles, usually about 20 ft high, and is being piled to 40 ft or even 50 ft high. Arriving coal has to be spread out layer upon layer and packed down tightly to avoid oxygen penetration. It should be watched for excessive heating. There is a spontaneous combustion hazard due to heating of the lower storage, especially with frequent high sulfur content. A bulldozer or yard scraper is used to level off the pile and to keep it cool. Where space is not available for a spread out storage pile, concrete or steel silos, 15–20 ft in diameter, and, say, 50–60 ft high, are used. Elevators feeding to and reclaiming the coal from the silos should be arranged to recirculate the coal and keep it cool. A rod thermometer attached to a lighting signal or belt should warn plant personnel of rising temperature in the silo. Coal in steel bins should be recirculated often to prevent fires that could distort the bins.

If coal is stored in several silos and then reclaimed, a gravity discharge elevator in combination with a conveyor, makes an ideal setup. This can be done with skip hoist and pivoted bucket carrier (the most expensive way of doing it), or by the gravity discharge elevator-conveyor combination that will take care of most operations for intermittent service with low operating and maintenance cost. This latter type of unit will handle nonabrasive, free-flowing materials such as bituminous coal, whether it is operating vertically or on an incline, and will function as a bucket elevator vertically and as a scraper flight conveyor horizontally. Steel buckets are placed either 24 in. or 36 in. apart and are rigidly attached to a double strand of steel chain usually 12- or 18-in. pitch. A chain of this type does not stretch much, and experience has indicated that the bucket should be of welded construction and thus keep the chains in position. There are no swivel joints on this type of chain. Centrifugal discharge elevators will be very noisy.

During the winter months, when sized bituminous coal (say −1¼ in.) is stored outdoors and packed down by bulldozers, even with alternate freezing and thawing, the moisture will seldom penetrate more than 2 ft. The coal can be handled easily by a shovel crane.

The top crust will break fairly easily into 3- or 4-in.

round lumps, and the crusher will simply squeeze the lumps apart. Heating the frozen coal in the discharge hoppers of railroad cars by blow torches will move it through the restricted hopper and start the flow. The railroad companies frown on the use of blow torches, claiming that heat distorts the steel body of the freight car. Small plants will not use torches due to fire hazard and insurance. Hammering the sides of the car or using a car shaker will release the coal stuck to the sides. A car unloader is available, which breaks up the surface of the coal and pushes it out at the bottom of the car. To reduce the size of the lumps of coal for stokers or pulverizers, especially run-of-mine coal, or coal that is lumpy, crushers are used (refer to section 10). The crushers used are a single-roll type, two-roll crushers, hammermills and, in very large operations, the Bradford Breaker, As new surfaces are formed by the crushers, the moisture already on some of the lumps will tend to distribute itself. When dry lumps are mixed with the wet fines, serious size and moisture segregation may occur in the coal bunkers or silos. Fine crushing reduces the difficulty considerably.

Large coal lumps, supplied to a grate, ignite slowly and, for a given stoker speed on traveling grates, will drop into the ash pit, high in carbon. Finely sized coal that stays on the grate burns out more completely. If the coal is too fine, it will fall through the grates, or burn out in spots, allowing excessive quantities of air to pass through. Operating efficiencies have resulted when the ash pit and excessive air losses have been reduced.

Bituminous coal, when shipped from the mine, has some fines minus $\frac{1}{8}$ in. The fines when wet are sluggish and will not flow easily. Chutes should be at least on a 60° angle to take care of any sluggish coal. If this is impossible, the chutes should have a minimum of a 50° slope, and the bottom should be lined with replaceable 12-gauge stainless-steel plates. The smoothness of this plate usually takes care of the 10 degrees lost in the slope. Sometimes it is necessary to change from slack bituminous coal ($-1\frac{1}{4}$ in.) lumps to run-of-mine (-10 in.). It is well to provide for possible future use of a crusher under the discharge of the apron feeder, or install the crusher and provide a bypass.

Some boiler plants use pulverized coal. The coal is pulverized to 70–80% through a 200-mesh screen and fed into the boiler by primary air fans blowing temperate air through the pulverizer and discharging air-coal mixture into the furnace at about 150°F. Secondary air is fed to the burner at around 600–800°F. Pulverizing does provide an opportunity for using less expensive, low-grade bituminous coal. As most of the ash goes up through the boiler, an electrostatic precipitator, or bag house, is needed before the flue gas is discharged to the stack. Fine pulverized ash must then be discharged from the hoppers of the dust collectors to the ash system, either by screw conveyors or pneumatically.

Debris and refuse found in the coal will vary with the source. Sulfur balls, slate, rock, tramp iron, timber, magnetic metal lumps, rags, and straw may be found with the coal. A magnetic separator should be used to remove iron particles and protect the crusher.

To keep accurate records of coal consumption, a traveling weigh larry is used to deliver and weigh the coal as it is fed to the stoker. With a long suspension overhead bunker, the larry permits close spacing of discharge gates, and drawing coal from any part of the bunker, which is impossible with chutes. A larry distributes coal uniformly in the stoker hopper, prevents segregation of lumps and fines, and thus increases efficiency. Stoker extension hoppers with a 2- to 3-hour supply should be used where the stokers are fed by larries, to save trips of the larry. Some boiler plants use the weigh larry, while in some large power plants, automatic scales on belt conveyors in the coal system are used. Small hoppers can be supported by the stokers, while large ones should be supported from the floor (see paragraph 2.12 and figure 2.10).

Figures 14.9 and 14.10 show a layout of a Typical Coal Handling System. Run-of-mine (-10 in.), or slack bituminous coal ($-1\frac{1}{4}$ in.) is received on siding in 70- to 100-ton cars. The coal is discharged to the track hopper through the steel grating, and conveyed by an apron feeder to a single roll crusher, if run-of-mine, where it is reduced to $-1\frac{1}{2}$ in. If slack coal ($-1\frac{1}{4}$ in.) is received, it can bypass the crusher, flowing to the boot of the bucket elevator (at some locations -8 in. is considered run-of-mine). Skip hoists work well on run-of-mine coal.

A popular type of apron used for handling coal is shown on figure 14.11 with apron pans. An apron of this type will safely operate up to a period of 10–12 years without much attention and maintenance. It requires the placing of heavy oil on the chain joints occasionally, and used engine oil is satisfactory.

A bucket elevator lifts the coal high enough to discharge into a two-way chute with flop gate feeding either silo. The overhead storage is filled first and the overflow will then go to the bottom storage. Coal can be reclaimed from bottom storage through gate and chute leading to boot of bucket elevator, or gate and chute can reclaim coal from bottom of silo and discharge it to floor of boiler room in an emergency. Coal from the overhead storage in silo can be fed to stoker in front of boiler by a gate and chute on face of silo, then to stoker spout or chute. From the two-way chute at head of elevator, another chute can discharge coal to reclaim pile on ground. Coal in the silo at times must be recirculated to prevent overheating due to depth of silo.

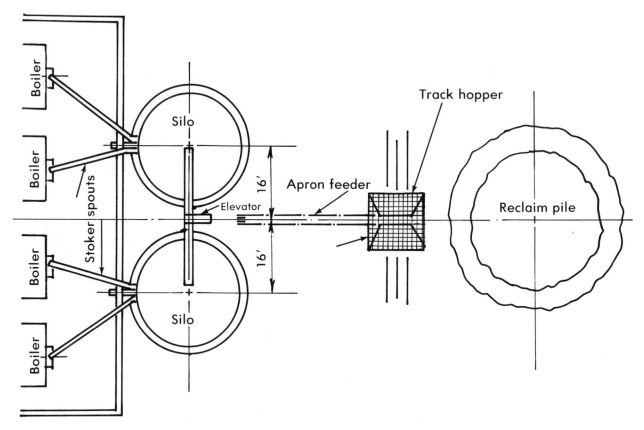

Figure 14.9   Plan of a coal-handling plant.

Figures 6.7 and 6.8 show a bar scraper feeder used to take the coal from car hopper to an elevator. An apron feeder could be used instead of a bar scraper feeder. It would have to carry the load of coal on the top side, greatly reducing the slope of the hopper sides to maintain a depth of 4 ft for foundations, where a 4-ft frost depth has to be considered. This apron would cost considerably more for machinery and installation. Belt conveyors can be used operating at 400 fpm for 18 in. belts, up to say 800–1000 fpm for 42-in. and over belts.

An 18-in. wide bar feeder consists of two strands of either C-188 or C-131 combination malleable iron and steel chain and 1½-in. steel flights welded to chain attachments every 10 in. Operating at a speed of 20 fpm (30 fpm would be about maximum to avoid considerable maintenance) results in a capacity of approximately 12 tons of coal (50 lb/ft^3) per hour. When the weather is fairly dry, this coal flows easily, so that a 50-ton car could be unloaded in about 5 hours, barring any unforseen delays.

Feeders should be used to control the rate of feeding to crushers and thus improve their operation. Single reciprocating feeders give a pulsating flow which can be improved by use of twin units that discharge alternately. Apron feeders give a more continuous flow. Mechanical and electromagnetic feeders screen fines ahead of crusher, and have an easily adjusted feed rate. Controls must be provided to stop feeder when crusher stops. A bar scraper feeder should not be used on run-of-mine coal, as the clearances are not enough for large lumps.

A typical inclined screw feeder receiving fine bituminous coal from storage room and delivering to boiler stoker, is shown on figure 14.12. Fine coal, when dry, will flow by itself out of storage area until it reaches its angle of repose. Bar grating over hopper does not normally interfere with flow unless coal is very damp or wet. In this event, it may become necessary to stop the screw and poke down through the grating to get the coal moving. Grating is also in place to possibly catch any pieces of iron or wood hidden in the coal, and to prevent a laborer from falling into the screw if the coal is low and a person is in the storage room shoveling to the hopper.

A 9-in. screw operating at 60 rpm handled about 8 tons per hour of minus 1¼-in. bituminous coal from coal storage bins. The motor was 2 hp for a 30-ft long con-

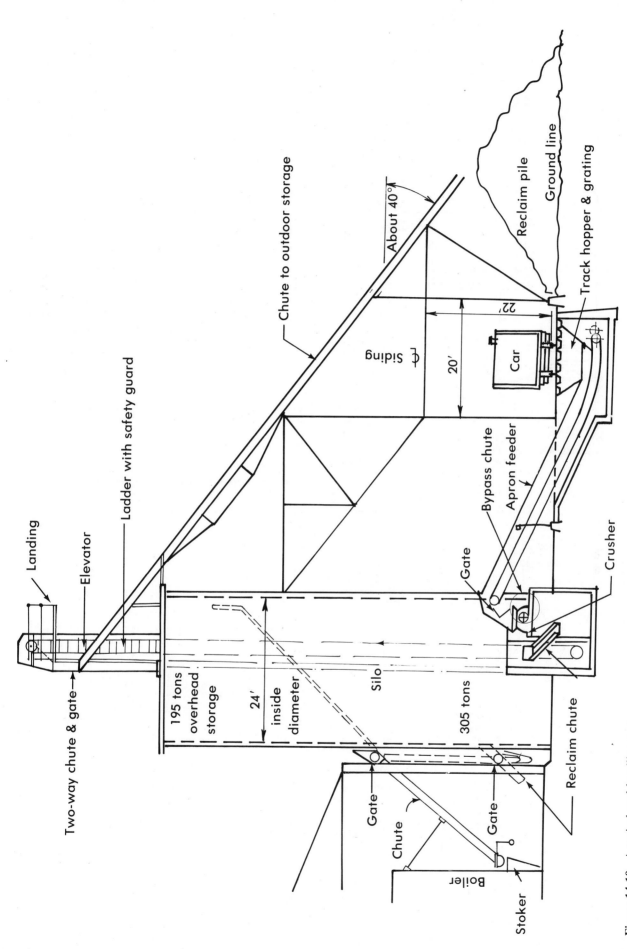

Figure 14.10 A typical coal-handling system.

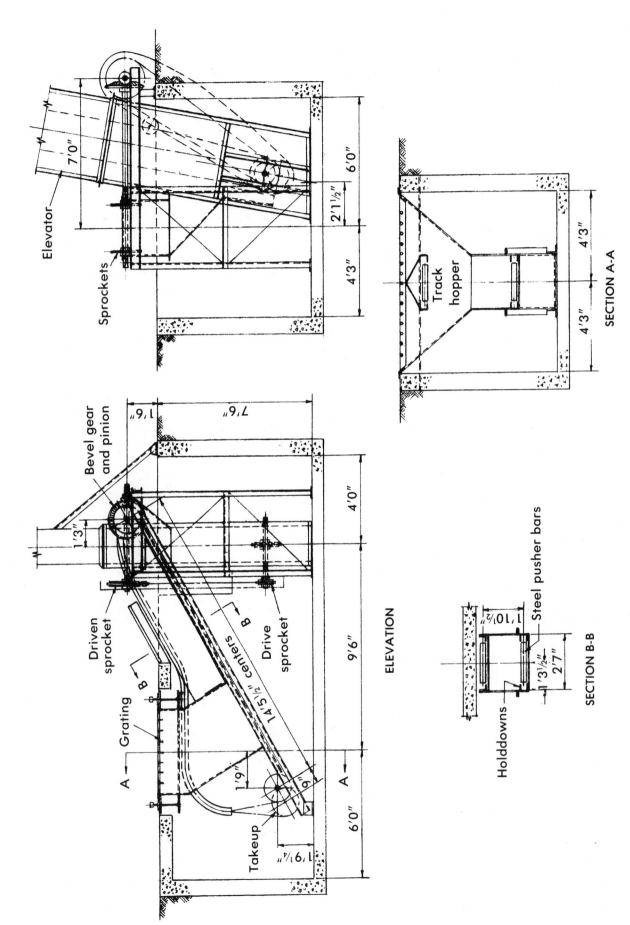

Elevator

7'0"

Sprockets

6'0"

2'1½"

4'3"

SECTION A-A

Track hopper

4'3"

4'3"

Bevel gear and pinion

1'6"

7'6"

1'3"

Driven sprocket

Grating

B

A

Drive sprocket

B

14'5½" centers

4'0"

9'6"

Takeup

1'9"

9"

A

1'9¼"

6'0"

ELEVATION

Holddowns

Steel pusher bars

1'10½"

1'3½"

2'7"

SECTION B-B

Figure 14.11   An apron conveyor at a coal hopper.

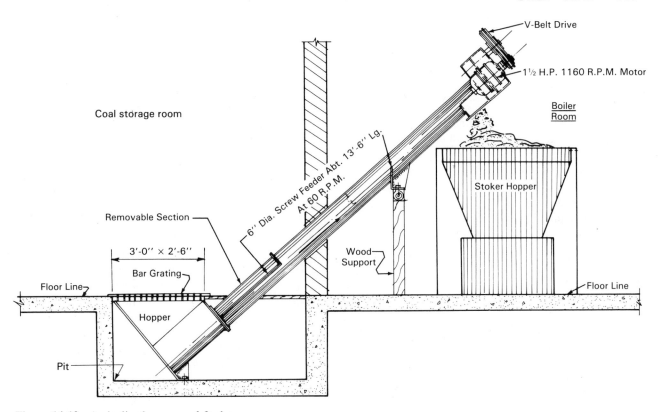

Figure 14.12 An inclined-screw coal feeder.

veyor. In using side inlets to feed the screw, the opening farthest from the discharge end would feed first and the intermediate feed openings did not feed until the farthest opening from the discharge had exhausted its portion of the coal. Coal should be fairly dry, free-flowing and, during winter, not bunched up by ice crystals. If the coal comes to storage 2 in. and under, then a 12-in. diameter reclaiming screw should be used, as a 9-in. diameter screw is too small to handle anything larger than minus 1¼-in. lumps. For small plants (20,000–50,000 lb of steam/hr), coal can be brought into the storage area adjacent to firing floor by truck and dumped on the floor. Until about 1950, the boiler operator would use a wheelbarrow to move the coal into the stoker hopper. Now a small one-person fork-lift truck with a front-end bucket scoop would be used to move both the coal to the stoker, and the ash out to the disposal hopper or truck.

14.10 COKE

Weight/ft^3: 25–35 lb
Angle of repose: 30–40°
Very abrasive

The volatile parts, hydrogen and oxygen compounds, of

bituminous coal may be driven off by heating the coal in closed ovens, and the residual mass is coke, almost pure carbon. This is distillation, and the ovens in which this is done without trying to save the volatile products are called bee-hive ovens, while the more modern ovens, which save the byproducts, are called byproduct ovens. A ton of bituminous coal treated in a typical byproduct oven will yield on the average 1410 lb of coke, 7.1 gal of tar, 18.9 lb of ammonium sulfate (fertilizer), 2.4 gal of light oils, and 10,440 ft^3 of illuminating gas, about half of this gas being used to furnish the heat for the distillation. The coal-tar dye industry is built on the tar thus produced. Toluol, benzol, etc, come from the light oils, and half of the gas produced is available for household illumination.

The coke thus produced is hard, clean, and abrasive. Fine-grained coke dust with rough edges will move slowly in a vibrating conveyor, but when dropped only 18 in., will mat together on hitting the steel plate, and form a mass that cannot be broken down by any vibration. A screw conveyor designed to keep the load low in the trough (15% capacity) is best for the purpose.

Elevator buckets should be ¼-in. steel, with continuous welds on the outside. Plastic buckets have given satisfactory performance.

14.11 CRUSHED ICE

Weight/ft^3: 35–45 lb
Angle of repose: 19°
Nonabrasive

Combination ice crushers and slingers are used for producing fine ice to cover perishable food while in transit. After a cake of ice is fed to a crusher, the resulting crushed ice can be handled in a vertical screw lift or by a centrifugal discharge bucket elevator, equipped with a single strand of no. 830 or no. 844 Ley steel bushed malleable chain, with malleable iron enameled round-bottom buckets (style A or AA), at standard spacing, and operating at standard speed. The chain pins can be made of Everdur bronze with S-shaped stainless steel cotter pins.

Corrosion does not affect malleable chain that is equipped with an Everdur pin and stainless steel cotter. The S-shaped cotter will not become loose in the pin. Ice has an affinity for a plain malleable iron bucket and, therefore, these should be enameled. This coating is extremely durable, and will not allow any ice particles to build up. The attachment bolts and washers for buckets-to-chain should be of stainless steel, to prevent corrosion.

In a continuous bucket elevator, the ice settles in the V, or bottom, of the bucket, congeals rapidly, forms gobs of ice and, at times, will not come out of the bucket. These are slow-operating elevators, running at a speed of about 100–150 ft/min, and not throwing the ice out of the bucket, like a centrifugal discharge elevator does at a speed of around 230–260 ft/min.

The boots of ice elevators should be equipped with ^{3}/₁₆-in. stainless or ^{3}/₁₆-in. steel galvanized plate (heavy dipped). These are made by most conveyor manufacturers. Bearings should be adjustable, and screws should be stainless steel. These boots outlast a steel boot three to one. Under normal operation, these boots have lasted almost 10 years.

Screw conveyors are used for the distribution of crushed ice. The trough loading capacity chart of 30% should be used. All parts should be fairly heavy to withstand corrosion. It is good practice to use not less than a 12-in. diameter screw, because if the screw was stopped for a long time and the trough had much ice in it, the ice would congeal and form rather large lumps. Then, when the conveyor was started, these lumps would get past the hangers, at times get caught, and cause a breakdown. Then again, the working parts of a 12-in. screw are larger and stronger to take almost any overload shock. Experience has indicated that the 12-in. diameter screw operates much better than a conveyor of smaller diameter.

14.12 CRUSHED STONE (LIMESTONE—DOLOMITE)

Weight/ft^3: crushed limestone: 85–90 lb; crushed dolomite: 80–100 lb
Angle of repose: 38° for limestone; 30–44° for dolomite
Semiabrasive

Up to a depth of, say, 50 ft, it appears more economical to remove the overburden by scrapers and/or shovels, and truck it to a disposal area. Beyond that depth, at the present time, mining appears more economical. If a mixture of sand or clay is present in the overburden, and it becomes damp or wet, the stone becomes covered with a coating which is very difficult to remove. In view of the rigid requirements for stone in highway and building construction, and the difficulties caused by clay in conveying equipment, provision must be made to scrub the rock or stone.

The rock or stone is brought to the plant in 3- or 4-ft blocks and dumped into hoppers made of heavy structural sections and plates, at times lined with renewable wooden stringers to absorb the shock. The opening in the bottom of the hopper should be made large enough so that the large pieces will not jam in dropping on the apron feeder. The valley angle of the hopper should be at least 48°, so no fines will hang up in it. With this angle, the hopper sides will come close to 55°. The apron-feeder side plates on the supporting steel structure should be made to flare out a little, from the dump hopper opening to the discharge point of the feeder, to prevent the rock binding en route to the primary crusher.

A heavy-duty, slow-operating apron feeder will take the load from the hopper. The chain should be heavy. If a feeder is not provided, the load in the hopper might arch over, and then let go, and flood the elevator or belt conveyor. The openings in the bottoms of the hopper are made large when handling large capacities.

When buying a crusher, it is best to try one or more on actual samples of the rock. Usually a gyratory crusher on a vertical shaft, but sometimes a jaw crusher, is used as the primary crusher to handle the very large pieces, and a jaw crusher as the secondary crusher.

At a typical plant (figure 14.13), the primary crusher will break the material down to, say, −8 in. and, under an apron feeder, will take it to the secondary crusher where it will be reduced to −2½ in. and under, and then it will discharge on belt conveyor no. 1 (inclined not more than 15°). If necessary to make incline as much as 18°, care should be taken to prevent loose rock from falling off conveyor while moving up incline. The head shaft of the conveyor should have a band brake to stop the loaded belt conveyor from running backward in case

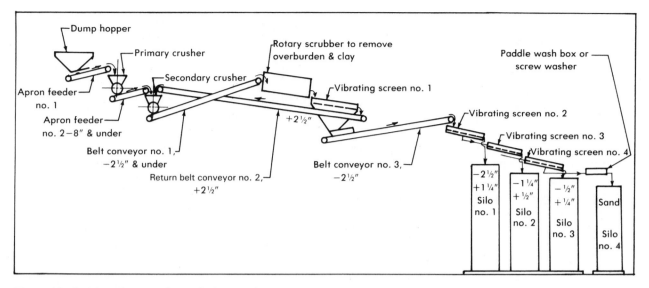

Figure 14.13 Flow diagram of a crushed stone plant.

of power failure. A band brake will allow conveyor to drift backward only about 2–3 ft when power fails, and will automatically assume its safety position when power is restored. If a small amount of overburden sticks to the rock, several return beater idlers should be installed, to help clean off the overburden.

A rotary scrubber is located to receive the rock from belt conveyor no. 1, if it is felt necessary. The scrubber is a heavy affair, 5–6 ft in diameter and 12–15 ft long, with steel partitions lengthwise and crosswise to keep the rock in motion and soaked with water as it passes through the scrubber. The dirty water is bypassed to a pond. From the scrubber, the rock discharges onto vi-

brating screen no. 1, where the oversize drops onto reclaiming belt conveyor no. 2, which returns these large pieces to the secondary crusher for reduction to −2½ in. If larger pieces (over −2½ in.) do get through the secondary crusher, they are fed back on return belt conveyor no. 2.

The −2½-in. and under that passes through vibrating screen no. 1 drops onto belt conveyor no. 3, which discharges onto vibrating screen no. 2 located above silo no. 1, which receives −2½ in. to +1¼ in., from the top deck of vibrating screen no. 2 with −1¼ in. and under discharging onto vibrating screen no. 3. All this is a wet, or water, operation at about 90 psi, to clean

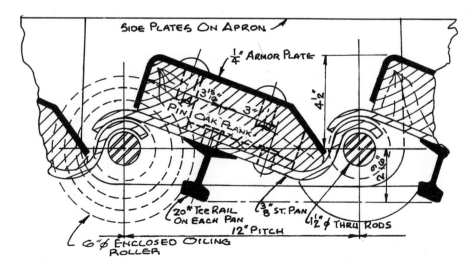

Figure 14.14 A cross-section of a 60-in. wide heavy-duty rock apron feeder.

Figure 14.15 Diagonal spacing of rail.

the rock. The vibrating screens over the silos classify the material for various users.

The so-called rock sand ($-\frac{1}{4}$ in. and under) is not produced in large quantities and can be reclaimed in a wash box or sand screw (silo no. 4).

If the sand has to be screened again, it should be sprayed, as sand must be either completely wet or completely dry to screen properly on a vibrating screen. If the scrubber and vibrating feeder do not remove the clay film, inclined conical screens should be used. These conical screens are mounted on a central shaft turning about 10 rpm, and with water sprays and water under heavy pressure (100 psi), the clay film will be removed. When using screw conveyors, keep load below hangers. Use the 30% chart for capacity.

The steel pans should be at least $\frac{1}{2}$ in. thick. and reinforced with replaceable oak planks to absorb the shock (figure 14.14). On the bottom of each pan is welded continuously a T rail which slides on a $\frac{3}{4}$-in. wear plate that, in turn, is welded to the supporting structure frame. These wear plates are spaced diagonally under the opening of the dump hopper only (figure 14.15).

Figure 5.12 shows the $\frac{1}{4}$-in. clearance between the bottom of the chain roller and the 25-lb. rail. This clearance, at the dump hopper opening, is provided so that when the stone is dumped, the shock would be absorbed by the T rails attached to each pan, with the $\frac{3}{4}$-in. wear plate thus saving the chain joints from shock. The hardened head of the rails attached to the pans will eventually wear enough to require replacement.

The apron feeder side plates on the supporting steel structure should flare out a little from the dump hopper opening to the discharge point of the feeder. This prevents any tendency of the rock to bind on the way to the primary crusher (figure 14.16).

Many elevators are equipped with belts, and others are equipped with chains, because some limestones contain a certain amount of silica, which is abrasive, while other limestones have a high percentage of calcium and very little silica. If much silica is present, the belt is to be preferred. Limestone will produce a cloud of dust in crushing. Dolomite will not.

At the quarry, conveying can be done by belts, which can handle the material, whether wet or dry. At the processing mill, continuous bucket elevators, with their V-shaped troughs and screens, will work well if the material is dry and the percentage of clay or overburden is low.

The speed of belt conveyors should be under 600 fpm for 24-in. belts, and may go up to 1000 fpm for 72 in. and over. Stones crushed to $\frac{3}{8}$ in. and under will travel on a belt, wet or dry, up to a maximum angle of 20°.

A heavy-duty rock feeder is shown on figure 14.17.

14.13 EXPLOSIVE MATERIALS

When handling materials that could cause an explosion, special precautions must be taken in the selection of the elevator parts. The inside of the elevator boot should be lined with 12-gauge replaceable aluminum sheet or sparkproof material. A traction wheel (sprocket without teeth) made of aluminum can be used on the foot shaft. The chain, which can be made of malleable iron, should be equipped with pins made of Everdur bronze. To prevent any possible spark, the malleable iron or steel buckets should be attached to the chain with nonsparking materials, such as bronze bolts and aluminum washers. Plastic buckets, made of reinforced phenolic plastic, are spark- and static-proof. The discharge chute at the head or top of the elevator should be lined with aluminum or other sparkproof material to prevent sparking from the discharge of material from the buckets into the discharge chute. The speed of such elevators should be slowed down to avoid dust. Motors and equipment should be grounded. Sometimes the buckets and the traction wheel are made of cast aluminum. The use of continuous bucket elevators travelling at about 100 fpm

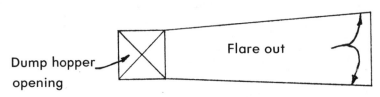

Figure 14.16 Flare of a feeder at a dump hopper.

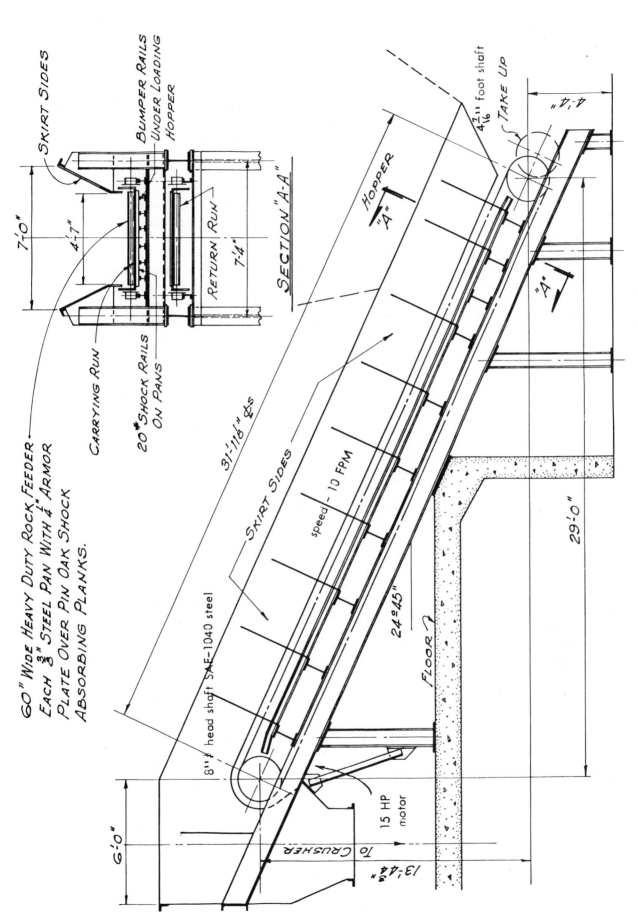

Figure 14.17 A heavy-duty rock feeder.

60" WIDE HEAVY DUTY ROCK FEEDER
EACH 3/8" STEEL PAN WITH 1/4" ARMOR
PLATE OVER PIN OAK SHOCK
ABSORBING PLANKS.

CARRYING RUN

20 # SHOCK RAILS
ON PANS

SKIRT SIDES

BUMPER RAILS
UNDER LOADING
HOPPER

7'-0"

4'-7"

RETURN RUN
7'-4"

SECTION "A-A"

HOPPER

"A"

"A"

4 7/16" foot shaft

TAKE UP

4'-7"

29'-0"

31'-1 1/8" ℄s

SKIRT SIDES

speed - 10 FPM

8" ⌀ head shaft SAE-1040 steel

24° 45'

FLOOR

6'-0"

15 HP motor

TO CRUSHER

13'-4 3/8"

is preferred. A belt elevator with staticproof or conductive-rubber belting will help reduce the hazard of sparking.

The inspection doors can be covered with material similar to muslin that ruptures on a slight increase in pressure within the casing. The top cover is usually equipped with a 6-in. diameter vent stack with weather cap (see figure 14.18).

The width of the elevator casing for explosion hazard should be at least 4 in. wider than standard, for elevators on over 50-ft centers, and 2 in. wider for those on less than 50-ft centers.

A gravity-type takeup (see figure 3.12) with aluminum angle guides will help avoid sparks. The sprocket or traction wheel can be made of cast aluminum, if patterns are available. A traction wheel (without teeth) is most desirable since it can be made of welded aluminum plates, and the absence of teeth helps prevent sparks. The traction wheel runs loose on the fixed steel shaft, and is kept in place by safety collars on each side of the hubs of the traction wheel. The diameter of the hubs is made larger than necessary so that, when in time, the bore of the traction wheel becomes a sloppy fit, the wheel can be bushed with an aluminum bushing in the plant's shop. No lubrication is provided, since the shaft is pinned and does not rotate. The shaft does not extend through the casing. Another reason for using a gravity takeup is that the pit is always clean. Usually, if the elevator is on over 35-ft centers, there is enough weight in the takeup frame to keep it in position but, if necessary, additional weight can be placed on the frame. Sprockets on head and foot shafts should be nonsparking

bronze, case aluminum, or welded aluminum. Suggestions have been made to make the casing dusttight, so that CO_2 can be introduced to kill sparking. In practice, this is very difficult to do. Grounding the casing will help minimize danger from sparks.

Screw conveyors have proved satisfactory since they are slow moving, and allow air to circulate through the material. The breaking up of the lumps is all to the good. On long screw conveyors, say, over 50 ft, it is advisable to break the flighting into several sections for 3 in. or 4 in. to kill any sparks. If the top cover of the screw conveyor is made dusttight, the casing must be ventilated to the outside of the building.

Figures 14.19, 14.20, and 14.21 show an explosive material being processed. The material is brought in drums lightly loaded and is put in the drum conveyor cage. Enough loading positions are located at floor level to avoid lifting and dropping the container, and the conveyor cannot be moved, except from a control house a suitable distance away, and then only after the truck and men delivering the material are out of the area. Figure 14.19 shows the container-dumping device. Figure 14.20 shows the drum holder in detail. Figure 14.21 shows the plastic conveyor moving the material to the incinerator. Note the oversize hopper and the metal detector. Oversize material is dropped through a hinged gate to a hopper, while the metal detector helps remove metallic objects. The velocity of the sludge determined the 4-ft length of the hinged gate. The sludge is introduced just before the drums reach the area shown on figure 14.19. Figure 14.22 shows the electrical control wiring.

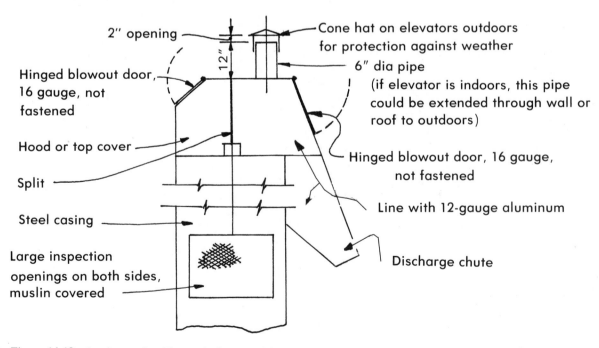

Figure 14.18   An elevator handling explosive material.

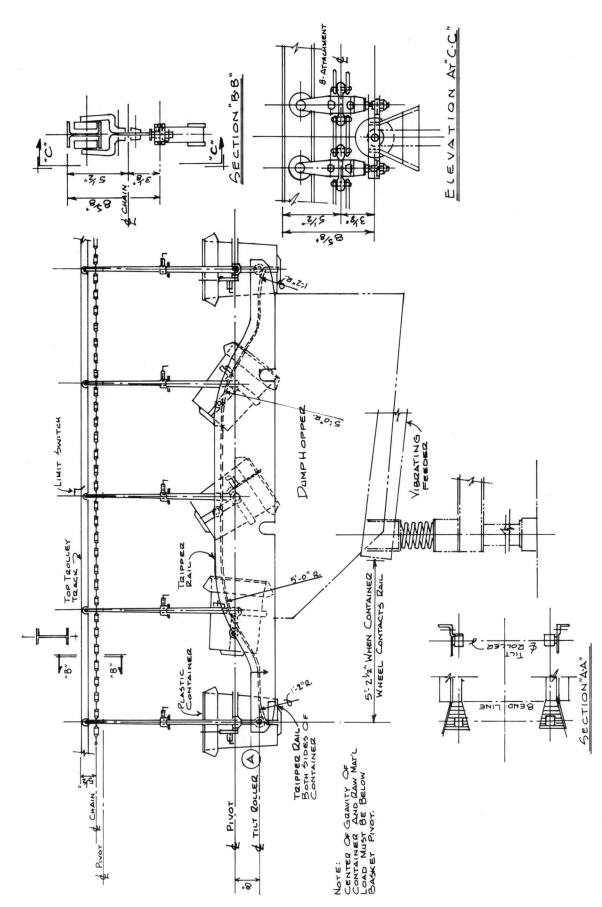

Figure 14.19 A container-dumping guide. For smooth operation, the dumping cycle is based on multiples of 15° tilt for 6 in. of trolley travel; that is, 15°, 30°, 45°, 60°, 75°, 90°, 105°, 120°, 135°, 150°, 165°, 180°.

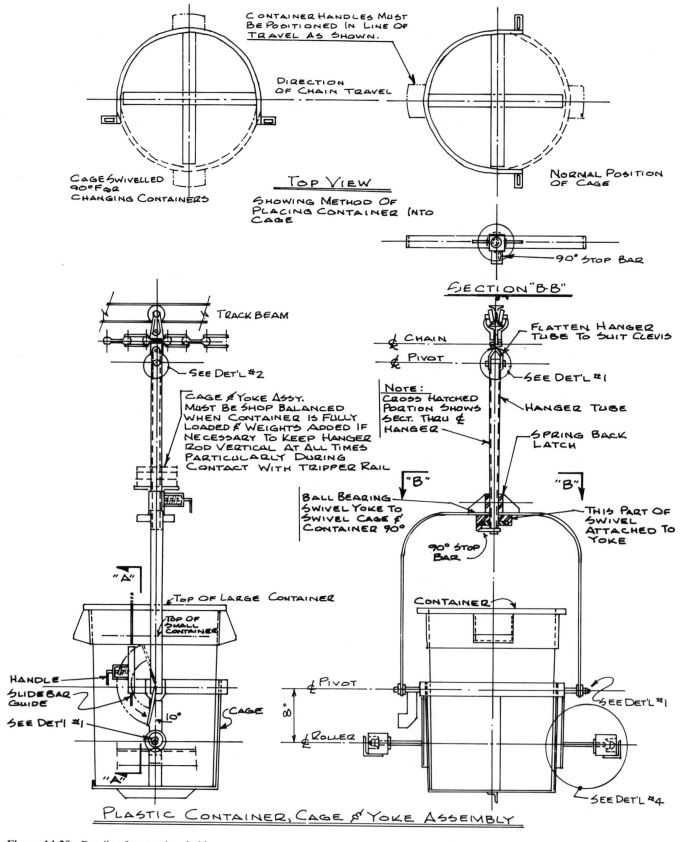

Figure 14.20   Details of a container holder.

FIGURE 14.20 347

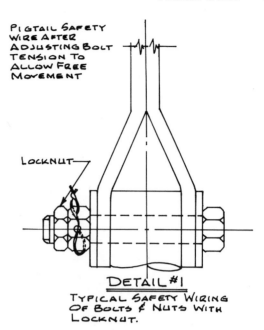

PIGTAIL SAFETY WIRE AFTER ADJUSTING BOLT TENSION TO ALLOW FREE MOVEMENT

LOCKNUT

DETAIL #1

TYPICAL SAFETY WIRING OF BOLTS & NUTS WITH LOCKNUT.

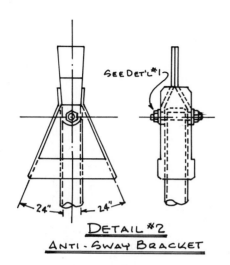

SEE DET'L #1

24°   24°

DETAIL #2

ANTI-SWAY BRACKET

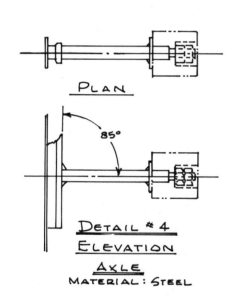

PLAN

85°

DETAIL #4

ELEVATION

AXLE

MATERIAL: STEEL

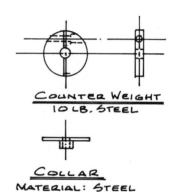

COUNTER WEIGHT

10 LB. STEEL

COLLAR

MATERIAL: STEEL

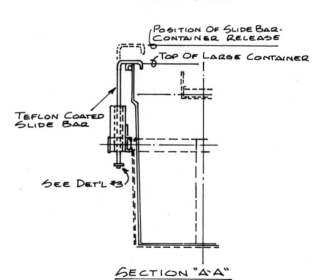

POSITION OF SLIDE BAR-CONTAINER RELEASE

TOP OF LARGE CONTAINER

TEFLON COATED SLIDE BAR

SEE DET'L #3

SECTION "A-A"

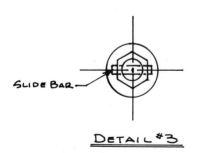

SLIDE BAR

DETAIL #3

Figure 14.20 (*Continued*)

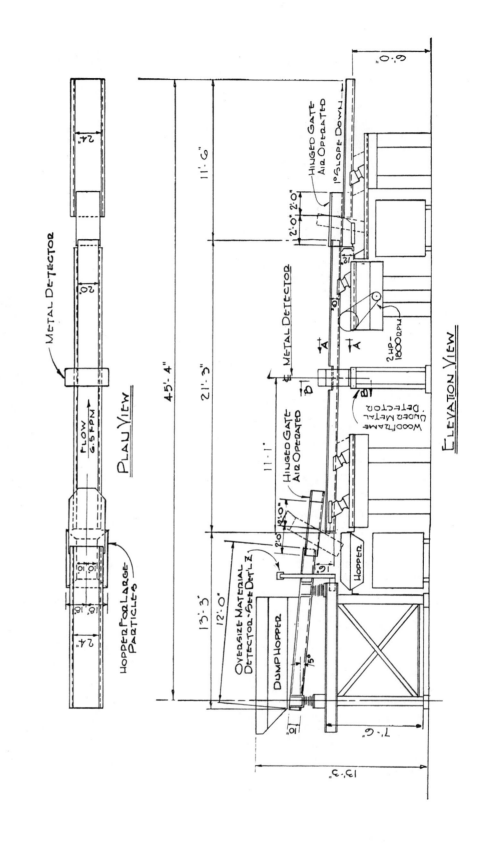

Plan View

Metal Detector

Flow
6.5 FPM

Hopper for Large Particles

24"

20"

18"

10"

10"

45'-4"

Elevation View

Metal Detector

₡ Metal Detector

Hinged Gate
Air Operated

1° Slope Down

2'-0" 2'-0"

Wood Frame
Under Metal
Detector

2 HP-
1800 RPM

Hinged Gate
Air Operated

Oversize Material
Detector - See Det'l. Z

Dump Hopper

Hopper

11'-6"

21'-3"

11'-1"

13'-3"

12'-0"

2'-0"

8'-0"

6'-0"

7'-6"

13'-3"

6"

6"

10"

10"

15°

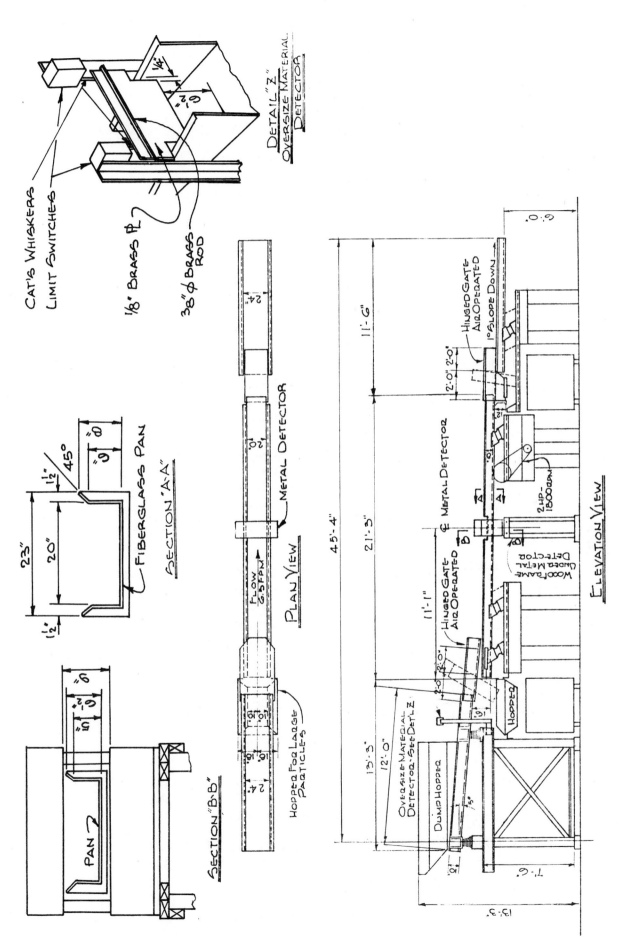

Figure 14.21 A vibrating conveyor for plastic.

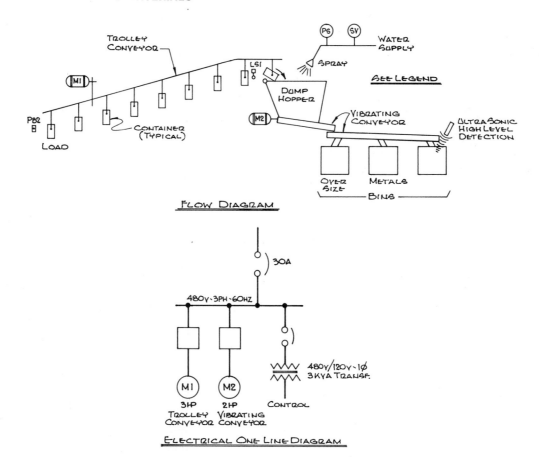

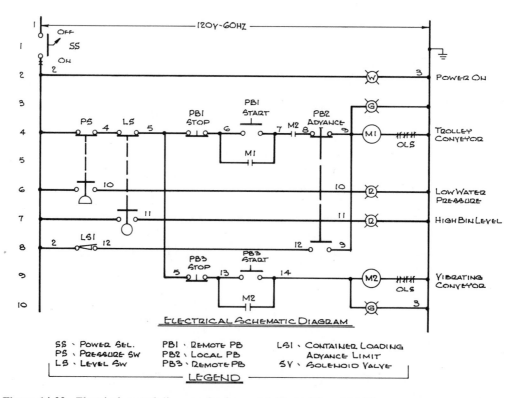

Figure 14.22 Electrical control diagrams for figures 14.19, 14.20, and 14.21.

While the design of the equipment housing is not within the scope of this text, it may be worthwhile to add these notes:

The Underwriters Laboratories, Factory Mutual, National Fire Code, and various municipalities impose standards for construction of buildings subject to explosions. These regulations include the amount of vent space per square foot of building, types of vents, pressures at which vents will open, introduction of inert gases (refer to paragraph 11.10), various controls, and so on. Quick relief of pressure may be obtained by release of window closers or roof and wall vents. It is important to check the weight and type of the closers to make certain they will open when necessary. Pressure-relieving valves may be too slow.

In general, one square foot of vent space is required for every 30 ft^2 of floor space (OSHA). The release area can be obtained by roof vents and/or wall release areas. Roof vents can be either lightweight covers over openings held down by clips, or covers hinged on one side and clamped on the other, that will release when the pressure exceeds a predetermined value. Hinges on one side will prevent the cover from flying loose and doing damage. In areas subject to snow or freezing, the covers should not be insulated, but should be open to a heat source. Wall areas can be made to release pressure by using windows hinged at the top or side, and latched lightly at the other side or bottom, as the case may be, or by steel siding arranged to blow out.

Commercial fasteners for steel siding are available that are designed to fail at predetermined values, so that by taking the tributary area and the desired pressure at failure, the proper fastener can be selected. The fasteners should have washers and be of stainless steel.

One word of caution in the design of ventilation systems of plant laboratories seems worth mentioning. When slightly noxious or toxic fumes are involved, that are not vented to outside of collectors, most laboratory rooms are designed for slightly lower pressures than the corridors or office spaces; when an explosion takes place, the raised pressure in the laboratory will tend to force the gases into the entire building. Controls should be provided to close the vents to the hallways when the pressure is raised.

14.14 FEED, FLOUR, AND GRAIN (SEE EXPLOSIVE MATERIALS)

Weight/ft^3: 40 lb
Angle of Repose: 45°

When grain arrives by ship at a grain storage facility, it is usually unloaded by a marine elevator leg. The ma-

rine elevator is supported in a structure on the dock alongside the storage facility. The marine leg is lowered and raised by a mechanism in the supporting structure for lowering the elevator into the grain in the hold of the ship, and lifting and retracting it up into the structure for movement of the ship.

The ship is equipped with drag scrapers or shovels in the hold or hull of the ship to move the grain to the marine leg. The marine leg discharges onto a belt conveyor which transports the grain to the headhouse or workhouse of the storage facility. It is then elevated by bucket elevator to the upper floor of the headhouse and discharged into garner bins located above a scale. The weighed grain is garnered in holding bins and spouted to a receiving separator where a rough scalp is taken to remove foreign objects before distribution and storage. A magnetic separator usually precedes the separator to remove tramp metal (wire, nails, bolts, etc.). The weighed grain is transferred to the various silos by means of belt conveyors on the bin top floor. Trippers in the belt conveyors discharge the grain into the silo selected by the operator.

Marine elevator legs are constructed to withstand severe conditions both in service and weather. They are constructed in one casing of heavy durable metal. The cups or buckets are usually in double row and properly spaced to give the capacity required. Marine legs are known to have a capacity of up to 60,000 bushels per hour.

The belting must be rubber-coated fabric capable of withstanding moisture, heat, and wear. Because of its remote location and exposure, it is often neglected in periodic maintenance until some damage is done to the buckets or belting. Some rubber manufacturers, such as Goodyear and Goodrich, produce special belting for this purpose (refer to paragraph 2.4.1). Proper belting in these elevators is important since moisture, heat, oil, or fat all have definite effects on the belting.

The bucket elevators vary in construction, depending on capacity required, type of grain handled, and other factors developed by the engineering departments of the various companies from their field experiments. The buckets are constructed of heavy durable metal or plastic and are manufactured by firms such as Screw Conveyor Corporation and others. Marine legs as complete units can be supplied by such firms as Buehler-Miag, Henry Simon, Ltd, and Sprout Waldron, who have furnished this equipment worldwide.

Bucket elevators in a mill or food processing plant are similar to, but much smaller in size and lighter in contruction, than those used for raw grain handling. There are several methods to eliminate or greatly reduce contamination: using a floating boot airswept; a fixed boot with wing-type pulley and curved bottom plate, which then would have the takeups in the head; or a conventional boot selection with wing-type tail pulley and cleanout on both sides, wherein the boot can be cleaned out periodically (see figure 14.23).

The type of cup or bucket is selected for the particular product to be handled and the capacity required. There are several types of buckets used specifically for fine material versus granular product. Belt speeds in these cases are critical, to prevent blowing. Too high a belt speed impedes discharge at the elevator head; too low reduces capacity.

Inspection doors are installed on both up and down legs for periodic inspection and servicing when necessary. The buckets occasionally will become loose because of some obstruction foreign to the product which may hinder operation or cause damage, sometimes of a serious nature. Usually, a telltale knocking in the casing or the leg will indicate some minor difficulty. This can be remedied by heeding the warning before serious damage occurs. Bucket elevators in the mill have aspiration or dust control on both the boot and the head at the discharge at the elevator. An inspection door should be provided on the side of the head to allow close inspection of the head pulley, which often is a source of infestation.

Pulleys are up to 7 ft in diameter, depending on centrifugal speed. Speed depends on the rpm of the shaft and how the grain is thrown out. The speed of travel is about 600 fpm. With this type of elevator, no cobwebs can accumulate in the casing. With a single-leg elevator, there would be ample room for this to happen. Double-leg elevators cost about 25–30% more than single-leg units.

Material is often delivered in hopper-bottom cars that are arranged for pneumatic unloading or loading onto mechanical conveyors and, in some cases, for both types of loading. Pneumatic conveying offers the advantage of protection against contamination, and ease of delivery to storage silos and, through diverting valves, directly to other areas. Controls of the operation, including weighing and packing, are available. The extent to which automatic devices can be justified depends primarily on the capacities involved. The more automated a system, the more expensive it is in original cost, breakdowns, and maintenance (refer to paragraph 18.4).

Dust control is a must at every discharge point of an elevator or conveyor and on the receiving separator in the headhouse. This is critical to prevent dust explosions, to reduce air pollution, and to safeguard working conditions for the operators. Sufficient data and engineering information is available from various manufacturers to size a marine leg properly for the capacity required. Data is available for bucket size, type, and spacing; belt width, thickness of belt, type of construction and tensile strength, and speed of belt; pulley sizes; power required; and casing fabrication, to enable any manufacturer of this type of equipment to tailor make it to specific requirements.

The Salem Steel elevator bucket, made by most conveyor manufacturers, has been used for many years, handling all kinds of feeds and grains. The Screw Conveyor Corporation has developed the Nu-Hy grain bucket from research on grain elevators, claiming that the shape and contour works with the natural grain flow in an elevator. Perforated and vented buckets are especially designed for the handling of flour, bran, shorts, and other soft stocks. They also handle cottonseed efficiently, one of the most difficult materials. The perforations and vents eliminate vacuum suction between the bucket walls and contents at discharge. This permits buckets to discharge completely since the material will not cling. The perforations and vents enable the buckets to scoop up a full load with the least amount of puffing or blowing. Air currents in the leg are reduced, eliminating suction of stock down the back leg. Perforated buckets have $5/32$-in. or $1/4$-in. holes in body and ends for handling mill feeds, bran, small grain, and so on. The buckets, perforated with $1/4$ in. $\times$ 1 in. slotted holes in body and ends, are used for handling flour and similar soft stocks. This grain elevator bucket offers features that permit continuous spacing and clean discharge with no side spill. It has a wide bottom and high-positioned front lip, together with high, sweeping sides raised above the strike line for increased capacity. The shape of this bucket with the proper contour brings the load up over the head pulley without premature discharge.

Flour and feed mills producing edible products are equipped with electromagnets to remove any foreign materials, such as steel buckets breaking off, bolts, nuts, and nails.

Handling flour in bucket elevators is difficult unless the bottoms of the buckets have four or five $1/4$-in. diameter holes, and the sides of the buckets have $1/2$ in. diameter holes near the top. These holes break the suc-

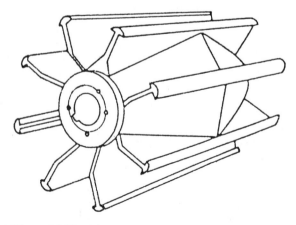

Figure 14.23 A lagged wing pulley.

tion created by the centrifugal force of the buckets picking up the load in the boot. Without these holes, considerable flour stays in the bucket and falls down to the boot, thus overloading the elevator. With these holes in the bucket, there is no apparent loss of capacity.

These high-speed type elevators should be equipped with explosion-proof doors and vents, and piping leading to outside of building. High-speed belt conveyors operating at 500–600 fpm on top of silos equipped with trippers, should be vented to outside when stationary over a silo. The vent hose is arranged to move with the tripper. The tripper shafts operate on shafts supported by fully enclosed antifriction bearings. The inside of the chutes and housing should be lined with light-gauge copper plate.

Flour can be stored in concrete bins if provision is made to avoid condensation. The flour comes to the bin at 90°F, and no problem should be encountered if the bins are placed inside a heated building, or double walls with an airspace are provided. Air can be circulated in the airspace.

14.15 FOUNDRY SAND—GREEN SAND MOLDING

Weight/ft^3: 100 lb
Angle of repose: 40°

When dry foundry sand is stored in large diameter bins of several hundred tons capacity, the diameter of the bottom opening can be made somewhat smaller than the diameter of the bin, if the side is equipped with compressed air jets at 90 lb pressure. These jets go on and off automatically to break any so-called vacuum created between the sand particles and the inside of the bin. If these are not installed, the sand will arch and hang on bin plates, forming a hard crust that tends to build up. Directional vanes can be installed in the bin to prevent segregation of particle sizes.

Small storage bins over a molder are emptied so fast that the prepared sand does not get a chance to arch or cling to the sides. The sides of the bins, usually rectangular or square, flare out at the bottom a little to prevent arching.

Newly bonded or reconditioned molding sand usually contains 3–5% moisture and, together with binder and sea coal (carbon black), the sand is very likely to be on the sluggish side enough so it is not free-flowing. Normally a large rotary table feeder, anywhere from 5 ft to 8 ft in diameter is fitted at the bottom of the bin so as to move the sand above it. This tends to prevent clogging of the sand at this point. Using a simple duplex undercut gate (see figure 13.1e), the opening in the bottom of the bin would be too small to get the sand out of

the bin. The sand would arch and get hard, and become difficult to clean out. Storage bins should have a large opening in the bottom and use a rotary table feeder if possible. Small storage hoppers over molding machines have been troublesome at times, when made with vertical sides. The sand tends to hang up on the vertical sides, but this is usually overcome by tapering the sides out slightly and adding a small vibrator (figure 14.24).

Bucket elevators in a foundry are usually of the centrifugal discharge type, with the malleable iron buckets mounted on a belt. The elevator may be called on to handle hot reclaimed sand, so a heat-resistant plylon belt manufactured by the Goodyear Tire and Rubber Company can be used. The rubber-belt manufacturers have also put on the market a rubber-covered belt that has given good results.

On shakeout belt conveyors handling very hot sand and sprues from the castings, a heat-resistant, canvas-impregnated belt is used. Here again, the rubber-belt manufacturers offer a rubber-covered belt to withstand high temperatures. When the sprues are knocked off the castings and fall through a grating at floor level to the belt conveyor below, they could be glowing red hot. They lose their heat rapidly but, even at that, they may cause a slight smudge on the belts, whether Sahara brand or rubber belt is used, but the sprues will not set anything on fire, just smoke a little for a short time. Even a thin layer of hot shakeout sand on the belt will act as a protective insulator and an oxygen barrier.

Usually at the discharge end of the shakeout conveyor, an overhead stationary magnetic pulley is used to separate the sprues from the sand, preventing the

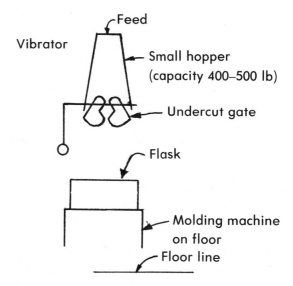

Figure 14.24 A foundry sand hopper over a molding machine. Use an air or electric vibrator, interlocked to vibrate only when the hopper gate is open.

sprues from getting in the bucket elevator. To prevent the electromagnetic field from getting into the supporting steel around the pulley, it may be necessary to line some of the steel of chutes with a nonmagnetic material like sheet aluminum. This is important, as the magnetic field is considerably reduced at this point, and the magnetic pulley will lose its effectiveness to remove any iron.

In large production foundries, the conventional shakeout belt conveyor has been replaced by an oscillating (vibrating) conveyor, heavy-duty type. When hot sands and hot sprues are dropped onto this unit, the vibrating feature keeps the sand and sprues moving forward; thus, a good amount of cooling is accomplished. These oscillating conveyors are set in a pit below the shakeout station, and are very clean in their operation, requiring almost no cleanup in the pit, saving labor at this location in the overall picture. They fit in a foundation trench, and the trench or pit requires very little cleanup of spilled sand. The belt conveyor in the pit is supported on steelwork, with very little clearance provided under the return belt run for a person to clean up spilled sand. Hence, in many cases, the sand piles up and spreads out, making a mess in the pit. The speed of belt conveyors is about 350 fpm. (Refer to paragraph 6.7 for use of drag conveyor in foundry.)

Molding sand contains enough clay to move on an incline of up to 24°, if the moisture content is under 4%.

14.16 GRAPHITE

Graphite is found in nature, often mixed with impurities. Broken graphite from pits, weighing 95–100 lb/ft^3 is considered very abrasive, like any rock. Graphite flake weighing 40 lb/ft^3 is granular, usually ½ in. and under, and is considered nonabrasive. The angle of repose of graphite fines is 32.1°. It varies from 29.4° to 34.9°.

Graphite flour (100 mesh) is very fine, weighs 28 lb/ft^3, is considered nonabrasive, aerates, and becomes fluid. For certain uses, such as foundry facings, pencil graphite, paints, and stove polishes, at least 99% must pass the 200 mesh screen. The Hardinge ball mill was found more efficient for producing foundry facings. With very soft material, where a very fine-grained product is required, horizontal Rock Emery mills were found more efficient. In both cases, the products are ground dry. For sizing the finished product, Raymond Mills systems are in use.

In the processing mill, centrifugal units with rounded bottom buckets, or continuous bucket elevators with standard V-shaped buckets can be used. Even if wet or damp, this rock is sufficiently slippery to prevent buildup

in V-shaped buckets, and usually empties the buckets clean unless, of course, there is overburden in the material.

Production of finished graphite is usually inside a mill or building where it is dry. It does not seem to make any difference if the building is heated or cold, as long as it is dry. Finished graphite handles very easily in a centrifugal discharge elevator or continuous bucket elevator, due to the slippery nature of the graphite. The continuous bucket elevator, operating at about 125–150 fpm to reduce any dusting conditions, can be used if desired.

If a centrifugal discharge elevator is used, with the speed of chain reduced to about 185 fpm and malleable iron style A or AA round-bottom buckets spaced very closely together, the actual capacity is as much as 40% over the rated capacities of standard centrifugal bucket elevators, as listed in manufacturer's catalogs.

The buckets usually empty fairly clean, unless a percentage of clayey type overburden gets mixed in with the rock. If this happens, a buildup will result in time in the bottom of the buckets, but the buckets can be hosed out with water.

Screw conveyors handle graphite rock or finished product well, if you keep the loading capacity down to 50%. Vibrating conveyors should not be used if the material passes through a 10-mesh screen, as the capacity is considerably reduced for finer mesh. The graphite rock as mined may contain a larger percentage of fines and could be damp or wet, depending on the weather.

14.17 GYPSUM

Weight/ft^3: 100 lb ($-1''$)
            150 lb (large, solid blocks)
            180 lb (anhydrite in large, solid blocks)
Angle of repose: 30–40° (in lumps)
Abrasive in rock form (¼ in. and up); and ¼ in. to dust; not abrasive

This industry uses chain on bucket elevators instead of belts, and their maintenance costs seem to indicate chain is better to use throughout a gypsum plant.

In many locations, gypsum is deposited 20–40 feet below the surface. After the overburden is removed, it is mined in pieces up to 3 × 3 × 5 ft and down to small pieces; then trucked to a primary crusher, where it is reduced to -8 in. to dust. If clay or shale stick to the gypsum in wet weather, the handling of the larger pieces becomes difficult. Secondary crushing is usually required. The material is then fed to a rotary kiln dryer.

Belt conveyors for transporting the gypsum from the mines to outdoor storage areas work well, although care must be exercised to keep the return of the belt fairly clean with double scrapers and water jets. Trippers on these belt conveyors must be made with extra clearance where the head or top pulley discharges to the chute to prevent a jam at this location. Chutes must have at least 60° slope. From outdoor storage, the rock is handled by belt conveyors through the reclaiming tunnels, to inclined belt conveyors, to secondary crushing and screening, and to the mill.

Bucket elevators are not used in handling the rock if any amount of clay is present, since the continuous type buckets or type A centrifugal discharge buckets tend to fill up with fines, eventually getting hard as rock.

In the mill, the material is calcined (reduced to a powder and cooked). The dryer revolves slowly and tumbles the rock at a temperature of about 150°F. This removes the surface moisture that might cause clogging of the crushing and grinding equipment. If, during calcining, the gypsum is heated to a point where the water of crystallization is below 0.5% down to bone dry, the gypsum changes back to anhydrite, which cannot be processed and must be discarded. The crushing is done in a hammermill that feeds onto the screen. The final grinding is done in roller mills. Before calcining, gypsum contains 20.93% of chemically combined water. The final product is stored in suspension bins or silos.

For handling "gypsum rock" −4 in. to dust (free from clay or mud), a super-capacity continuous elevator can be used, having 18-in. wide × 8¾-in. projection × 11⅝-in. deep, ¼-in. steel fully welded buckets on a double strand of 12-in. pitch strap steel chain, operating at about 90–100 fpm. This elevator should be equipped with a gravity takeup in the boot to keep the chains tight at all times. Usually the weight of the steel parts and machinery parts of a gravity takeup (figure 3.12) is sufficient to hold the elevator chains tight, although additional counterweight can be added. This takeup rides loosely up and down in steel guides inside the steel casing and always compensates for any slight stretch in the chain, or temperature changes that would affect the steel parts of an elevator. The gravity takeup shaft does not go through the casing, is stationary, and the traction wheels run loose on the shaft with no openings in the boot, to avoid any spill and keep pits clean. If necessary to place screw-type takeup on head shaft, the centers of bucket elevator should not exceed 90 ft, because the total weight of chain (or belt) plus buckets and load in buckets, on up or carrying side, is hanging on the takeup screws in tension, and this is not good engineering practice. The use of traction wheels on the head shaft has not gained favor in the industry, but it is used on the foot shafts.

Some machinery manufacturers use shorter pitch steel elevator chains, with bucket attachments spaced every other pitch to replace the long pitch chains, which do have a tendency to cause vibration in an elevator when passing over the head sprocket.

Head sprockets have also been designed with an odd number of teeth, with the idea that the same point of chain contact with sprocket does not come in the same place on the sprocket. In a slow-moving (like a bucket) elevator, with the sprockets on head shaft turning very slowly, this does not seem worthwhile. The shorter pitch steel elevator chains instead of the long pitch, say 9-in. pitch instead of 18-in. pitch, may have some merit on these elevators. They should be tried out over a period of time, however, to see if excessive maintenance occurs, as there are twice as many chain joints in the short-pitch chains.

Land plaster, stucco moulding, gaging and dust are handled successfully in centrifugal discharge elevators. Material like stucco (gypsum plaster) is very fine, minus 140 mesh, and may be hot or cold when placed in a storage bin. It weighs about 55 lb/ft^3. A single strand of either C-188, C-102B, C-102 or C-110 combination malleable iron and steel chain, or all-steel chain, can be used with malleable iron style A buckets spaced close together as for a continuous bucket, and at a speed of about 185 fpm.

Hot gypsum, above 150°F, flows like water, has no angle of repose, and experiments show these elevators give about 50% more capacity than the advertised standards in manufacturers' catalogs. The discharge chutes at the top of the elevators are lowered about 1 ft over standard elevators to give the close spaced bucket a chance to discharge (see figure 14.25).

When figuring capacity of these elevators, say minus 10 mesh, use 60% capacity of a bucket, 70% from plus 10 mesh to minus 1, and 75% from +1 in. to −4-in. material. The use of continuous bucket elevators to handle calcined gypsum is not advisable because of so much carryover down the casing on the return side.

Gypsum can also be transported after calcining by using air-activated gravity slide conveyors. Air requirements are generally in ounces of pressure and relatively low cfm, resulting in low horsepower for work performed. The air passes through a fluidizing membrane, causing the gypsum to be suspended and flow like water at inclines of 8°.

Both hot and cold calcined gypsum have a tendency to stick to steel, and in discharging from one continuous bucket to the back of the forward bucket, a whirling action takes place, actually building up a bead of gypsum on the back of the forward bucket, and preventing the gypsum from sliding off the back of the forward bucket to the discharge chute, thus carrying a considerable amount of gypsum down the elevator casing to the boot. Figure 14.6 shows why continuous bucket el-

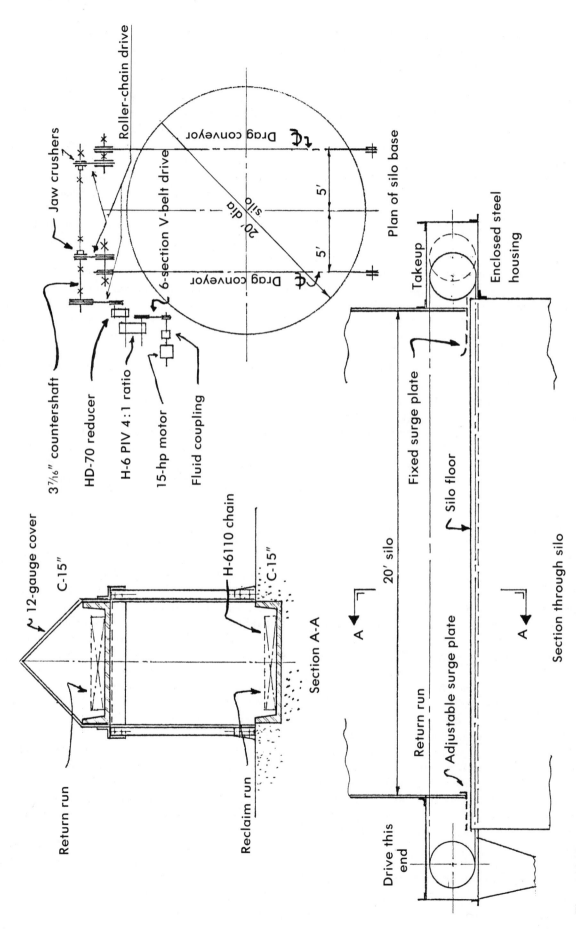

Figure 14.25 Layout of two drag conveyors in the flat bottom of a silo. (Gypsum temperature is 180°F.)

evators have been discarded in favor of slower running centrifugal elevators using style A round-bottom buckets, spaced close together.

Storage of gypsum and cement give rise to about the same problems and solutions (refer to paragraph 14.6).

When reclaiming stucco from a bin, the carrying or reclaiming run is located on the bottom of the bin. The return run can be located under the bin and hung from it, as shown on figure 14.7, or supported on a frame resting on the bottom of the bin and protected by a continuous inverted V-shaped plate cover, as shown on figure 14.25. This figure shows two H-6110 drag chains located 5 ft each side of the silo center line. Each silo is equipped with a single strand of chain operating at a variable speed of 36–143 fpm, resulting in capacities of 18–30 tons per hour of hot stucco, 100 mesh, weighing 60 lb/ft^3 at a temperature of 180°F. A 15-hp motor is arranged to drive the two conveyors, operating one at a time.

A 60-in. wide apron feeder was used to convey mined gypsum rock with maximum sized pieces measuring 36 in. (cubes) and weighing about 2,000 lb. An average size is 15 in. and under, weighing 100 lb/ft^3. Design capacity to crusher is 100 tons per hour. This apron feeder must always be stopped before dump hopper is completely empty, to assure a bed of rock on apron to cushion the impact of the next truck or car load. It is much better to have apron at rest when dumping into hopper, to allow the large pieces to find their place and settle. Otherwise, large pieces could get into such a position and wedge together if apron is operating, causing possible jamming against feeder skirt sides.

14.18 LEAD AND ZINC ORES

Lead ore
   Weight/ft^3: 200–270 lb
   Angle of repose: 30°
   Abrasive, mildly corrosive
Zinc ore
   Weight/ft^3: 160 lb (3 in. and under)
   Angle of repose: 38°
   Abrasive
Zinc oxide (heavy type)
   Weight/ft^3: 30–35 lb
   Angle of repose: 45–55°
   Nonabrasive
Zinc oxide (light type)
   Weight/ft^3: 10–15 lb
   Angle of repose: 45°
   Nonabrasive

Zinc oxide (heavy type) is very fine, minus 100 mesh. Zinc oxide (light type) is fluffy, minus 100 mesh, and

can be handled in a centrifugal discharge bucket elevator with round-bottom, malleable iron buckets spaced close together and operating at a chain speed of 185 fpm. Continuous bucket elevators should not be used, as this material packs under pressure, and will not discharge properly out of V-shaped buckets. Screw conveyors can handle the fine ore (minus 10 mesh) if loaded only to 15% (in an emergency, to 30%) of capacity, and will work well with little maintenance. The ore should be kept below the hanger bearings. Vibrating conveyors can usually handle the ore larger than 10 mesh.

Belt conveyors are used whenever possible in handling these ores from the quarry and in the preparation plant, since the belts can be cleaned of any fine dust, wet or dry, on the return run, with double-edged stationary rubber-edged scrapers and, if necessary, with high-pressure water sprays.

Graded ore, say from ¼ in. and up, can be handled on continuous bucket elevators operating at 125–150 fpm. The ore must be dry and free flowing.

14.19 LIME

Weight/ft^3: 50–60 lb (hot, burnt)
Angle of repose: –40°

When calcium is burned in a kiln at about 3200°F, it comes out at the discharge end with a low glow. It is hot and difficult to handle. Commercially built bucket elevators with malleable iron or steel buckets will not keep their shape very long under this heat.

Hot lime can best be handled in a pivoted bucket carrier, with the cross rod supporting the bucket and covered by a loose-fitting pipe, to allow air to circulate between rod and pipe. By operating at a maximum speed of 40 fpm, and loading each bucket with only about 2–3 in., a length of run can be used that will cool the material in transit, and avoid distortion of buckets. This unit has proved highly successful, operating 24 hours a day, stopping only for monthly inspections. Maintenance is extremely low, and a very low operating cost per ton of lime handled results. The chain should be malleable iron with ample wide links. Use links with ⅞-in. wide bearings on the chain bushing and 1⅛-in. heat-treated nickel chromium pin, alemite lubricated through pin, to case-hardened bushing of single-flange flint rim rollers. Strap steel links may require replacement after a while, because of elongation of chain pin holes caused by pulsating action in going around the drive sprocket.

Cold lime can be handled in bucket elevators, with buckets spaced close together. The speed should be about 75 fpm. No-leak apron or skip hoists can be used, although they are expensive.

In the production of hydrated lime, flat steel apron

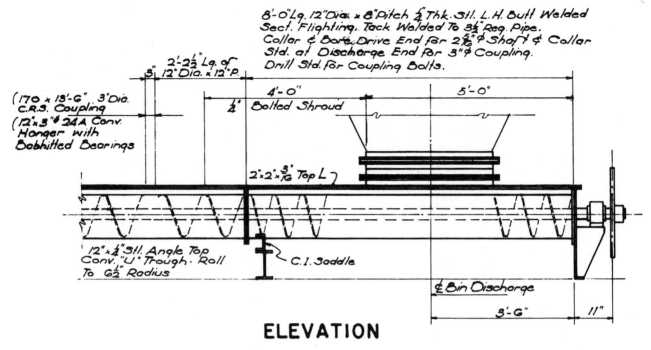

8'-0"Lg. 12"Dia × 8"Pitch ¼"Thk. Stl. L.H. Butt Welded
Sect. Flighting. Tack Welded To 3½" Reg. Pipe.
Collar & Bore. Drive End for 2⅝"φ Shaft & Collar
Std. at Discharge End for 3"φ Coupling.
Drill Std. for Coupling Bolts.

ELEVATION

Figure 14.26 Details of a conveyor for lime hydrate.

pans with curved ends at chain joints work well. The lime may be cold or hot (maximum 450°F). See figure 14.26 for details of a lime hydrate screw conveyor.

Sideplates are usually 3–4 in. high. No. 1130 malleable iron chain (6-in. pitch, 3750 lb allowable pull) are attached to the pans outside of the sideplates. The apron does not have to be leakproof type, the offset end plates are welded to the pans, and the two strands of steel chain are attached underneath the ends of the pans, where they are protected from grit and abrasive materials which may spill over the conveyor sides. The through rods stiffen the conveyor, and at the ends of these rods are placed single flanged chilled tread rollers to carry apron on angle or T-track.

14.19.1 Apron Conveyor Design

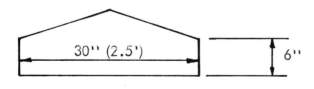

The capacity can be determined by

$$2.5 \text{ ft} \times 0.5 \text{ ft} \times 0.75 = 0.9375 \text{ ft}^2$$

where the value 0.75 allows 25% reduction of the cross section area because of lumps. Then,

$$0.9375 \text{ ft}^2 \times 1 \text{ ft} = 0.9375 \text{ ft}^3 \text{ per ft of feeder}$$

$$\text{at } 32 \text{ lb/ft}^3 = 30 \text{ lb of lime}$$

where 1 ft is the length of the pan. The cross-section shows capacity peaked above the sides of the apron. While moving, the apron often settles the material to almost level, while filling in the voids between lumps.

To determine tons per hour,

$$30 \text{ lb} \times 50 \text{ fpm} \times 60 = 90,000 \text{ lb} = 45 \text{ tph}$$

where 30 lb is the amount of lime per foot of feeder.

14.19.2 Horsepower at Head Shaft

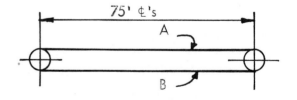

First, determine the total weight per ft of conveyor by adding the load of lime on the feeder per ft to the approximate weight of the apron with 3/16-in. steel pans, 6-in. high sides of 10-gauge steel and two strands of no. 1130 roller chain with all attachments, respectively: 30 lb + 55 lb = 85 lb.

Then determine carrying run A by

75 ft × 85 lb × 0.20 = 1275 lb

where the value 0.20 is the one-fifth friction factor for roller chain. Figure return run B by

75 ft × 55 lb × 0.20 = 825 lb

(55 lb is used because the feed return run is empty). Adding the two results gives total pull on the head shaft: 1275 lb + 825 lb = 2100 lb.

The allowable working strength of no. 1130 chain is 3750 lb per strand (the ultimate strength is 28,000 lb) and therefore the chain is safe.

To determine theoretical horsepower:

$$\frac{2100 \text{ lb} \times 50 \text{ fpm}}{33,000} = 3.2 \text{ hp}$$

Taking into consideration about 10% friction for chain drive and 15% friction for a speed reducer, a 5-hp motor is ample to drive this feeder.

14.20 LOGS

Paragraphs 6.11 and 6.12 refer to drag and scraper flight conveyors in pulp and paper mills. It should be noted that the standard method of handling logs is as described in paragraph 6.12. Figure 6.12 shows a method of unloading logs for storage by using two strands of drag chain. Similar details would apply for a single-strand chain.

The general method is by a single-strand of drag chain only. For very large capacities, two strands of chain operating as one drag conveyor are used. The steel trough necessary to stand the abuse of the shock of logs rolling down to the conveyor chain requires heavy construction. In order to avoid the logs building up between the skirt sides, the width between the skirt sides should not be close to the length of the logs. The chain should be heavy duty. Some manufacturers use manganese steel with manganese steel attachments. Some Canadian mills use belt conveyors, but they have not been generally accepted by the southern mills.

Mr. Richard Ludwig, Hammermill Paper Company, writes that pulpwood lengths vary considerably. In the Lake States (Michigan, Wisconsin, and Minnesota), 100 in. is the standard. In the south, rail pulpwood is normally 5 ft 2 in. long, with truck pulpwood in 8–9-ft lengths. The trend in the industry is to long logs (18 ft up to tree length), which are either "merchandized" to veneer and/or lumber lengths or slashed to 5–8 ft lengths.

Mr. Ludwig prefers "that no idlers be located under pulpwood transfer impact points, but to have the belt absorb the impact of the logs. Turning the logs at transfer point creates bad problems of jamming. Instead, logs are aligned on one belt, and then sort of rolled over the

end of that conveyor pulley onto the next belt, without changing the orientation of the logs." This will reduce the height of the fall considerably.

If the transfer point must be located where a change of direction is required, the design of the transfer chute will depend on the angle between the two conveyors. For a 90° turn, the height between the two conveyors for a 4-ft log would have to be at least 10 ft, to prevent any log from stopping in the chute. A cord of wood is 4 × 4 × 8 ft = 128 ft³ (with average 8-in. diameter logs a cord = 64 logs). When these logs go through the chipper the chips, because they are green and have considerable moisture, could weigh up to 22 lb/ft³, but when dry, may weigh only 10 lb/ft³.

The rubber-belt manufacturers produce a tough-carcass belt with a top-rubber cover that will stand the rough handling of logs when discharged from a ship or railroad car, even the plowing off the belt where the logs discharge to the barking drums. The rough edges of a sawed log are too sharp for any material to stand up day after day with the constant pounding.

The accepted practice in the design of loading chutes for belt conveyors is to locate the chute on the side of the belt and roll the logs over lengthwise. The logs should not hit the surface of the belt with the rough, sawed edges of the logs.

14.21 METALLURGICAL COKE

Weight/ft³: 35 lb*
Angle of repose: 30–40°
Very abrasive

Metallurgical coke is received by truck or hopper-bottom car, and is unloaded into a receiving hopper that discharges onto a belt conveyor, vibratory feeder, or into a helical screw conveyor that feeds a bucket elevator. The coke is discharged from the bucket elevator into a surge bin that feeds a screen to separate the coke into fines, required screened size and tailings, or oversized pieces. The screened coke and fines are discharged into their respective storage silos. The tailings are discharged into a double-roll type crusher, such as a Gundlach, Pennsylvania, or Allis Chalmers, recrushed and recycled back into the receiving bucket elevator, and rescreened along with the incoming coke.

Since the coke is generally received wet, because of previous ground bulk storage or inclement weather in transit, in those northern areas where the screened coke and fines will freeze in the storage silos, a coke drying system is generally installed between the receiving

_____
*The weight of metallurgical coke varies with the source and the weather.

bucket elevator and screen, to remove moisture. Figure 16.2 shows an FMC dryer that works well on this coke. The extra heat in the exhaust can be used for heating the discharge of the elevator in cold weather areas. Space is always at a premium, and some can be saved by putting the motor and drive under the rotary dryer.

The receiving bucket elevator should have additional capacity, as much as 25%, to allow for the recrushed oversized tailings. The speed of the belt conveyors and bucket elevators should be about 125 fpm. The bucket elevator should be a continuous type. The buckets should be nylon or other abrasion-resistant material. The belt should be a 4-ply synthetic carcass. The receiving hoppers should be lined. For want of better data on the particular coke used, assume 0.265 Btu/lb/°F specific heat, and maximum moisture content of 18%, when designing drying equipment for the coke. These values are considered on the high side, but are possible in areas such as the northeastern section of the U.S.

14.22 PENCIL PITCH

Weight/ft^3: 50 lb
Nonabrasive

Pencil pitch is lumpy and free flowing. It is mildly corrosive and degradable. The particles are $1\frac{1}{4} \times \frac{1}{4}$ in. Discharging any fine or lumpy material from a hopper-bottom car to a screw conveyor requires a screw feeder. If a standard-pitch screw or even a short-pitch screw is used under the hopper, the load is likely to plug the conveyor, and it may be necessary to dig the material out by hand to free the conveyor. Using a screw feeder where the screw section is short-pitched and tapered, together with a shroud plate, works well. The combination of the tapering and short pitch acts just the opposite as when using a wood screw. It gradually pulls the load along and does not get jammed from an overload.

14.23 PERLITE AND EXPANDED PERLITE

Weight/ft^3: 72–75 lb (some, 80 lb)
            8–10 lb (expanded)
Angle of repose: 40°
Very abrasive in ore state

Commercial perlite comes about $\frac{1}{32}$ in. in diameter. Most of it will pass 20-mesh cloth. Perlites that weigh 80 lb/ft^3 in the ore state are used with gypsum in the manufacture of wallboard.

Perlite is shipped in closed, drop-bottom railroad cars and is dry and free flowing (see figure 14.27). In the process, the small spheres of perlite are fed to a rotary dryer, where most of the moisture is driven off, resulting in considerable expansion of the perlite.

A centrifugal discharge elevator will handle this light product if the speed of the chain is reduced to about 185 fpm, and round bottom buckets, style A or AA, spaced close together, are used. The discharge chute should be lowered to 18 in. as shown on figure 14.6*, to give the buckets a good chance to discharge properly. Screw conveyors will also handle this material without difficulty. With screw conveyors, the 30% capacity values should be used to arrive at a resonable capacity per hour.

Pneumatic conveying is commonly used for transporting perlite after expansion. The perlite is introduced into the blower discharge pipe through an eductor receiving the perlite from a feed hopper. Long-radius elbows are recommended because of abrasion and possible plugging with changes in velocity.

On account of the lightness of the expanded perlite, using a vibrating (oscillating) conveyor involves risk. Some particles of this expanded perlite will move on a vibrating conveyor, while others will hesitate and pile up in the trough.

14.24 PETROLEUM COKE

Weight/ft^3: 48–52 lb
Angle of repose: 30–40°
Abrasive

Petroleum coke is shipped in closed freight cars and is generally unloaded through a track hopper to a belt conveyor or an apron feeder that, in turn, feeds a bucket elevator. The apron feeder should handle about 75 tons per hour to unload a car in one hour. Travel speed for apron conveyor should be about 20 fpm. The coke comes in sizes −2 in. to dust and is generally free flowing in chutes. If it is delivered in larger sizes, and if a centrifugal discharge elevator is used to raise the material, it should be put through a crusher. A crusher with teeth similar to the Gundlach operates well. If the larger sizes are only occasional, the feed to the elevator should be regulated so that the larger pieces can be picked up. At any rate, the size should not exceed 3–$3\frac{1}{2}$ in. If the size cannot be regulated, a continuous bucket elevator operating at 125–150 fpm should be used. It does have some tendency to mat together and cause jams or hang up in track hoppers. While it feels greasy to the touch, it is abrasive in sizes −1 in. to dust. There is a difference of opinion on the use of chain versus belt for mounting the elevator buckets. The pin joints of the chain catch dust and are subject to wear. Belts, espe-

---

*12-in. dimension on figure will be 18 in. for perlite.

FIGURE 14.27 361

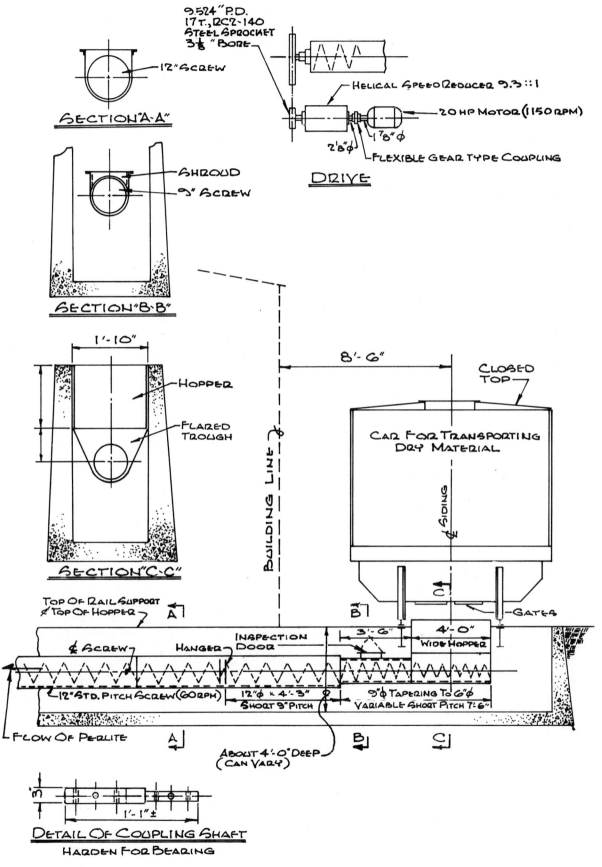

Figure 14.27 Arrangement of a screw feeder under track (typical for fine granular material).

cially in outdoor elevators, tend to collect moisture on the underside and rot. Slippage at the head pulley will occur in time. But with good maintenance, both have proved satisfactory. With plastic buckets, belts must be used.

A slow-operating continuous bucket elevator will require less maintenance and is recommended in spite of its 15–25% higher cost, if the size of the incoming coke cannot be controlled. Using lumps up to 2 in. puts a heavy strain on buckets and conveyor belts, if traveling at high speed. Operating at the lower speed, a continuous bucket elevator should cause no problems. The buckets should be ¼-in. steel, with continuous welds on the outside. In northern states where extreme changes in weather occur, the catalog ratings for capacities should be field checked.

A continuous bucket elevator with the buckets mounted on a single strand of chain gives the best results, maintenance-wise, per ton handled. These elevators are slow operating at 125 fpm when coke is fed to the buckets (buckets do not dig coke in the boot). If a centrifugal discharge elevator operating at about 260 fpm is required, higher maintenance costs can be expected. If super-capacity continuous bucket elevators are used, a double strand of 9–12-in. pitch will work satisfactorily. All-steel chains, such as SBS-188, SBS-102B, or SBS-110, are good selections.

For screw conveyors, the 15% loading chart is advisable, so that the travel of the coke is below the hangers, but 30% seems to work well. In either case, the drive should be designed for 30% capacity or even higher, to allow for overloads.

Petroleum coke handles well in vibrating conveyors (oscillating units). Petroleum coke dust, however, is very fine, possibly as fine as 200 mesh, and it does not handle well in a vibrating conveyor. In falling any distance to the vibrating conveyor, the dust seems to settle densely with next to no air in the dust. What little air remains in the dust tries to work its way up through the dense mass and bubble out at the top of the dust. This dust will travel along the conveyor but very slowly, resulting in very small capacity. A screw conveyor designed to keep the dust low in the trough (15% capacity rating) would give better results.

14.25 PLASTIC (FINE) MATERIALS

Cellulose cotton, synthetic resin (bakelite)
Weight/ft^3: 35–45 lb
Angle of repose: 45°
Nonabrasive

Raw materials such as resins and cellulose products are handled in screw conveyors and bucket elevators, the same as for wood flour. Some of the resin products,

when handled, tend to become slightly warm, and have a tendency to act somewhat like tacky material. A continuous bucket elevator operating at higher speeds may cause the resin to heat up and cause trouble in discharging. If the product is dusty, the heat makes the situation worse.

The products are generally −½ in. to dust. If a centrifugal discharge elevator is installed, and it is desired to change it to a continuous, it is necessary to respace the chain bucket attachments, and to raise the inlet chute in the boot, or to let the continuous bucket dig into the fine product. If the boot is left alone, the buckets will not get a full load, but the elevator can operate. Any dust present, where cellulose sheets are broken into small pieces, can be highly explosive.

14.26 RICE

Weight/ft^3: 36 lb
Angle of repose: 35°
Abrasive (paddy rice)

Rice must be dried to about 13% moisture to be milled, and have hulls removed. Kernels of rice are then polished in a wooden barrel tumbler. It is handled like grain by screw conveyors, double-leg bucket elevators, and belt conveyors. Separation of grades of rice is done on an inclined movable screen with perforated plate. Dryers used are mostly vertical, baffle type.

14.27 SALT (MINERAL-HALITE)

Weight/ft^3: 70–80 lb (fine)
        45–50 lb (coarse)
Angle of repose: 25° for fine
        35° for coarse
Mildly corrosive; abrasive

Solar evaporation of sea water is the simplest extraction method of obtaining salt, but is employed only in regions that lack rock-salt deposits or subterranean brines. Salt is obtained in two other ways, mostly in the central and eastern parts of the US, either by evaporation of brine, or by mining of rock salt. Brine from wells can be raised by pumping water into wells drilled into deposits of rock salt. Two evaporation methods are used, the vacuum method and the Grainer method. The vacuum refining method produces salt in hermetically sealed vacuum pans, in the form of minute, crystal-like cubes. This type of salt is most familiar as table salt. The Grainer refining method produces salt in Grainer pans at atmospheric pressure. This type of salt is produced for many industrial users who require or prefer a flake salt. Both evaporated salt and rock salt may be produced from the same rock-salt deposit. In one process, water dissolves the salt in a well to produce brine; in the other, the solid rock salt is mined more or less like coal. Rock salt comes from underground in large lumps, which are crushed, screened, and graded to various commercial sizes needed.

Where handled in large quantities, rock salt arrives in hopper-bottom (sometimes box) cars in sizes $-\frac{1}{4}$ in. It should be dry and free flowing. If moisture is present, the salt cakes and gets hard and lumpy, and must be broken up before being handled in a bucket elevator. It may be necessary at times to use large hammers to break the salt crust down, when it arrives in a leaky car, so it will pass through a 4-in. grating to get to the boot of the elevator.

Pure salt (sodium chloride) does not absorb water from the air, but common salt, containing several impurities, such as magnesium chloride, makes it appear so. To prevent caking of common salt due to the presence of magnesium chloride, various chemicals, such as starch or sodium bicarbonate, are added. Sodium chloride melts at 801°C.

Belt conveyors are used to move rock salt indoors or outdoors, but the fines tend to build up on the carrying and return idlers. Keeping the belt conveyor idlers clean can become quite a problem. Brushes are not always effective, as they may get filled up. Scrapers do not keep the fine particles from hanging on the belt.

Salt stored in suspension bins can receive the material without problems, but reclaiming the salt has often been difficult, as the material has arched over 18-in. square openings, so that it was necessary to rod it to start flowing, and even this was very difficult at times. When stored in vertical cylindrical bins with conical bottoms, the same difficulty was encountered. It is best not to have the conical bottom with limited gate opening. A very large opening with a slow-operating rotary table gives good results (see figures 7.11 and 7.12). A system of air jets attached to the vertical sides and conical bottom of the bins, and operated for a fraction of a minute at 90 psi, will loosen the material at the sides and bottoms.

Belt conveyors are used to handle either damp or dry salt in outdoor storage and reclaim systems. Usually, a belt cleaner must be provided to keep the return belt clean. Damp salt must not be allowed to remain on the belt overnight, or the belt can get stiff.

Centrifugal discharge bucket elevators with round-bottom buckets are used mostly in handling processed or refined common table salt. These buckets travel at a speed of over 225 fpm. A single-strand of chain, type 830 or 844 Ley bushed, and closely spaced buckets are generally used. A continuous bucket elevator will handle the material, but is more expensive. A continuous bucket elevator should be equipped with a single strand of chain, with steel or iron buckets, spaced one after another, and operating at a speed of 125–150 fpm. The V-shaped bucket, even with a filler in the bottom, will

fill up with salt and cake, if the salt is damp. To produce this salt in a processing plant, care must be taken in the selection of various parts of machinery, as this salt is used as a food product as well as for industrial use.

Use bucket elevators with enameled buckets, as salt will adhere to malleable iron or steel buckets and corrode the chain. Boots should have cast iron sides. Plastic buckets have proved satisfactory, but with chain, they should be reinforced on the backs with steel. Malleable iron buckets can be granitized (baked with several coats of enamel, the outside coat being a very shiny material, making for a slippery surface). This process tends to keep the buckets clean. Enameled inside surfaces of discharge chutes may allow for a 5° reduction in the slope of the chutes. With close clearance, this may be a big advantage. Where the relative humidity is kept low (by providing heat), standard bucket elevators have operated for years with little trouble in maintenance and operation.

Salt should be dry to be handled in any bucket elevator, but there are times during processing when this salt can become somewhat damp. Malleable iron chain, processed to give better wearing qualities, such as Promal by Link-Belt or Z-metal by Chain-Belt, give a very good account of themselves. It has been found that a little extra clearance in the chain pin joint of 0.0015 in. (between pin and barrel of link) is necessary so any dry caked salt (that has been allowed to stand in an elevator boot too long) would have a chance of working itself out of the chain joint, preventing the stiffening of the joint.

The chains should be equipped with Everdur bronze pins, instead of steel pins, as the Everdur pins are corrosion resistant and have lubricating qualities. No oils should be used on these chains. Instead of using the conventional steel cotter pin in the chain, which has been known to become loose and fall out into the product, most of these chains are equipped with S-shaped stainless-steel cotters made from welding rod, by simply taking a straight piece of rod and bending over the ends in opposite directions. These are very effective, do not corrode, and do not become loose. Bolts should be of alloy steel.

Screw conveyors are generally used for handling any grade of salt. If the system is shut down, the screw conveyor should be run for about 10 minutes after the feed stops, to clean it out as much as possible. The salt will get as hard as rock if allowed to stay in the open any length of time. Where no electrical equipment is provided to shut off a screw conveyor when the trough is filled due to the last discharge chute being plugged, it is best to extend the trough a foot beyond the last discharge opening, and install an opening with a pipe leading to a receptacle to catch the overflow. This is not foolproof, since the overflow chute can become filled or plugged up. A very successful method used in screw

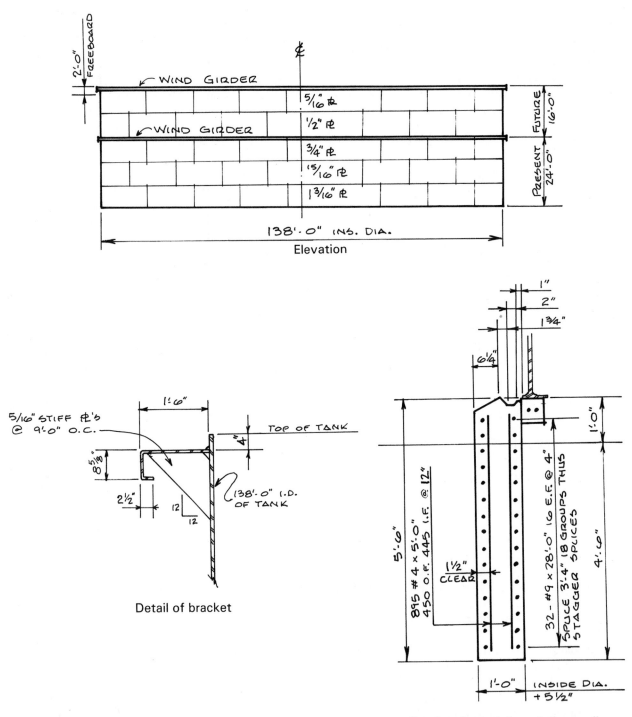

Elevation

Detail of bracket

Section through foundation wall

Figure 14.28 An open storage tank for brine. *Specifications:* 1. Paint: not required; 2. Steel: all plates to meet ASTM specification serial A-283 for Grade C steel. All other steel to meet requirements for serial A-7; 3. Welding: bottom, ⁵/₁₆ in. fillet top side only. Shell to bottom, ⁵/₁₆ in. fillet inside and outside. All circumferential and vertical shell joints to be butt double welded. All work to meet API specification 12-C for All-Welded Storage Tanks; 4. Testing: after work is completed, the tank shall be filled by the contractor and tested for leakage. Repairs are to be made by the contractor in accordance with the American Welding Society's Rules for Field-Welding of Storage Tanks; 5. Fittings: one exterior spiral stairway with an outside platform at the top, one 10-in. diameter outlet at the bottom, and one 24-in. diameter manhole 4 ft 0 in. above the bottom; 6. Fill: remove silt and organic material from the ground surface and replace it with 1 ft 0 in. of compacted sand saturated with oil.

conveyor casings consists of a swinging baffle, as shown on figure 4.15 (refer to paragraph 4.20).

If a screw conveyor is used, take the next larger size than the one required, and increase the required horsepower by 25%. The ½-in. clearance between the screw flighting and the trough fills up quickly and may freeze up if salt is slightly damp, and the conveyor is stopped for a period. The trough loading should not exceed 30%.

Vibrating conveyors can handle rock salt, say ¼ in. and larger, if dry. They should not be used for fine salt, as it tends to mat and form a dense mass.

In storing brine, the tanks should be rubber lined. In figuring the thickness of steel plates required, it is best to add ¹/₁₆ in. to the computed value. If stored in unlined open steel tanks, it is best to put a slight cover of oil on top of the brine, then, when the brine level is lowered, the oil coats the side of the tank and acts as a further protection. Covered tanks must be vented. When rock salt has been dissolved at the plant in hot water, it may be advisable to insulate and weatherproof the exterior of the tank to conserve heat (see figure 14.28).

Rubber-lined steel pipe with full-faced rubber-lined flanges or fiberglass-reinforced polyester pipe with plastic flanges are suitable for brine. Rubber-lined butterfly valves with teflon-coated discs are satisfactory. Pumps should also be rubber lined. Specific gravity can be controlled with a Dynatrol from Automation Products, Inc, or a Densart system from the Ohmart Corporation; however, the Ohmart system uses radioactive cesium-137, and may require a license. Flow can be controlled with the use of a Magnetic Flow Meter by Fischer-Porter Company. This instrument has the advantage that the brine does not come into physical contact with any metal, except for two electrodes made of a suitably corrosion-resistant metal, in its flow through the instrument. If not run every day, it is advisable to turn over equipment for at least five minutes daily to prevent freezing up.

14.28 SAND AND GRAVEL

Weight/ft³: 100 lb
Angle of repose: 26.6° (varies between 25.1° and 28.6°) for sand; 40° for gravel
Abrasive

Sand and gravel are taken primarily from open pits. In some areas of the country, dredging and water transportation are used. Except for extremely large operations, dry excavation is generally considered much more economical. It depends a good deal on the cost of the disposal of the wet soil from the barges. Pumping has proved very costly in maintenance and fuel, and has proved difficult to handle because of increase in length of lines, as distance to supply is increased. If ground water level is high, pumping to an inclined dredging elevator of heavy construction (¼-in. steel buckets with manganese lips, manganese chain links, heavy band brake) will work economically.

Where dragline operations are used for wet materials, difficulty has been encountered in maintenance of the serving roads and track beds. Serving roads and track beds are subject to damage by drainage from equipment. It is expensive to move the equipment to take full advantage of variation in supply. Dry plants are flexible, since variations in requirements can be supplied by moving excavators or, where economical, to use two or more excavators, in different areas, that can meet the specifications. A movable dredge can perform to meet varying requirements, but not as easily. Where a dry pit may become wet because of ground-water level, a study should be made, based upon relative depths of dry and wet supply.

In most so-called dry pit operations, the raw sand and gravel is reclaimed by a gasoline- or diesel-operated crawler shovel, discharging into large capacity trucks which, in turn, discharge into a common dump hopper, then are moved by a belt conveyor system to a preparation plant. This method has proved very economical, and maintenance is at a low level. In many dry pits, the raw material may contain a small percentage of clay or shale which is very objectionable as far as highway specifications go. Clay and shale must be removed because they may cause flaking of concrete.

Many sand and gravel preparation plants are equipped with vibrating screens for sizing the finished products. Some plant operators feel that where clay and shale are encountered, vibrating screens are too fast in operation and do not allow the material a chance to get clean.

Scrubbers at the top of the preparation plant are most desirable in the form of a cylinder with lifting vanes and plenty of water to put clay in solution and soften the shale so it can be slushed out at the end to a disposal pile on the ground. Then, the raw material continues to pass through screens, where the material gets a continuous bath of water to rid it of any clay or foreign material in solution, and these screens, operating slowly, have a chance to do their intended work by allowing those pieces of gravel that go through the perforated plates to fall into different bins below, holding various-sized finished material. Sand continues on with dirty wash water until it usually reaches "wash boxes" or "dewatering" conveyors. These units consist of two strands of chain with steel flights attached every 12 in., operating in a wood or steel tank into which the dirty water with sand flows. Much of the dirty water flows out the back of the tank and the sand, being heavier, falls to the bottom. Fresh water is added, and the flights, moving at approximately 50 fpm, bring a fairly clean sand out to be discharged into a bin, but the sand is not sized. Wet screening by vibrating screens can produce almost any combination of specifications as to size and percentages of mesh.

Sometimes, in combination with the wash boxes, and depending on the condition of the sand as it discharges from the tanks, screw washers are used to clean up the sand if it is still discolored with objectionable material. At times, a good percentage of fine sand goes out to the waste pond with the dirty water, and this sand can be reclaimed, so it can be sized and made a profitable material. There is usually a large demand for a very fine sand, in industry and filter beds.

If the raw material deposit is mostly sand with very little gravel, then the sand can be scrubbed and washed very easily at the top of the plant and flumed to classifiers operated entirely by water, to give a definite grain size. These sand separators will allow the sand to settle in the bottom, and the dirty water to flow out, but no actual or accurate grain size can be produced in these units, as they are not designed to do so. Sand coming out of the separators will contain 25–35% water.

Some elevator manufacturers suggest an elevator chain with 6-in. pitch, instead of the usual standard of 12-in. pitch (on double-strand elevators) and with the bucket attachments every other pitch, so as to make the chain run more smoothly over the head sprocket. The same idea can be used on standard 18-in. pitch chains, using 9-in. pitch to get smoother action on head sprockets. Shortening the chain pitch allows the chain to follow closer to the circumference of the sprocket, but the action of a 12-in. pitch chain is not objectionable, and a 6-in. pitch chain is not necessary. When an elevator takes an 18-in. pitch chain, a 9-in. pitch can be considered. Elevators using a 12-in. pitch chain have proved satisfactory. Many super-capacity continuous bucket elevators having a double strand of 12-in. pitch steel strap chain have been installed and have very little maintenance, if the elevators operate at 100 fpm or under. Sand and gravel should be handled by belt conveyors. Joint pins in a chain will wear out. It is best to limit the angle of belt conveyors to 15°. If it is absolutely necessary to go to 18°, it is best to place a layer of fine sand on the belt, before the gravel is fed. Moist, sluggish sand, with 8–10% moisture, will move on an incline of 20°, but it is best to limit the angle to 15°.

14.29 SHALE, KAOLIN (CLAY)

Weight: 63 lb/ft^3 for kaolin
85–90 lb/ft^3 for crushed shale
Angle of repose: 35° for kaolin
39° for crushed shale

If the shales or clays are handled from the quarry during damp or wet weather, they are apt to break down and tend to stick on steel plates. V-shaped buckets tend to fill up quickly and must be cleaned to maintain capacity. A centrifugal discharge elevator equipped with plastic or enamelled malleable iron style A or AA (round bottom) buckets should be used, if an elevator is a must. It is better to avoid elevators. The clay or shale will tend to wear out the enamel and require replacement.

Because of the generally large quantities being handled, super capacity continuous bucket elevators equipped with a double strand of steel chain and flat bottom continous buckets should be considered. Even these buckets will need occasional cleaning. Enamelling the buckets will help.

If the clay or shale has been dried in a rotary kiln and the large pieces crushed to −2 in. and under, a centrifugal discharge elevator can be used, but a continuous bucket elevator running slower would serve better and require less maintenance. If the shale or clay has been crushed and screened to give −10 mesh material, a centrifugal discharge elevator, with malleable iron buckets spaced every other link, and with operating speed reduced to about 185 fpm, will prove very satisfactory.

Screw conveyors should not be used on damp or wet clay or shale. If other reasons make it necessary, the trough loading should be only about 15%. Dry material will handle easily, and trough loading of 30% can be used.

Vibrating conveyors can be used for material up to −10 mesh. Material beyond that may become aerated and not move along. Very fine-grain clays do not handle well in vibrating conveyors.

Belt conveyors have proved the best to use if the material is damp. If material clings to the belt, it can be cleaned by rubber brushes, sometimes even followed by steel scrapers, on the return run. It can be hosed off further along the return run. Fuller's earth is usually mined by the open pit method. At the mill, it is often placed in ventilated storage bins, allowed to dry, and then crushed, elevated, and fed to rotary dryers, crushed again, and screened to required size.

14.30 SOAPSTONE, TALC, ASBESTOS, AND MICA

Weight/ft^3 and angle of repose:
soapstone ⅛ in. and under: 40–50 lb
and 20–29°
soapstone and 100-mesh talc: 40–60 lb
and 20–29°
shredded asbestos: 20–25 lb and 45°
shredded mica: ⅛ in. and under: 13–15 lb and
34°
Mildly abrasive

Soapstone, talc, asbestos, and mica in rock form may be found with a shallow or deep overburden. If shallow, drag scrapers can remove a large percentage of the overburden. If pockets of overburden are present, then these must be cleaned with hand labor, but here again it must be the decision of the mining superintendent as to whether or not machinery can be used to keep the costs down.

Any of these products, wet or dry, can be handled by a belt conveyor system from the quarry to the mill. If, by chance, some overburden goes along with the rock then, somewhere along the line, there will be trouble. If it is necessary to use a bucket elevator after the rock is quarried wet or dry, use a super-capacity bucket elevator with flat-bottom continuous buckets. With this type of bucket, and the greasy feel of the material, there will be less chance of the bucket filling up with fines. The continuous elevator with V-shaped buckets should not be used, as these buckets would gradually fill up with fines that would get hard as rock, making it almost impossible to remove without a lot of labor and downtime.

In reducing soapstone and talc to powder, these products aerate and become fluid when dry. They handle well in screw conveyors and centrifugal discharge bucket elevators with very closely spaced round-bottom malleable iron buckets, style A or AA, and the chain operating at a reduced speed of 185 fpm.

Asbestos ore or rock weighs about 81 lb/ft³ and is considered abrasive by some operators; others call it mildly abrasive. During the crushing cycle, the asbestos may give off dust or fumes that could be harmful to life. The design must provide for the removal of the dust and fumes.

Asbestos, when shredded, handles well in centrifugal discharge elevators with closely spaced round-bottom buckets and the chain speed reduced to 185 fpm. Screw conveyors are used in the mill to distribute the finished product, using the normal capacity carrying chart. Vibrating conveyors (oscillating type) may or may not handle the shredded asbestos. Consult with the manufacturers of these conveyors, and try to make a test run. These conveyors will handle the asbestos rock well.

Mica handles satisfactorily in a screw conveyor if the load is kept low in the trough to avoid a cutting action on the hanger couplings. A centrifugal discharge bucket elevator with standard spacing of buckets on chain, and operating at standard speed, produces good results. A standard vibrating conveyor (oscillating type) usually works well on this size product.

Screw conveyors will handle pulverized mica by using the 15% trough capacity chart, keeping the load low in the trough. Centrifugal discharge elevators will give good results, if the round-bottom buckets, style A or AA are spaced every other link (close spacing) on chain and speed of chain is reduced to 185 fpm. A continuous type bucket elevator does not give good results due to fine mesh product filling the V-shaped buckets and packing in the bottom of the buckets. The slippery nature of mica does not help clear a continuous bucket. A vibrating conveyor (oscillating type) should not be used, as there is not enough weight to the mica to move it along the trough to get the desired capacity.

In handling any of these mesh size materials that tend

to aerate when handled, it is well to lower the discharge chute to 18″ from the head shaft of the elevator to a point where the discharge chute is attached to the elevator casing.

14.31 SODA ASH

Weight/ft³: 55–65 lb (dense ash)
20–35 lb (light ash)
Angle of repose: 32° (heavy ash)
37° (light ash)
Mildly abrasive

Material should be unloaded from hopper-bottom cars in dry weather (it has a tendency to pick up moisture), or in an enclosure. Canvas connections are generally used between the cars and the receiving conveyor. Centrifugal discharge bucket elevators handle the dense ash well. The light ash can be moved on centrifugal discharge elevators at a reduced speed of 175–185 fpm. Buckets are closely spaced, malleable iron, type A or AA. Pivoted bucket carriers can also be used. Full volume or Bulk-Flo elevators work successfully, and with little maintenance, on this product.

Screw conveyors operating at 15% loading will work well. Vibrating conveyors will not work well on grain sizes below −10 mesh. As it becomes aerated, it will not move along the trough. If stored in tanks and left to stand for a while, it gets somewhat hard next to the steel plates of the tank or bin, and air jets at 90 lb pressure should be provided.

Soda ash is handled in about the same way as salt. It should be dry and free flowing to be properly conveyed. It is usually shipped in closed gondola cars with tight slide gates in the bottom of the car. Conveying equipment should be as dusttight as practical. The system should be vented through a dust collector.

14.32 STEEL OR CAST IRON CHIPS

Weight/ft³: 100–150 lb
Angle of repose: 30–40°
Very abrasive

Chips about ½ in. long, without curls, are free flowing. Chips larger than ½ in. should be put through a crusher and reduced to ½ in. They are usually saturated with oil.

A centrifugal discharge bucket elevator with round-bottom, standard-spaced, malleable buckets, and operating at standard speeds, probably is the best. Continuous bucket elevators, with V-shaped buckets, have been found to get clogged and packed with chips, and do not discharge properly. For oil-soaked, cast-iron chips, the continous bucket elevator will not work properly.

An apron feeder will handle the material if it is cut up into small pieces and, if mixed with wood chips, the wood chips should be 2 in. or under. With the apron feeder moving at, say 20 fpm, if the hopper is loaded with curls of steel mixed with wood chips, a rolling action of the mass of material results, and the curls form into a ball about 12 in. in diameter that keeps turning around in the hopper and causing a heavy pressure against the front plate of the hopper. To relieve this pressure, the opening in the front plates of the hopper should be made large and equipped with a free-swinging, counterweighted gate plate to prevent jamming of these balls of steel. The wood chips with slivers are less troublesome to handle, but if the percentage of slivers is large, the swing-gate feature is an asset. Most plants provide space under the machines to take away the chips that are produced in the machine operations. Often these chips are soaked with oil, and this adds to the handling problems.

A screw conveyor with close clearances can be used under production machines. If a screw conveyor is used with standard clearance between the flighting and the trough, usually ½ in., the small chips are likely to get jammed and cause a breakdown. It is better to use a close clearance of ¼ in. This means the flighting should be mounted on a stronger pipe to prevent deflection of the screw section due to its own weight. Formula for deflection of pipe is given in table 4.1.

For a long distance, say 150–200 ft or more, the H class drag chains will do a good job. H-110 chain, operating in a steel or concrete trough under the production machines at 50 fpm or under, is in popular use. Maintenance is rather low on these chains if they are made of Promal or Z-metal, both with a Brinell hardness of about 190. Promal or Z-metal is not a skin-hardening process. The complete metal is changed by heat treating, and with a Brinell hardness of about 190 against a Brinell of 120 for standard malleable iron, a tough wear-resisting chain results.

Refer to paragraph 6.8 for the design of a drag conveyor, to paragraph 6.10 for a flight conveyor handling bales of steel scrap, and to paragraph 14.34 for the handling of a combination of steel ships and sawdust.

The accompanying sketch shows a 36-in. wide apron feeder, 13 ft long, and equipped with a double strand of SS-942 steel strap roller chain with cross rods every 18

in. It has 5-in. diameter single-flange rollers to support the apron on T-rail track (L-track could have been used). The pans are ½-in. thick with 4-in. high sides that are ¼-in. thick. The speed is 20 fpm. Its capacity is about 30 tph with the average weight of material being 75 lb/ft³.

The cross-section area of the feeder is

2.75 ft × 2.0 ft × 1.0 ft = 5.5 ft³/ft of feeder

Because of such loose materials, figure 70% voids or 30% solids, giving

5.5 × 0.3 × 75 lb = 123.75 lb/ft

(say, 124 lb/ft)

124 × 20 fpm × 60 min = 148,800 lb/hr

= 74 tph

Some steel turnings in raw material restrict the flow of smaller scrap to such an extent that discharge from the hopper is very slow and the balls of turnings have to be removed by hand. In that case, capacity may be temporarily reduced by as much as another 50%, thus giving only 35 tons per hour. If everything is operating correctly with good material, this feeder can deliver 70 tons per hour.

To determine the total pull on the head shaft, first figure the rolling friction (*RF*) of the 5-in. diameter rollers (*D*) on 1-in. diameter cross rods (*d*):

$$RF = X \times \frac{d}{D}$$

$$= 0.25 \times \frac{1 \text{ in.}}{5 \text{ in.}} = 0.05$$

(but use 0.10 due to fine dust on the track or lack of lubrication)

where the value 0.25 (*X*) is the coefficient of steel pins on bored, greased rollers. Next, figure the approximate weight of the apron with ½-in. thick plates, ¼-in. steel, 4-in. high sides, two strands of SS-942 chain, 1-in. cross rods 18-in. on centers with rollers at 130 lb/ft. Add the weight of the material on the apron, 124 lb/ft:

130 + 124 = 254 lb/ft

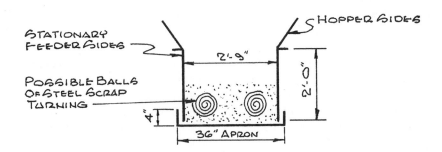

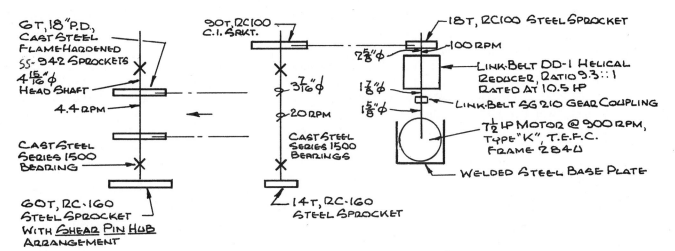

Figure 14.29 Arrangement of a drive for a steel-chip conveyor.

Figure the carrying run at

$$13 \times 254 \times 0.10 = 330 \text{ lb}$$

and the return run at

$$13 \times 130 \times 0.10 = 169 \text{ lb}$$

giving the total pull on the head shaft as

$$330 + 169 = 499 \text{ lb} \text{ (say 500 lb)}$$

Due to the possibility of steel having trouble getting through the opening in the hopper, use 1000 lb.

To figure the theoretical horsepower:

$$\frac{1000 \text{ lb} \times 20 \text{ fpm}}{33000} = 0.6 \text{ hp}$$

It was found experimentally that so much friction is developed between the shredded steel pieces turning over and over and forming balls with the larger pieces, and along the hopper sides and skirt boards, that a 7.5-hp motor was required. A 5-hp motor and drive proved too light to operate under all conditions.

A shear-pin hub arrangement was used on the roller chain sprocket mounted on the head shaft as an extra safety valve.

The accompanying diagram shows the size of the headshaft.

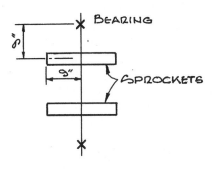

The bending moment is

$$\frac{1000 \text{ lb} \times 9 \text{ in.}}{2} = 4500 \text{ in.-lb}$$

The torsion moment is

$$1000 \text{ lb} \times 9 \text{ in.} = 9000 \text{ in.-lb}$$

Figure 9.3 shows a $3^7/_{16}$-in. diameter shaft required for heavy shock (stopping and starting under a loaded hopper).

Based on experience, the head shaft should not be less than $4^{15}/_{16}$ in. in diameter.

Figure 14.29 shows the arrangement of the drive for a steel-chip conveyor.

14.33 SULFUR

Weight/ft³: 75 lb
Angle of repose: 40°
Mildly abrasive, highly explosive

Sulfur is obtained by drilling holes into the ground and melting it by the introduction of hot steam. The sulfur is then pumped to the surface into large storage vats with removable sides. When the mass is solidified, the sides are removed and the huge blocks (over 99% pure sulfur) are blasted into fragments and loaded into railroad cars by cranes for shipment to the user, where it is reclaimed from open storage piles.

If sulfur is damp or wet, a belt conveyor will best serve the purpose. If dry and free-flowing and not over 10% of the material is 2-in. lumps, a centrifugal discharge elevator with round-bottom buckets, operating between 225 fpm and 260 fpm, will properly discharge the material. Fine material could cake to the bucket, especially if the elevator is shut down and the sulfur is allowed to dry in the bucket.

If the material has, say 20% large lumps, a continuous bucket elevator operating at 100–125 fpm can be considered. If lumps are 5–6 in., a double-strand chain super-capacity continuous bucket elevator should be used, operating at a speed of 75 fpm to, say, 120 fpm. Buckets should be flat-bottom, continuous, style SC. Plastic or cast aluminum buckets can be used with the elevators to help reduce explosion hazard. The chains on any of these elevators should preferably be malleable iron, equipped with Everdur bronze pins with stainless steel S cotter pins. Enameled malleable iron buckets can be used and attached to chain with aluminum or bronze washers and bronze bolts. Dura buckets are shatterproof, noncorrosive, and self-cleaning. The boot and the feed chute leading to the boot of the elevator should be lined on the inside with 12-gauge replaceable aluminum plates (refer to paragraph 14.13).

A 14-in. screw conveyor handling dry sulfur weighing 75 lb/ft^3, 2½ in. and under, with a possible overload capacity of 45 tons per hour, is shown on figure 4.18. The top cover of a screw conveyor casing is not dusttight (unless ordered that way), and any trapped dust has a chance to leak out. If the cover must be made dusttight, then the cover must be ventilated at about 15-foot intervals, and connected by piping to the outside of the building.

Use aluminum bushings. If a spark hit and travelled along the pipe, the coupling would kill it in the bushing in the hanger. Ground all equipment. White metal and babbit bearings in the hangers give good service. If sulfur has, say, 50% of capacity in lumps 1½ in. to 2½ in., use oversized trough. In storing fine sulfur outdoors, it is advisable to heat the surface to the melting point, thus forming a coating over the storage pile and avoiding polluting the neighborhood.

Vibrating conveyors with removable linings and dusttight covers generally have been used for dry and free-flowing sulfur. If placed over storage bins, the supporting structure must be designed to dampen the vibration.

14.34 WOOD: CHIPS, FLOUR, AND SAWDUST

Weight/ft^3:   10–13 lb (sawdust)
15 lb (dry wood flour)
18 lb (cellulose)
10–30 lb (chips)*
Angle of repose: 36° (sawdust)
45° (chips)

Sawdust will vary in texture and weight, depending on from what part of the country it originates. Many lum-

*A cord of southern hardwood, gum, or baywood (4' 0" × 4' 0" × 8' 0" = 128 ft^3) produces about 219 ft^3 of chips when 50% wet. Southern pine will produce 200 ft^3 of chips per cord.

ber mills produce a large volume of sawdust, but they are usually located far from transportation, making it too costly to prepare and ship it. In this event, the sawdust is usually burned right at the mill for quick disposal. Other mills, close to shipping points, find it worthwhile to size the sawdust, bag it, and sell it as such. True sawdust, clean and without any curls or slivers of any kind, has a market value. Dry sawdust is delivered to an industrial plant by covered truck and discharged into a dump hopper, which is also covered, to keep the sawdust dry.

Oscillating conveyors or feeders and centrifugal discharge bucket elevators, operating at a speed of 185 fpm or less, with malleable iron or steel, round-bottom buckets spaced very closely together, work well. Continuous bucket elevators may find the V-bottoms of the buckets filling up and hardening, thus reducing capacity. A continuous bucket elevator operating under 100 fpm will reduce the dust problem and is often used. It will handle well in vibrating feeders in spite of its light weight. The top of the elevator should be vented through the roof to a dust collector. Elevator casing should be dusttight. Sawdust can be stored in a bin with straight sides if it is dry, and the sawdust can be reclaimed with Live Bottom Reclaiming Screws. Figure 14.30 shows a series of 12-in. screw conveyors, 9-in. pitch, feeding a 14-in. cross-screw conveyor of standard pitch.

Wood flour is sawdust broken down or crushed to pass 100–200 mesh screens. Wood flour can be explosive around machinery, but is not considered as explosive as sulfur. Chain should be equipped with Everdur bronze pins and stainless steel S-shaped cotter pins. Buckets should be attached to chain with bronze bolts and brass washers. Sprockets can be cast iron and chutes carbon steel. Two cloth blowout doors should be provided, and the top of the elevator vented through the roof. All motors should be explosion proof. It handles well in screw conveyors, but not in vibrating or oscillating conveyors, as the flour does not have enough weight to travel along properly. A centrifugal discharge elevator works satisfactorily if the speed is reduced to about 185 fpm, and the round-bottom buckets are spaced very closely together. Elevator casings should be 2 in. wider than standard. Driving machinery should be explosion proof. There may be some dusting, which would require the top of the elevator to be vented through the roof of the building, where a trap could be installed to catch any dust. All elevator casings or screw conveyor casings should be made dusttight.

On Figure 14.31, the wood chips (no sticks), averaging about 1½ in. and under, are brought to the plant by truck and dumped into a large hopper. At the bottom of this hopper are two 12-in. diameter short-pitch screw sections, one right-hand and one left-hand, in a single-channel flat-bottom trough. Both screws are driven by a

FIGURE 14.30 371

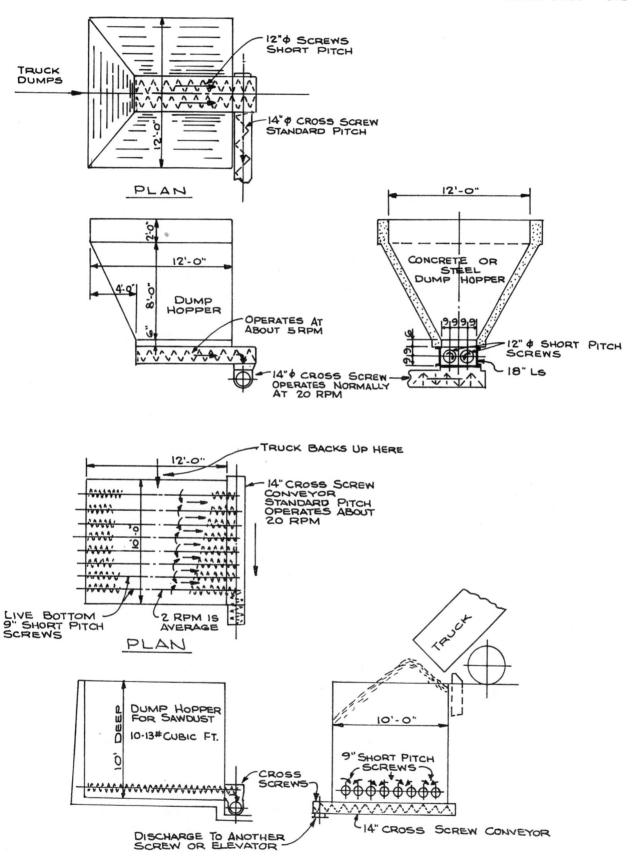

Figure 14.30 Sawdust hoppers. *Top:* arrangement of a dump hopper with sloping sides and two reclaiming screws handling sawdust. *Bottom:* arrangement of a dump hopper with straight sides and eight "live" bottom reclaiming screws handling sawdust.

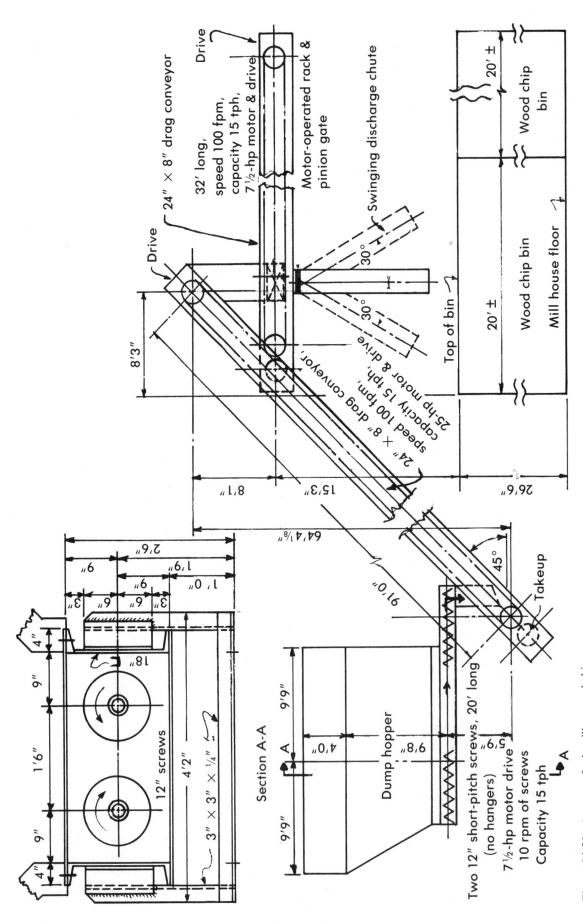

Figure 14.31 A system for handling wood chips.

FIGURE 14.32 373

Figure 14.32 An apron feeder for handling a mixture of steel scrap and wood chips.

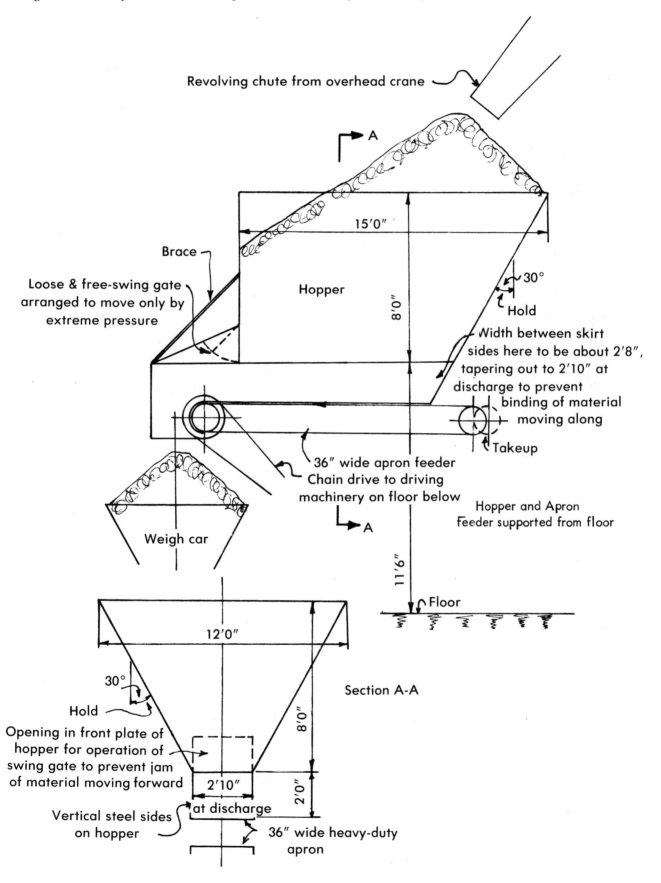

Revolving chute from overhead crane

15'0"

Brace

Loose & free-swing gate arranged to move only by extreme pressure

Hopper

8'0"

30°

Hold

Width between skirt sides here to be about 2'8", tapering out to 2'10" at discharge to prevent binding of material moving along

Takeup

36" wide apron feeder Chain drive to driving machinery on floor below

Hopper and Apron Feeder supported from floor

Weigh car

11'6"

Floor

Section A-A

12'0"

30°

Hold

8'0"

Opening in front plate of hopper for operation of swing gate to prevent jam of material moving forward

2'10"

2'0"

at discharge

Vertical steel sides on hopper

36" wide heavy-duty apron

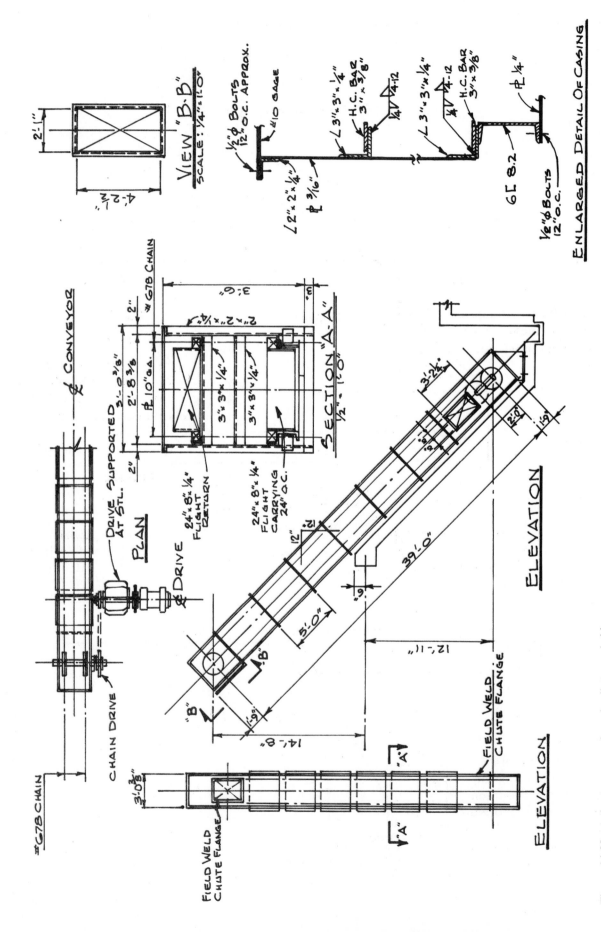

Figure 14.33 A drag conveyor for unloading wood chips.

pair of cut-tooth steel spur gears, reducer, RC-160 roller chain drive. To prevent any possible stalling of the screws under a load of wood chips such as this, the screws are made short pitch of 9 in., which gives a more vertical grab on the chips. Usually 12-in. standard pitch on 12-in. diameter screw is too flat for operating under a load like this. (See also figure 14.30.) No hangers are used in the screw length of 20 ft, so the screw pipe is made 5 in. (4.063 in. ID) to give minimum deflection. The flighting is ⅜-in. sectional butt welded, and welded to the pipe continuously both sides. The drive shaft is 3¹⁵/₁₆-in. diameter with three-bolt connection. A pair of steel spur gears, with 36 cut teeth, 2 DP 18 in. PD, 4½ in. F, reverses the motion of these screws. The screws discharge directly to a double-strand inclined drag flight conveyor at a 45° angle on about 91 ft centers. A stair walkway and platform runs the entire length of the conveyor.

This inclined conveyor consists of a double strand of no. 678 rivetless chains to which are attached ¼-in. steel flights 24 in. wide × 8 in. deep every 24 in. The 3¹⁵/₁₆-in. diameter head shaft is equipped with 6-tooth sprockets, 23.18 in. PD plate center and roller bearing pillow blocks.

The conveyor operates at 100 fpm and is driven with RC-180 roller chain drive, speed reducer equipped with a backstop and 25-hp motor. This backstop is effective if conveyor is loaded or empty when the power source is interrupted and conveyor tends to run backward. When power comes back on the line, the brake automatically releases.

The inclined flight conveyor discharges to a horizontal flight conveyor of the same cross-section equipped with 7½ hp motor, speed reducer and RC-140 roller chain drive. Capacity of the conveyor is 15 tons per hour. In the bottom of this conveyor trough, the openings to storage bins are equipped with motor-operated rack-and-pinion gates. Pinion shaft operates at 10 fpm, driven by a ¾-hp motor, speed reducer direct-connected. Under rack-and-pinion gate, a universal type swing chute is located to fill storage bins.

This sytem works well, but the chips cannot contain any sticks or slivers of wood.

In handling a mixture of shredded steel turnings and wood chips, the wood chips should not be over 2 in. long, and the shredded steel should not be long curls of steel shavings. If a hopper is loaded with a mixture of steel curls and wood chips (figure 14.32), the motion of the apron feeder tends to cause a rolling action of the mass of material above it, and the curls may form into a 12-in. ball, and keep turning in the hopper, causing severe strains in the hopper plates. If this is likely to

happen, increase the opening in the front plates, and add a free-swinging counterweighted gate plate to prevent jamming of these balls of steel. This will also help if a large percentage of wood slivers is in the mixture.

Pins should be Everdur bronze, and buckets should be attached to chain by bronze bolts and brass washers. Sprockets can be cast iron. Chutes can be carbon steel. Belts should not be used, as the flour gets between belt and pulley even if pulley is slotted. Malleable iron 730-lb chain is satisfactory. Plastic will help avoid explosion.

It may be well to vent screw conveyor covers every 15 ft or so to relieve buildup of dust pressure. If there is any possibility of the entrance of foreign matter, such as steel, the coupling shafts and bolts should be of bronze.

Wood chips, say −1½ in., can be moved in a screw conveyor, but it is best to reduce the standard 12-in. pitch to 9 in., to obtain a more vertical bite into the chips or sawdust. If there is the possibility of small sticks or blocks of wood getting into the sawdust, offset screws have worked well. Wood chips, damp or dry, and mostly 1–2 in. in size (no sticks) tend to mat together and, if their movement is restricted, will usually tend to arch and not move at all. If the bottom of the bin opening is made almost the same size as the diameter of the bin, rotary table feeders will work satisfactorily. Wood chips will travel on a belt conveyor up to 27° slope, wet or dry, but it is best to keep the angle under 20°.

A wood-chip unloading drag conveyor is shown on figure 14.33.

Logs ground down into small pieces are taken up in bucket elevators (round-bottom buckets) at high speed. The chips are slightly curled and have sharp edges. If using belt or chain conveyor, use 45° idlers. Plows will distribute the chips to bins or silos. Most plants prefer to use chain instead of belt because the cost is lower and the life is longer.

Cellulose is very fine and abrasive. Bucket elevators should be slowed down about 25% because of the dust they create. The elevator should be vented through roof to dust collector. Buckets should have holes drilled to break vacuum. Cellulose handles well in screw conveyors.

14.35 ZIRCONIUM

Zirconium sand is very abrasive. A screw conveyor handling it should have ¼-in. trough clearance, and the flighting should be ⅜ in., tapered down to ³/₁₆ in. next to pipe. The load should be kept below the hangers.

Overhead and Portable Car Shakers

15.1 GENERAL

Bituminous coal is screened to size and washed at the mine. If the coal cars are sent in transit immediately and they encounter cold weather the coal will freeze to some depth. To prepare for such conditions, many plants have thawing pits at unloading locations where men with torches try to thaw the coal at the discharge openings in the cars. Pounding with a sledge hammer does not accomplish too much. Torches with a very hot flame distort the steel body of the car, a practice frowned upon by car manufacturers but allowed, against their better judgement, by the railroads. Larger industrial plants use a car shaker to help unload cars and, if coal is badly frozen, also use torches.

15.2 NOISE

The car shakers are heavy and rugged units. They produce a series of heavy hammer-like blows on the top of the car and usually loosen the coal even if it has a 2-ft frozen crust at the top. Any of these units is noisy and normally cannot be tolerated when used near a factory or office building located within 100 ft of an operating car shaker. The acoustical properties of the location and the sounds produced at night may keep everyone in the area awake.

Allis-Chalmers Co. makes a heavy, self-contained shaker that is operated by a 20-hp motor. The manufacturer claims it takes two to five minutes to unload a car. This assumes summer operation and free-flowing coal. Using manual methods, in the summer, it takes three to six men from 15 minutes to nearly an hour to unload coal from a hopper-bottom car.

15.3 HEATING CARS

In the winter months, even with the use of a car shaker, intense heat may have to be applied to a single car for an hour or so to break the ice (adhesion) crystals, so that when the car shaker operates, it can perform its duty in a shorter period of time. This heating of car hoppers is called ''toasting,'' and is usually performed by blow torches. Once this bond is broken, the intense vibration of a car shakeout quickly discharges the load. Sometimes cars are frozen almost solid. It becomes necessary to shuttle these cars into warm areas to thaw out for 24 hours before they can be unloaded. Even after the 24-hour period, unloading may be difficult. Sometimes, a heavy guillotine blade can be used to cut through the semithawed coal in the car.

15.4 HEWITT-ROBINS SHAKERS

The design of the Hewitt-Robins HDV car shakeout is such that it vibrates the cars sideways on the track (see figure 15.1). These car shakeouts weigh approximately 3000 lb. When operating, a heavy vibration is set up in the entire car structure. The rigid steel frame of the shaker is stress-relieved after welding and is a very husky unit. Bearings are well protected against contaminants. The vibrator motor is 20 hp, tefc. It requires an overhead 6-ton electric hoist, usually supported by an

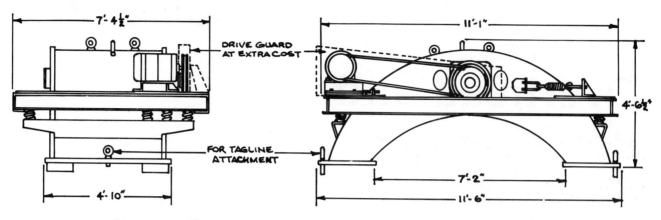

HEWITT-ROBINS HEAVY DUTY CAR SHAKEOUT

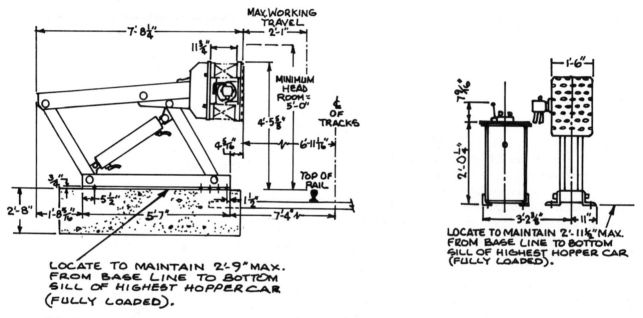

HEWITT-ROBINS TRACKSIDE CAR SHAKEOUT

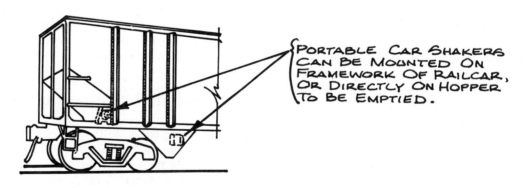

TYPICAL PORTABLE CAR SHAKERS

Figure 15.1 Car shakers.

A-frame to span one or two tracks. Hoist weight is 2500 lb; required motor is 7½ hp. Remote pushbutton control station for both shakeout and hoist operation, and automatic controls for limiting upper and lower travel of hoisting blocks, are included.

Hewitt-Robins also makes a trackside car shaker that is placed alongside the car. This transverse shaking action produces virtually no undesirable sidewise movement. It can handle up to 10 cars per day, but is best recommended for operations averaging three to five cars daily.

Vibrating a car vertically appears to cause less damage to the car because the car springs are located to take vertical vibration.

The HDV is an overhead-type car shakeout that is controlled from a hoist with a 6-ton capacity. It is recommended by the manufacturer for up to 100 car-per-day operation although, at the upper range, car turnovers might be possible. The HDV car shakeout imparts a vertical downward vibration to hopper cars.

15.5 ALLIS-CHALMERS CAR SHAKERS

The Allis-Chalmers Car Shaker is a selfcontained unit, with the motor located inside the body of the car shaker. The motor is supported on four rubber mountings, for isolation of vibration. Total weight of car shaker with motor and drive is 10,500 lb. The car-shaker body is a one-piece steel fabrication that is stress-relieved after welding.

The car shaker is normally supported by a special

5½-ton hoist with pendant control which contains all the required operating controls and safety interlocks (figure 15.2). All motor starters and controls are included with the hoist. The car-shaker drive motor is 20 hp. The hoist is 7½ hp and the trolley motor, if required, is ½ hp.

The entire weight of the car shaker is working weight, imparting approximately 27,000 lb force to the rail car.

15.6 PORTABLE VIBRATORS

The National Air Vibrator Co. produces a portable heavy-duty hydraulic vibrator unit known as the Navco HCP Model. The hydraulic clamp assembly can be attached to structural members of hopper cars to receive the vibrator unit, and both can be moved anywhere along the car. They are made of stainless steel. This unit requires moving along the car by hand to vibrate the length of a car (average 50 ft long), and it may take two to three hours to unload a car in extreme winter weather. If a company uses only, say, one or two cars a week, the handling of a unit of this kind may meet the company's requirements. Other portable units are also produced by Cleveland Vibrator Co., Martin Engineering Co., and the Aldon Co., to mention a few other producers. These units are usually available for pneumatic or electric operation. Special application precautions must be taken, however, with the pneumatic units, regarding filter-oiler units as well as special antifreeze lubricants used in climates where applicable. Cleveland Vibrator Co. furnishes a dry solid film-lubricant treatment for the bore of the pneumatic piston vibrators which eliminates the need for an oiler. This optional treatment is designed to withstand temperatures ranging from −360°F to 1200°F, and also is beneficial for applications where it is difficult to refill the oiler reservoir. Always refer to manufacturers' recommendations for all installations.

The National Air Vibrator Co. produces a pneumatic shaker known as their Model ROS that works on top of the car. It is operated by two 8-in. vibrators.

15.7 LARGE-SCALE CAR UNLOADERS

For industries moving thousands of tons of coal, special car unloading devices are available which, though expensive, achieve low unit costs—for example, side dump cars (usually captive) where the entire car body is raised approximately 60–90° to the pulling axis. These cars can be used to haul sludge and materials very difficult to remove by normal unloading procedures. Navco supplies permanently mounted vibration for these cars.

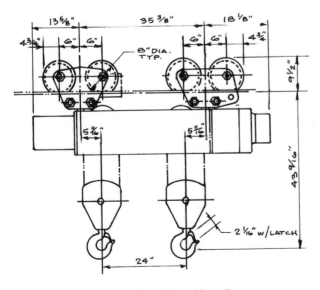

Figure 15.2 A-C car-shaker hoist with trolley.

Heating, Cooling, and Ventilating Furnace Buildings

16.1 GENERAL—HEATING AND COOLING

Sometimes it is necessary to either heat or cool material as it is conveyed or stored. The material itself can be heated or cooled, or an air or water stream can be introduced in a separate duct that is made part of the system.

Infrared dryers are used for drying drum containers for bulk material. The drums are placed on a conveyor, and are moved through a housing fitted with infrared lamps. The length and speed of travel is adjusted to time required for drying the drum paint.

Paragraph 10.9 shows the design of a system introducing inert gasses to avoid explosions. Paragraph 16.3 shows design of a lime cooling system, and 16.7 shows design for heating (or cooling) of storage bin. The calculations explain the processes involved.

The use of apron conveyors for conveying and cooling of materials is described in paragraph 5.14 and shown on figures 5.18 and 5.19.

16.2 TYPES

The two types of dryers/coolers most often used in handling of bulk materials are the rotary and the fluid bed.

The Roto-Louvre®, made by FMC, is essentially an inner shell that increases in diameter from the input to the discharge end, and an outer uniform-diameter shell. The inner shell consists of overlapping plates that form full-length louvres. The entire drum revolves at 90–150 ft per minute. Heated air is passed through the moving material. There is very little degradation of product,

moisture control is good, and resistance to wear and abrasion is good. It occupies less space than other types, and by putting the motor and drive under the dryer, space requirements may be reduced further.

In the fluid-bed process, enough air is introduced to separate and suspend the particles (refer to section 18). Material should be fairly uniform in size, and of small particles. Degradation will be at a minimum. It is a high-capacity type. Dust collectors (figure 16.1) are required with both types of dryers, but are larger with the fluid-bed type.

FMC also makes a ''Roto-Fin''® type for high-volume conduction drying or cooling of fine and granular free-flowing solids. The fins or tubes act as heat exchangers.

16.3 COOLING OF LIME

16.3.1 Data

Kiln capacity, 120 tons/day = 167 lb/min
Lime leaves kiln at   2000°F
Temperature of cooling air = 70°F
Altitude   = sea level
Fuel ratio   = 5 : 1
Heat value of coal   = 13,500 Btu
Btu/ton of lime

$$= \frac{13,500 \times 2000}{5.0} = 5,400,000 \text{ Btu}$$

Average specific heat of lime:
   range 450°F to 2000°F = 0.22 Btu/lb/°F
Specific heat of air:
   range 70°F to 500°F   = 0.23 Btu/lb/°F

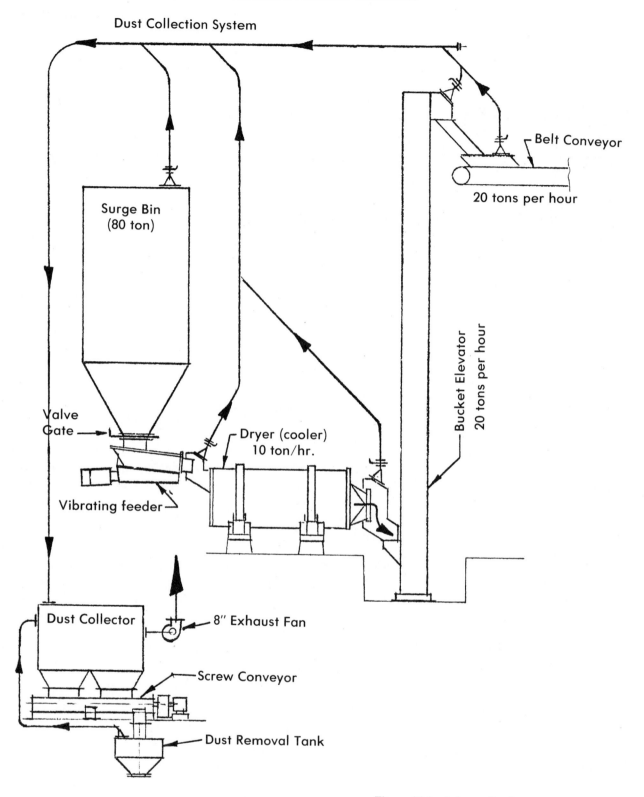

Dust Collection System

Surge Bin
(80 ton)

Valve
Gate

Vibrating feeder

Dryer (cooler)
10 ton/hr.

Belt Conveyor

20 tons per hour

Bucket Elevator
20 tons per hour

Dust Collector

8″ Exhaust Fan

Screw Conveyor

Dust Removal Tank

Figure 16.1 A dust collection system at a cooler.

Air required for combustion of
1 lb coal
(pulverized, soft)   = 10.8 lb
Pounds of coal consumed

$$= \frac{120 \times 2000}{24 \times 5.0} \qquad = 2,000 \text{ lb/hr}$$

$$= 33.3 \text{ lb/min}$$

Cubic feet of air per lb
at 70°F   = 13.33 ft^3
Lime cooled to   150°F
Excess air leaving cooler at   200°F

16.3.2 Cooling Air Requirements

Sensible heat of lime =

$$167 \times 0.22(2000° - 150°)$$

$$= 67,970 \text{ Btu/min}$$

Air required for combustion =

$$33.3 \times 10.8 = 360 \text{ lb/min at 70°F}$$

$$360 \times 13.33 = 800 \text{ cfm at 70°F}$$

With highly volatile bituminous coals, 35% of the air should be primary, thus 0.35(4800) = 1680 cfm, 126 lb/min.

Air from the cooler =

$$360 - 126 = 234 \text{ lb/min} = 3120 \text{ cfm}$$

Sensible heat content of recirculated air, assuming a temperature of 500°F for the secondary air from cooler to kiln =

$$234 \times (500°F - 70°F) \times 0.23 = 23,100 \text{ Btu/min}$$

Assuming a temperature of 400°F for air from the cooler entering the mill, the sensible heat content of air to the coal mill =

$$126(400° - 70°) \times 0.22 = 9100 \text{ Btu/min}$$

The temperature of the air and coal leaving the mill should not exceed 150–160°F. Since the amount of heat required in the mill will depend on the moisture content, ice conditions, and temperature of the coal, the amount of outside air mixed with the air from the cooler should be controlled by a temperature regulator at the outlet of the mill.

Radiation losses from the cooler (assumed as 3½% of total heat removed from line) =

$$167 \text{ lb/min } (2000°F - 150°F) \times 0.22 \times 0.035$$

$$= 2380 \text{ Btu/min}$$

Total sensible heat content of excess air required for cooling =

$$67,970 - 23,100 - 9,100 - 2380$$

$$= \text{aprx } 33,400 \text{ Btu/min}$$

Amount of excess air required =

$$\frac{33,400}{(200 - 70°) \times 0.22} = 1168 \text{ lb/min}$$

$$= 15,570 \text{ cfm at 70°F}$$

and, allowing 25% for safety, we will require 19,500 cfm at 70°F at sea level.

16.4 REQUISITION FOR COKE DRYER

Required, one coke dryer to dry 10 tons per hour of coke having a maximum moisture content of 18% to an allowable residual content of 1%.

  Maximum exhaust temperature: 220°F
  Inlet temperature: 800–1000°F
  Ambient temperature of coke: 60°F
  Weight of coke: 43 lb/ft^3
  Type of Dryer: rotary
  Electric power available: 230/460 V, 3 ph, 60 Hz
  Installation to meet FIA approval
  Space available: approximately 10'-0" × 25'-0"
  Bidder to state power required
The specifier should note

1. Since the dry material will be handled on a belt, the exit temperature should not exceed that recommended for the particular belt or for safety, but should be high enough to be of value in heating storage bins in cold weather areas, or for other use of the heat.
2. Check ability of dryer to physically transport material through it, and period of retention of product in dryer.

16.5 DESIGN OF COKE DRYER

16.5.1 Data

Material: metallurgical coke
  weight: 43 lb/ft^3
Feed to dryer = 10 tons/hr
       = 20,000 lb/hr
Moisture content = 18%
       = 0.18 × 20,000
       = 3600 lb
Ambient temperature
       = 60°F (material)
Solid weight = 20,000 − 3600
       = 16,400 lb

Residual moisture = 1%
$$= 0.01 \times 16{,}400$$
$$= 164 \text{ lb}$$

Discharge from dryer
$$= 16{,}400 + 164 = 16{,}564 \text{ lb}$$

Water evaporated
$$= 3600 - 164 = 3436 \text{ lb/hr}$$

Average specific heat
$$= 0.265 \text{ Btu/lb/}^\circ\text{F}$$

Ft^3 of air per pound
$$= 13.34 \text{ at } 70^\circ\text{F}$$
$$= 14.4 \text{ at } 212^\circ\text{F}$$

Inlet air temperature = 805°F

Exhaust air temperature
$$= 220^\circ\text{F (material out at } 200^\circ\text{F)}$$

Differential = 630°F

Specific heat of air
$$= 0.241 \text{ Btu/lb/}^\circ\text{F}$$

$0.241 \times 60 \text{ min} \times 630^\circ\text{F} = 9110 \text{ Btu}$

Enthalpy of exhaust vapor
$$= 1153.4 \text{ Btu/lb (from steam chart)}$$

Enthalpy of feed liquid
$$= 28.06 \text{ Btu/lb (from tables) or}$$
specific heat $\times T = 1.0(60^\circ - 32^\circ)$
$$= 28^\circ$$
$$= 200^\circ - 60^\circ$$
$$= 140^\circ$$

16.5.2 Internal Heat Required

Coke:

$16{,}400 \text{ lb} \times 0.265 \times 140^\circ\text{F} = 608{,}440 \text{ Btu/hr}$

Water:

$164 \text{ lb} \times 1.0 \times 140 \qquad = 22{,}960 \text{ Btu/hr}$

Evaporation:

$3{,}436 \times (1153.4 - 28.06)$

$$\begin{array}{ll} & = 3{,}866{,}668 \text{ Btu/hr} \\ & \overline{\quad 4{,}498{,}068 \text{ Btu/hr}} \\ 10\% \text{ heat losses} & = 449{,}807 \\ \text{Total:} & \overline{\quad 4{,}947{,}875 \text{ Btu/hr}} \end{array}$$

The 3,866,668 figure can also be obtained thusly:

$$\begin{array}{lr} 3436 \times 1.0(212 - 60) & = \quad 522{,}272 \\ 3436 \times 970.3 & = 3{,}333{,}951 \\ 3436 \times 0.45(220 - 212) & = \quad\ 12{,}370 \\ \hline \text{Total:} & \quad 3{,}868{,}593 \end{array}$$

16.5.3 Required Air

$$\frac{4{,}947{,}875}{9110} = 543 \text{ air/min}$$

$$543 \times 13.34 \text{ ft}^3/\text{lb} \times \frac{460^\circ + 220^\circ}{460^\circ + 70^\circ}$$
$$= 9294 \text{ cfm air}$$

$$\frac{3436}{60} \times 26.8 \text{ ft}^3/\text{lb} \times \frac{460^\circ + 220^\circ}{460^\circ + 212^\circ}$$
$$= 1553 \text{ cfm vapor}$$

Total cfm through dryer =

$$9249 \text{ cfm air} + 1553 \text{ cfm vapor}$$
$$= 10.847$$

with assumed through-bed velocity of 100 ft/min (108 ft^2 required). Selected bed size (L-B data) = 104.9 ft^2; bed velocity =

$$\frac{10{,}847}{104.9} = 103.4 \text{ ft}$$

A velocity of up to 120 fpm is recommended for metallurgical coke. The velocity is figured at exhaust temperature.

16.5.4 Thermal Efficiency

Allowing 10% for fans and 10% for miscellaneous losses,

$$543 \text{ lb} \times 14.4 \text{ ft}^3/\text{lb} \times (850^\circ\text{F} - 60^\circ\text{F})$$
$$\times 1.1 \times 1 = 7{,}474{,}373 \text{ Btu/hr}$$

The thermal efficiency is found by dividing the total heat required by the total heat input:

$$\frac{4{,}947{,}875}{7{,}474{,}373} = 66\%$$

16.5.5 Inlet Air

With the required air being 543 lb/min, at 70°:

$$543 \times 13.34 \text{ ft}^3/\text{lb} = 7244 \text{ ft}^3$$

Using a 10% factor for the safety of the fan sizing gives a figure of 724. Thus the total acfm at 70°F is

$$7244 + 724 = 7968 \text{ (say, } 8000)$$

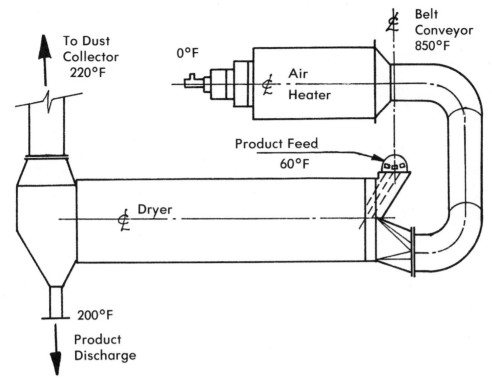

Figure 16.2 Layout of a coke dryer.

16.5.6 Exhaust Air

With the exhaust air being 10,847 acfm at 220°F and given a 10% factor for fans:

10,847 + 1,085 = 11,932 acfm at 220°F

16.5.7 Furnace Heat Required (see figure 16.2)

The air comes in at 0°F* and goes out at 850°F. Allowing 10% for fans and 10% for miscellaneous losses:

543 × 0.241 × (850°F − 0°F)

× 1.1 × 1 × 60 min = 8.08 × 10^6 Btu/hr

where 0.241 is the specific heat of air/lb/°F.

16.6 OPERATION OF THE ROTO-LOUVRE

The main unit of the Roto-Louvre (figure 16.3) is a cylindrical drum arranged to rotate on a horizontal axis

---

*0°F is used as a safety factor for furnace sizing.

and enclosing a series of overlapping louvres (figure 16.4). Louvres extend the full length of the drum and form a conical inner shell that increases in diameter toward the discharge end. Louvre supports are continuous radial partitions that form longitudinal passages extending the entire length of the drum.

16.7 HEATING MATERIAL IN BINS

16.7.1 General

Note that all nomenclature and equation numbers used in paragraph 16.7 are those used in Gebhardt's "Heat Transfer," and are different from those in the rest of this book. The values for any two projects will be different because of the moisture content of the coke and its temperature. It may be necessary to refer to the original Gebhardt text.

Figure 16.5 shows a bin storing 80 tons of coke that arrives wet or frozen in cakes. Ten tons of coke and moisture is drawn from tank every 7 hours. The temperature of the coke, the moisture in the coke, and the moisture leaving the tank should be 45°F. Heat is available from a near dryer. It was decided to use this heat

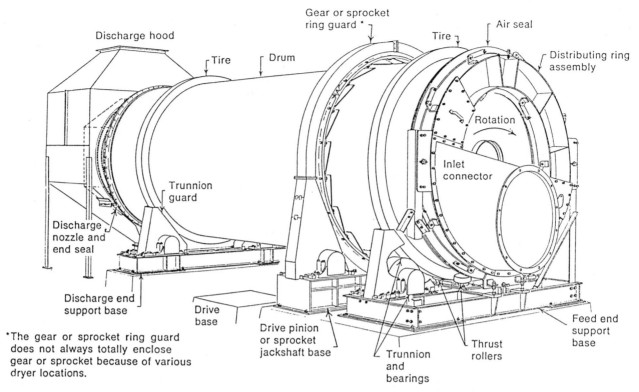

Figure 16.3 The Roto-Louvre. (*Courtesy FMC*)

*The gear or sprocket ring guard does not always totally enclose gear or sprocket because of various dryer locations.

to avoid freezing of the material in the bin. The pipes are also intended to prevent the stored material from arching, and to help produce an even flow of hot air. The figure defines sections and temperatures for heat-transfer analysis.

For purposes of this design, it is accurate enough to assume that the temperature of any horizontal section is uniform at any instant, and that the properties of the material are independent on temperatures. The volume of the six pipes is neglected.

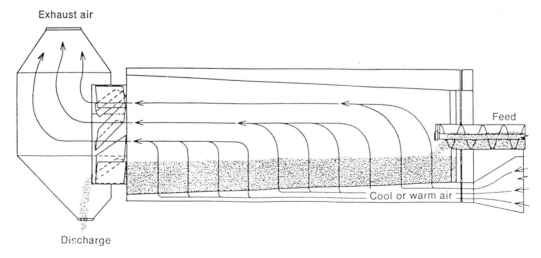

Figure 16.4 Operation of the Roto-Louvre. (*Courtesy FMC*)

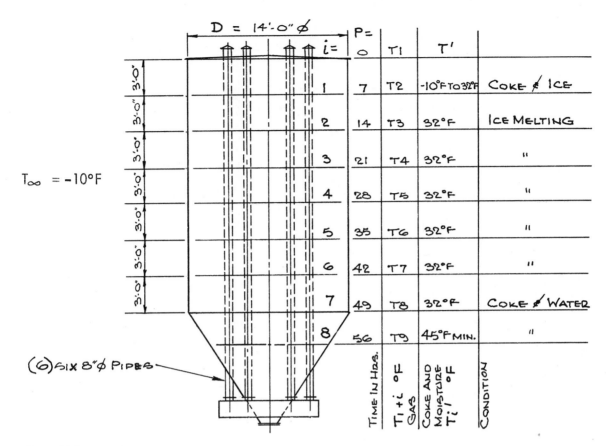

Figure 16.5 Heat diagram.

16.7.2 Volumes and Areas

Since the cylindrical section of the bin has a capacity of 70 tons, it is divided into seven sections, with each section 3 ft high.

The cross-section area is $A_r = 153.93$ ft². The volume of a 3-ft high cylinder is $153.93 \times 3 = 461.8$ ft³, and the total capacity of the cylindrical section is 3233 ft³, or 9.95 (say about 10) tons per section.

The surface area of the cylindrical part of the tank is 923.6 ft², or $A_c = 131.9$ ft² per section.

The cone area is $A_{co} = 256.5$ ft², the volume is 478.6 ft³, and capacity is 10.3 tons.

The outside surface area of the six 8-in. pipes 30 ft 4 in long, is 407.8 ft². The inside surface area of the pipes is 377.4 ft² and the inside surface area of a 3-ft long section of the pipes is $A_r = 37.33$ ft². The cross-section of the area of the six pipes is 2.08 ft² inside. The inside volume of the pipes = 62.7 ft³.

Weight per 3-ft section, with moisture at 18% = 10 × 2000 × 0.18 = 3600 lb of moisture plus the coke = 10 tons × 2000 × 0.82% = 16,400 lb of coke.

16.7.3 Properties

16.7.3.1 Heat Capacities Per 10 Tons.

Coke: $C_c = 16,400 \times 0.2$
$= 3280$ Btu/°F

Ice: $C_i = 3600 \times 0.476$
$= 1717$ Btu/°F

Water: $C_w = 3600$ Btu/°F

Heat of fusion:

$$h_{sf} = 144 \text{ Btu/lb}$$

The value of C_c varies. The assumption of 0.2 is the author's. C_i is from Marks' *Mechanical Engineers' Handbook*, page 300 (courtesy of McGraw-Hill). The coefficient of heat conductivity of coke and moisture is

$$k = 6 \text{ Btu/hr/ft}^2/°F$$

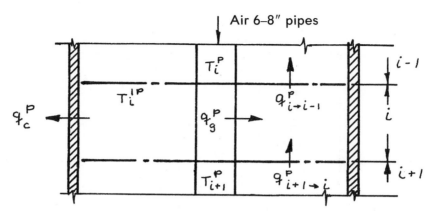

Figure 16.6 Heat transfer.

16.7.3.2 Convection Coefficients. For the outside bin surface:*

Top (roof): h_r = 0.165
(Gebhart eq. 8-73)

Conical part: h_{co} = 0.84
(Gebhart eq. 8-72)

Cylindrical
    section: h_c = 0.84
(Gebhart eq. 8-63)

All three figures represent Btu/hr/ft²/°F.

Based on Gebhart's equation 7-48, forced convection inside the pipe at 75°F film temperature and with a velocity of 24 fps is h = 4.47 Btu/hr/ft²/°F. The velocity is assumed, based on an experienced guess.

16.7.3.3 Gas Flow Rate.

$$V = A_{pc} \times V_p \times 60 \text{ min}$$

$$= 2.08 \times 24 \times 60$$

$$= 2995 \text{ cfm}$$

16.7.3.4 Insulation. One inch of fiberglass with a thermal resistance of R = 3.62 Btu/hr/ft²/°F is used on the tank.

16.7.4 Heat Transfer Equations*

Figure 16.6 defines temperatures and heat transfer quantities for heat balance analysis of any section i at time

---

*Source: B. Gebhart, 1971, *Heat Transfer*, 2nd ed., New York: McGraw-Hill, by permission.

p. The heat losses from the bin surface are determined by:

1. losses through the roof

$$q_r^p = \frac{A_r(T_i'^p - T_\infty)}{R + (1/h_r)}$$

$$= \frac{153.93(T_i'^p - T_\infty)}{3.62 + (1/0.165)}$$

$$= 15.9(T_i'^p - T_\infty) \text{ Btu/hr}$$

where A_r is the area of the roof (from paragraph 16.6.2), R is the thermal resistance of the insulation (from paragraph 16.7.3.4), and h_r is the convection coefficient of the top of the roof (from paragraph 16.7.3.2).

2. losses through conical section

$$q_{co}^p = \frac{A_{co}(T_8'^p - T_\infty)}{R + (1/h_{co})}$$

$$= \frac{257.5(T_8'^p - T_\infty)}{3.62 + (1/0.84)}$$

$$= 57.6(T_8'^p - T_\infty) \text{ Btu/hr}$$

where A_{co} is the area of the conical section (from paragraph 16.6.2) and h_{co} is the convection coefficient of the conical part (from paragraph 16.7.3.2).

3. losses through cylindrical 3-ft section (refer to Gebhart, p. 460)

$$q_{ic}^p = \frac{A_c(T_i'^p - T_\infty)}{R + (1/h_c)}$$

$$= \frac{131.9(T_i'^p - T_\infty)}{3.62 + (1/0.84)}$$

$$= 27.4(T_i'^p - T_\infty) \text{ Btu/hr}$$

where A_c is the area of the 3-ft cylindrical section (from paragraph 16.6.2) and h_c is the convection coefficient of the cylindrical section (from paragraph 16.7.3.2).

4. losses from air in six 8-in. pipes for 3-ft section (refer to Gebhart, p. 13)

$$q_g^p = \frac{h_g A_p (T_i^p - T_{i+1}^p)}{ln\left[(T_i^p - T_i'^p)(T_{i+1}^p - T_i'^p)\right]}$$

where h_g is the forced convection coefficient for air (in the range of 2–15 Btu/hr/ft²/°F) that, for the given conditions (air velocity of 24 fps), equals 4.47 Btu/hr/ft²/°F, and where A_p is the inside area of the pipes equalling 37.6. Therefore, $h_g A_p = 168.07$ Btu/hr/°F.

Applying the first law of thermodynamics to the air in the pipes, we get

$$q_g^p = C_p \zeta V (T_{i+1}^p - T_i^p)$$

$$= 0.24 \times 0.074 \times 2995 (T_{i+1}^p - T_i^p)$$

$$= 3191 (T_{i+1}^p - T_i^p) \text{ Btu/hr}$$

where C_p is the specific heat capacity of air in Btu/lb/°F, ζ is the air density in lb/ft³, and V is the cfm (from paragraph 16.7.3.3).

5. heat conduction from sections adjacent to section i

The expressions $q_{i \to i-1}^p$ and $q_{i+1 \to i}^p$ represent the heat conducted from sections $i - 1$ to i and from $i + 1$ to i, respectively, and are equal to

$$A_r k \frac{\Delta T}{\Delta x} \text{ Btu/hr}$$

where A_r is the cross-section area for conduction (from paragraph 17.6.2), k is the coefficient of heat conductivity of coke and moisture (from paragraph 16.7.3.1), ΔT is the temperature in °F between two adjacent sections, and Δx is an effective length in ft. Therefore,

$$q = A_r k \frac{\Delta T}{\Delta x}$$

$$= 153.93 \text{ ft}^2 \times 6.0 \text{ Btu/hr/ft}^2/°F \times \frac{\Delta T}{2}$$

$$= 462 \Delta T \text{ Btu/hr}$$

Finally,

$$q_{i \to i-1}^p = 462 (T_i'^p - T_{i-1}'^p) \text{ Btu/hr}$$

and

$$q_{i+1 \to i}^p = 462 (T_{i+1}'^p - T_i'^p) \text{ Btu/hr}$$

6. sum of heat transfer (energy-gain) rates at time p

The sum of heat transfer rates at time p is set equal to the time rate change of stored energy for section i.

$$q_g^p - q_r^p - q_{co}^p - q_{ic}^p + q_{i+1 \to i}^p - q_{i \to i+1}^p$$

$$= (C_c + C_i + C_w)(T_i'^{p+1} - T_i'^p) + h_{sf} M_{ice}$$

where M_{ice} is the ice melted in lb/hr. Not all of the terms in this equation make a contribution. Whether they do or not depends on the section and the time.

16.7.5   Sample Computations for Section 1 (see figure 16.5)

The time cycle is from $p = 0$ to $p = 7$ hr. At $p = 0$, the coke and ice are at $-10°F$. The temperatures are calculated at the beginning of each 1-hr interval and are considered constant during the 1-hr period.

Using equations and data in paragraphs 16.7.3 and 16.7.4, the air temperature leaving section 1 is

$$T_2^p = T_1'^p + \text{Exp}\left[ln(T_1^p - T_1'^p) - 0.053\right]$$

The heat loss is through the roof and the 3-ft cylindrical section.

$$q^p \text{ loss} = q_{ic}^p + q_r^p = 43.4 (T_1'^p + 10) \text{ Btu/hr}$$

The heat transfer to section 1 from section 2 (see figure 16.5) is

$$q_{2 \to 1}^p = 462 (T_2'^p - T_1'^p) \text{ Btu/hr}$$

The heat transfer from the air to the coke in section 1 is

$$q_g^p = 3191 (T_1^p - T_2^p) \text{ Btu/hr}$$

The net heat transfer to the coke and ice of section 1 is

$$q_{net}^p = q_g^p + q_{2 \to 1}^p - q^p \text{ loss Btu/hr}$$

The net energy gain of section 1 may increase its temperature to

$$T_i'^{p+1} = T_i'^p + q_{net}^p / 4994 °F$$

or melt part of the ice when $T_i'^p = 32°F$ is reached. The ice melted is calculated by

$$M_{ice} = q_{net}^p / 144 \text{ lb/hr}$$

Assuming that $T_i^p = 155°F$ and $T_\infty = -10°F$, and the initial temperature of the coke and ice in section 1 is $T_1' = -10°F$, then all the quantities given by the seven equations above at time $p = 0$ can be determined except for $q_{2 \to 1}^p$, which requires knowledge of the temperature of the adjacent section, T_2'. Since it is not known, it must be assumed. In this case we will assume $T_2'^p = 32°F$. When $T_2'^p$ is later calculated, the calculations for section 1 must be recalculated. This iteration must continue until a satisfactory result is obtained. Table 16.1 shows the results for section 1 for $p = 0$ to $p = 7$ hr. Three iterations are performed and final results are given in table 16.2 for $T_1 = 155°F$.

The total mass of ice melted is 3682 lb, which is greater than the 3600 lb required. The final temperature of the 10-ton batch before it is drawn is 49°F, which satisfies the temperature requirement of 45°F.

The total heat transfer from the air to the coke and moisture is maximum

$$Q_{c+w} = C_p \zeta V(T_1 - T_q) = 3191(155 - 111.1)$$
$$= 140,000 \text{ Btu/hr}$$

16.7.6 Details of Bin Design

Some details of the bin design are shown on figure 16.7. Figures 16.1 and 16.2 show a diagram of a dryer sys-

Table 16.1 Tabulation of Calculations for Section 1

$P(hr)$	0	1	2	3	4	5	6	7
T_2 (°F)	146.5	147.0	147.4	147.7	148.0	148.0	148.5	
q_{loss}^p (Btu/hr)	0	405	755	1,060	1,326	1,558	1,758	
$q^p\ 2 \to 1$ (Btu/hr)	19,404	15,093	11,365	8,109	5,279	2,820	683	
q_g^p (Btu/hr)	27,178	25,642	24,312	23,151	22,142	21,266	20,504	
q_{net}^p (Btu/hr)	46,582	40,330	34,922	30,199	26,095	22,528	19,429	
$T_1'^p$ (°F)	−10	−0.67	7.4	14.4	20.6	25.9	30.5	32
M_{ice} (lbm/hr)	0	0	0					82.9
Assumed $T_2'^p$	32	32	32	32	32	32	32	32

Note: Similar calculations are performed for the other sections. After 7 hr, 10 tons is drawn from the bin and section 1, above, drops down and becomes section 2. A 10-ton batch will stay in the bin 56 hr, going through the 7-hr cycle eight times.

Table 16.2 Tabulation of Final Results

Time (hr)	1	2	3	4	5	6	7
Comb. T_1 °F	155	155	155	155	155	155	155
Gas T_q °F	111.1	111.4	111.4	111.5	111.5	111.6	111.7
Coke + ice T_1' °F	−0.67	7.4	14.4	20.6	25.9	30.5	34.5
Melting $T_2' = \ldots = T_6'$ °F	32	32	32	32	32	32	32
Coke T_7' °F	34.3	36.8	38.9	40.8	42.4	43.7	44.9
+ Water T_8' °F	45.8	46.3	46.7	47.2	47.7	48.4	49.0
Mass of $M_{ice\,1}$							87.3
Ice $M_{ice\,2}$	−7.6	15.	45.	67.9	88.	105.3	120.4
Melted in $\sum_3^5 M_{ice}$				2328.5			
lbm $M_{ice\,6}$	101.4	107.2	113.5	119.3	124.9	130.3	135.6

FIGURE 16.7 391

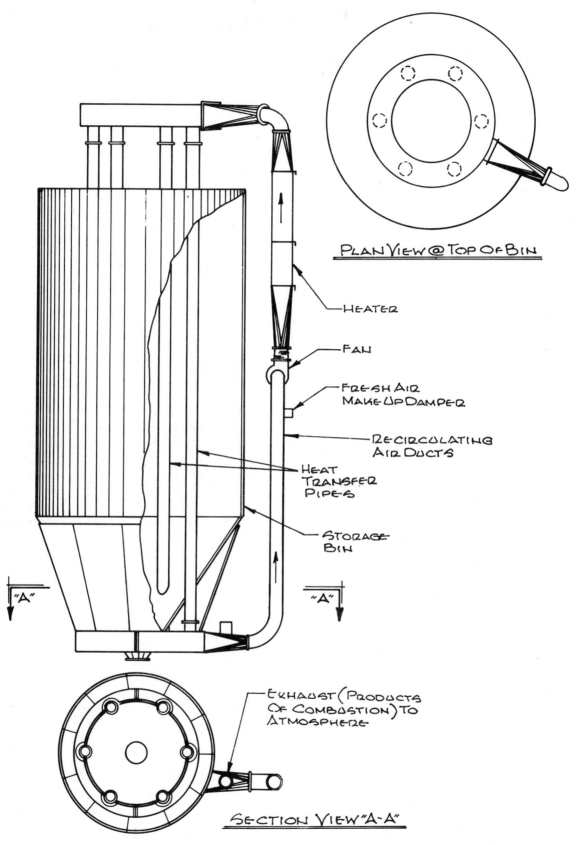

PLAN VIEW @ TOP OF BIN

HEATER

FAN

FRESH AIR
MAKE-UP DAMPER

RECIRCULATING
AIR DUCTS

HEAT
TRANSFER
PIPES

STORAGE
BIN

"A"   "A"

EXHAUST (PRODUCTS
OF COMBUSTION) TO
ATMOSPHERE

SECTION VIEW "A-A"

Figure 16.7 Details of a bin design.

tem. Note that the number of vertical pipes and their spacing as shown in the drawings should be reduced for larger particles, and should be eliminated for lumps over 2 in.

16.7.7   Nomenclature for Paragraph 16.7

A_{co} = area of conical section in ft^2
      = 256.6 ft^2
A_c = area of cylindrical section in ft^2
      = 131.9 ft^2
A_p = area inside the six 8-in. pipes in ft^2
      = 37.6 ft^2
A_{pc} = cross-section of the six pipes
      = 2.08 ft^2
A_r = area of the roof in ft^2
      = 153.9 ft^2

C = material heat capacity in Btu/°F/lb

C_c = coke = 0.20
C_i = ice   = 0.476
C_w = water = 1.00

h = convection coefficient in Btu/hr/ft^2/°F

h_c = cyclindrical section = 0.84
h_y = top (roof)   = 0.165
h_{co} = conical part   = 0.84

Forced convection inside the pipe at 75°F and at velocity of V_p = 24 fps is h_{sf} = 4.47 Btu/hr/ft^2/°F

p = index for time in 1-hr increments
i = index for sections

q = heat losses or transfer to adjacent sections in Btu/hr

q_r^p = roof of the bin
q_{co}^p = surface of the conical section of the bin
q_{ic}^p = cylindrical 3-ft section of the bin
q_g^p = from the gas in the six 8-in. pipes (3-ft long) to the coke

T = temperature in °F
T_∞ = outside air temperature at -10°F
$T_i^{'p}$ = section i at time p
$T_{i+1}^{'p}$ = section $i+1$ at time p
T_i^p = entering air at time p
T_{i+1}^p = outgoing air at time p
T' = coke

Exp = exponent to base "e"
ln = natural logarithm
M_{ice} = ice melted in lb/hr

16.8   GENERAL—VENTILATION

To study the average temperature and the air changes in the work-level area under different directions and velocities of wind, calculations were made and summarized in table 16.3. Because of the repetitive nature of the design, most of the computations were originally made on a TI-SR51A calculator. To speed up the work, the program described in paragraph 16.11 was developed. Three tabulations (items 2, 4, and 6 in table 16.3) were checked against the original computations.

Table 16.3   Summary of Calculations of Air Flow

Item No.	Heating Load 10^6 (Btu/h)	Wind Direction and Velocity (mph)	t_o (°F)	Δt	Airflow Rate × 1000 (cfm)	Air Changes per Hour
1	70.4	SW, 6	125	38	1893.5	34.7
2	80.3	SW, 6	129	42	1970.0	36.1
3	59.6	W, 8	123	36	1697.0	31.1
4	79.55	W, 8	130	43	1822.8	33.4
5	67.5	S, 6	123	36	1921.0	35.2
6	82.7	S, 6	129	42	2028.0	37.2
7	77.66	0	129	42	1904.0	34.9
8	91.17	0	129	42	2010.0	36.8

Notes: Wind direction and velocity are assumed, using the best available data, and the heating load is from other data.

16.9 CHECK OF BUILDING VENTILATION

16.9.1 Object

The object of this exercise is to evaluate the possible rate of air change and the average temperature in the work-level area of the building.

16.9.2 Data (see figure 16.8)

Volume = length × width × average height

= 575 ft × 110 ft × 51.76 ft

= 3,273,734 ft³

Heat produced during the process = 23,900,000 Btu/hr.

16.9.3 Ventilation Arrangement

16.9.3.1 Inlet Air Through Doors and Louvers.

South elevation
vent doors:

21 × 10' × 11.5'	= 2415 ft²

overhead doors:

1 × 14' × 14'	= 196 ft²

south wall
subtotal (A_s): 2611 ft²

West elevation
overhead door:

1 × 14' × 12'	= 168 ft²

louvers:

2[18.72(4.17 + 5.42)] × 0.5
= 179.5 ft²

louvers:

2[12.38(4.17 + 5.42)] × 0.5
= 118.7 ft²

west wall subtotal (A_w): 466.2 ft²

North elevation
overhead doors:

2 × 14' × 14'	= 392 ft²

vent doors:

5 × 12' × 12'	= 720 ft²

louvers:

10 × [18.72(4.17 + 5.42)]
× 0.5 = 897.6 ft²

north wall subtotal (A_n): 2009.6 ft²

East elevation
overhead door:

1 × 25' × 23'	= 575 ft²

louvers:

4[12.38(4.17 + 5.42)] ×
0.5 = 237.4 ft²

louvers:

2[18.72(4.17 + 5.42)] ×
0.5 = 179.5 ft²

east wall subtotal (A_e): 991.9 ft²

Total effective inlet area = 6078.1 ft².

16.9.3.2 Engineering Data.
The engineering data has been taken from Buffalo Forge Company's "Fan Engineering" (B); ASHRAE Handbook "Fundamentals" (A); and "Ventilation of Industrial Plants," Baturin (VB). These sources will be referred to in the text by the initials following them.

16.9.3.3 Roof Monitor.
The roof monitor consists of two symmetrical streamlined vents 4 ft wide by 573.5 ft long. The total outlet area of the roof vents is

$$A = 2 \times 4 \times 573.5$$

$$= 4588 \text{ ft}^2$$

The ratio of inlet area to outlet area γ is

$$\gamma = \frac{\Sigma A \text{ in}}{\Sigma A \text{ out}}$$

$$= \frac{6078.1}{4558}$$

$$= 1.325$$

16.9.4 Heat Content of the Discharged Air

The rate of ventilation can be checked using the following equation:

$$H = 3600 G_m C_p \Delta t \qquad \textbf{(Eq. 1)}$$
$$= 3600 Q \gamma (t_o - t_a)$$

where

G_m = rate of ventilation, lb/sec
C_p = specific heat of air
= 0.24 Btu/lb/°F (B)
t_o = exit air temperature, °F
t_a = outside air temperature
= 87°F
Q = volume of air discharged from the building
$\gamma = P/RT$

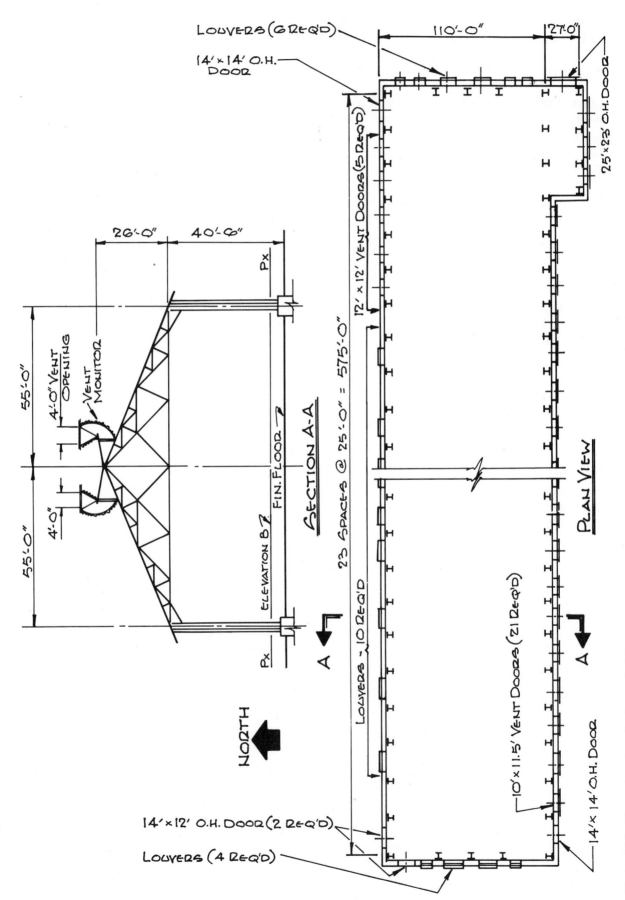

Figure 16.8 Plan and section of a furnace building.

For $\gamma = P/RT$, P is the absolute atmospheric pressure in lb/ft^2 at the top of the roof monitor (the elevation = 700 + 70 = 770 ft above sea level = 29.1 inches of mercury) (B); R is the gas content for air; and T is the absolute temperature. P is calculated by

$$P = \frac{29.1 \times 13.61 \times 62.4}{12}$$

$$= 2059.5 \text{ lb/ft}^2$$

Therefore,

$$\gamma = P/RT$$

$$= 2059.4/53.3T \qquad (\textbf{Eq. 2})$$

$$= 38.639/T$$

16.9.5 Evaluation of the Pressure Differential for Air Flow

The air flow rate through an opening of area A due to the pressure differential ΔP (in lb/ft^2) between the indoors and outdoors is calculated using the equation:

$$G = AC\sqrt{2g\gamma\Delta P} \text{ (VB)} \qquad (\textbf{Eq. 3})$$

where G is the rate of air flow in lb/sec, C is the discharge coefficient of the opening, and g is gravitational acceleration. G is always less than unity. Its value is a function of the flow conditions, geometry of the opening, viscosity, and density of the gas. The air density in lb/ft^3 is represented by γ.

In the following analysis we assumed:

1. steady-state conditions, that is, all factors affecting the ventilating rate remain constant,
2. air pressure varies linearly with the elevation,
3. infiltration of air through the building enclosure is neglected,
4. internal resistance along the path traveled by the air inside the building is neglected, and
5. the amount of air passing through the inlet openings is equal to the amount leaving the building.

Consider vent openings in the building as shown on section A-A of figure 16.8. Let the internal pressure of the air on the level of the vent openings be P_x and the outside pressure exerted by the wind on the inlet opening be P_1 (vent no. 1). The existing pressure for vent no. 1 is

$$\Delta P_1 = P_1 - P_x$$

This same relationship exists for each inlet vent at elevation B (section A-A, fig. 16.8).

$$\Delta P_2 = P_2 - P_x$$

$$\Delta P_3 = P_3 - P_x$$

$$\Delta P_4 = P_4 - P_x \qquad (\textbf{Eq. 4})$$

where P_1 is wind pressure on the south wall; P_2, on the west wall; P_3, on the north wall; and P_4, on the east wall. The wind pressure in lb/ft^2 can be calculated using the equation

$$P_w = k\gamma x\left(\frac{5280V_w}{360}\right) \times \frac{1}{2g} \qquad (\textbf{Eq. 5})$$

where

P_w = wind pressure in lb/ft^2
k = coefficient, depending on the direction of the wind and the geometry of the building enclosure (see table 16.4 and figures 16.9 and 16.10*)
γ = outside air density for $t = 87°$F
V_w = wind velocity (mph)
g = gravitational constant = 32.17 ft/sec

Using Equation (2) gives

$$\gamma_1 = \frac{P}{RT}$$

$$= \frac{2059.4}{53.3(460 + 87)}$$

$$= \frac{38.639}{460 + 87}$$

$$= 0.07064 \text{ lb/ft}^2$$

Substituting values in Equation (5)

$$P_w = k \times 0.07064\left(\frac{5280V_w}{3600}\right)^2 \times \frac{1}{2 \times 32.17}$$

$$= 0.002362kV_w^2. \qquad (\textbf{Eq. 6})$$

We will compare this calculation with the ASHRAE formula $P = 0.000482(V_w)^2$ inches of water, where V = mph, and γ is given as 0.075. To convert inches of water to lb/ft^2, divide by 5.2 (value at 4°C).

$$P_w = 0.075 \times \left(\frac{5280}{3600}\right)^2 \times V_w^2 \times \frac{1}{2 \times 32} \times \frac{1}{5.2}$$

$$= 0.000484V^2$$

We see that it agrees with Equation (6). The factor k is added to Equation (5).

---

*The data on table 16.4 and figures 16.9 and 16.10 are based on actual tests made in Russia and Poland. The values of k used in these computations are interpolated from these data.

Table 16.4   Air Change: Pressure Coefficients on Building Surface

Wind Direction	Position									
	1	*2*	*3*	*4*	*5*	*6*	*7*	*8*	*9*	*10*
0°	−0.26	−0.68	−0.34	−0.75	−0.17	−0.60	+0.96	+0.95	+0.93	+0.90
15	−0.15	−0.09	−0.16	−0.09	−0.30	−0.37	+1.00	+0.70	+1.00	+0.78
30	−0.04	+0.16	−0.04	+0.25	−0.39	−0.27	+0.87	+0.46	+1.00	+0.45
45	+0.17	+0.41	+0.10	+0.52	−0.32	−0.29	+0.66	+0.14	+0.70	+0.09
60	+0.49	+0.70	+0.37	+0.82	−0.57	−0.69	±0	−0.06	±0	−0.11
75	+0.70	+0.93	+0.57	+0.93	−0.61	−0.93	−1.00	−0.15	−1.10	−0.36
90	+0.85	+0.98	+0.72	+0.96	−0.64	−0.90	−0.84	−0.64	−0.86	−0.64
105	+1.00	+0.92	+1.00	+0.98	−0.30	−0.95	−0.54	−0.62	−0.55	−0.59
120	+0.94	+0.64	+1.10	+0.79	−0.38	−0.62	−0.51	−0.54	−0.47	−0.64
135	+0.71	+0.41	+0.81	+0.50	−1.02	−0.32	−0.44	−0.49	−0.42	−0.48
150	+0.32	+0.15	+0.39	+0.21	−0.97	−0.31	−0.41	−0.54	−0.38	−0.50
165	−0.84	−0.04	−0.85	−0.08	−0.90	−0.34	−0.33	−0.45	−0.30	−0.41
180	−0.96	−0.88	−0.92	−0.73	−1.07	−0.73	−0.23	−0.23	−0.21	−0.23

Source: From V. Baturin, *Fundamentals of Industrial Ventilation*, International Series of Monographs on Heating, Ventilation, and Refrigeration, vol. 8, Oxford: Pergamon Press, by permission of Pergamon Press; copyright 1972.

The internal pressure in the building, at the level of the monitor exit, L ft from the center line of the inlet vent openings, is

$$P_m = P_x - L\gamma \ \text{lb/ft}^2 \qquad \textbf{(Eq. 7)}$$

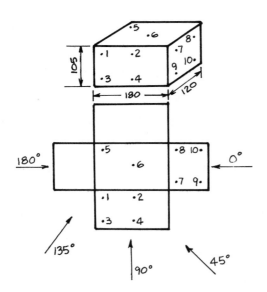

Figure 16.9   Investigation of wind pressures on a building. (Reproduced from V. Baturin, 1972, *Fundamentals of Industrial Ventilation*, International Series of Monographs in Heating, Ventilation, and Refrigeration, vol. 8, Oxford: Pergamon Press)

where

P_m = internal pressure on the level of the roof monitor exit in lb/ft^2

P_x = internal pressure at the inlet level of the building vents

L = height from elevation B to the top of the roof monitor in ft = 64 ft

γ_{av} = average density of the air inside the building for the stated height L

The outside atmospheric pressure at elevation L, due to wind pressure and change in elevation is

$$P'_m = P_r - L\gamma_1 \qquad \textbf{(Eq. 8)}$$

where P_r is the wind pressure at roof monitor level. It can be evaluated using Equation (6).

Finally, the existing pressure differential at the monitor is

$$\Delta P_m = P_m - P'_m$$

or, using Equations (7) and (8)

$$\Delta P_m = (P_x - L\gamma_{av}) - (P_r - L\gamma_1)$$
$$= P_x - [P_r - L(\gamma_1 - \gamma_{av})] \qquad \textbf{(Eq. 9)}$$
$$= P_x - P_f \qquad \textbf{(Eq. 10)}$$

FIGURE 16.10 397

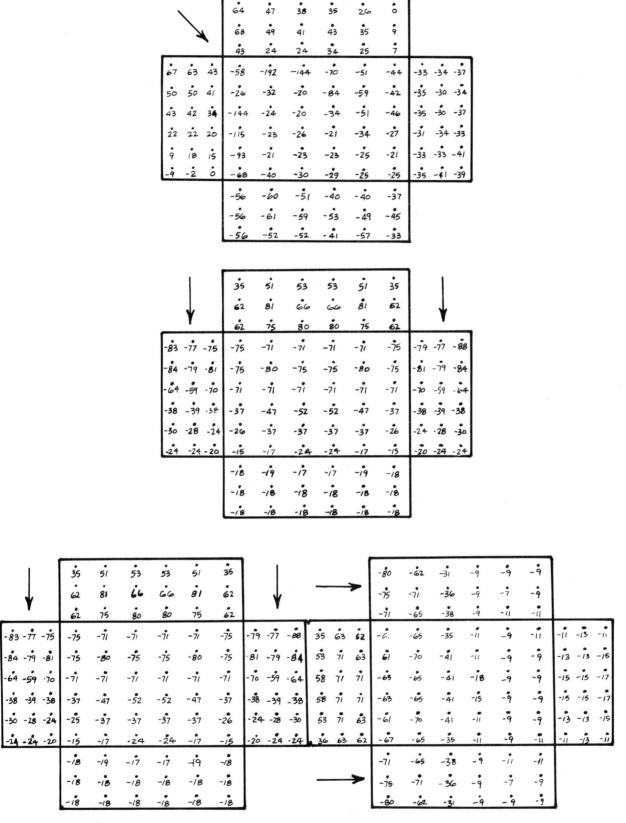

Figure 16.10 Values of *k* in percentage of Pw. (Values from and reprinted by permission of Prof. Jan Ferencowicz, *Ventilation and Air Conditioning*, Warsaw: Arkady Budownistwo-Szpuka-Architektura, in Polish)

where

$$P_f = P_r - L(\gamma_1 - \gamma_{av})$$

$$= L(\gamma_1 - \gamma_{av}) - \text{(pressure due to}$$

temperature difference)

(Eq. 11)

To calculate the available pressure differential for natural ventilation for the inlet and outlet vents of the building, we will evaluate two unknown variables:

1. P_x = the internal pressure at the inlet level of the building vents and
2. γ_{av} = the average density of the air inside the building for height L.

A precise evaluation of the average density of air, γ_{av}, is not possible. Experience has shown that the density should be evaluated for hot process buildings, with a large ratio of furnace area to the building floor area at average temperature.

$$t_{av} = b + t_a + \frac{\Delta t}{2} \qquad \textbf{(Eq. 12)}$$

where b varies from 0°F to 15°F, t_a is the outside temperature in °F, and Δt is the temperature rise of the air in the building that can be calculated if the heat load is known. In the following calculations, t_{av} is calculated assuming that $b = 5$°F.

16.10 CALCULATION OF THE FLOW RATE

16.10.1 Case 1: NW Wind at 6 mph

The wind pressure coefficients for the building vents are assumed to be

$$\text{south wall } (A_s) = 2611 \text{ ft}^2,$$
$$k_1 = -0.41$$
$$\text{west wall } (A_w) = 466.2 \text{ ft}^2,$$
$$k_2 = 0.33$$
$$\text{north wall } (A_n) = 2009.6 \text{ ft}^2,$$
$$k_3 = 0.36$$
$$\text{east wall } (A_e) = 991.9 \text{ ft}^2,$$
$$k_4 = 0.33$$
$$\text{roof monitor } (A_m) = 4558 \text{ ft}^2,$$
$$k_m = 0.21$$

Applying Equation (6), calculate the wind pressure for each vent

$$P_w = 0.002362 \, kV_w^2$$

substituting $V_w = 6$ mph in the equation gives

$$P_w = 0.002362 \, k \times 6^2 = 0.085 \, k$$

$$P_1 = 0.085 \times k_1$$
$$= 0.085 \times (-0.41)$$
$$= -0.035 \text{ lb/ft}^2$$

$$P_2 = 0.085 \times k_2$$
$$= 0.085 \times 0.33$$
$$= 0.028 \text{ lb/ft}^2$$

$$G_1 + G_2 + G_3 + G_4 = G_m \qquad \textbf{(Eq. 13)}$$

Applying Equation (3) for each of the flow components, we obtain for the wall inlets:

south wall

$$G_1 = A_s C_1 \sqrt{2g\gamma_1(P_1 - P_x)}$$
$$= A_s C_1 \sqrt{2g\gamma_1} \sqrt{P_1 - P_x}$$

west wall

$$G_2 = A_w C_2 \sqrt{2g\gamma_1(P_2 - P_x)}$$
$$= A_w C_2 \sqrt{2g\gamma_1} \sqrt{P_2 - P_x}$$

north wall

$$G_3 = A_n C_3 \sqrt{2g\gamma_1(P_3 - P_x)}$$
$$= A_n C_3 \sqrt{2g\gamma_1} \sqrt{P_3 - P_x}$$

east wall

$$G_4 = A_e C_4 \sqrt{2g\gamma_1(P_4 - P_x)}$$
$$= A_e C_4 \sqrt{2g\gamma_1} \sqrt{P_4 - P_x}$$

where

$A_s = 2661 \text{ ft}^2$
$A_w = 466.2 \text{ ft}^2$
$A_n = 2009.6 \text{ ft}^2$
$A_e = 991.4 \text{ ft}^2$
C_1 = discharge coefficient for rectangular vent doors
$\quad = C_2 = C_4 = 0.64$
C_3 = industrial louvers coefficient
$\quad = 0.61$
$g = 32.17 \text{ ft/sec}^2$
$\gamma_1 = 0.07064 \text{ lb/ft}^3$
$P_1 = -0.035 \text{ lb/ft}^2$
$P_2 = 0.028 \text{ lb/ft}^2$
$P_3 = 0.03 \text{ lb/ft}^2$
$P_4 = -0.028 \text{ lb/ft}^2$

Substituting these values in the equations above, we obtain

$$G_1 = 2611 \times 0.64 \sqrt{2 \times 32.17 \times 0.07064}$$
$$\cdot \sqrt{-0.035 - P_x}$$
$$= 2611 \times 0.64 \times 2.1319 \sqrt{-0.035 - P_x}$$
$$= 3562.5 \sqrt{-0.035 - P_x}$$

$$G_2 = 466.2 \times 0.64 \times 2.1319 \sqrt{0.028 - P_x}$$
$$= 636.1 \sqrt{0.028 - P_x}$$

$$G_3 = 2009.6 \times 0.61 \times 2.1319 \sqrt{0.031 - P_x}$$
$$= 2613.4 \sqrt{0.031 - P_x}$$

$$G_4 = 991.1 \times 0.64 \times 2.1319 \sqrt{-0.028 - P_x}$$
$$= 1353.4 \sqrt{-0.028 - P_x}$$

For the roof monitor, the flow rate is

$$G_m = A_m C_m \sqrt{2g\gamma_{av}\Delta P_m} \qquad \textbf{(Eq. 14)}$$

where

$$\Delta P_m = P_x - (P_r - P_t)$$
$$= P_x - P_f; \qquad P_f = P_r - P_t$$

Substituting the computed values for P_r and P_t gives

$$\Delta P_m = P_x - (-0.10725)$$

where

$$P_f = -0.018 - 0.089247$$
$$= -0.10725$$
$$\Delta P_m = P_x + 0.10725$$
$$A_m = 4558 \text{ ft}^2$$
$$\gamma_{av} = \gamma_{98}$$
$$= 0.069246 \text{ lb/ft}^3$$
$$C_m = \text{discharge coefficient}$$
$$= 0.61$$

Substituting in Equation (14) gives

$$G_m = A_m C_m \sqrt{2g\gamma\Delta P_m}$$
$$= 4558 \times 0.61$$
$$\cdot \sqrt{2 \times 32.17 \times 0.069246(P_x = 0.10725)}$$
$$= 5868.7 \sqrt{P_x + 0.10725}$$

The air balance Equation (14) gives

$$3562.5 \sqrt{-0.035 - P_x} + 636.1 \sqrt{0.028 - P_x}$$
$$+ 2613.4 \sqrt{0.031 - P_x} + 1353.4 \sqrt{-0.028 - P_x}$$
$$= 5868.7 \sqrt{P_x + 0.10725}$$

Solving this equation, we obtain $P_x = -0.04597$ lb/ft^3. Hence, the flow throughout the monitor will be

$$G_m = 4558 \times 0.61 \sqrt{2 \times 32.17 \times 0.0692455}$$
$$\cdot \sqrt{(-0.04597) + 0.10725}$$
$$= 1452.7 \text{ lb/sec}$$

Check the heat removed by the discharged air using Equation (1)

$$H = 3600 \times G_m C_p \Delta t$$
$$= 3600 \times 1452.7 \times 0.24 \times 14$$
$$= 17.57 \times 10^6 \text{ Btu/hr}$$

This is less than the design load $H_{des} = 23.9 \times 10^6$ Btu/hr, therefore, the assumed temperature rise Δt should be increased; try $\Delta t = 19°F$.

$$t_{av} = 4 + 87 + (19/2)$$
$$= 100.5°F$$
$$\gamma_{av} = \gamma_{100.5} = \frac{38.639}{400 + 100.5}$$
$$= 0.068937 \text{ lb/ft}^2$$
$$P_t = L(\gamma_i \gamma_{av})$$
$$= 64(0.07064 - 0.06894)$$
$$= 0.10901 \text{ lb/ft}^2$$
$$P_f = -0.018 - 0.10901$$
$$= -0.12701 \text{ lb/ft}^2$$
$$\Delta P_m = P_x - P_f$$
$$= P_x - (-0.12701)$$
$$= P_x + 0.1270.1$$

Substituting in Equation (14) gives

$$G_m = 4558 \times 0.61 \sqrt{2 \times 32.17 \times 0.06894}$$
$$\cdot \sqrt{P_x + 0.12701}$$
$$= 5855.7 \sqrt{P_x + 0.12701}$$

Using the air balance equation, we obtain

$$3562.5 \sqrt{-0.035 - P_x} + 636.1 \sqrt{0.028 - P_x}$$
$$+ 2613.4 \sqrt{0.031 - P_x} + 1353.4 \sqrt{-0.028 - P_x}$$
$$= 5855.7 \sqrt{P_x + 0.12701}$$

Solving the above equation where $P_x = -0.05192$ lb/ft^2 gives

$$-0.0554 \times 0.197 = 0.0102 \text{ in. water}$$

Hence

$$G_m = 5855.7 \sqrt{P_x + 0.12701}$$

$$= 5855.7 \sqrt{-0.0554 + 0.12701}$$

$$= 1604.6 \text{ lb/sec}$$

Check the heat content, applying Equation (1)

$$H = 3600\, G_m C_p \Delta t$$

$$= 3600 \times 1604.6 \times 0.24 \times 19$$

$$= 26.34 \times 10^6 \text{ Btu/hr}$$

This is larger than the indicated design load of 23.9×10^6 Btu/hr. Recalculate assuming larger temperature difference. Try $\Delta t = 18°F$.

$$t_{av} = 4 + 87 + (18/2)$$

$$= 100°F$$

$$\gamma_{av} = \gamma_{100} = \frac{38.639}{460 + 100}$$

$$= 0.06900 \text{ lb/ft}^3$$

$$P_t = 64(0.07064 - 0.069)$$

$$= 0.10507 \text{ lb/ft}^2$$

$$P_f = -0.018 - 0.10507$$

$$= -0.12307 \text{ lb/ft}^2$$

$$\Delta P_m = P_x - P_f$$

$$= P_x - (-0.12307)$$

$$= P_x + 0.12307$$

Substituting in Equation (14) yields

$$G_m = 4558 \times 0.61 \sqrt{2 \times 32.17 \times 0.069}$$

$$\cdot \sqrt{P_x + 0.12307}$$

$$= 5858.3 \sqrt{P_x + 0.12307}$$

Using the air balance equation, we obtain

$$3562.5 \sqrt{-0.035 - P_x} + 636.1 \sqrt{0.028 - P_x}$$

$$+ 2613.4 \sqrt{0.031 - P_x} + 1353.4 \sqrt{-0.028 - P_x}$$

$$= 5858.3 \sqrt{P_x + 0.12307}$$

Solving the above equation, where $P_x = -0.0507$ lb/ft^2, we obtain

$$G_m = 5858.3 \sqrt{-0.0507 + 0.12307}$$

$$= 1576 \text{ lb/sec}$$

Check the heat content, applying Equation (1)

$$H = 3600\, G_m C_p \Delta t$$

$$= 3600 \times 1576 \times 0.24 \times 18$$

$$= 24.5 \times 10^6 \text{ Btu/hr}$$

This fits very close to the design load. The difference is only

$$100 \times \frac{24.5 \times 10^6 - (23.9 \times 10^6)}{23.9 \times 10^6}$$

$$= 2.5\%$$

Thus, the ventilation rate of the building will be

$$G_m = 1576 \text{ lb/sec}$$

For volume per minute,

$$Q_m = \frac{60\, G_m}{\gamma_{av}}$$

$$= \frac{60 \times 1576}{0.069}$$

$$= 1,370,434 \text{ cfm}$$

Anticipate air changes per hour in the building

$$n = \frac{60\, Q_m}{Q_f}$$

$$= \frac{60 \times 1370434}{3273734}$$

$$= 25.1$$

cfm per square foot of floor area is

$$q = \frac{Q_m}{A_f}$$

$$= \frac{1370434}{63250}$$

$$= 21.66 \text{ cfm/ft}^2$$

Check the volume of air passing through the vents of the south wall in the rectifier area:

$$G_1 = G_n = 3562.5 \sqrt{-0.035 - P_x}$$

$$= 3562.5 \sqrt{-0.035 - (-0.0507)}$$

$$= 446.4 \text{ lb/sec}$$

$$Q_1 = \frac{60\, G_1}{\gamma_1}$$

$$= \frac{60 \times 446.4}{0.07064}$$

$$= 379,162 \text{ cfm}$$

Calculate the volume of air admitted to the furnace area from the west wall vent openings:

$$G_2 = G_w = 636.1\sqrt{0.028 - P_x}$$

$$= 636.1\sqrt{0.028 - (-0.0507)}$$

$$= 178.5 \text{ lb/sec}$$

The volume of air is

$$Q_2 = Q_{\text{west}} = \frac{60 \times 178.5}{0.07064}$$

$$= 151,570 \text{ cfm}$$

Calculate the volume of air passing to the furnace area through the north wall vent doors:

$$G_3 = G_n = 2613.4\sqrt{0.031 - (-P_x)}$$

$$= 2613.4\sqrt{0.031 - (-0.0507)}$$

$$= 747 \text{ lb/sec}$$

The ventilation volume of air admitted direct to the furnace area from the north wall is:

$$Q_n = \frac{60G_3}{\gamma_1}$$

$$= \frac{60 \times 747}{0.07064}$$

$$= 634,485 \text{ cfm}$$

The velocity of the air through the north vents is

$$V_n = \frac{Q_n}{A_n}$$

$$= \frac{634,485}{2009.6}$$

$$= 315.7 \text{ fpm}$$

16.10.2 Other Cases

In a similar manner, seven other cases were developed and tabulated in table 16.3.

16.11 COMPUTER PROGRAM FOR CALCULATIONS OF VENTILATION

The program shown as figure 16.11 is for an IBM PC computer, using BASIC language. It is intended to determine temperature increases, air flow rate, and the number of air changes per hour.

The program developed uses an iterative procedure to determine the temperature rise to fit the heat content of the air, as calculated in Equation (1). The program follows an average of 8–12 iterations before it readies the accurate solution. It must be mentioned that the computations require 45 seconds for the solution of one case of wind load.

The program has been developed for a user-friendly operation where the engineer is questioned by the computer for data and the computation process prints the result. The program starts with "Input of Data," and follows with data for "Wind Calculations." The results, after analysis, are printed in detail. The results shown in this table are different from those shown in the summary of calculations (of hand computations, table 16.3) because of the precise iteration performed by the computer (compare 10 cycles by the computer and 2 cycles by hand).

If one wants to analyze another set of wind conditions, then only wind data should be entered in the program. The example included two more cases of wind conditions as also analyzed by hand computations. Also, in these cases, the results are different due to reasons specified above. The results computed by the program are done accurately and quickly in a user-friendly fashion.

16.12 DESIGN OF A CUPOLA EXIT GAS SCRUBBER

The principal features of construction, dimensions, and data of the Cupola Exit Gas Scrubber are shown on figure 16.12.

The scrubber requires no vanes, cover plates, or other appurtenances, and can be assembled from steel plates and structural shapes. The shell weight varies from about 1,500 pounds for a one-ton-per-hour cupola to about 7,500 pounds for one having a capacity of 25 tons per hour. This generally permits it to be supported directly from the top of most cupolas with a minimum of reinforcement of the shell. In some cases, it may be necessary to use rubber-lined or alloy-steel sections for parts of the scrubber. A study of the composition of the gases, and the temperatures involved, should be made before any decision is reached.

In operation, the natural draft from the cupola induces a flow of secondary air between the scrubber shell and the cupola, cooling the cupola exit gas and providing oxygen to burn the carbon monoxide present. The portion of the scrubber above the top of the cupola acts as an expansion chamber, reducing the gas velocity, thus settling out coarser dust particles. The water spray in this upper portion of the scrubber further cools the gas, absorbs the soluble fumes, and removes a large portion of the finer dust particles. The flow of secondary air and

```
10 REM *************** VENTILATION OF BUILDING ******************
20 REM *    PROGRAM DEVELOPED FOR J. FRUCHTBAUM CONSULTING ENG.   *
25 REM *         BY  ANDREI REINHORN PH.D P.E. BUFFALO, NY.        *
30 REM ************************************************************
50 DIM A(5), G(5), P(5), PRED$(5), K(5)
60 CLS: LPRINT CHR$(15):
70    LOCATE  2,20,0:PRINT "    ANALYSIS OF VENTILATION SYSTEM"
80    LOCATE  3,21,0:PRINT "            INPUT OF DATA           ":PRINT:
90    LOCATE  5,20,0: INPUT "WHAT IS THE NAME OF THE BUILDING";     NAM$
100    LOCATE  6,20,0:INPUT "WHAT IS THE VOLUME OF THE BUILDING";    VOL
110    LOCATE  7,20,0:INPUT "ABSOLUTE BAROMETRIC PRESSURE (INCH)";   BAR
120    LOCATE  8,20,0:INPUT "OUTSIDE TEMPERATURE (DEG. FARENHEIT)"; TOUT
130    LOCATE  9,20,0:INPUT "LEVEL OF ROOF MONITOR (FEET)";         LEVM
140    LOCATE 10,20,0:INPUT "AREA OF OPENINGS ON SOUTH WALL"; A(1)
150    LOCATE 11,20,0:INPUT "SOUTH WALL OPENINGS ARE D(OORS) OR L(OUVERS)"; PRED$(1)
160    LOCATE 12,20,0:INPUT "AREA OF OPENINGS ON EAST WALL"; A(2)
170    LOCATE 13,20,0:INPUT "EAST WALL OPENINGS ARE D(OORS) OR L(OUVERS)"; PRED$(2)
180    LOCATE 14,20,0:INPUT "AREA OF OPENINGS ON NORTH WALL"; A(3)
190    LOCATE 15,20,0:INPUT "NORTH WALL OPENINGS ARE D(OORS) OR L(OUVERS)"; PRED$(3)
200    LOCATE 16,20,0:INPUT "AREA OF OPENINGS ON WEST WALL"; A(4)
210    LOCATE 17,20,0:INPUT "WEST WALL OPENINGS ARE D(OGRS) OR L(OUVERS)"; PRED$(4)
220    LOCATE 18,20,0:INPUT "AREA OF OPENINGS ON ROOF MONITOR"; A(5)
230    LOCATE 19,20,0:INPUT "ROOF MONITOR OPENINGS ARE SQUARES OR LOUVERS"; PRED$(5)
250    CLS: LOCATE 2,25,0: PRINT "DATA FOR WIND CONDITIONS ":
260    LOCATE 5,20,0:INPUT "WIND DIRECTION "; DIRW$
270    LOCATE 6,20,0:INPUT "WIND VELOCITY MILES/HOUR"; VELW
280    LOCATE 7,20,0:INPUT "WIND COEFICIENT FOR SOUTH WALL"; K(1)
290    LOCATE 8,20,0:INPUT "WIND COEFICIENT FOR EAST WALL"; K(2)
300    LOCATE 9,20,0:INPUT "WIND COEFICIENT FOR NORTH WALL"; K(3)
310    LOCATE 10,20,0:INPUT "WIND COEFICIENT FOR WEST WALL"; K(4)
320    LOCATE 11,20,0:INPUT "WIND COEFICIENT FOR ROOF MONITOR"; K(5)
340 Q = 1400000#
350 CP = .241
360 TI = 130
370 FOR I = 1 TO 5
380 IF PRED$ (I) = "D" THEN C(I) = .64  ELSE C(I) = .61
390 NEXT I
400 T=TI
410 GOSUB 850
420 GAR130 = GAMA
430 H = 60 * Q * CP * GAR130 * (130 - 70)
440 GR= 32.17
450 T=TOUT
460 GOSUB 850
470 GOUT=GAMA
480 PWO = GOUT * (5280 * VELW/3600)^2 /(2 * GR)
490 REM **** ASSUME INITIAL TEMPERATURE RAISE *****
500 DELT = 36
510 TAV = 5 + TOUT + DELT/2
520 T = TAV
530 GOSUB 850
540 GAMAV = GAMA
550 PT = LEVM * (GOUT - GAMAV)
560 FOR I = 1 TO 5
570 P (I) = K (I) * PWO
580 NEXT I
590 CN =1
600 PMIN = P(1): FOR I = 2 TO 5: IF P(I) < PMIN THEN PMIN=P(I): NEXT I
```

Figure 16.11 A computer program for building ventilation requirements. (Prepared by Dr. Andrei Reinhorn, University of Buffalo)

FIGURE 16.11   403

```
610 PX = PMIN
620 GOSUB 910
640 IF ABS(FUN) < .5 THEN GOTO 730
650 IF FUN/FUN1 < 0! THEN GOTO 670 ELSE CN=1
660 IF CN=1 THEN CN=CN+1:PX1=PX:FUN1=FUN:PX=PX-.1*ABS(PX):GOTO 620:ELSE GOTO 670
630 IF CN=1 THEN FUN1 = FUN
670 PX2=PX
680 PX=PX-((PX-PX1)/(FUN-FUN1))*FUN:FUN2=FUN1:FUN1=FUN:REM PRINT PX
690 GOSUB 910
700 IF ABS(FUN) < .5 THEN GOTO 730
710 IF FUN/FUN1 <0  THEN PX1=PX2 :GOTO 670 ELSE GOTO 720
720 IF FUN/FUN1 >0  THEN FUN1=FUN2:PX1=PX1: GOTO 670 ELSE PRINT "IMPOSIBLE"
730 HM = 3600 * G(5) * CP * DELT
740 DELH = HM - H
760 IF ABS(DELH) > .1 THEN DELT= DELT*(1-DELH/H): GOTO  510   ELSE GOTO 770
770 QM= 60 * G(5) / GAMAV
780 FOR I = 1 TO 5
790 IF I = 5 THEN GOUT =GAMAV
800 V(I) = G(I) * 60/ (GOUT * A(I))
810 NEXT I
820 N = 60 * QM / VOL
830 GOSUB 1010
840 END




850 REM **** SUBROUTINE GAMA ****
860 PO = 70.77201*BAR
870 R = 53.3
880 TA = 460 + T
890 GAMA = PO / (R * TA)
900 RETURN
910 REM **** SUBROUTINE FUN ****
920 FUN = 0
930 FOR I = 1 TO 4
940 G(I) = A(I)*C(I)*SQR(2*GR*GOUT*(P(I)-PX))
950 FUN=FUN+G(I)
960 NEXT I
970 G(5) = A(5)*C(5)*SQR(2*GR*GAMAV*(PX-P(5)+PT))
980 REM FOR I=1 TO 5 :PRINT "G(";I;")=";G(I): NEXT I
990 FUN=FUN-G(5)
1000 RETURN
1005 REM  ***** PINTING RESULTS ********
1010 CLS:
1020 LOCATE  2,20,0:PRINT "    ANALYSIS OF VENTILATION SYSTEM"
1030 LOCATE  3,20,0:PRINT "        FOR ";NAM$;" BUILDING"
1040 LOCATE  5,10,0:PRINT "Analysis made for wind direction: ";DIRW$
1050 LOCATE  6,10,0:PRINT "                Wind velocity:";VELW;"MPH"
1060 LOCATE  7,10,0:PRINT "          Outside temperature:";TOUT;CHR$(248);"F"
1070 DELT = INT(DELT): TEXT=TOUT+DELT
1080 LOCATE  8,10,0:PRINT "     Temperature of exhaust air:";TEXT;CHR$(248);"F"
1090 LOCATE  9,10,0:PRINT "   Change of temperature of air:";DELT;CHR$(248);"F"
1100 HM=HM/1000000!
1110 LOCATE 10,10,0:PRINT USING "                    Heating load:###.## MILLIONS BTU/H
1120 QM=QM/1000
1130 LOCATE 11,10,0:PRINT USING "                 Air flow rate: ####,.# THOUSANDS CFM"
1140 N=INT(N)
```

Figure 16.11 (*Continued*)

```
1150 LOCATE 12,10,0:PRINT USING "          Number of air changes: ##. PER HOUR";N
1160 LOCATE 13,10,0:PRINT USING "      Air velocity thru south wall: ###.# FPM";V(1)
1170 LOCATE 14,10,0:PRINT USING "       Air velocity thru east wall: ###.# FPM";V(2)
1180 LOCATE 15,10,0:PRINT USING "      Air velocity thru north wall: ###.# FPM";V(3)
1190 LOCATE 16,10,0:PRINT USING "       Air velocity thru west wall: ###.# FPM";V(4)
1200 LOCATE 17,10,0:PRINT USING " Air velocity thru roof monitor: ###.# FPM";V(5)
1210 LOCATE 19,10,0:INPUT "     Do you want another analysis";AN$
1220 IF AN$ = "Y" GOTO 240
1230 RETURN
```

                    ANALYSIS OF VENTILATION SYSTEM
                           INPUT OF DATA

```
        WHAT IS THE NAME OF THE BUILDING? Example
        WHAT IS THE VOLUME OF THE BUILDING? 3273734.0
        ABSOLUTE BAROMETRIC PRESSURE (INCH)? 29.1
        OUTSIDE TEMPERATURE (DEG. FARENHEIT)? 87.0
        LEVEL OF ROOF MONITOR (FEET)? 64.0
        AREA OF OPENINGS ON SOUTH WALL? 2611.0
        SOUTH WALL OPENINGS ARE D(OORS) OR L(OUVERS)? D
        AREA OF OPENINGS ON EAST WALL? 991.9
        EAST WALL OPENINGS ARE D(OORS) OR L(OUVERS)? D
        AREA OF OPENINGS ON NORTH WALL? 2009.6
        NORTH WALL OPENINGS ARE D(OORS) OR L(OUVERS)? L
        AREA OF OPENINGS ON WEST WALL? 466.2
        WEST WALL OPENINGS ARE D(OORS) OR L(OUVERS)? D
        AREA OF OPENINGS ON ROOF MONITOR? 4558.0
        ROOF MONITOR OPENINGS ARE SQUARES OR LOUVERS? L
```

                        DATA FOR WIND CONDITIONS

```
        WIND DIRECTION ? SW
        WIND VELOCITY MILES/HOUR? 6
        WIND COEFICIENT FOR SOUTH WALL? .36
        WIND COEFICIENT FOR EAST WALL? -.33
        WIND COEFICIENT FOR NORTH WALL? -.41
        WIND COEFICIENT FOR WEST WALL? .33
        WIND COEFICIENT FOR ROOF MONITOR? -.21
```

                    ANALYSIS OF VENTILATION SYSTEM
                         FOR Example BUILDING

```
Analysis made for wind direction: SW
                   Wind velocity: 6 MPH
              Outside temperature: 87 F
        Temperature of exhaust air: 128 F
     Change of temperature of air: 41 F
                     Heating load: 79.55 MILLIONS BTU/HOUR
                    Air flow rate: 1,960.8 THOUSANDS CFM
            Number of air changes: 35 PER HOUR
       Air velocity thru south wall: 379.5 FPM
        Air velocity thru east wall: 255.4 FPM
       Air velocity thru north wall: 225.7 FPM
        Air velocity thru west wall: 374.9 FPM
      Air velocity thru roof monitor: 430.2 FPM
```

Figure 16.11 (*Continued*)

DATA FOR WIND CONDITIONS

WIND DIRECTION ? S
WIND VELOCITY MILES/HOUR? 6
WIND COEFICIENT FOR SOUTH WALL? .42
WIND COEFICIENT FOR EAST WALL? -.35
WIND COEFICIENT FOR NORTH WALL? -.18
WIND COEFICIENT FOR WEST WALL? -.35
WIND COEFICIENT FOR ROOF MONITOR? -.37

ANALYSIS OF VENTILATION SYSTEM
FOR Example BUILDING

Analysis made for wind direction: S
Wind velocity: 6 MPH
Outside temperature: 87 F
Temperature of exhaust air: 127 F
Change of temperature of air: 40 F
Heating load: 79.55 MILLIONS BTU/HOUR
Air flow rate: 2,018.8 THOUSANDS CFM
Number of air changes: 37 PER HOUR
Air velocity thru south wall: 388.5 FPM
Air velocity thru east wall: 251.0 FPM
Air velocity thru north wall: 273.6 FPM
Air velocity thru west wall: 251.0 FPM
Air velocity thru roof monitor: 442.9 FPM

DATA FOR WIND CONDITIONS

WIND DIRECTION ? W
WIND VELOCITY MILES/HOUR? 8
WIND COEFICIENT FOR SOUTH WALL? -.31
WIND COEFICIENT FOR EAST WALL? -.14
WIND COEFICIENT FOR NORTH WALL? -.31
WIND COEFICIENT FOR WEST WALL? .46
WIND COEFICIENT FOR ROOF MONITOR? -.10

ANALYSIS OF VENTILATION SYSTEM
FOR Example BUILDING

Analysis made for wind direction: W
Wind velocity: 8 MPH
Outside temperature: 87 F
Temperature of exhaust air: 131 F
Change of temperature of air: 44 F
Heating load: 79.55 MILLIONS BTU/HOUR
Air flow rate: 1,847.4 THOUSANDS CFM
Number of air changes: 33 PER HOUR
Air velocity thru south wall: 268.2 FPM
Air velocity thru east wall: 326.2 FPM
Air velocity thru north wall: 255.6 FPM
Air velocity thru west wall: 477.7 FPM
Air velocity thru roof monitor: 405.3 FPM

Figure 16.11 (*Continued*)

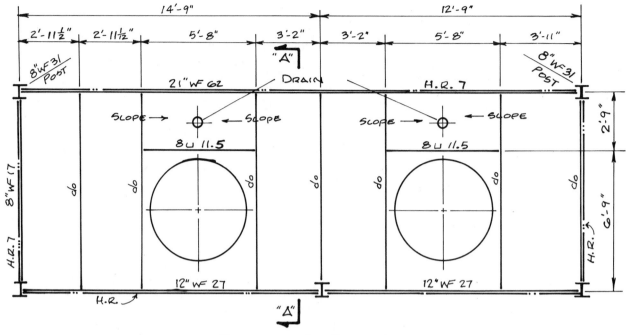

PLAN OF PLATFORM

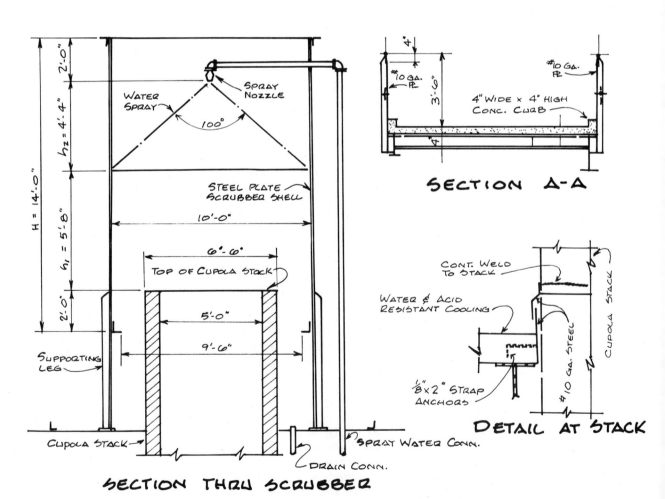

SECTION THRU SCRUBBER

SECTION A-A

DETAIL AT STACK

Figure 16.12   Scrubber details.

water over the inside surface of the scrubber shell maintains a temperature sufficiently low so that ordinary carbon steel, properly protected, may be used.

The scrubber, being open at the top and bottom, generally places less back pressure on the cupola than those of the enclosed type, thus minimizing charging door operational difficulties. It is also notable that the scrubber will generally continue to function without damage in the event of water spray failure, since the secondary air provides the greater portion of cooling to the scrubber shell.

The quantity of secondary air is self-regulating, as an increase or decrease in the quantity of primary exit gas increases or decreases the cupola draft which, in turn, increases or decreases the aspirating effect on the scrubber, with a corresponding increase or decrease in the velocity at which secondary air enters the scrubber.

The spray water requirements are low compared to other cupola exit gas scrubbers of this type. Except for the amount lost by evaporation, this water serves to carry the dust particles to waste, and may also be used for slag cooling or similar uses. The water can also be recirculated in the scrubber after proper treatment and removal of solids.

16.12.1 Data

Cupola dimensions:
external diameter:	6' 6"
external cross-section:	33.18 ft^2
internal diameter:	5' 0"
internal cross-section:	19.63 ft^2
height:	40' 0"

Iron melting rate:	13 tph
Iron/coke ratio:	10 : 1
Carbon content of coke:	88%
Heating value of coke:	12,750 Btu/lb

Cupola air supply from blower:
40,000 cfm = 30 lb/min at 70°F

Field analysis of exit gas by volume (from cupola):
CO_2:	16.20%
CO:	1.82%
O_2:	3.46%
N_2:	78.52%
	100.00%

16.12.2 Physical Constants

Specific volume of air
at 70°F:	13.33 ft^3/lb
Average specific heat of air:	0.25 Btu
Specific heat of exit gas:	0.30 Btu

Temperature of air
entering scrubber:	70°F
Average specific heat of iron (70–2900°F):	0.167 Btu
Latent heat of fusion of iron:	40.0 Btu

16.12.3 Cupola Exit Gases

In table 16.5, the value of 30.734 would correspond to the molecular weight of the exit gas if it were a chemical compound. For the specific volume of the cupola exit gas, since 359 ft^3 of any gas weighs its molecular weight in pounds, one pound of exit gas will occupy:

$$\frac{359}{30.734} = 11.68$$

say, 11.7 ft^3 at 70°F.

16.12.4 Air Required

Since the iron/coke ratio is 10 : 1, coke consumption will be

$$\frac{13}{10} \times \frac{2000}{60} = 43.33 \text{ lb/min}$$

and carbon consumption will be
$$0.88 \times 43.33 = 38.1 \text{ lb/min}$$

Oxygen required for combustion
Of the 38.1 lb of carbon required per minute, the amount that will burn to CO_2 is

$$\frac{16.20 \times 38.1}{16.20 + 1.82} = 34.25 \text{ lb}$$

The amount that will burn to CO is

$$38.1 - 34.25 = 3.85 \text{ lb}$$

Table 16.5 Percent of Exit Gas Components by Weight

Constit- uent	V Percent of exit gas	M Molecular weight	M × V	% of M × V
CO_2	16.2	44	7.128	23.2
CO	01.82	28	0.509	01.7
O_2	03.46	32	1.107	03.6
N_2	78.52	28	21.990	71.5
	100.00		30.734	100.0

The oxygen required for combustion is

$$34.25 \times (32/12) = 91.33 \text{ lb/min}$$

$$3.85 \times (16/12) = \underline{5.13 \text{ lb/min}}$$

$$94.46 \text{ lb/min}$$

Free oxygen equals

$$\frac{0.036}{0.017} \times (3.85 + 5.13) = 19.02 \text{ lb/min}$$

Therefore, the total oxygen supplied to the cupola is
$$96.46 + 19.02 = 115.48 \text{ lb/min}$$

Total air supplied to the cupola
The nitrogen supplied is

$$\frac{71.5}{3.6} \times 19.02 = 377.75 \text{ lb/min}$$

The ratio of nitrogen to oxygen by weight is

$$\frac{377.75}{115.48} = 3.27:1$$

which checks closely (it should be $3.32:1$).

Air required
The total air required is

$$377.75 + 115.48 = 493.23 \text{ lb/min}$$

Air furnished by the blower is 300 lb/min, so air supplied through the cupola door is

$$493 - 300 = 193 \text{ lb/min}$$

Amount of exit gas

$$493.23 + 38.1 = 531.33 \text{ lb/min}$$

$$531 \times 11.7 = 6210 \text{ cfm at } 70°\text{F}$$

16.12.5 Temperature of Cupola Exit Gas

The heat released by the coke is

$$43.3 \times 12,750 = 552,100 \text{ Btu}$$

assuming all carbon burns to CO_2 before its release to the atmosphere. To find the heat absorbed by iron (assume the temperature to be 2900°F):

$$\frac{13 \times 2000 \times 0.167 \times (2900 - 70)}{60}$$

$$= 204,800 \text{ Btu/min}$$

$$\frac{13 \times 2000 \times 40}{60} = 17,300 \text{ Btu/min}$$

$$204,800 + 17,300 = 222,100 \text{ Btu/min}$$

To find the heat absorbed by gas:

$$552,100 - 222,100 = 330,000 \text{ Btu/min}$$

$$\text{less 1\% loss through walls} = \underline{3,300}$$

$$\text{heat loss in stack} = 326,700 \text{ Btu/min}$$

To find the temperature of the exit gas:

$$\frac{326,700}{531 \times 0.30} + 70°\text{F} = 2121°\text{F}$$

16.12.6 Size of Scrubber Shell

The volume of the exit gas is

$$6210 \times \frac{2121 + 460}{70 + 460} = 30,250 \text{ cfm at } 2121°\text{F}$$

The velocity of the exit gas is

$$\frac{30,250}{19.63} = 1541 \text{ fpm}$$

The velocity of the outside air at the bottom of the scrubber should be kept high enough so as not to disturb the operation under expected wind conditions, and low enough so as not to carry the spray outside the scrubber. In one particular problem, this value was determined to be about 700 fpm.

Secondary air required
Since kinetic energy varies as the square of the velocity, the amount of outside air required (assuming 700 fpm) is

$$\frac{531 \times (1541)^2}{700^2} - 531 = 2042 \text{ lb/min}$$

Although we are figuring 700 fpm, the actual velocity probably will be nearer 650 fpm, after allowing for friction.

Volume of outside air at 70°F

$$2042 \times 13.33 = 27,200 \text{ cfm}$$

Cross-sectional area of scrubber

$$\frac{27,200}{700} = 38.86 \text{ ft}^2$$

$$38.86 + 33.18 = 72.04 \text{ ft}^2 \text{ (inside bottom flange)}$$

Diameter of scrubber
Use a 9½-ft diameter clear opening, making the inside diameter of the scrubber 10 ft. This gives a net scrubber area of 70.88 ft².

Temperature of air-gas mixture
With no water, and assuming T is the temperature of the mixture,

$$531 \times 0.30 \times (2121 - T)$$

$$= 2042 \times 0.25 \times (T - 70)$$

$$= 558°F$$

Height of scrubber

The dimension h_1 (see figure 16.12) should allow for the distribution of the exit gas over the entire scrubber area. We have found that making h_1 6–8 in. larger than the difference between the inside diameters of the scrubber and cupola will produce satisfactory results. Thus,

$$h_1 = 8'' + (10' - 5')$$

$$= 5'\ 8''$$

The dimension h_2 depends on the type of spray nozzles. Using 100° spread, h_2 will equal 4 ft 4 in.

Thus, the total dimension is

$$H = 2'\ 0'' + 5'\ 8'' + 4'\ 4''$$

$$= 14'\ 0''$$

16.12.7 Water Requirements

Since the temperature of the air-gas mixture is low enough for discharge, the purpose of the water becomes merely a matter of trapping the dust. For properly designed cupolas and scrubbers of this type, 0.4 gpm per ton of coke burned per hour will produce satisfactory results, even in areas with very rigid dust collection regulations.

The pressure of the water should have a definite relation to the velocity of the air-gas mixture. For the velocities in this problem, pressures between 35 lb/in.^{2} and 50 lb/in.^{2} (normally available) give good results.

With the slight amount of water used, and the large amount of heat lost in the scrubber, no danger exists in winter operation. In some other processes where similar scrubbers are used, this should be checked.

Assuming a good distribution, about 15 gpm will evaporate as steam.

Package-Handling Conveyors

17.1 GENERAL

Package-handling conveyors are either gravity type, roller (nonpowered); belt drive, roller; chain driven, roller; belt; or slat. Belt conveyors in general are covered in section 2. Drives are covered in section 9.

The packages may vary in weight from a few ounces to as much as 20–25 tons. The heavier loads are encountered in special processes such as steel mill operations. This section covers medium packages and pallets.

The bottom of the package must be reasonably flat and hard enough to stand up in the rolling process. All rollers are equipped with ball bearings at both ends, mounted on either round or hexagonal shafts.

CEMA has adopted a series of minimum standards for design and fabrication of light and medium package conveyors. Limits are placed on length and speed of these conveyors, that can be exceeded with proper design. The CEMA tables apply within the given limits.

17.2 ROLLER CONVEYORS

The frames supporting the rollers are generally either channels or angles. The rollers may be set high so the top of the roller is above the top of the angle or channel, or set low so the top of the roller is below the top of the steel support, thereby using the frame as a guard (see figure 17.1). Because of the cover over the chain and sprocket, the rollers cannot be set high in chain-driven roller conveyors. The rollers should be 2–3 in. longer than the largest width of package to be handled.

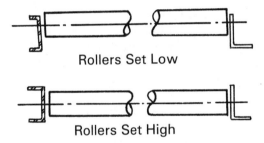

Figure 17.1 Setting of rollers on angle frame rails.

Tapered containers or cartons with open flaps may overhang the conveyor width and foul the supports, so this condition should be watched. Tapered tote boxes and containers with protruding handles, if accumulated under pressure or shock conditions, will tend to buckle. The handles will lock or jam.

17.3 BASIC CONVEYOR BEARING LOAD RATING*

The Basic Conveyor Bearing Load Rating is defined as that constant, radial bearing load which a group of apparently identical bearings can endure for one million revolutions with 90% of the bearings surviving.

---

*Paragraph 17.3 and Tables 17.1, 17.4, and 17.5 are adapted from CEMA Standard No. 401-1985, with permission.

The Basic Conveyor Bearing Load Rating can be calculated by the following equation:

$$C = f \times Z^{2/3} \times D^{1.8} \qquad \text{(Eq.1)}$$

where C = the basic conveyor bearing load rating in pounds; f = a factor that depends on the geometry of the bearing components, the accuracy to which the bearing components are made, and the material; Z = the number of balls in the bearing; and D = the ball diameter in inches. The values of f, Z, and D for commonly used ball sizes and numbers are given in table 17.1.

Equation (1) was developed empirically in the English system of measurements; therefore, calculations must be performed in the English system and answers converted to SI. In those instances where results of equation (1) differ significantly from the catalog rating, consult manufacturer for the basis of the rating. In applications where the need for greater carrying capacity and extended life require the use of Precision bearings, the load ratings should be established by reference to ANSI/AFBMA Standard 9-1978, Load Rating and Fatigue Life for Ball Bearings.

The rating life of Commercial grade conveyor bearings operating at loads other than the Basic Conveyor Bearing Load Rating (C) can be predicted by the following equation:

$$L_{10} = (C/P_e)^2 \times 10^6 \text{ revolutions} \qquad \text{(Eq.2)}$$

where L_{10} = the rating life, in revolutions; C = the basic conveyor bearing load rating in pounds; and P_e = the applied radial load on a bearing, in pounds (calculation follows). Note that when the applied load (P_e) is equal to the Basic Conveyor Bearing Load Rating (C), the Rating Life (L_{10}) is equal to one million revolutions.

Thus, the Load/Life relationship expressed above establishes a Rating Life that is based on laboratory conditions of continuous running and constant load. Roller conveyors seldom operate at constant load and constant speed. Reduced loads and speeds frequently result in Service Life that is 30–40 times the Rating Life calculated above (equation 2).

Life in hours of operation (t) for conveyor rollers may be calculated as shown in the following example.

$$t = \frac{L \times \pi \times D}{12 \times 60 \times S} \qquad \text{(Eq.3)}$$

where

C = 2258 (table 17.2)
L = service life (L_{10})S, in revolutions
 = $(2258/2330)^2 \times 1,000,000$
 = 939,000

Table 17.1 Values for Equation 1

No. of Balls Z	$Z^{2/3}$	Ball Dia. D (in.)	$D^{1.8}$	Factor f
7	3.66	1/8	0.024	180
8	4.00	5/32	0.036	185
9	4.33	3/16	0.049	190
10	4.64	7/32	0.065	210
11	4.95	1/4	0.083	230
12	5.24	9/32	0.102	250
13	5.53	5/16	0.123	270
14	5.81	11/32	0.147	300
15	6.08	3/8	0.171	335
16	6.35			
17	6.61	7/16	0.225	440
18	6.87	1/2	0.288	630
19	7.12	9/16	0.355	1150
		5/8	0.430	1580

Source: Courtesy CEMA

D = roller diameter (in.)
 = 4.25 in. × 0.438 in. thick
S = roller surface speed (fpm)
 = 20

Therefore,

$$t = \frac{939,000 \times \pi \times 4.25}{12 \times 60 \times 20}$$

$$= 870 \text{ hr (aprx)}$$

17.4 SPACING OF ROLLERS

Rollers 1.9 in. in diameter are generally spaced 3, 4, 6, 8, and 12 in. on centers. Rollers 2.5, 2.75, and 3.5 in. can be spaced at the diameter of the roller plus 1/2 in. on centers, and in increments of 1 in. beyond that. Rollers 4 in. and larger can be spaced at roller diameter plus 1 in. and at increments of 1 in. beyond that. Rollers may be spaced closer, provided mill tolerances on straightness, concentricity, shape, etc., are considered. Straightening may be required when rollers are long.

17.5 WEIGHT OF ROLLERS

Defining the width as the dimension inside to inside of conveyor frame rails, Table 17.3 gives the weight of the rollers that are most often encountered in bulk material

Table 17.2 Values of C

Dia of Roller (in.)	Thickness	Bearings[a]	$Z^{2/3}$	$D^{1.8}$	f	C	Max Length bf
1.9	12 ga	11⁹⁄₃₂	4.95	0.102	250	126	39
2.5	11 ga	12³⁄₈	5.24	0.171	335	300	72
2.75	¼ in.	13½	4.00	0.198	630	499	99
3.5	9 ga	12³⁄₈	5.24	0.171	335	300	90
3.5	0.300 in.	13½	4.00	0.198	630	499	120
4.0	0.500 in.	9⁷⁄₁₆	4.33	0.225	440	429	120
4.25	¼ in.	13½	4.00	0.198	630	499	120
4.25	0.438 in.	13⁹⁄₁₆	5.53	0.355	1150	2258	120
5.0	0.71 in.	13⁵⁄₈	5.53	0.429	1580	3748	120

Note: The values of C that are most often involved in bulk material handling are tabulated here. For other sizes or number of balls, C can be computed by using equation (1).

[a]For increased values of P, increase number and/or size of bearing balls.

Table 17.3 Average Weight of Rollers (lb)

Dia of Roller (in.)	Thickness	Width of Roller (in.)[a]								
		18	20	24	30	36	42	48	54	60
1.9	12 ga	5.3	5.8	6.9						
2.5	11 ga	8.5	9.2	10.7	12.1	13.6				
2.75	¼ in.			23.2	28.6	33.9	39.2	44.6		
3.5	9 ga				20.1	23.5	27.0	30.4		
3.5	0.300 in.				38.4	45.5	52.5	60.0		
4.0	0.500 in.					79.0	91.0	103.0	115.0	127.0
4.25	¼ in.				42.0	49.5	57.0	64.5	72.0	79.5
4.25	0.438 in.					74.5	86.2	97.9	109.6	121.3
5.0	0.71 in.						146.8	208.6	270.4	332.2

[a]Weights for other bf (between faces) dimensions can be interpolated.

handling. These weights may vary for different manufacturers.

Data on other sizes can be found in manufacturers' catalogs. Rollers are available in diameters up to 7⁵⁄₈ in.

17.6 CURVES

The radius of curves and the distance between the guard rails will depend upon the maximum width and length of packages handled (see figure 17.2).

Allow a minimum of 2-in. clearance at outside curve. The grade should be increased at curves.

17.7 TRANSFERS

Ball transfers and turntables are used when a load must be rotated or deflected to another conveyor at right angles. Ball transfers are very flexible when used with hard, flat-bottom loads up to 200 lb. For heavier loads, or when the bottom is soft and rough, gravity-roller turntables will perform a similar service. Standard ball transfers are designed to fit into the frame of a roller conveyor at any point and do not require separate supports.

Turntables are made with a center pivot, but the weight is supported by casters mounted on a heavy structural frame. In most applications, the carrying sur-

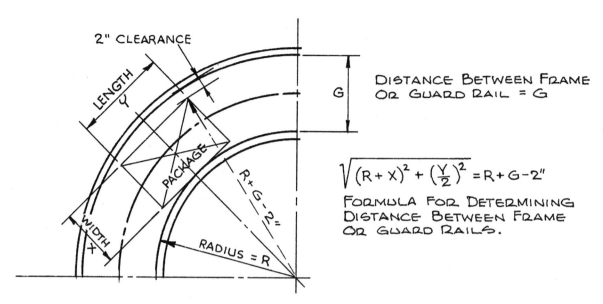

Figure 17.2 Curves.

face of the turntable is a roller conveyor and the turning action is manual.

For automated systems or very heavy loads, the carrying surface may be powered and the turning action effected by motor or air cylinder.

Conveyors can be designed to transfer from one system to another, to elevate packages, and to load pallets. (Figure 17.12 details of a right-angle transfer to a palleting machine.)

17.8   BEARINGS

Bearings are furnished as free-running, grease-sealed, or greased bearings with Alemite fittings.

Free-running bearing rollers are used on many industrial projects and provide the strongest and liveliest bearings in the conveyor industry. For corrosive, dusty, and severe operating conditions, the bearings for ½- and ¾-in. diameter shafts may be furnished with all metal labyrinth seals. Grease may be added at the final assembly of the roller to form a grease-sealed bearing, or Alemite fittings may be furnished for periodic lubrication, but care should be taken that these bearings are not overgreased. Normally, greased bearings are used on high-speed powered conveyors, for both long life and quietness. Too much grease may slow up the turning of the ball race; therefore, grease is not used on gravity lines.

17.9   GRAVITY ROLLERS (NONPOWERED)

17.9.1   General

Gravity conveyors carry packages on rollers by gravity, aided by power at curves, or to control speed on steep grades, or to help at loading points.

In handling heavy packages, the runs should be limited or brakes provided. Where possible, at least three rollers should form the riding surface of packages.

Gravity wheel conveyors are used in combination with gravity roller conveyors or power-driven belt conveyors. These commercial units consist of a series of wheels approximately $1^{15}/_{16}$-in. diameter × ⅝ in. face and are equipped with seven ¼-in. diameter high-carbon, polished and ground steel balls that revolve in two case-hardened steel races. Wheels up to 6 in. in diameter are made to suit various problems. They are designed to act as transfers and switches. The steel outer case is plated to resist rust and corrosion. Some wheels are grease-packed for life, but most are oiled for gravity service.

17.9.2   Grade

CEMA recommends a grade of ½-in. per foot under average conditions. This might have to be increased for packages weighing 25 lb or less, and decreased for packages weighing 50 lb or more. It is always best to experiment with the packages to arrive at a proper slope. When a wide variation between the lightest and heaviest load on the same line results in the setting of an excessive grade in order for the lighter articles to run, the use of retarders may be indicated.

Cover Surface Finish	Angle
Friction surface, or smooth	10°
Fabric impression (use rough top belting)	
light	25°
heavy	35°
Molded design with pushers	45°

17.9.3 Curves

Where the load has to be turned or transferred to another conveyor, turntables can be used. Ball transfers can be used for hard-bottom packages weighing up to 150–200 lb.

17.10 PACKAGE BELT CONVEYORS (See also section 2)

17.10.1 General

CEMA includes roller and slider types under the heading of Package Belt Conveyors.

17.10.2 Roller Type

The belt runs on spaced rollers mounted in channel or angle frames. Horizontal conveyors require at least two rollers under each package at all times, and vertical conveyors require three or more. Roller beds require a minimum of power and will give longer belt life.

17.10.3 Slider Type

The belt and commodities are carried by sliding on a metal bed. This type is used for impact loading (as at airport baggage loaders) and where the package shapes or size might interfere with roller operation.

17.10.4 Inclined Package Belt Conveyor

The RMA recommends the following maximum angles of incline under dry conditions:

Note that the weight, size, shape, and surface finish of the package may reduce the maximum angle.

CEMA adds that the values above are not generally used under average conditions: "These figures are based on using the best grades of rubber compound, and should be used only under ideal conditions with new, clean, dry cartons. . . . The angles of incline for damp conditions are not listed due to the many variables. . . ."

Generally, the maximum angle of incline or decline is limited to 25°, with a few installations slightly exceeding this value, provided the coefficient of friction between the belt and the packages will prevent sliding.

17.10.5 Package Configuration

The relationship of the height of the package to its base length is important in determining the maximum slope. A safe rule to follow is to make the slope such that a perpendicular line drawn through the center of gravity of the package will fall within the middle one-third of its base length (see figure 17.3).

17.10.6 Transfers

Quoting from CEMA Standard No. 402-1985:

Probably the most universally used transfer between belt conveyors is the gravity curve, either roller or wheel type. For intermediate receiving or feeding, a spur section of roller or wheel conveyor is used at an angle of approximately 30 degrees to the direction of travel of the through conveyor [see figure 17.4a]. The tapered portion of the transfer is fitted with varying-length rollers or with a series of wheels or casters. The rollers in this section may also be power driven.

Another form of power is the right-angle one, where the feed line is brought in at a right angle to the main line and a turning wheel or roller is used to assist the load in negotiating the turn [see figure 17.4b]. Such an arrangement works very well if there is always sufficient space between unit loads to permit one to negotiate the turn before the next one arrives.

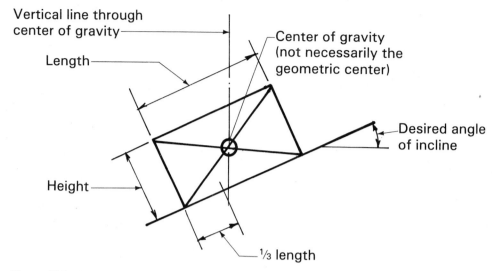

Figure 17.3 Slope of packages.

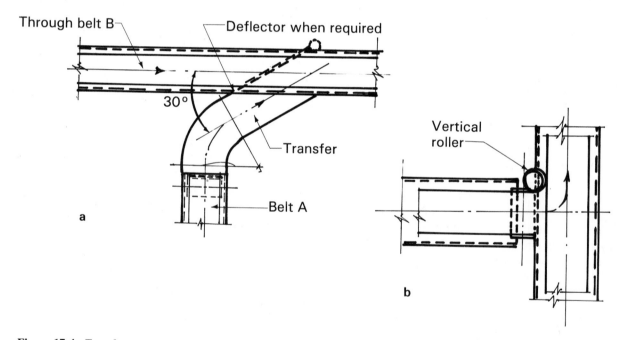

Figure 17.4 Transfer.

17.10.7 Frames

Formed channel or angle frames are furnished by various manufacturers with single or double adjustable or fixed steel or pipe supports to floors or steel rod hangers for attaching to steel and wood ceilings by bolts or to concrete ceilings by expansion shields. When there is a long line of gravity conveyor actually blocking through traffic in certain aisles, hinged-gate sections 3–4 ft long can be installed and operated by hand to provide a passageway for personnel traffic.

17.11 BELT- AND CHAIN-DRIVEN ROLLER CONVEYORS

17.11.1 General

Both belt- and chain-driven roller conveyors are used for controlled movement of a wide variety of package weights and shapes. They can be inclined or declined slightly, and can be run at a speed suitable to the operation of which they are a part. Belt-driven conveyors should be used where accumulation is permitted. Chain-

Table 17.4 Package Conveyor Chains

Chain	Allowable Chain Pull[a] (lb)	Average Ultimate Strength (lb)	Max Unit Load (lb)
SR-420	1800	22,000	500
SR-620	1800	21,000	500
SR-625	1800	21,000	500
SS-933	9200	56,000	1250
H-78	2380	16,000	
C-131	3220	24,000	
SM-12	2200	12,000	
348 Mod	1600	24,000	
X-458	3200	48,000	
468	4700	70,000	

Source: Courtesy Rexnord

[a]Up to 80 fpm.

Table 17.5 Percent Rolling Friction, Steel Roller Chain, Clean Track

Roller Dia (in.)	Dry (%)	Slight Lub (%)
1½	25	20
2	20	15
2½	15	12
3	12	9
4	11	8
5	10	7
6	9	6
Rolls frozen	30	20

Source: Courtesy Mathews Conveyor

Note: The use of 15% on straight track and 25% on curved track for 2½-in. rollers and over is frequently recommended.

driven conveyors should be used for heavy loads, particularly those on skids or pallets. Table 17.4 gives data on package conveyor roller chains. Tables 17.5 and 17.6 give information on rolling and sliding friction.

17.11.2 Cross-Sections, Chain Arrangements, and Ball Transfer Beds

Belt- and chain-drive cross-sections, types of chain arrangements, and a ball transfer bed are shown on figures 17.5, 17.6, and 17.7, respectively.

Table 17.6 Percent Sliding Friction

Condition	Dry (%)	Slight Lub (%)
Unblocked line	30	20
Blocked line	50	40
Curves	25	Exit chain pull-friction guides
Curves	10	Exit chain pull-wheel guides

Source: Courtesy Mathews Conveyor

Note: Add 30% of the weight of the heaviest article to be deflected, unless coef/frict of the article is greater (such as a rubber tire).

17.11.3 Pressure Rollers for Belt-Driven Conveyors

Pressure rollers are normally spaced at every third carrying roller on horizontal runs, and every second roller on inclined runs, at transfer deflections and at loading points.

17.11.4 Design of Chain-Driven Live Roller (Roller-to-Roller Chain Drive) Conveyor

To convey four 4000-lb loads on 5-ft ctr:

$$\text{Weight} = 4 \times 4000 \text{ lb} \times 0.05 \text{ (coef frict)}$$
$$= 800 \text{ lb (for conveying 4000-lb loads on 4 ft} \times 4 \text{ ft} \times 1\tfrac{1}{2} \text{ in. thick plywood pallets)}$$

Conveyor length = 20 ft

Clear conveying surface = 51 in.

Rollers = 3½-in. dia on 12-in. ctr, 57 in. long, weighing 1060 lb × 0.05 (wt × coef/frict)
= 53 lb

Speed = 30 fpm

Sprockets = 20 t, no. 60 weighing 88 lb × 0.05 (coef/frict)
= 5 lb

Chain = no. RC-60 weighing 75 lb × 0.05 (coef/frict)
= 4 lb

Total weight of material, pallets, rollers, sprockets, and chain
= 862 lb

Roller-to-roller condition = 862 lb × 1.2 = 1034 lb

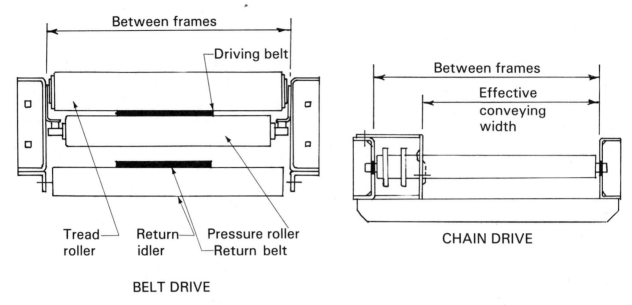

Figure 17.5   Belt- and chain-drive cross-sections.

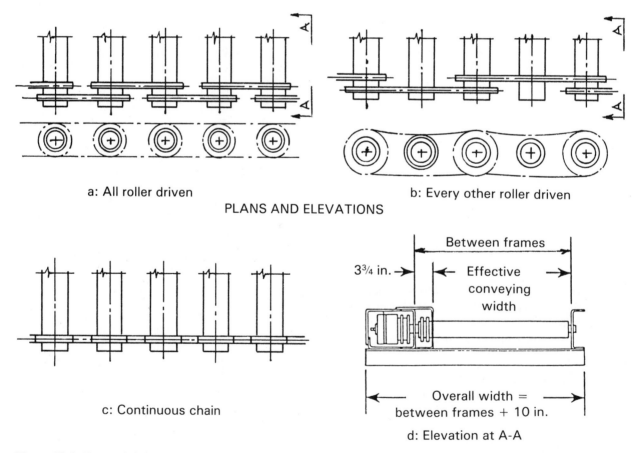

Figure 17.6   Types of chain arrangements.

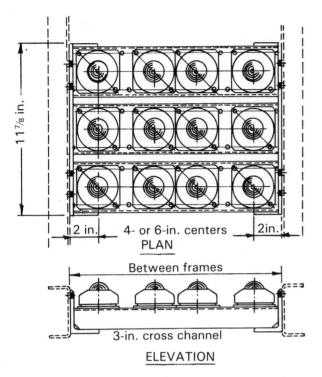

Figure 17.7 Ball transfer bed (Mathews). The ball bed must be positioned so that there is no interference from cross braces in the frame rails, or the cross braces must be relocated.

Dirty condition = 1034 × 1.2 = 1241, say 1300 lb
Effective pull speed
  (1300 × 30)/33,000 = 1.18 (effective hp)
Drive friction losses
      1.18/0.95 (chain ineff)
    = 1.24 × 1.3 (safety)
    = 1.62 hp
  rpm = (30 × 12)/10.99 = 32.75
      where 10.99 is the roller cir-
      cumference (in.) based on a di-
      ameter of 3½ in.

Use a 2-hp, 35-rpm output right-angle gearmotor good for 1.62 hp output. Put the drive in the center and use no. RC-60 chain (figure 17.8).

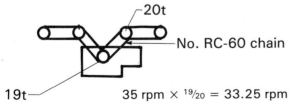

Figure 17.8 Gear reduction.

If we had standstill condition on our 4000-lb load,

4000 × 0.35 (plywood) = 1400 lb extra pull

and so no. RC-80 chain must be used and the extra hp required would be

$$\frac{1400 \times 30}{33,000} = 1.2$$

Net hp would then be 1.2 + 1.18 = 2.38/0.95 = 2.5. With 5 hp at 35 rpm, the output = 4.26 hp. With 3 hp at 35 rpm, the output = 2.5 hp (too close). Consequently, use a 5-hp motor. (Note that the allowance for loading variations is 25%.)

The properties of hexagonal and round shafts are given in tables 17.7 and 17.8, respectively.

17.12 ROLLER AND SHAFT DEFLECTION

Shaft deflection through the bearing cone (figure 17.9) should not exceed 1° on commercial bearings, much less than that on precision bearings (refer to manufacturers' recommendations).

Table 17.7 Section Properties of Common Hexagonal Shafts

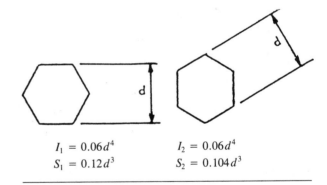

$I_1 = 0.06d^4$	$I_2 = 0.06d^4$
$S_1 = 0.12d^3$	$S_2 = 0.104d^3$

Hexagonal Shaft (across flats, in.)	$I_1 = in^4$	$S_1 = in^3$	$I_2 = in^4$	$S_2 = in^3$
¼	0.00023	0.00188	0.00023	0.00163
5/16	0.00057	0.00367	0.00057	0.00318
3/8	0.00119	0.00633	0.00119	0.00549
7/16	0.00220	0.01005	0.00220	0.00871
5/8	0.00920	0.02930	0.00920	0.02540
11/16	0.01341	0.03900	0.01341	0.03380

Table 17.8 Section Properties of Common Round Shafts

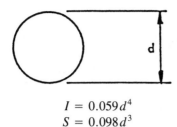

$$I = 0.059 d^4$$
$$S = 0.098 d^3$$

Round Shaft (dia, in.)	$I = in^4$	$S = in^3$
1/4	0.00019	0.00153
5/16	0.00047	0.00299
3/8	0.00970	0.00520
1/2	0.00306	0.01230
3/4	0.01550	0.04130

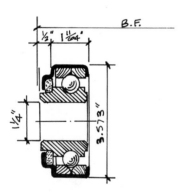

Figure 17.9 Labyrinth steel and felt front seal, Mathews No. 632 bearing. Data: straight-face conveyor bearing, 1¼-in. hex bore, 2250 lb load rating, raceways carefully hardened, 13 hardened steel balls 9/16-in. diameter, pressed steel outer shell.

The angle of deflection (θ) is determined by the following equation:

$$\theta = \frac{57.3\,Pab}{2EI}$$

$$= \frac{57.3 \times 1245 \times 1.09 \times 75.82}{2 \times (30 \times 10^6) \times 0.1465}$$

$$= 0.67° \ (<1.0)$$

where

P = allowable load on each bearing
   = 2490 lb/2 = 1245 lb
   = (1.172/2) + 0.5 in. = 1.09 in.
a = 1.09 in. (see figure 17.10)
b = 75.82 in. (see figure 17.10)
E = 30×10^6 psi for structural steel
I = moment of inertia (in^4)
   = $0.06d^4 = 0.1465$

When θ is one degree,

$$P = \frac{2EI}{57.3\,ab}$$

(see figure 17.10).

$$\text{Roller tube bm} = 1245 \times \frac{78}{4} = 24{,}277 \text{ in-lb}$$

$$S = 0.0983 d^3 = 0.0983 \left[(4.25)^3 - (3.374)^3 \right]$$

$$= 3.77 \text{ in}^3$$

$$I = 8.01 \text{ in}^4$$

$$\Delta = \frac{5WL^3}{384EI}$$

$$= \frac{5 \times 2490 \times 78^3}{384 \times (30 \times 10^6) \times 8.01}$$

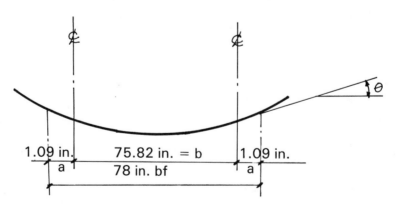

Figure 17.10 Deflection diagram.

$$= 0.064 \text{ in.} < (78/360 = 0.22 \text{ in.})$$

$$f = \frac{24,277}{3.77} = 6440 \text{ psi}$$

17.13 AIRPORT BAGGAGE CONVEYORS

Stearns Airport Equipment Company makes two baggage conveyors consisting of coated 0.350-in. rotating level steel plates designed to carry 200 lb per linear foot at a speed of 90 fpm. The conveyor can make 90° and 180° turns on a minimum radius of 21 in. for the standard unit, and 24 in. for the oversize unit. By making closed loops, baggage can be removed from the conveyor if missed the previous time around (see figure 5-15).

17.14 BAG FILLING

Streeter Richardson supplies bag-filling and handling scales and systems. Equipment ranges from low-cost manual bag fillers to fully automated, high-speed systems for filling, closing, and pelletizing bags of all types, valve or open-mouth, sewn or heat-sealed.

Special feeders, product densifiers and deaerators allow highest accuracy and speed in effectively bagging a wide variety of products, from pellets to flours and powders of many characteristics.

The baggers, made by the Chantland Company are either auger-operated or air-type valved for filling open-mouth bags or drums with powdered, granular, or pelleted materials. The machine can be caster-equipped for mobility.

It is very important to provide dust collection at each filling station. Refer to section 19 for dust collection design information.

17.15 HANDLING OF PALLETS

Figures 17.11 and 17.12 show roller conveyors feeding pallets loaded with bags for shipment by car. Pallets are placed manually on a platform at an area similar to the Transfer Section shown on Figure 17.12. The bags are brought to this point by a belt conveyor or in lift trucks, and the bags are loaded into the pallet in 2 rows of three each. The pallet support is rotated 90°, and 6 more bags are loaded. The pallet is raised 4 in. (the height of the bag) by a foot control every time a row is completed. When the pallets have 12 rows of bags, a motorized

roller moves the pallet to its destination, from whence it is moved by lift truck to car.

17.16 CHAIN PULL*

17.16.1 Level or Inclined Slat, Apron, or Sliding Conveyors (figure 17.13)

$$\text{Eff chain pull} = (A + B + C) \times F$$
$$+ (A \times D) + E$$

(also max chain pull on level)

$$\text{Max chain pull} = \text{Eff chain pull} + (B + C) \times D$$

(on inclines) compute B + C

for inclined length of

top run only

17.16.2 Vertical Chain or Suspended Tray Conveyors (figure 17.14)

$$\text{Eff chain pull} = A \times (1 + F)$$

$$\text{Max chain pull} = \text{Eff chain pull} + \frac{B + C}{2}$$

17.16.3 Allowable Chain Pull—Max Unit Load

Allowable chain pull must be compared with maximum chain pull and not with effective chain pull.

Mathews standard SR type chains can be furnished with heat-treated side bars for working loads up to 2750 lb. All drop-forged chain are furnished heat-treated.

17.17 DRIVES

All chain conveyors should have overload protection. Either a torque-limiting device or a shear-pin sprocket should be used. Inclined conveyors should have a safety

---

*The chain engineering data in the remainder of this section is courtesy of Rexnord, Inc.

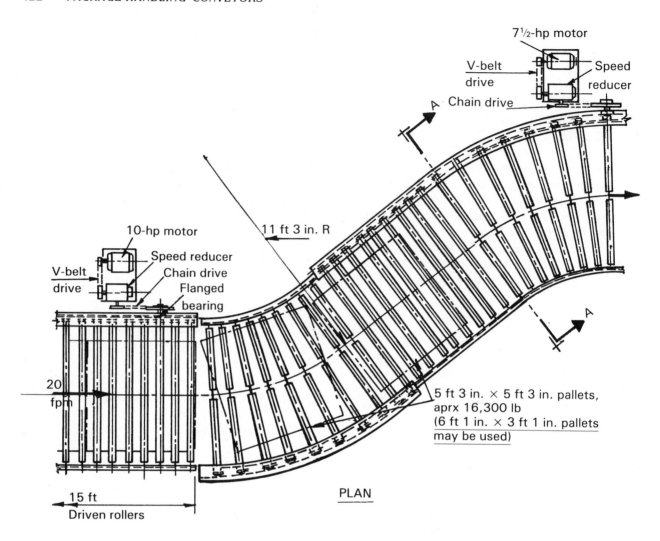

7½-hp motor

V-belt drive

Speed reducer

A

Chain drive

11 ft 3 in. R

10-hp motor

Speed reducer

V-belt drive

Chain drive

Flanged bearing

A

20 fpm

5 ft 3 in. × 5 ft 3 in. pallets, aprx 16,300 lb
(6 ft 1 in. × 3 ft 1 in. pallets may be used)

15 ft
Driven rollers

PLAN

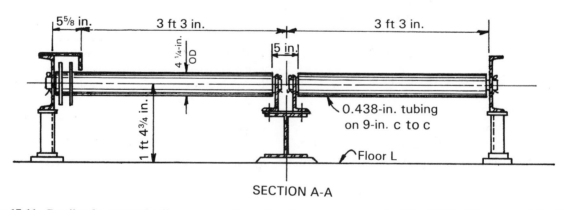

5⅝ in.

3 ft 3 in.

3 ft 3 in.

4 ¼-in. OD

5 in.

1 ft 4¾ in.

0.438-in. tubing on 9-in. c to c

Floor L

SECTION A-A

Figure 17.11   Details of a powered roller conveyor. Open chain boxes are not permitted by OSHA. For approved detail, see figure 17.6.

FIGURE 17.12 423

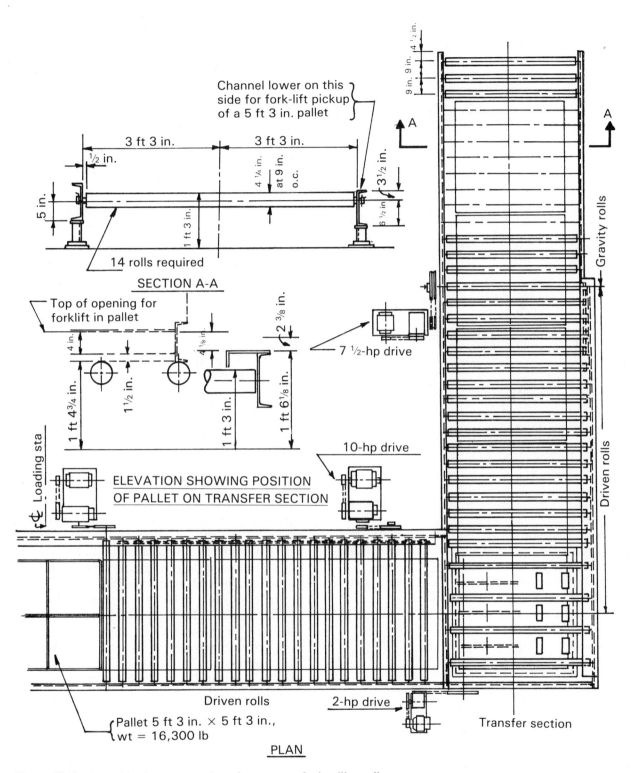

Figure 17.12 A combination power and gravity conveyor for handling pallets.

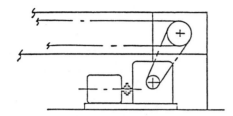

Figure 17.13   Slat, apron, and sliding-chain conveyor.

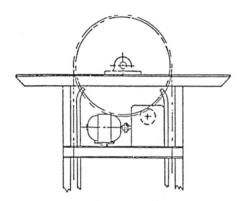

Figure 17.14   Suspended-tray conveyor.

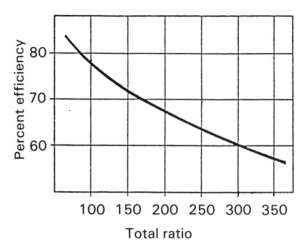

Figure 17.15   Efficiency of double worm-gear reducers.

holdback if a power failure or the breaking of a shear pin would cause the conveyor to back up.

When computing drives, add 5% to effective chain pull for head and tail shaft loss when mounted on anti-friction bearings, and 10% for friction bearings.

Add approximately 5% to the effective chain pull for each open spur gear or chain reduction.

For single worm-gear efficiency, use the formula 100 minus half the ratio.

Example: using ratio 30:1

$$100 \text{ minus } \frac{30}{2} = 85$$

For single worm gears with helix attachment, use the formula above for the worm ratio and deduct 10 for the helix.

The efficiency of double worm-gear reducers can be determined from the chart on figure 17.15.

SR chain data and attachments are shown on figure 17.16.

17.18   SPECIFICATION FOR VERTICAL LIFT DRUM CONVEYOR

The equipment covered by this specification consists of a Swinging Carriage for installation in _____ _____.

The equipment is to convey empty steel drums from first level to second level and also return them from second level to first level. The drums are of one size only: _____ outside diameter × _____ long, weighing _____ each, and are of the all-steel type with three chimes, one at each end and one in the center. The drums are handled on their side with the axis across the face of the carriages, so they may be rolled to and from the loading and unloading stations.

The conveyor is to be set in a pit at the first level, with the first-level loading station at floor level so the drums may be conveniently rolled into the loading station. Purchaser may elect to use runways or skids from cars to station which, if so elected, are to be furnished and installed complete by purchaser, with any manual assistance provided, if and as required, at the loading station, to ensure that the drums will be delivered at the proper time to be properly picked up by the ascending carriages. Once in the station, the drum will be picked up by the next ascending carriage, conveyed to the top of the machine, around the head sprocket, and down the descending side, where it is delivered to the second-level discharge station, being automatically combed from the carriage as the fingers pass through the station fingers, and rolling off of its own weight onto the floor.

FIGURE 17.16 425

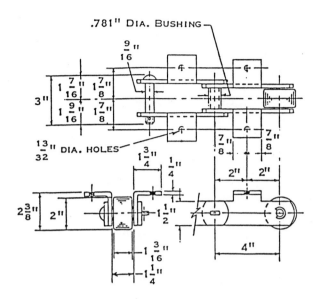

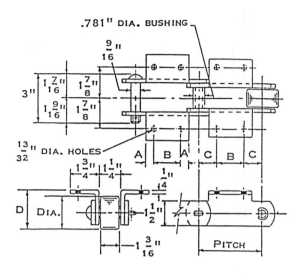

K - 1 ATTACHMENTS

FOR SR - 420 CHAIN ONLY

K-2 ATTACHMENTS

FOR SR-420, SR-620 AND SR-625 CHAIN

CHAIN	PITCH	DIA.	A	B	C	D
SR-420	4"	2"	1/2"	1 3/4"	1 1/8"	2 3/8"
SR-620	6"	2"	3/4"	2 1/4"	1 7/8"	2 3/8"
SR-625	6"	2 1/2"	3/4"	2 1/4"	1 7/8"	2 5/8"

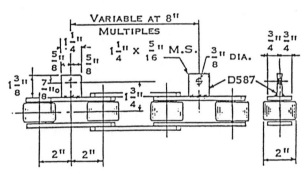

TWO MATCHED STRANDS OF SR-420 CHAIN.

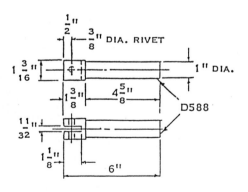

PUSHER BAR ATTACHMENTS
D587 AND D588

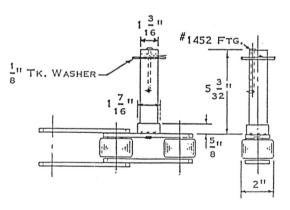

TRAY CONVEYOR ATTACHMENT- D583

TWO MATCHED STRANDS OF SR-420 CHAIN.

ATTACHMENTS WELDED TO INSIDE BARS OF CHAIN ON CENTERS ADEQUATELY SPACED FOR TRAY CLEARANCE.

Figure 17.16 SR chain data and attachments. *(Courtesy Rexnord, Inc.)*

In returning the drums from second to first level, the same procedure is used, but the second-level discharge station is manually thrown out of position by means of the control levers provided. The second-level loading station also is equipped with the same type of lever arrangement, so it may be thrown out of position when service from first to second level is required. It will be impossible, therefore, to provide the two types of service simultaneously. The capacity of the conveyor is to be a maximum of seven drums per minute.

17.18.1 Work to Be Performed By Others

First: Preparation and finishing of all necessary floor openings and pits, as required, together with any protection that may be necessary for same, such as guard-rails, doors, etc.

Second: Clearing of the site, including removal of any obstructions such as piping, conduits, machinery, etc, which might interfere with the work of installation.

Third: Performing of all field wiring, including mounting of starters and pushbutton control stations. The Contractor, however, is to mount the motor unit and the limit switches that comprise the electric safety devices, and to provide a wiring diagram to show the hookup of these limit switches in the control circuit.

17.18.2 Equipment to Be Provided by Contractor

One Swinging-Carriage Vertical-Lift Continuous-Chain Conveyor, with one fixed and one hinged loading station with manual feed, and one fixed and one hinged automatic discharge station, and eight carriages.

Speed: Approximately 45 ft per minute.
Capacity: Approximately seven drums per minute.
Drive Mechanism: To consist of a 1½ hp, 1200 rpm, 440 V, 60 Hz, 3 ph motor, direct-connected to an enclosed worm-gear reducer by means of a flexible coupling. The motor unit is to be furnished by Purchaser, but the reducer and couplings by the Contractor. The slow-speed shaft of the reducer unit drives the conveyor head shaft through a 4-dp spur-gear reduction with 5-in. pitch diameter, 20-tooth 2½-in. face-cut steel pinion, and 38-in. pitch diameter, 152-tooth, 2½ in. face-cut cast-iron gear. The motor and reducer unit shall be mounted on a structural steel stand built integral with the conveyor head framing. The motor which Purchaser is to furnish is to be provided with an extended shaft to permit the applying of a solenoid brake, which also is to be furnished by the Contractor.

Control: Automatic starter with necessary Stop and Start pushbutton control stations, to be furnished complete by Purchaser.

Framework: Structural steel throughout, approximately 25 ft 8½ in. high overall, of the 4-post type, with main members of 5 in. × 3½ in. × ⁵⁄₁₆ in. angles and auxiliary members of 1½ in. × 1½ in. × ³⁄₁₆ in. carriage-guide angles, rigidly braced and crossbraced as required. (Auxiliary guides to be provided around the head sprockets that engage with offset guidewheels mounted on carriage arms and which ensure that the carriages remain in a true horizontal position in passing from the ascending to the descending side around the head sprocket.)

Chain: Two endless strands of No. C-188 malleable combination type 2.069 in. pitch of 14,000-lb ultimate strength per strand, with each strand approximately 48 ft 9 in. long.

Carriage: To be of the fingered type with three V-shaped, or cradle fingers, of heavy structural steel and designed so as to intermesh with the loading and unloading station fingers spaced on approximately 6 ft 1 in. centers with 3 in. channel hangers 24 in. long, guide wheels, arms, etc., complete as required.

Stations: To be of the double-leaf fingered type, of structural steel construction, with two sets of fingers in each leaf. The loading and unloading stations on the second level are to be provided with a hinge arrangement and hand-lever control, permitting them to be thrown out of position when not in use.

Head and Tail Sprockets and Shafts: Head shaft to be 2³⁄₁₆ in. dia with babbitt pillow-block bearings, and with 23 gap-tooth (normal 46-tooth), 38.23 in. pd cast-iron sprockets mounted on the shaft. Tail shafts to be of the stub type, 1⁵⁄₁₆ in dia; four are required, on each of which is mounted a 7 gap-tooth (normal 14-tooth), 11.72 in. pd cast-iron sprocket, with the cross angles on which the four stub shafts are mounted provided with a hand-screw takeup mechanism, and the necessary framework to permit vertical adjustment of the carrying chain, as required to keep chain at proper operating tension.

Safety Device: Both unloading stations are to be provided with electric-type automatic safety devices consisting of two limit switches for each station, with arms and mounting so connected with the control circuit that, in the event a drum fails to discharge properly or is held on the station for any reason, the conveyor will automatically come to an instantaneous stop, and cannot be started again until the offending drum has been cleared. *Alternative No. 1:* If, in lieu of the drive mechanism specified, with the motor furnished by Purchaser and the balance of the drive by Contractor, a geared-head motor, complete with brake and controls, of 440 V (nominal), 60 Hz, 3 ph power description is provided, to be furnished complete by Contractor, including an automatic enclosed starter with two pushbutton control stations—the balance of the equipment is to remain as specified.

Shipment: Approximately _____ weeks after receipt of order, with the work of installation estimated to require approximately _____ days after arrival of material.

Drawing _____ attached.

17.19 COMPARISON WITH BELT CONVEYORS

In comparing belt conveyors (section 2) with package conveyors (section 17), the following should be noted:

1. For belt conveyors, the speed will run anywhere from approximately 200 fpm to 800 fpm. The belt is always troughed. The idlers are spaced 4–5 ft on center.

2. For package conveyors, we are concerned mostly with large boxes that contain small packages and pallet conveyors. Because of the close spacing of idlers, only $\frac{1}{2}$-in. clearance is required between any two idlers, and they may have no cover at all. In handling small packages, metal sliding plates are used and the speeds will vary from 10 ft to 200 ft.

3. The loads are uniform in belt conveyors, but concentrated in handling pallets or large packages.

Pneumatic Conveying

18.1 GENERAL

While pneumatic conveying uses more power per pound of material carried, it offers advantages under certain conditions. It occupies little floor space, makes possible moving material through crowded areas, is dust-free, except perhaps at loading and unloading points, and offers more safety. It requires less maintenance and less operating labor. It can run only from one definite point to another, except that, with switching devices, material can be diverted to any number of delivery points. There are two general types. The low-pressure type, commonly called Dilute Phase systems, uses the concept of air to carry a product in an airstream which is generally low pressure (under 15 psig), high velocity.

The second type, commonly called Dense Phase systems, uses the concept of air to push a slug of material, and is generally a high-pressure (over 15 psig), low-velocity system. The velocity and pressure will vary with the manufacturer, type of product, and distance the product will be conveyed.

In either case, power must be supplied to float the material in the pipe and then to move it either by suction or pressure. Table 18.1 gives the recommended average velocity, and the volume of air required for the operation of Dilute Phase systems, assuming average conditions. When either the gas or the material is hot, the ability of the gas to transport the material will be reduced, all other conditions remaining the same, and the values given in table 18.1 should be modified. When handling finely ground material, it may be necessary to introduce an inert gas into the system to avoid explo-

sions (refer to paragraph 10.9). At present, there is no exact method for figuring pneumatic conveyors. It is always best to send samples to manufacturers, and to consult with them. Some design work should be done to check recommendations, and to write specifications for obtaining bids.

18.2 DESIGN OF SYSTEM

18.2.1 General Data

Material: sand (moisture $<1\%$)
Capacity: 10 tph
Weight of material: 100 lb/ft^3
Weight of air: 0.075 lb/ft^3
Velocity: 7,000 fpm (table 18.1)

18.2.2 Buffalo Forge Company's Fan Engineering Method

This method is outlined in *Fan Engineering*, 7th ed., chapter on "Conveying."

18.2.2.1 Floating Velocity V_f. The floating velocity V_f for any shape particle in a vertical stream of air is

$$V_f = \sqrt{\frac{2g}{f_D} \times \frac{P_p}{P_a} \times \frac{\text{Volume}}{\text{Frontal area}}}$$

429

Table 18.1   Average Velocities and Air Volumes Required for Conveying Materials at Low Pressures

Material	Weight of Material (lb/ft^3)	Average Velocity to Convey Material (fpm)	Ft^3 Air per Lb Material (aprx)	Fr_s^*
Ashes, coal	30	5500	35	0.8
Cement	94	7000	15	0.8
Coal, powdered	30	4000	30	0.8
Corn, shelled	45	5500	40	0.4
Cotton seed (dry)	5	3500	75	0.7
Grinding dust	30	4500	25	0.9
Iron oxide	25	6500	45	0.8
Lime, hydrated	30	5500	35	0.5
Limestone, pulverized	85	5000	15	0.9
Malt	35	4800	35	0.6
Oats	26	4500	40	0.4
Paper cuttings	20	5000	45	0.6
Powders, dry dust	20	2500	40	0.6
Rags, dry	30	4500	25	0.8
Salt, fine	80	6000	20	0.6
Salt, coarse	50	5500	30	0.6
Sawdust, dry	12	3500	65	0.7
Sand	100	7000	10	0.6
Shavings, light	9	3500	75	1.0
Shavings, heavy	24	4000	45	1.0
Wheat	46	5800	30	0.4
Wood flour	19	2500	35	0.6

Source: Some data courtesy of Buffalo Forge Co., Chicago Blower Corp.
*Coefficient of friction of material sliding on steel pipe.

which, for spherical particles, reduces to

$$V_f = \sqrt{\frac{4}{3} \times \frac{g}{f_D} \times \frac{P_p}{P_a} \times d_p}$$

where

V_f = floating velocity in ft/sec
g = acceleration due to gravity in ft/sec/sec (32.2)
d_p = diameter of particle in ft = 0.003417
P_p = bulk weight in lb/ft^3 = 100
P_a = weight of air in lb/ft^3 = 0.075
f_D = average coefficient of drag = 0.5 for spheres and 1.0 for cylinders
W_a = air rate in lb/min

The diameter of an individual sphere of sand was obtained by screening and was found to be 0.041 in. or 0.003417 ft.

$$V_f = 60 \times \sqrt{\frac{4 \times 32.2 \times 100}{3 \times 0.5 \times 0.075} \times 0.003417}$$

$$= 19.78 \times 60$$

$$= 1187 \text{ fpm (say, 1190 fpm)}$$

18.2.2.2   Relative Velocity (V_r) and Losses.   In a vertical pipe, the relative velocity, V_r, of material and air is equal to the floating velocity. For a horizontal pipe,

Fan Engineering, quoting Gasterstadt, gives:

$$V_r = V_f(0.18 + 0.65 \times 10^{-4} \times V_{air})$$

$$= V_f(0.18 + 0.65 \times 0.7)$$

$$= 0.635 \times 1190$$

$$= 756 \text{ fpm (say, 750 fpm)}$$

Material Velocity V_m

$$V_m = 7000 - 1190 = 5810 \text{ fpm (for vertical run)}$$

$$= 7000 - 750 = 6250 \text{ fpm (for horizontal run)}$$

Material flow $= (10 \times 2000)/60 = 333$ lb/min
Cross-section 8-in. pipe $= 50.2$ in.$^2 = 0.349$ ft^2

$$\text{10-in. pipe} = 78.5 \text{ in.}^2 = 0.545 \text{ ft}^2$$

$$W_a = 7000 \times 0.349 \times 0.075$$

$$= 183 \text{ lb/min for 8-in. pipe}$$

$$W_a = 7000 \times 0.545 \times 0.075$$

$$= 286 \text{ lb/min for 10-in. pipe}$$

Material Loading R

$$R = \frac{333}{183} = 1.82 \text{ for 8-in. pipe}$$

$$R = \frac{333}{286} = 1.16 \text{ for 10-in. pipe}$$

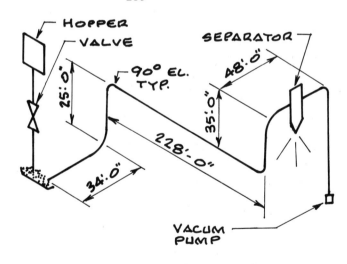

Cubic Feet of Air per Pound of Material

$$\frac{7000 \times 0.349}{333} = 7.34 \text{ for 8-in. pipe}$$

$$\frac{7000 \times 0.545}{333} = 11.46 \text{ for 10-in. pipe}$$

With weight of water $= 62.4$ lb/ft^3
weight of 1 ft^2, 1-in. deep

$$= \frac{62.4}{12} = 5.2 \text{ lb}$$

equivalent height of air for one in. of water

$$= \frac{5.2}{0.07495} = 69.4 \text{ ft}$$

Losses Due to Material Flow

height raised $= 60$ ft

$$\text{lift losses} = 1.82 \times \frac{60 \text{ ft}}{69.4}$$

$$= 1.57 \text{ in. of water for 8-in. pipe}$$

$$= 1.16 \times \frac{60 \text{ ft}}{69.4}$$

$$= 1.00 \text{ in. of water for 10-in. pipe}$$

Horizontal Losses (310 lin ft—horizontally)

$$\begin{array}{r} 228 \text{ ft (fig. 18.1)} \\ 34 \text{ ft} \\ \underline{48 \text{ ft}} \\ 310 \text{ ft} \end{array}$$

Average coefficient on steel $= 0.6$ (table 18.1)

$$\frac{310}{69.4} \times 1.82 \times 0.6$$

$$= 4.88 \text{ in. of water for 8-in. pipe}$$

$$\frac{310}{69.4} \times 1.16 \times 0.6$$

$$= 3.11 \text{ in. of water for 10-in. pipe}$$

Acceleration Losses (see figure 18.1). With the weight of air taken as 0.075 lb/ft^3, the relation between the velocity head (VP) in inches of water, and the velocity (V) of the air stream in feet per minute, is given by

$$VP = \left(\frac{V}{4005}\right)^2$$

Quoting from *Fan Engineering* for a 90° ell, leaving velocity is assumed to be $0.8 \times$ entering velocity; thus, after the first elbow, $V_m = 0.8(6250) = 5000$ fpm (for a long-radius bend with smooth pipe, 0.85 might be used instead of 0.8).

The material starts from rest and reaches a velocity of 6250 fpm on the horizontal run. After the first elbow, its velocity drops to 5000 fpm and, in its vertical 25 ft, reaches a velocity of 5810 fpm. At the end of the second elbow its velocity drops to 0.8(5810) or 4650 fpm.

During its 228-ft horizontal run, it will again reach a velocity of 6250 fpm and, after the third elbow, it will drop to 5000 fpm.

In its 35-ft vertical run, its velocity will increase to 5810 fpm, and drop to 4650 after the elbow.

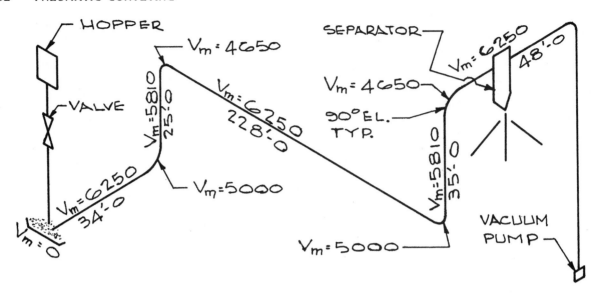

Figure 18.1 V_m diagram.

In the 48-ft horizontal run, it again reaches 6250 fpm.

(1) VP for V_m of 6250 $= \left(\dfrac{6250}{4005}\right)^2 = 2.44$ in. of H_2O

(2) VP for V_m of 5000 $= \left(\dfrac{5000}{4005}\right)^2 = 1.56$ in. of H_2O

(3) VP for V_m of 5810 $= \left(\dfrac{5810}{4005}\right)^2 = 2.10$ in. of H_2O

(4) VP for V_m of 4650 $= \left(\dfrac{4650}{4005}\right)^2 = 1.35$ in. of H_2O

ΔVP (34-ft run) $= 2.44 - 0 \quad = 2.44$ in.
ΔVP (25-ft rise) $= 2.10 - 1.56 = 0.54$
ΔVP (228-ft run) $= 2.44 - 1.35 = 1.09$
ΔVP (35-ft rise) $= 2.10 - 1.56 = 0.54$
ΔVP (48-ft run) $= 2.44 - 1.35 = 1.09$
$$\Delta VP \text{ total} = \overline{5.70} \text{ in.,}$$
$$\text{say, } 6.0 \text{ in. of } H_2O$$

$TP_A = 6.0 \times 1.82$ for 8-in. pipe $= 10.92$ in.
$TP_A = 6.0 \times 1.16$ for 10-in. pipe $= 6.96$ in.

Elbow Losses
   Average velocity

   in 1st and 3rd elbows $= \dfrac{6250 + 5000}{2}$

$$= 5625 \text{ fpm}$$

   Average velocity

   in 2nd and 4th elbows $= \dfrac{5810 + 4650}{2}$

$$= 5230 \text{ fpm}$$

VP (1st ell) $= \left(\dfrac{5625}{4005}\right)^2$

$$= 1.97 \text{ in. of } H_2O$$

VP (2nd ell) $= \left(\dfrac{5230}{4005}\right)^2$

$$= 1.71 \text{ in. of } H_2O$$

Average coefficient of friction on steel $= 0.4$
   Total for 2 elbows $= 3.68$ in., say, 3.70 in.
   Total for 4 elbows $= 3.70 \times 2 = 7.40$ in.
   TP (4 elbows) $\quad = 7.40 \times 1.82 \times 0.4\pi$
   $\quad\quad\quad\quad\quad\quad = 16.93$ for 8-in. pipe
   $\quad\quad\quad\quad\quad\quad = 7.40 \times 1.16 \times 0.4\pi$
   $\quad\quad\quad\quad\quad\quad = 10.79$ for 10-in. pipe

Losses Due to Air Flow
   Total length of pipe $\quad = 370$ ft
   Four elbow equivalents
   $\quad\quad\quad = 4 \times 10 \text{ ft} = \underline{40 \text{ ft}}$
   Length (L) $\quad\quad\quad\quad = 410$ ft

VP for air velocity of 7000 fpm $= \left(\dfrac{7000}{4005}\right)^2$

$$= 3.05 \text{ in. of } H_2O$$

$$TP_D = \text{Duct loss} = \dfrac{(N_D + 10n_e) \, VP_a}{N \times F_c}$$

where

   N_D = number of pipe diameters in length L
   D = diameter of pipe = 10 in. or 8 in.
   N = number of pipe diameters for a loss of one
   velocity head = 55 for 10-in. pipe and 52 for
   8-in. pipe

n_e = number of 90° elbows = 4

VP_a = velocity pressure of air = 3.05 in.

F_c = correction factor for N = 1.52 in. for 10-in. pipe and 1.53 for 8-in. pipe

$$N_D + 10n_e \times VP = \left(\frac{370 \times 12}{10} + 4 \times 10\right) \times 3.05$$

$$= 1476 \text{ for 10-in. pipe}$$

$$N_D + 10n_e \times VP = \left(\frac{370 \times 12}{8} + 4 \times 10\right) \times 3.05$$

$$= 1815 \text{ for 8-in. pipe}$$

$$TP_D = \frac{1476}{55 \times 1.52}$$

$$= 17.66 \text{ in. of } H_2O \text{ for 10-in. pipe}$$

$$TP_D = \frac{1815}{52 \times 1.53}$$

$$= 22.81 \text{ in. of } H_2O \text{ for 8-in. pipe}$$

Values of N are taken from figure 18.2 and F_c is taken from figure 18.3.

Other Losses

TP_e: 0.0 in. entry losses

TP_x: 1.0 in. exit losses (assumed)

TP_s: 2.0 in. separator loss (assumed)

TP_t: 3.0 in. total losses

If entry is accomplished by suction, the entry losses will vary from, say, 1½ in. to as much as 5 in. of water for sand. If the system is hopper fed, the entry loss can be taken as zero.

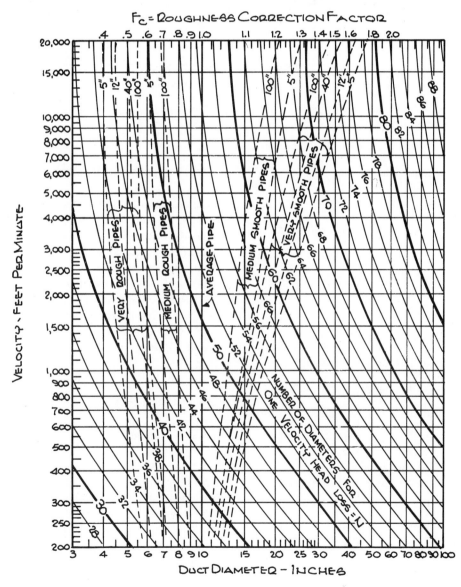

Figure 18.2 Duct friction for velocity head loss (N) and roughness correction factors for ducts (F_c). (*Courtesy Heating Piping and Air Conditioning magazine*)

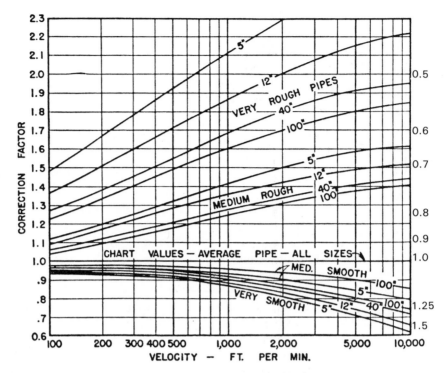

Figure 18.3 Correction factors for pipe roughness. To correct for pipe roughness, multiply the friction loss obtained from figure 18.5 or 18.6 by the correction factor obtained from this figure. (*Courtesy ASHRAE*)

18.2.2.3 Fan Requirements.

Capacity

$$\text{cfm} = \frac{183}{0.075} = 2440 \text{ cfm std air for 8-in. pipe}$$

$$\text{cfm} = \frac{286}{0.075} = 3810 \text{ cfm std air for 10-in. pipe}$$

Pressure

$$1.57 + 4.88 + 10.92 + 16.92 + 3.0 + 22.81$$

$$= 60.10 \text{ in. of } H_2O \text{ for 8-in. pipe}$$

$$1.00 + 3.11 + 6.96 + 10.79 + 3.0 + 17.66$$

$$= 42.52 \text{ in. of } H_2O \text{ for 10-in. pipe}$$

Horsepower

Assume 70% efficiency, which is reasonable. Air horsepower for 1 cfm at 1 in. H_2O = 0.0001573 = 1/6356.

$$\text{hp} = \frac{2440 \times 60.10}{6356 \times 0.70} = 32.96 \text{ for 8-in. pipe}$$

$$\text{hp} = \frac{3180 \times 42.52}{6356 \times 0.70} = 36.41 \text{ for 10-in. pipe}$$

18.2.3 Analysis of Results

The design, as shown, is the original design for a project. A vacuum system was contemplated, to keep the material from being blown around the work area in case of leakage in the ducts. But the size of the required fans and motors would make the system too expensive to operate and too large for a vacuum system. Two alternatives were considered: (1) divide the system in two, or (2) use a high-pressure (dense-phase) system. The second alternative proved more economical and more efficient. The diagram on figure 18.4 shows the revised layout.

18.3 HIGH-PRESSURE SYSTEMS

18.3.1 General

High-pressure (over 15 psig), low-velocity systems use a higher pressure source of supply and generally push the material through the conveying line. The length of the slug that can be pushed through the conveying line will vary with the material handled. Figure 18.4 shows diagrammatically a high-pressure system design for the problem treated in paragraph 18.2.

Typical of most high-pressure systems is the ability to convey materials with heavier bulk densities, because

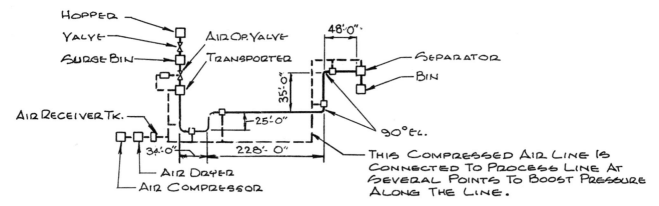

Figure 18.4 High-pressure system diagram.

of the higher pressures at which these systems convey. Bulk densities of 100 lb/ft³ up to and including 300 lb/ft³ are common within the industry. Practical distance limitations include systems 5000 ft long. Generally, they range somewhere between 10 lb and 300 lb of material per pound of air, depending upon the material and distance conveyed. Capacities exceeding 400 tph are possible. This is a function, however, of blow tank size, line size, conveying pressure, distance conveyed, and material conveyed.

18.3.2 Friction

To solve the friction problem, three basic approaches are used.

The first approach is to fluidize the material in the blow tank itself, and add enough air to compensate for the conveying line friction (see figures 18.5 and 18.6). When more air is added to fluidize the material, the exit velocity at the blow tank as well as the terminal velocity will increase.

The second approach is to push the material into the conveying line without fluidizing, in slugs, either mechanically or by pulsing higher volumes of air. The concept is to limit the length of the slug, to prevent high back pressure and thus a plugging condition, but yet to convey at higher densities to improve efficiency.

The third approach is again to push the material into the conveying line without fluidizing, for maximum efficiency, but in a continuous flow closer to extruding. Once into the conveying line, air is added to the conveying line by means of booster fittings, which are spaced evenly to add air along the conveying line where and when it is needed, to overcome the frictional loss as it occurs.

18.3.3 Efficiency

The greatest factor affecting rate and efficiency is the size of the blow tank itself. The larger the blow tank,

the greater the efficiency and the conveying rate. In addition, maintenance is lower, since less purging of the line reduces wear, and fewer cycles mean less wear and tear.

The velocity can be controlled to operate at just about any range required to optimize efficiency, rate maintenance, and product degradation.

For instance, a degradable material such as granulated sugar or salt may be conveyed at a velocity of 400 fpm terminal velocity. This is as compared with conveying a nondegradable powder, such as 325-mesh coal, at a velocity of 2000 fpm. No scientific formulae exist with which to calculate rate relative to a known material; most manufacturers rely on product testing with full-scale equipment to determine conveying characteristics.

18.4 AIRSLIDE

Fuller Company's Airslide™ conveying system uses low-pressure air to fluidize fine dry bulk material, and gravity to move it. The flow depends on the angle of repose and bulk density of the material.

The conveyor duct consists of an upper and a lower compartment separated by a special porous medium. The material flows down in the upper part of the duct, fluidized by air entering from the lower part of the duct through the porous medium. Since the ducts are sloped, the conveying distance is limited by the amount of headroom. Air inlets should be provided at least every 100 feet. Material at temperatures up to 350°F can be moved with conveyors supplied with the ordinary polyester medium. Fuller uses woven glass fabric for temperatures up to 800°F. Capacities up to 50,000 ft³/hr are possible.

Airslide conveyors are in use loading and unloading special cars or trucks with cement, gypsum, flour, and similar products. Fuller Company also makes an Airslide pump conveying system, with a rated conveying

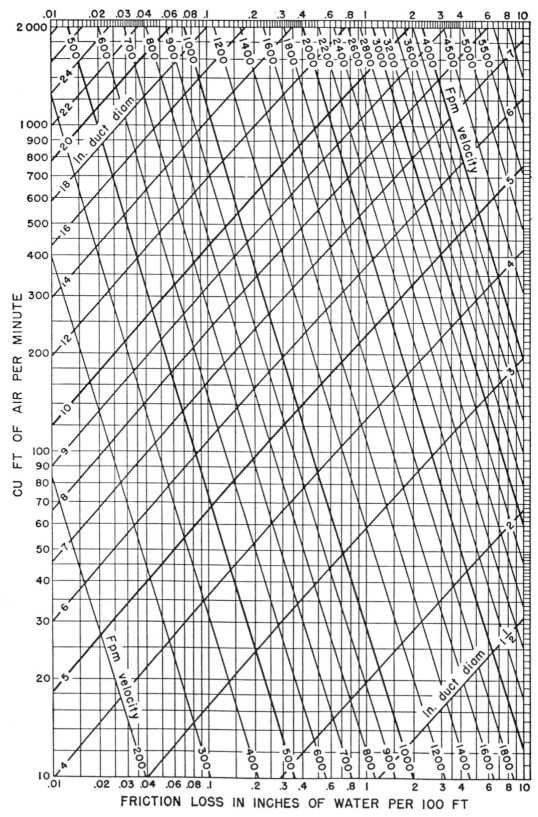

Figure 18.5 Friction of air in straight ducts for volumes of 10–2000 cfm, based on Standard Air of 0.075 lb/ft³ density flowing through average, clean, round, galvanized metal ducts having approximately 40 joints per 100 ft. Caution: Do not extrapolate below the chart. (*Courtesy ASHRAE*)

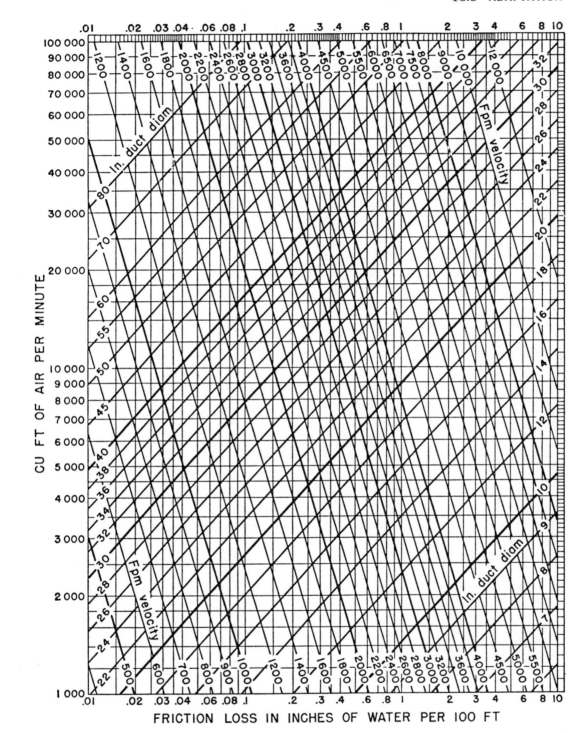

Figure 18.6 Friction of air in straight ducts for volumes of 1000–100,000 cfm, based on Standard Air of 0.075 lb/ft³ density flowing through average, clean, round, galvanized metal ducts having approximately 40 joints per 100 ft. (*Courtesy ASHRAE*)

capacity of up to 40 tph. The pump delivers materials into a pneumatic conveying line. The material flows through the tank toward the discharge line moved by one or more Airslide conveyors installed at the bottom of the tank.

18.5 ADAPTATION

Pneumatic conveying can be adapted to work with other methods of conveying, or can be arranged to provide mixing, and/or weighing and, by introduction of inert

Table 18.2 Classification of Pneumatic Conveyor Systems

System:	Dilute Phase Fan	Dilute Phase Blower	Medium- Dense Phase Pump	Dense Phase Blow Tank	Air-Activated Gravity Conveyor Airslide
Pressure range	±20 in. H$_2$O	±7 psi	15–35 psi	30–125 psi	fan type 0.5–1 psi (closed) 4–5 psi (open)
Saturation ft^3 air/lb matl	vac: 10–30 pres: 4.5–13	vac: 3–5 pres: 1–3.5	0.35–0.75	0.1–0.35	3–5 cfm/ft^2
Matl loading lb matl/lb air	vac: 1.3–0.45 pres: 3–1	vac: 4.5–2.5 pres: 13–3.8	45–18	135–45	—
Air velocity (fpm)	6000	4000–8000	1500–3000	200–2000	10 through dia- phragm
Max capacity (tph)	100	300	300	400	500
Practical distance limits (ft)	vac: 100 pres: 200	vac: 200 pres: 500	3000	8000	100 ft 6 ft drop/length 3–10° slope

gases, can improve the safety of handling explosive, dusty material.

18.6 COMPARISON OF METHODS

Table 18.2 is from an address by Mr. James R. Steele, and is reproduced here with his permission.

18.6.1 Advantages of Dense Phase over Dilute Phase

Ability to convey over longer distances.
Ability to convey materials with higher bulk densities.
Higher material-to-air ratios.
Ability to convey at lower conveying-line velocities.
Ability to convey abrasive materials with less wear.
Ability to convey fragile materials with less degradation.
Ability to start and stop in the conveying line.
Ease in controlling desired conveying-line velocity.
Ability to convey hot materials.

Higher conveying capacities.
Quiet operation.
Little or no segregation during conveying.

18.6.2 Limitations of Dense Phase

Inability to convey fibrous materials (tendency to bridging in the blow tank). Thumb rule is: length-to-width ratio should be less than 6:1.
Inability to convey large particles (inefficiency and tendency to plug). Thumb rule is: particle size to conveying-line ratio 1:12.
Inability to convey materials that pack, smear, or break down under pressure (tendency to plug).
Conveying-line slugging.
Rigid conveying-line supports.
Heavy-duty design that conforms to ASME Code requirements and/or National Board approval.
Minimum blow-tank size at feed end.
Little or no cooling effect during conveying (due to lower air-to-material ratios).
Cost considerations: fan, dilute, or dense phase.

Dust Collection

19.1 GENERAL

There are departments, both Federal and State, that regulate fumes and dusts. The federal EPA (Environmental Protection Agency), and the state Department of Environmental Resources (Pennsylvania) or Department of Environmental Conservation (New York) (many other states have comparable agencies) regulate emissions from plant buildings or processes; the Occupational Safety and Health Administration (OSHA) of the U.S. Department of Labor regulates workplace standards.

Most states now require filing an application for a "Permit to Construct" when installing any equipment to be utilized in the control or collection of dust or fumes. The filing of this document is checked by the appropriate regulatory commission and, provided that design methods are satisfactory, a "Permit to Construct" is issued, which then allows construction and installation of the equipment to begin. After the equipment has been installed and the system is operating at normal production levels, the regulatory commission makes a site inspection, with the owner supplying documented results by an independent testing laboratory, indicating the efficiency of the system. If these results meet or exceed those on the original application, a "Permit to Operate" is issued. This permit is renewed annually. In some states, Pennsylvania, for example, an application to dispose of the waste stream from a dust collector to an approved landfill is also submitted with the "Permit to Construct" application for approval. The regulatory commission can and does make various return site checks to ensure that these systems are maintaining their projected cleanup levels.

Various methods of cleanup are employed by modern industry but, by far the most common and popular, is the dust collector. This piece of equipment generally involves a series of hoods at the dust sources, at a negative pressure induced by a fan on the discharge side of the collector; a series of interconnecting ducts that convey dust to the dust collector; a dust collector that generally utilizes cloth bags to trap airborne dust, in order to discharge approximately 99.9% clean air to the atmosphere or back into the work area.

Figure 19.1 illustrates what might be termed a typical example of a cloth bag dust collector. This particular style of collector uses cloth bags, which collect dust particles on the outer surface, allowing the filtered air to be carried vertically upward into the clean-air section for exhaust to the atmosphere by the induced-draft fan. The heavy dust particles will fall into the hopper at the base section of the collector, due to the velocity pressure drop upon entering the dust collector chamber. The finer particles become trapped in the cloth bags, and are eliminated from the bag fabric via periodic blasts of clean air down into each bag. This air blast creates a shock wave and bag flexing action that causes the fine dust particles to become dislodged and fall down into the hopper at the base of the collector. This air-blast cleaning can be accomplished during normal dust collection in the system, because the collector is divided into sections, permitting one section at a time, at specific timed intervals based on the dust loading of the bags, to be air-blast

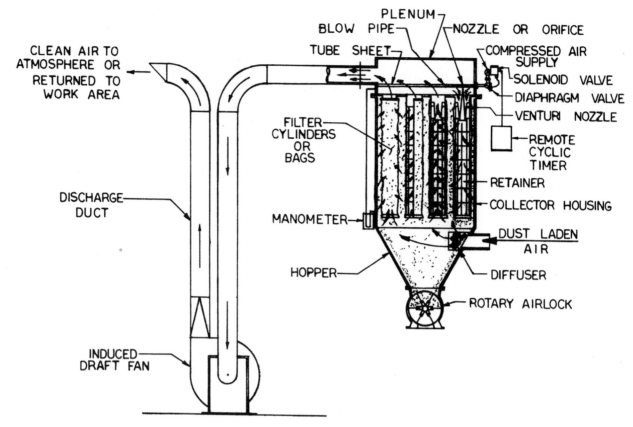

Figure 19.1 A typical dust collector.

cleaned. This type of dust collector requires a reliable dry compressed-air source at a minimum of 90 psi. The collector, as illustrated, should utilize a rotary feeder or air lock at the base of the hopper section, which permits continuous dust discharge during operation of the dust collection system.

When sizing a dust collector based on total system cfm requirements, close contact should be maintained with various reliable dust collector manufacturers' representatives, because of the various air-to-cloth ratios used. These air-to-cloth relationships will vary dramatically between mechanical shaker collectors and reverse air collectors, as well as the judgment factor on the material being collected. At this point, it might be well to mention that only reliable and proved dust collectors be considered because of the importance of a viable system, and the consequences that could result from a poor selection. In addition to the mechanical shaker collectors and reverse air collectors, there are pulse-jet collectors. The reverse air collectors use blowers to clean the bags; the pulse-jet collectors use compressed air.

When handling dust at elevated temperatures, insulation may be required for the collector, to avoid condensation on the bags.

Figure 19.2 illustrates a simple dust collection system, made up of a dust collector, a fan, a system of dust piping, and collection hoods at the dust source.

Assuming that the total cfm requirements of the system have been established based on the individual collection hood requirements, hood and duct design must now follow. An excellent aid to this solution can be found in "Industrial Ventilation," which is a manual of recommended practice and basic design. The most important segment of this design is capture velocity and carrying velocity of the dust in question. Ducts must be sized to maintain minimum carrying velocities of the dust, in order to convey the material to the collector without fallout occurring prior to entry into the collector. If ductwork is improperly sized on the large side, the material being collected will fall out prematurely, building up in the ductwork until one of two things occurs: either the ductwork will collapse from the excessive weight of the dust, or the velocities will increase, causing excessive and premature wear on the ductwork.

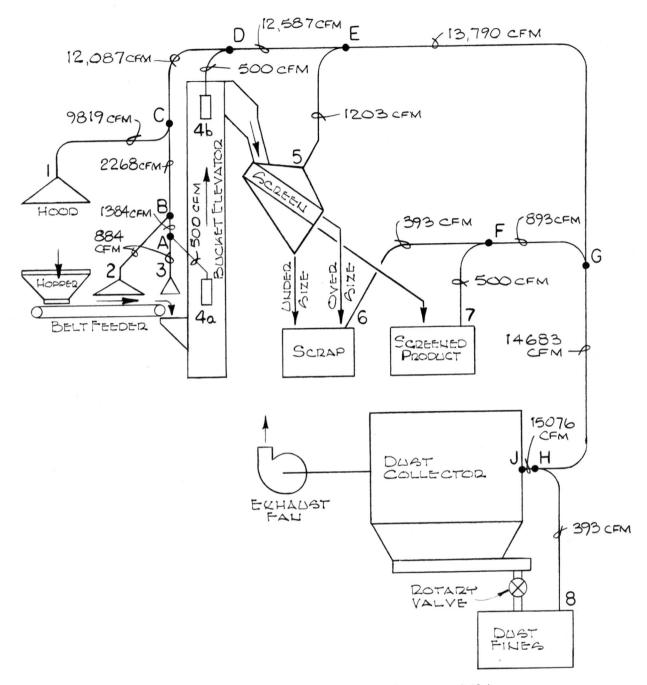

Figure 19.2 A system to collect graphite dust. For design computations, refer to paragraph 19.4.

After installation of the properly designed systems, periodic capacity readings should be taken at the collection hoods and the system ductwork. There may be slight changes in the system capacity. If the system is subsequently modified by the addition or deletion of collection hoods, however, it may not perform as designed.

Each individual hood in a dust collection system must be carefully designed, and all dust at the point of generation carefully analyzed to ensure proper application of good dust collection practice. Careful attention to hood design can greatly reduce the amount of air flow required at each load source. This reduction is a direct function of sizing the dust collector and the induced draft fan required to generate this air movement. Prompt attention to, and solution of, dust and fume collection problems in today's industrial-related life style lead to good positive relationships with the personnel employed by industry, as well as the more informed sector of the community that has become involved with the desire to obtain the cleanest air possible for coexistence with industry.

19.2 SOUND TESTING AND OCCUPATIONAL NOISE EXPOSURE

In an effort to control fugitive dust and fumes in today's workplace, the type of equipment utilized often generates sound levels that are not only irritating, but damaging to personnel working in the vicinity of this equipment. AMCA (Air Movement and Control Association) has issued Standards 300-85, "Reverberant Room Method for Sound Testing of Fans" and 303-79, "Application of Sound Power Level Ratings for Fans."

The following requirements from OSHA Bulletin No. 2206, Section 1910.95 must be implemented:

Protection against the effects of noise exposure shall be provided when the sound levels exceed those shown in OSHA Table G-16 [table 19.1], when measured on the A-scale of a standard sound-level meter at slow response.

When employees are subjected to sound levels exceeding those listed in the table, feasible administrative or engineering controls shall be utilized. If such controls fail to reduce sound levels to within those allowed by the table, personal protective equipment shall be provided and used to reduce sound levels to within the levels of the table.

If the variations in noise level involve maxima at intervals of one second or less, it is to be considered continuous.

In all cases where the sound levels exceed the values shown herein, a continuing, effective hearing conservation program shall be administered.

Table 19.1   Permissible Noise Exposures

Duration/Day (hr)	Sound Level (dBA, slow response)
8	90
6	92
4	95
3	97
2	100
1½	102
1	105
½	110
¼ or less	115

19.3 DETAILS OF DUST COLLECTION INLETS

Figures 19.3, 19.4, and 19.5 give ventilation details for several types of situations; figures 19.6 through 19.10 cover hoods; figures 19.11 and 19.12 give design data for ductwork; figure 19.13 covers screens; figure 19.14 presents hood entry loss calculation data; and figures 19.15, 19.16, and 19.17 offer additional dust-collection information. Table 19.2 (page 454) lists hood design formulas.

Inside duct diameters are shown on the sketches of the industrial hood examples, figures 19.7 through 19.10. The outside diameters of the ducts will be deter-

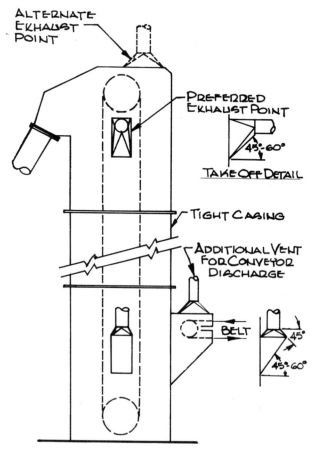

Figure 19.3 Bucket elevator ventilation. See figures 3.3 and 3.4 for inspection doors. For the casing only, $Q = 100 \text{ cfm/ft}^2$ cross-section; duct velocity = 3500 fpm minimum; and entry loss = 1.0 *VP*, or calculate from individual losses. Takeoff at top for hot materials, at top and bottom if elevator is over 30 ft high, otherwise optional. With a belt speed of less than 200 fpm, the volume will be 350 cfm/ft of belt width and not less than 150 cfm/ft of opening. With a belt speed of over 200 fpm, the volume will be 500 cfm/ft of belt width and not less than 200 cfm/ft of opening. *(Courtesy Industrial Ventilation, 15th ed.)*

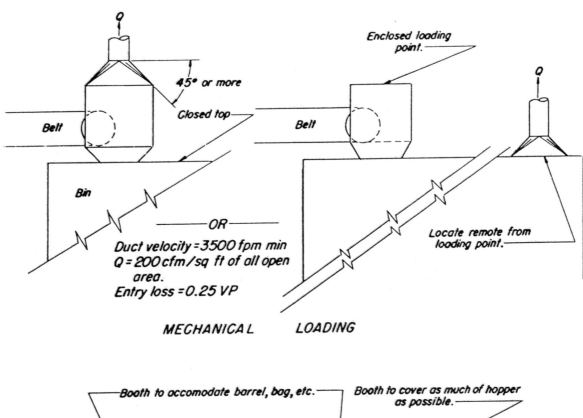

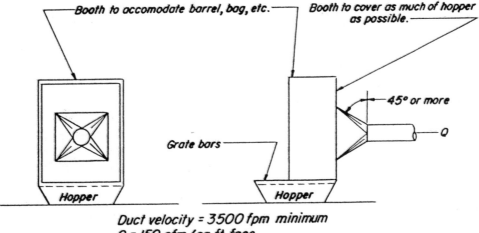

Figure 19.4 Bin and hopper ventilation. For mechanical loading, with a belt speed of less than 200 fpm, the volume will be 350 cfm/ft of belt width and not less than 150 cfm/ft of opening. With a belt speed of over 200 fpm, the volume will be 500 cfm/ft of belt width and not less than 200 cfm/ft of opening. *(Courtesy American Conference of Governmental Industrial Hygienists)*

mined by the gauge of the material used to fabricate the duct.* The gauge of the material will vary depending on the duct diameter, duct velocity, and abrasive action of

the material collected. It will also depend on the material used (for example, carbon steel, USX gauge, galvanizing sheet steel gauge, or stainless steel gauge) and on the method of fabrication, such as spiral lock seam, longitudinal seam, round duct fittings, or welded ducts. The gauge of the material can be increased to two to

---

*See R. Jorgenson, *Fan Engineering*, 8th ed., Buffalo Forge Co., pp. 3-20–3-23.

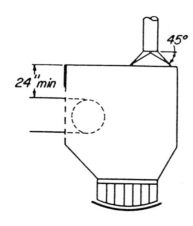

1. Conveyor transfer less than 3' fall. For greater fall provide additional exhaust at lower belt. See 3 below.

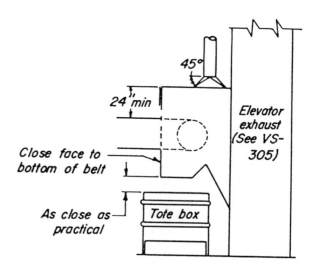

2. Conveyor to elevator with magnetic separator.

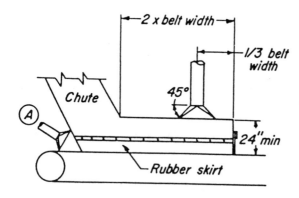

3. Chute to belt transfer and conveyor transfer, greater than 3' fall.
Use additional exhaust at (A) for dusty material as follows:
  Belt width 12"-36", Q=700 cfm
  above 36", Q=1000 cfm

Detail of belt opening

DESIGN DATA

Transfer points:
  Enclose to provide 150-200 fpm indraft at all openings.
  Minimum Q=350 cfm/ft belt width for belt speeds under 200 fpm
      = 500 cfm/ft belt width for belt speeds over 200 fpm and for magnetic separators
  Duct velocity = 3500 fpm minimum
  Entry loss = 0.25 VP

Conveyor belts:
  Cover belt between transfer points
  Exhaust at transfer points
  Exhaust additional 350 cfm/ft of belt width at 30' intervals. Use 45° tapered connections.
  Entry loss = 0.25 VP

Note:
  Dry, very dusty materials may require exhaust volumes 1.5 to 2.0 times stated values.

Figure 19.5 Conveyor belt ventilation. (*Courtesy American Conference of Governmental Industrial Hygienists*)

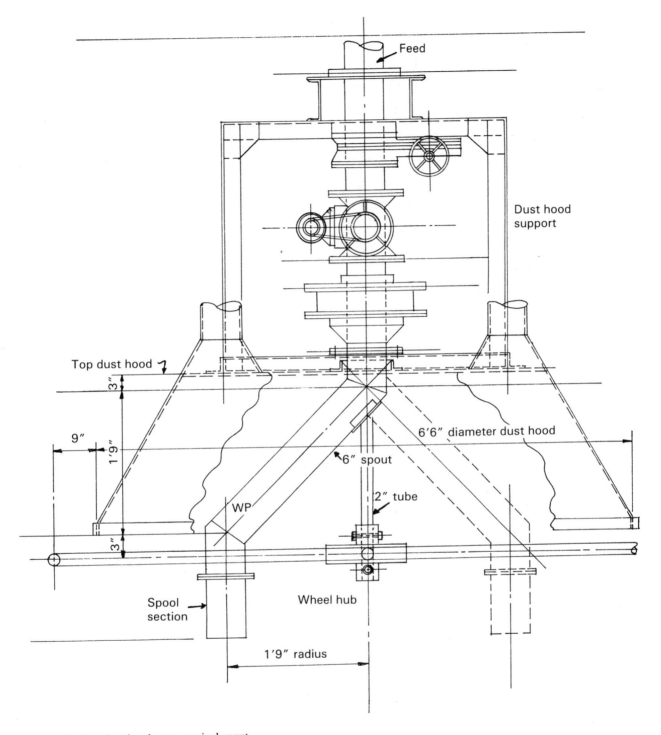

Figure 19.6 A dust hood over a swivel spout.

four times heavier to allow for degree of abrasive action.

The standards for duct construction have been established by the Sheet Metal and Air Conditioning National Association.

Figure 19.15 illustrates the details of a discharge gate and dust hood that have been designed for a specific application that was determined by several criteria: the nature of the dust to be collected; ambient conditions in the immediate vicinity of the hood; necessary capture

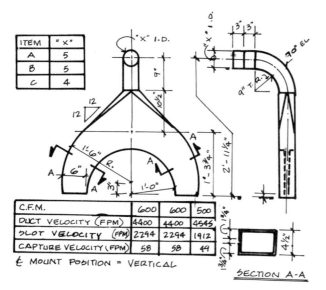

ITEM	"X"
A	5
B	5
C	4

C.F.M.		600	600	500
DUCT VELOCITY (FPM)		4400	4400	4545
SLOT VELOCITY (FPM)		2294	2294	1912
CAPTURE VELOCITY (FPM)		58	58	44

₵ MOUNT POSITION = VERTICAL

SECTION A-A

Figure 19.7 A barrel hood is used to collect dust or fume emissions from material discharged into an open-topped round container; for example, a barrel or drum. (For typical data, see *Industrial Ventilation*, 1977 ed., p. 5-30.)

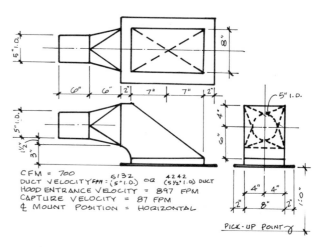

CFM = 700
DUCT VELOCITY (FPM) = 5132 (5" I.D.) OR 4242 (5½" I.D.) DUCT
HOOD ENTRANCE VELOCITY = 897 FPM
CAPTURE VELOCITY = 87 FPM
₵ MOUNT POSITION = HORIZONTAL

PICK-UP POINT

Figure 19.8 A conveyor hood is installed at the point of discharge onto a belt conveyor. (For typical calculations, see figure 19.5, item 3.)

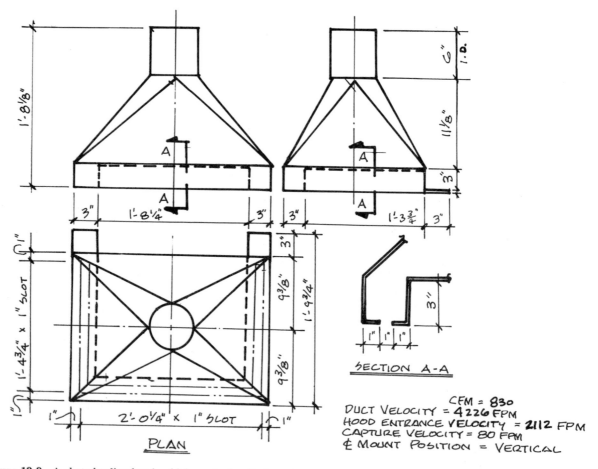

PLAN

SECTION A-A

CFM = 830
DUCT VELOCITY = 4226 FPM
HOOD ENTRANCE VELOCITY = 2112 FPM
CAPTURE VELOCITY = 80 FPM
₵ MOUNT POSITION = VERTICAL

Figure 19.9 A chute loading hood, which can be installed at a hopper loading station to collect dust emissions. The slot design on the three sides of the hood will contain the emissions within the hood to be exhausted through the duct.

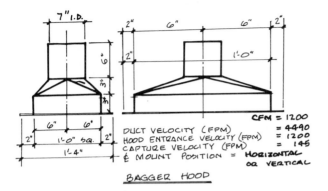

Figure 19.10   This bagger hood is a standard wide-flanged unit that can be located at a bag-loading operation to collect dust emissions away from the bag-filling point. (See *Industrial Ventilation*, 1977 ed., p. 5-28.)

Pipe D	90° Elbow *Centerline Radius			Angle of Entry	
	1.5 D	2.0 D	2.5 D	30°	45°
3"	5	3	3	2	3
4"	6	4	4	3	5
5"	9	6	5	4	6
6"	12	7	6	5	7
7"	13	9	7	6	9
8"	15	10	8	7	11
10"	20	14	11	9	14
12"	25	17	14	11	17
14"	30	21	17	13	21
16"	36	24	20	16	25
18"	41	28	23	18	28
20"	46	32	26	20	32
24"	57	40	32		
30"	74	51	41		
36"	93	64	52		
40"	105	72	59		
48"	130	89	73		

For 60° elbows —— 0.67 x loss for 90°
45° elbows —— 0.5 x loss for 90°
30° elbows —— 0.33 x loss for 90°

Figure 19.11   Duct design data giving equivalent resistance in feet of straight pipe. (*Courtesy American Conference of Governmental Industrial Hygienists*)

velocities required; and minimum carrying velocities required.

19.4   DESIGN OF PROBLEM ON FIGURE 19.2

19.4.1   Data for Design

On figure 19.2, the items are numbered 1 through 8. The identifications of these numbers, as well as exhaust cfm, are listed below.

No.	Exhaust cfm
1. Hopper exhaust hood	9819
2. Belt feeder	884
3. Belt discharge	884
4. Elevator	
4a. lower	500
4b. upper	500
5. Vibrating screen	1203
6. Scrap bin (disp. air)	393
7. Screened product	500
8. Dust fines (disp. air)	393

Note: Initial exhaust cfm is a basic assumption that may later require adjustment.

For dust collection data at elevators, see figure 19.3; for screens, figure 19.13; for bins and hoppers, figure 19.4; for conveyor belts, figure 19.5; and for hood entry, figure 19.14.

Table 19.3 (page 455) gives recommended air flows.

19.4.2   Capture Velocities*

Capture velocity is the velocity at any point in front of the hood necessary to overcome opposing air currents and to capture the contaminated air by causing it to flow into the exhaust hood.

Exceptionally high-volume hoods (for example, large side-draft shakeout) require less air volume than would be indicated by the high capture velocity values recommended for small hoods. This phenomenon is ascribed to:

1. The presence of a large air mass moving into the hood.
2. The fact that the contaminant is under the influence of the hood for a much longer time than is the case with small hoods.
3. The fact that the large air volume affords considerable dilution as described above. Table [19.4, page 455] offers capture velocity data.

(Text continues on page 453.)

---

*From *Industrial Ventilation*, 14th ed., pp. 4-4-4-5.

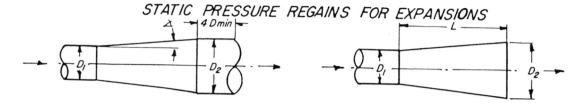

STATIC PRESSURE REGAINS FOR EXPANSIONS

Within duct

Regain (R), fraction of VP difference					
Taper angle degrees	Diameter ratios D_2/D_1				
	1.25:1	1.5:1	1.75:1	2:1	2.5:1
3 1/2	0.92	0.88	0.84	0.81	0.75
5	0.88	0.84	0.80	0.76	0.68
10	0.85	0.76	0.70	0.63	0.53
15	0.83	0.70	0.62	0.55	0.43
20	0.81	0.67	0.57	0.48	0.43
25	0.80	0.65	0.53	0.44	0.28
30	0.79	0.63	0.51	0.41	0.25
Abrupt 90	0.77	0.62	0.50	0.40	0.25
Where: $SP_2 = SP_1 + R(VP_1 - VP_2)$					

At end of duct

Regain (R), fraction of inlet VP						
Taper length to inlet diam L/D	Diameter ratios D_2/D_1					
	1.2:1	1.3:1	1.4:1	1.5:1	1.6:1	1.7:1
1.0:1	0.37	0.39	0.38	0.35	0.31	0.27
1.5:1	0.39	0.46	0.47	0.46	0.44	0.41
2.0:1	0.42	0.49	0.52	0.52	0.51	0.49
3.0:1	0.44	0.52	0.57	0.59	0.60	0.59
4.0:1	0.45	0.55	0.60	0.63	0.63	0.64
5.0:1	0.47	0.56	0.62	0.65	0.66	0.68
7.5:1	0.48	0.58	0.64	0.68	0.70	0.72
Where: $SP_1 = SP_2 - R(VP_1)$✖						

✖ When $SP_2 = 0$ (atmosphere) SP_1 will be (-)

The regain (R) will only be 70% of value shown **above** when expansion follows a disturbance or elbow (including a fan) by less than 5 duct diameters.

STATIC PRESSURE LOSSES FOR CONTRACTIONS

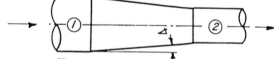

Tapered contraction
$$SP_2 = SP_1 - (VP_2 - VP_1) - L(VP_2 - VP_1)$$

Taper angle degrees	L(loss)
5	0.05
10	0.06
15	0.08
20	0.10
25	0.11
30	0.13
45	0.20
60	0.30
over 60	Abrupt contraction

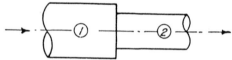

Abrupt contraction
$$SP_2 = SP_1 - (VP_2 - VP_1) - K(VP_2)$$

Ratio A_2/A_1	K
0.1	0.48
0.2	0.46
0.3	0.42
0.4	0.37
0.5	0.32
0.6	0.26
0.7	0.20

A = duct area, sq ft

Note:
In calculating SP for expansion or contraction use algebraic signs:
    VP is (+)
and usually
    SP is (+) in discharge duct from fan
    SP is (-) in inlet duct to fan

Figure 19.12 Ductwork design data. *(Courtesy American Conference of Governmental Industrial Hygienists)*

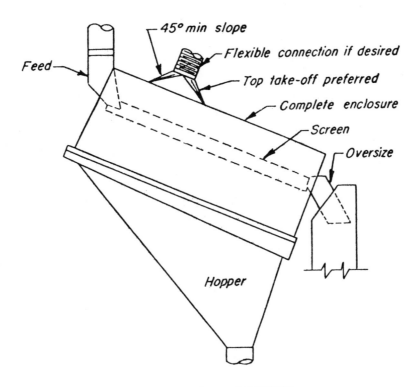

FLAT DECK SCREEN

Q = 200 cfm/sq ft through hood openings, but not less than
   50 cfm/sq ft screen area. No increase for multiple decks
Duct velocity = 3500 fpm minimum
Entry loss = 0.50 VP

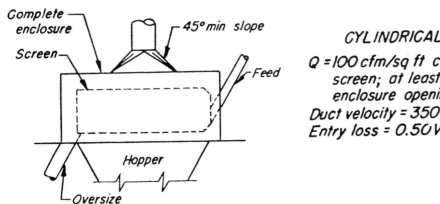

CYLINDRICAL SCREEN

Q = 100 cfm/sq ft circular cross section of
   screen; at least 400 cfm/sq ft of
   enclosure opening
Duct velocity = 3500 fpm minimum
Entry loss = 0.50 VP

Figure 19.13 Screens. *(Courtesy American Conference of Governmental Industrial Hygienists)*

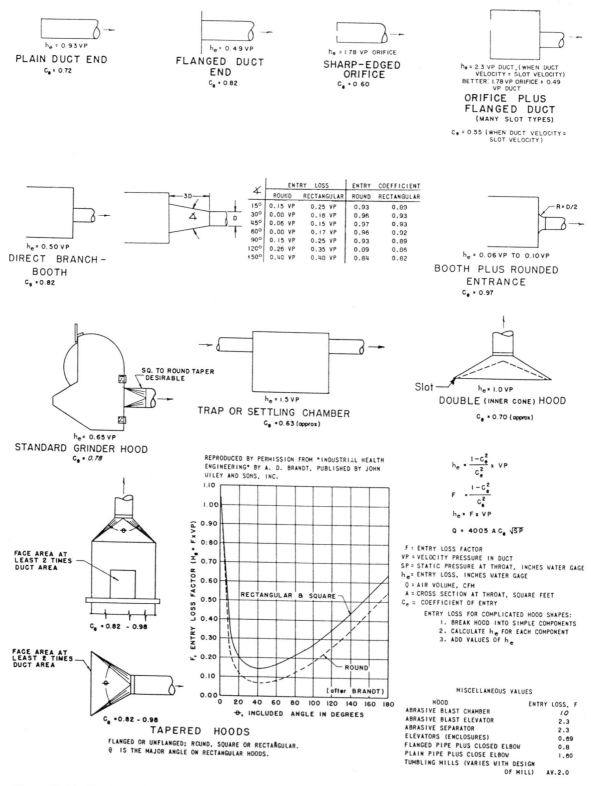

Figure 19.14 Hood entry loss. (*Courtesy American Conference of Governmental Industrial Hygienists*)

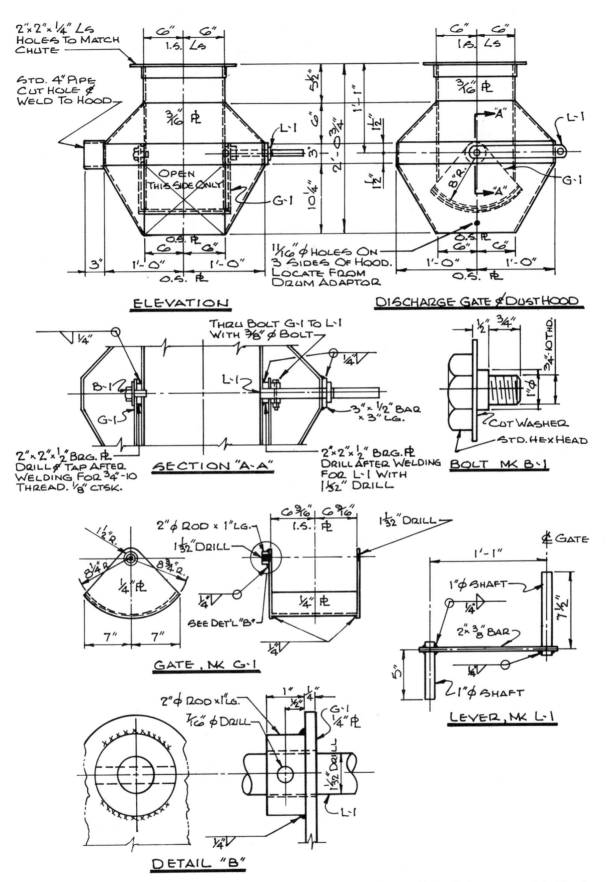

Figure 19.15 Detail of a discharge gate and dust hood.

451

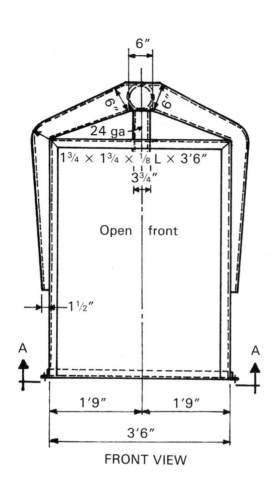

6"

6"   6"

24 ga

$1\frac{3}{4} \times 1\frac{3}{4} \times \frac{1}{8}$ L $\times$ 3'6"

$3\frac{3}{4}$"

Open   front

$1\frac{1}{2}$"

A   A

1'9"   1'9"

3'6"

FRONT VIEW

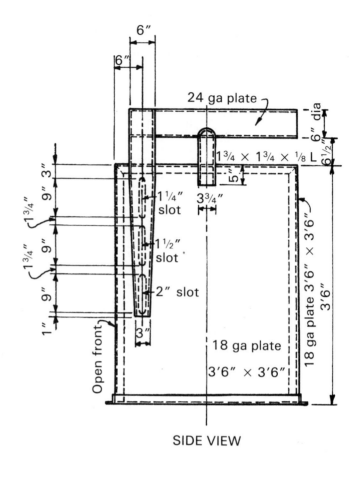

6"

6"

24 ga plate

6" dia

$1\frac{3}{4} \times 1\frac{3}{4} \times \frac{1}{8}$ L

$6\frac{1}{2}$

3"   3"

9"   3"

5"

$3\frac{3}{4}$"

$1\frac{3}{4}$"

$1\frac{3}{4}$"

9"

$1\frac{1}{4}$" slot

$1\frac{3}{4}$"

$1\frac{3}{4}$"

9"

$1\frac{1}{2}$" slot

18 ga plate 3'6" $\times$ 3'6"

3'6"

9"

2" slot

1"

Open front

3"

18 ga plate
3'6" $\times$ 3'6"

SIDE VIEW

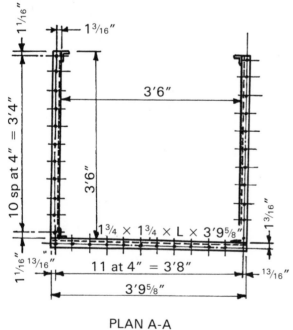

$1\frac{1}{16}$"

$1\frac{3}{16}$"

3'6"

10 sp at 4" = 3'4"

3'6"

$1\frac{3}{4} \times 1\frac{3}{4} \times$ L $\times$ 3'9$\frac{5}{8}$"

$1\frac{3}{16}$"

$1\frac{1}{16}$" $\frac{13}{16}$"

11 at 4" = 3'8"

$\frac{13}{16}$"

3'9$\frac{5}{8}$"

PLAN A-A

Entry Loss in Slots = 1.78 VP
Entry Loss in Duct = 0.49 VP
                      ───────
                     2.27 VP

Figure 19.16 Dust hood at a discharge hopper.

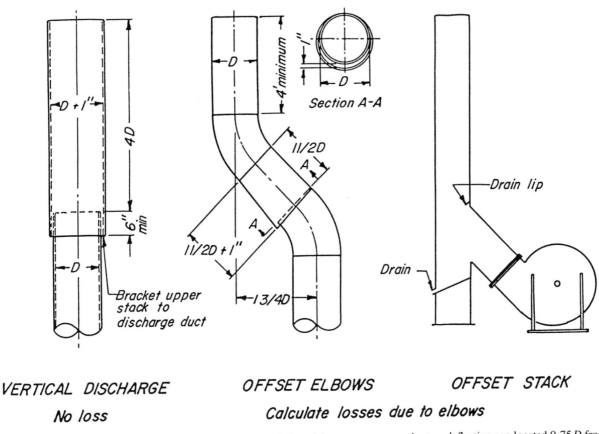

VERTICAL DISCHARGE

No loss

OFFSET ELBOWS

OFFSET STACK

Calculate losses due to elbows

Figure 19.17 Stackhead designs. Rain protection characteristics of these caps are superior to a deflecting cap located 0.75 D from the top of the stack. The length of the upper stack is related to rain protection. Excessive additional distance may cause "blowout" of effluent at the gap between the upper and lower sections. *(Courtesy American Conference of Governmental Industrial Hygienists)*

19.4.3 Cfm Required

The cfm required for the various parts of the system, and the lengths and composition of the ducts are given in table 19.5.

19.4.4 Design Procedure

The equations used in the calculations (see figure 19.2) are as follows:

1. Volume (cfm) = velocity (fpm) × duct area (ft²):
   refer to table 19.6, columns 2, 3, 4, 5, and 6.
2. Equivalent resistance in feet of straight pipe:
   refer to table 19.6, column 9. (90° elbow centerline radius, 2 OD)
3. Resistance to flow in inches of water gauge:
   see figure 18.2.
4. Velocity pressure (VP): $V = 4005\sqrt{VP}$:
   see "Acceleration Losses" in paragraph 18.2.2.2.

5. Entry loss:
   see figure 19.14 (table 19.6, column 14).

Assume collector resistance at 4 in.

Collector fan to be sized at 15,076 cfm and (5.89 in. + 4.00 in.) 9.89 in. sp.

All blast gates to be locked in place after the system is balanced.

Refer to table 19.6 (page 458) for tabulation of data related to the example in figure 19.2.

19.4.4.1 Explanation of Columns 1 Through 6 in Table 19.6 (figure 19.2). Starting at hood 3 and branch 3-A, assume 4500 fpm as adequate carrying velocity, and select a 6 in. duct. The resultant branch volume is 884 cfm.

0.1964 (area of 6 in. duct) × 4500 = 884 cfm

Continuing to hood 4a and branch 4a-A, take 500 cfm exhaust (paragraph 19.4.1) and (by calculation) a

Table 19.2   Hood Design Formulas

Hood Type	Description	Aspect Ratio (W/L)	Air Volume (Q)
	Slot	0.2 or less	$Q = 3.7 \, LVX$
	Flanged slot	0.2 or less	$Q = 2.8 \, LVX$ (figure 19.7)
$A = WL \, (sq. \, ft.)$	Plain opening	0.2 or greater and round	$Q = V(10X^2 + A)$ (figure 19.9)
	Flanged opening	0.2 or greater and round	$Q = 0.75V(10X^2 + A)$ (figures 19.8 and 19.10)
	Booth	To suit work	$Q = VA = VWH$
	Canopy	To suit work	$Q = 1.4 \, PDV$ (P = perimeter of work, D = height above work)

Source: Courtesy *Industrial Ventilation*, 14th ed.

$4\frac{1}{2}$ in. duct is required, resulting in a velocity of 4525 fpm. In exhaust main A-B, the total cfm (884 + 500) is 1384 cfm. With a $7\frac{1}{2}$ in. duct, the resultant velocity (1384/0.3068) is 4511 fpm. From hood 2 and branch 2B and 4500 fpm velocity, select a 6 in. duct. The resultant flow is 884 cfm. In exhaust main B-C, the total cfm is (884 + 1384) 2268 cfm and, with the use of a $9\frac{1}{2}$ in. duct, the resultant velocity is 4607 fpm. Continuing then to hood 1 and branch 1-C, with 4500 fpm velocity, select a 20 in. duct, and the resultant branch volume is 4500 × 2.182 = 9819 cfm.

In a similar manner, all the other values in columns 1–6 were obtained. Basically, the procedure is to determine the velocity needed to convey the dust. This should be based on experience of either project engineers or dust collector manufacturers. As a guide, table 18.1 can be used. For exhaust cfm of equipment, consult paragraph 19.4.1 or table 19.3. Select duct sizes to maintain

Table 19.3   Recommended Air Flows

Type of Equipment	Air Flow
Bucket elevator	100 cfm/ft^2 casing cross-section 3500 fpm duct velocity, minimum
Belt conveyors	
transfer points	350 cfm/ft belt width minimum, speeds < 200 fpm 500 cfm/ft belt width minimum, speeds > 200 fpm 3500 fpm duct velocity, minimum
between transfer points	exhaust additional 350 cfm/ft belt width at 30-ft intervals
transfer points > 3-ft fall	additional exhaust for dusty material: belt width 12-36 in., use 700 cfm; over 36 in., use 1000 cfm
Screens	
flat deck	200 cfm/ft^2 through hood openings, but not less than 50 cfm/ft^2 screen area 3500 fpm duct velocity, minimum
cylindrical	100 cfm/ft^2 circular cross-section of screen, at least 400 cfm/ft^2 of enclosure opening 3500 fpm duct velocity, minimum

Source: Courtesy *Industrial Ventilation*

Note: Dry, very dusty materials may require exhaust volumes 1.5–2.0 times the belt conveyor values stated in the table.

Table 19.4   Hood Design Data, Range of Capture Veolcities

Condition of Dispersion of Contaminant	Examples	Capture Velocity (fpm)
Released with practically no velocity into quiet air	Evaporation from tanks; degreasing, etc.	50–100
Released at low velocity into moderately still air	Spray booths; intermittent container filling; low-speed conveyor transfers; welding; plating; pickling	100–200
Active generation into zone of rapid air motion	Spray painting in shallow booths; barrel filling; conveyor loading; crushers	100–500
Released at high initial velocity into zone of very rapid air motion	Grinding; abrasive blasing, tumbling	500–2000

Source: Courtesy *Industrial Ventilation*

Note: In each category above, a range of capture velocity is shown. The proper choice of values depends on several factors:

Lower End of Range	*Upper End of Range*
1. Room air currents minimal or favorable to capture.	1. Disturbing room air currents.
2. Contaminants of low toxicity or of nuisance value only.	2. Contaminants of high toxicity.
3. Intermittent, low production.	3. High production, heavy use.
4. Large hood—large air mass in motion.	4. Small hood—local control only.

Table 19.5 Cfm and Duct Data

No. of Branch or Main	Cfm Required	Straight Run (ft)	Elbows	Entries
3-A	884	3	1 - 60°	1 - 30°
4a-A	500	5	—	—
A-B	1,384	6	—	—
2-B	884	10	1 - 60°	1 - 30°
B-C	2,268	20	—	—
1-C	9,819	13	1 - 90°	1 - 30°
C-D	12,087	20	1 - 90°	—
4b-D	500	5	1 - 90°	1 - 30°
D-E	12,587	10	—	—
5-E	1,203	10	1 - 90°	1 - 30°
E-G	13,790	30	1 - 90°	—
6-F	393	10	1 - 90°	—
7-F	500	10	1 - 90°	1 - 30°
F-G	893	15	—	1 - 30°
G-H	14,683	30	1 - 90°	—
8-H	393	15	1 - 90°	1 - 30°
H-J	15,076	5	—	—

Note: The conveying velocity of the system is taken at 4400–4500 fpm as a minimum.

the minimum velocity, exceeding it slightly as necessary. Where two ducts meet, add the cfm required for each.

19.4.4.2 Explanation of Columns 7 Through 19 in Table 19.6 (figure 19.2).

Table 19.5 lists the cfm required, the length of the straight runs, and the number of elbows and entries in each duct system. Taking the first horizontal row in table 19.6, hood 3 to branch 3A, and using data in table 19.5:

Col. 7: straight run = 3 ft

Col. 8: (from figure 19.11) one 60° elbow,
6-in. duct = 0.67 × 7 in. = 5 (approx.)
one 30° elbow, 5 in. duct = 5

Col. 9: equivalent length = 10 ft

Col. 10: 10 ft + 3 ft = 13 ft

Col. 11: (figure 18.5) Find the intersection of the 4500 fpm velocity and 6-in. duct and follow vertically down to read friction loss of 5.2 in./100 ft

Col. 12: for 13 ft = 0.13 × 5.2
= 0.676 in. water

Col. 13: entry loss is based on the velocity as shown in column 6

Formula $V = 4005\sqrt{VP}$, and $VP = (V/4005)^2$

where VP = velocity pressure in inches of water. Thus, $(4500/4005)^2 = 1.26$ in. water

Col. 14: taken from figure 19.14 and other figures as given below.

At point 3 entry loss = 0.25 VP (figure 19.5).

At elevator 4a and 4b, the entry loss is 1.0 in. (figure 19.3).

From 1 to C (figure 19.16), the entry loss is 2.27.

From 5 to E, the entry loss is 0.50 in. (figure 19.13).

From 6 to F, the entry loss is 0.25.

From 7 to F, the entry loss is 0.25.

From 8 to H, the entry loss is 0.25.

Column 15 is the result of adding 1 to column 14 to obtain the hood suction velocity pressure. Column 16 is the result of multiplying the unit velocity pressure in column 13 by the hood suction velocity pressure in column 15. Column 17 is the addition of duct resistance in column 12 and the hood suction resistance in column 16. Column 17 is the total resistance in inches of water in any one section of the system.

After the completion of all sections of the system in column 1, the final task is to select that portion of the system that offers the greatest resistance to flow, or the greatest static pressure. In the system illustrated here, the greatest resistance occurs starting with branch 1-C, flowing then through main sections C-D, D-E, E-G, G-H, H-J, and then into the dust collector, resulting in a total system resistance of 11.04 in.

19.4.5 Balancing System Without Use of Blast Gates

The same system can be balanced without the use of blast gates, as shown in the "Industrial Ventilation" manual. In bulk material handling, it is almost always necessary to use blast gates rather than to balance the pressure at piping junctions. Experience has shown that dust collection systems get overloaded rapidly. Extra connections to the system are made by plant employees without consulting engineering departments.

Besides, equipment is moved and changed. If the system is balanced without blast gates; that is, the static pressure is balanced at junctions of ducts, making changes in the system is very difficult.

In buying a dust collector, it is wise to allow 25% extra capacity to the calculated requirements.

19.5 DUST COLLECTORS AND SELECTION

19.5.1 Fabric Filter Collectors

Fabric filter collectors remove particulates from carrier gas streams by interception, impaction, and diffusion mechanisms. The "fabric" may be constructed of a variety of materials, and may be woven or nonwoven. Regardless of construction, the fabric represents a porous mass through which the gas is passed to separate out matter in suspension.

Fabric collectors are not 100% efficient, but well designed, adequately sized, and properly operated fabric collectors can be expected to operate at efficiencies in excess of 99%, often as high as 99.9+% on a weight basis. The inefficiency, or penetration, that does occur is greatest during or immediately after reconditioning. Fabric collector inefficiency is frequently a result of bypass due to damaged fabric, faulty seals, or sheet metal leaks, as opposed to penetration of the fabric. The combined mass becomes increasingly efficient as the fabric or filter mat accumulates a dust cake. At the same time, the resistance to air flow increases. Unless the air-moving device compensates for the increased resistance, the gas-volume flow rate will be reduced. Fabric collectors, as other dust collectors, are designed and intended for service on relatively heavy dust concentrations. The amount of dust collected on a single square yard of fabric may exceed five pounds per hour. In virtually all applications for fabric collectors, the amount of dust cake accumulated over several hours will represent sufficient resistance to flow to cause a reduction in air flow to a rate below an acceptable minimum.

In a well designed fabric collector system, the fabric, or filter mat, is cleaned or reconditioned before the loss of flow is critical. The cleaning is accomplished by fabric agitation, or motion, which frees accumulated dust cake from the fabric. After reconditioning, the fabric retains a residual dust cake and does not exhibit the same characteristics of efficiency or resistance to air flow that new fabric would.

Commercially available fabric collectors employ fabric configured as tubes or stockings, envelopes (flat bags), or pleated cartridge.

The variable design features of the many fabric collectors available are:

1. types of fabric (woven, nonwoven)
2. fabric configuration (tubes, envelopes, cartridges)
3. reconditioning (cleaning) capability (intermittent service, continuous service)
4. type of reconditioning (shaker, reverse air—low pressure, reverse jet—high pressure)
5. housing configuration (single compartment-multiple compartment)

At least two of these features will be interdependent. For example, nonwoven fabrics are more difficult to recondition and, therefore, require reverse jet—high pressure cleaning.

Fabric collectors are sized with the object of providing a sufficient area of filter medium to allow operation without excessive pressure drop. The amount of filter area required depends on many factors, including:

1. release characteristics of dust
2. porosity of dust cake
3. concentration of dust in carrier gas streams
4. type of fabric and surface finish, if any
5. type of reconditioning
6. reconditioning interval
7. air-flow patterns within the collector
8. temperature and humidity of gas stream

Because of the many variables and their range of variations, fabric collector sizing becomes a judgement based on experience. The sizing is usually made by the equipment manufacturer, but at times may be specified by the user or a third party where first-hand experience exists from duplicate or very similar applications. Where no experience exists, a pilot installation is the only reliable way to determine proper size. The sizing or range of a fabric filter is expressed in terms of air-volume rate versus filter-medium area. The resultant ratio is entitled "air-to-cloth ratio" with units of cfm/ft^2 of cloth. The ratio represents the average velocity of the gas stream through the filter medium. The expression, "filtration velocity" is used synonymously with "air-to-cloth ratio" for rating fabric collectors. For example, an "air-to-cloth ratio" of 3:1 (3 cfm/ft^2) is equivalent to a "filtration velocity" of 3 fpm.

19.5.2 Intermittent-Duty Fabric Collectors

These types may use either an envelope or tube configuration of woven fabric and will generally employ shaking or vibrating as a means of reconditioning.

With tube type, dirty air enters the open bottom of the tube and dust is collected on the inside of the tube. The bottoms of the tubes are attached to a tube sheet and the tops are connected to a shaker mechanism. Since the gas flow is from inside to outside, the tube tends to

Table 19.6 Tabulation of Cfm and Pressures

Blast Gate Method						Equivalent Foot Method			
1	2	3	4	5	6	7	8	9	10
									Col 7 Plus Col 9
			Air Volume (cfm)			Length of Duct (ft)			
No. of Branch or Main	Duct Dia (in.)	Duct Area (ft²)	in Branch	in Main	Vel (fpm)	Straight Run	One of each — Ell / Ent	Equiv Length	Total Length
3-A	6	0.1964	884		4500	3	60° / 30°	10	13
4a-A	4½	0.1105	500[a]		4525[c]	5	— / —	—	5
A-B	7½	0.3068		1384[b]	4511[c]	6	— / —	—	6
2-B	6	0.1964	884		4500	10	60° / 30°	10	20
B-C	9½	0.4923		2268[b]	4607[c]	20	— / —	—	20
1-C	20	2.182	9819		4500	13	90° / 30°	52	65
C-D	22	2.640		12087[b]	4578[c]	20	90° / —	36	56
4b-D	4½	0.1105	500[a]		4525[c]	5	90° / 30°	9	14
D-E	22½	2.76		12587[b]	4561[c]	10	— / —	—	10
5-E	7	0.2673	1203		4500	10	90° / 30°	15	25
E-G	23½	3.01		13790[b]	4581[c]	30	90° / —	40	70
6-F	4	0.0873	393		4500	10	90° / —	4	14
7-F	4½	0.1105	500[a]		4525[c]	10	90° / 30°	9	19
F-G	6	0.1964	893		4547[c]	15	— / 30°	5	20
G-H	24½	3.272		14683[b]	4487[c]	30	90° / —	42	72
8-H	4	0.0873	393		4500	15	90° / 30°	7	22
H-J	25	3.409		15076[b]	4422[c]	5	— / —	—	5

inflate during operation and no other support of the fabric is required.

With envelope-type collectors, gas flow is from outside to inside, with the effect of deflating or collapsing the envelope during operation. Support normally is achieved by inserting wire mesh or fabricated wire cages into the envelopes. The opening of the envelope, from which cleaned air exits, is attached to a tube sheet and, depending on design, the other end of the envelope may be attached to a support member or cantilevered without support. The shaker mechanism may be located in either the dirty air or cleaned air compartments. Periodically (usually at 3–6-hour intervals), the air flow must be stopped to effect reconditioning; thus, the classification of "intermittent." The time required for reconditioning seldom exceeds two minutes, but it must be done without air flow through the fabric. If reconditioning is attempted with air flow, it will be less effective, and the flexing of the woven fabrics will actually allow dust to be conveyed through to the cleaned air side.

				Calculation Sheet				
11	12	13	14	15	16	17	18	19
From Fig. 18.5	Col 10 × Col 11 / 100	From V = 4005 √VP	From Fig. 19.14	1.00 Plus Col 14	Col 13 Times Col 15	Col 12 Plus Col 16	At Junction	
Resistance (in.) Water Gauge					Resistance (in.) in Water			
Per 100	of Run	One VP	Entry Loss VP	Hood Suct VP	Hood Suct	Static Press	Gov SP	Corrected cfm
5.2	0.676	1.26	0.25	1.25	1.58	2.26		GATE
7.5	0.375	1.26	1.0	2.0	2.52	2.90		GATE
4.0	0.240	1.26						
5.2	1.04	1.26	0.25	1.25	1.58	2.62		GATE
3.2	0.640	1.32						
1.25	0.813	1.26	2.27	3.27	4.12	4.93		GATE
1.02	0.571	1.31					0.57	
7.5	1.05	1.26	1.0	2.0	2.52	3.57		GATE
1.0	0.100	1.30					0.10	
4.2	1.05	1.26	0.50	1.50	1.89	2.94		GATE
1.02	0.714	1.31					0.71	
8.5	1.19	1.26	0.25	1.25	1.58	2.77		GATE
7.5	1.425	1.26	0.25	1.25	1.58	3.01		GATE
5.2	1.04	1.29						
0.95	0.684	1.25					0.68	
8.5	1.87	1.26	0.25	1.25	1.58	3.45		GATE
0.90	0.045	1.23					0.05	

Notes: Total cfm = 15,076

Total governing resistance = 7.04 in.
Assume loss through collector = 4.00 in.

Total system resistance = 11.04 in.

[a]Refer to table 19.3.
[b]Air volume determined by addition.
[c]Velocity determined by required volume.

The rate of flow through the fabric seldom exceeds 6 fpm and normal selections are in the 2–4 fpm range. Lighter dust concentrations and the ability to recondition more often allow higher filtration velocities. Ratings are usually selected so that pressure drop across the fabric will be in the 2–5 in. Hg range between start and end of a cycle.

19.5.3 Multiple-Section, Continuous-Duty, Automatic Fabric Collectors

The disadvantage of stopping the air flow to permit vibration and variations in air flow can be overcome in several ways. Use of sectional arresters allows continuous operation of the exhaust system as automatic dampers periodically take one section out of service for reconditioning the fabric, while the remaining sections handle the entire gas volume. The larger the number of sections, the more constant the pressure loss. These types may use either envelope or tube configuration of the fabric, and will generally employ shaking or vibrating as a means of reconditioning. The most common designs are an extension of intermittent-duty collectors with the addition of internal walls and automatic dampers.

One variation on this design is the low-pressure, reverse-air type which does not use shaking to recondition. Due to a gas temperature in some applications, glass fabric may be employed. Because glass is fragile, service life may be reduced by shaking. The glass fabric is in tube shape, and a reversal of air flow and deflation is accomplished very gently so as to avoid damage to the glass fibers. A combination of shaking and reverse air flow has also been utilized.

When employing shakers as a means of reconditioning, rate of flow through the fabric will range from 1 fpm to 4 fpm. Reverse-air collapse-type reconditioning generally necessitates lower filtration velocities, since the reconditioning is not as complete. The air to cloth ratio or filtration velocity is based on net cloth area available when one compartment is out of service for reconditioning.

19.5.4 Reverse-Pulse, Continuous-Duty Fabric Collectors

These types of collectors may use envelope or tube configuration of nonwoven (felted) fabrics or a pleated cartridge on nonwoven mat (paper-like). They differ from the low-pressure reverse-air type in that they employ a brief burst of high-pressure air to recondition the fabric. The most common designs use compressed air at 80–100 psig while others use an integral pressure blower. Those using compressed air are generally called pulse-jet collectors, and those using pressure blowers are called fan-pulse collectors.

All designs, even the tube type, collect dust on the outside and have air flow from outside to inside the bag. All recondition the media by introducing the pulse of cleaning air into the opening where cleaned air exits from the tube, envelopes, or cartridges. In many cases, a venturi-shaped fitting is employed at this opening.

Reverse-jet collectors normally clean no more than 10% of the fabric at any one time. Because such a small percentage is cleaned at any one time, and because the induced secondary flow blocks normal flow or filtration during that time, reconditioning can take place while the collector is in service and without the need for compartmentation and dampers. The cleaning intervals are adjustable and considerably more frequent than the intervals employed by shaker-type collectors. A given piece of fabric may be pulsed and reconditioned as often as every one to six minutes.

Due to this very short reconditioning cycle, higher filtration velocities are possible with reverse-jet collectors. With all reverse-jet collectors, however, the fabric is in the dirty air compartment and when reconditioning, accumulated dust which is freed from one fabric surface may become re-entrained and redeposited on an adjacent or the same fabric surface. This phenomenon of redeposition tends to limit filtration velocity to something less than might be anticipated with cleaning intervals of a few minutes.

Laboratory tests have shown that, for a given collector design, redeposition increases with filtration velocity. Other test work indicates clearly that redeposition varies with collector design and, in particular, with normal air flow patterns in the dirty air compartment. EPA-sponsored research has shown that superior performance results from downward flow of the dirty air stream. This downward air flow reduces redeposition because it aids gravity in moving dusty particles toward the dust hopper.

Filtration velocities of 5–12 fpm are usual for tube- or envelope-type reverse-jet collectors. The pleated-cartridge type reverse-jet collectors are limited to filtration velocities of 1–2.5 fpm, because their pleated configuration produces very high approach velocities and greater redeposition.

19.5.5 Removal of Dust

The collected dust often has a resale value. Dust collectors are provided with a screw conveyor at the bottom, which can be extended to load a special truck through elephant ducts. The screw conveyor should be short pitch for materials under 10 mesh.

Layout of a Material Handling Plant

20.1 GENERAL

To consolidate the study of some of the sections, an analysis of the design of a plant is given below. Points to be considered at each stage of the operation are mentioned and are referenced to earlier sections.

Before beginning the design, the following information must be decided upon (usually by the plant engineer and his superiors):

1. desired quantity of finished product
2. quality and character of raw materials required per 8-hr day
3. sources of supply
4. probability of work stoppages and plant storage required
5. where different sizes of material are required, which are most required

The probability of using the calculations or details over again is almost nil, except for plants being duplicated in the same atmospheric conditions, using the same sources of supply, and requiring the same quantity and size of finished product.

The designs given are only guidelines of methods in standard use.

20.2 UNLOADING MATERIAL

The raw material arrives in cars and trucks. It may arrive frozen or wet. A shed may have to be provided for heating the car and its contents, or shaking the car (section 15) to help unload material that is frozen. In any case, it will be unloaded in a hopper where it may be easily broken up on the grating over the hopper, or arrive in sizes passing through the openings. Sometimes a ramp is built for the trucks, so that the material is unloaded above grade. If dust is created, the unloading should be done in a covered shed, with dust intakes at the sides or top of the shed (figure 19.16).

Freight cars must generally be unloaded within 24 hours to avoid demurrage charges. Trucks are given about an hour to unload the material. Thus, if a 20-ton truck delivers the material, the conveyor taking the material from the hopper must have a minimum capacity of 20 tons per hour.

For such materials as ores or rocks, it is best to use an apron feeder to receive the material from the hopper. For lighter material, a belt conveyer will generally prove satisfactory.

Material also may be delivered pneumatically or hydraulically, from some remote point. Much of the iron ore, coal, or stone is delivered thus from the mine direct to the plant. Materials such as grains or cements are delivered in cars and pneumatically conveyed directly to storage bins.

20.3 DRYERS AND COOLERS

The material, as it comes into the plant, often has to be dried to remove excess moisture before further processing, and has to avoid freezing up in the bins, especially if stored outdoors in cold climates.

Heating may be combined with the screening, or the transmission of the material in ducts or conveyors, or by separate kilns or dryers. Cooling of material may be required before final storage. Where the material in the

461

process is subject to heat, such as in furnaces, some heat recovery often is possible.

20.4   SCREENING

Screening, along with removing of tramp iron (refer to paragraph 2.10), should be done before the material has to be crushed, so that only the oversize is fed to the crusher. Tramp iron contaminates the product, sometimes causes explosions, and hinders screening.

20.5   STORAGE

The material is then fed to crushers (if necessary), to process (rarely), or to storage bins. The capacity of the crushers must equal at least the capacity of the conveyor. The kind and amount of storage will depend on:

1. distance to and reliability of supply source or sources
2. reliability and time of delivery
3. danger of shutdowns or labor disputes at either suppliers or plant being designed
4. indoor or outdoor storage
5. condition of material: does it have to be stored by size or do some shipments require different treatment
6. explosion hazard
7. dust problems

If bins are required, circular bins are more economical. Stave silos are generally more economical up to about 30 feet in diameter. Cast-in-place concrete or steel will provide better protection against dust, can be made watertight and, in large-diameter units, may even be more economical. If sliding forms can be used, that is, if the height is, say, over 30 ft, the poured-in-place concrete will often prove the most economical (refer to section 13).

The bins should have high- and low-level indicators, an emergency door near the bottom, manholes, feed holes, and a roof capable of supporting drives. Ladders are sometimes desirable both inside and outside (see Figures 13.17 and 13.18).

The bins may require heating or cooling or perhaps insulation. A bypass may be provided to take the material from the hopper or crusher directly to process.

20.6   ENTERING PROCESS

From the storage bins, the product may have to be taken to a secondary crusher and screens to give the sizes required. Oversize material is brought back to the secondary crusher. Dust collection will probably be required at the crusher and the entrance to the screen. The sized material will be fed to either storage bins, with or without bypasses, to conveyors, or direct for further processing.

20.7   DUPLICATING EQUIPMENT

Depending upon the amount of available storage, lead-time required in the production, the output required, and the estimated loss due to shutdowns, duplicate equipment may be required. Even where only a single set of equipment may be necessary, it may be advisable to provide duplicate equipment with only one-half or, preferably, three-quarters of the required capacity, where storage of finished product is limited, or involves costly inventory. Split production equipment will also help by enabling its efficient use, when required capacity is reduced. Where the plant operates only, say, 8 hours per day, another shift can be used in an emergency. This should be avoided where possible, because of decreased efficiency and other extra costs. It is necessary to duplicate equipment where a breakdown would seriously affect production and shipment.

20.8   BYPASSING EQUIPMENT

To avoid a shutdown, the design should provide bypassing equipment that is not duplicated. Before using a bypass, the flow of material should provide for processing the bypassed material. Thus, on figure 20.1, if screen VS-1 is bypassed, the +3-in. material will be returned to the gyratory crusher via belt conveyor BC-3, and all material remaining in the system will be −2 in., the same as if VS-1 were not bypassed. Occasionally, a bypass is inserted to make possible a larger size of material for shipping or further processing. Note that the sizes selected for the bypass are the ones that are most required by the market.

20.9   SAMPLING MATERIAL

On figure 20.1, note the sampling spouts at each screen. It is always advisable to place sampling spouts, so the operation can be checked and corrected.

20.10   INTERCONNECTION OF EQUIPMENT

On figure 20.1, the hammermills fed by screens VS-4 and VS-9 are interconnected, to provide for emergency when one of the hammermills is taken out of production. Throughout, a design provision should be made for continued operation. Where the equipment is very expensive, some parts are stocked, and storage in large

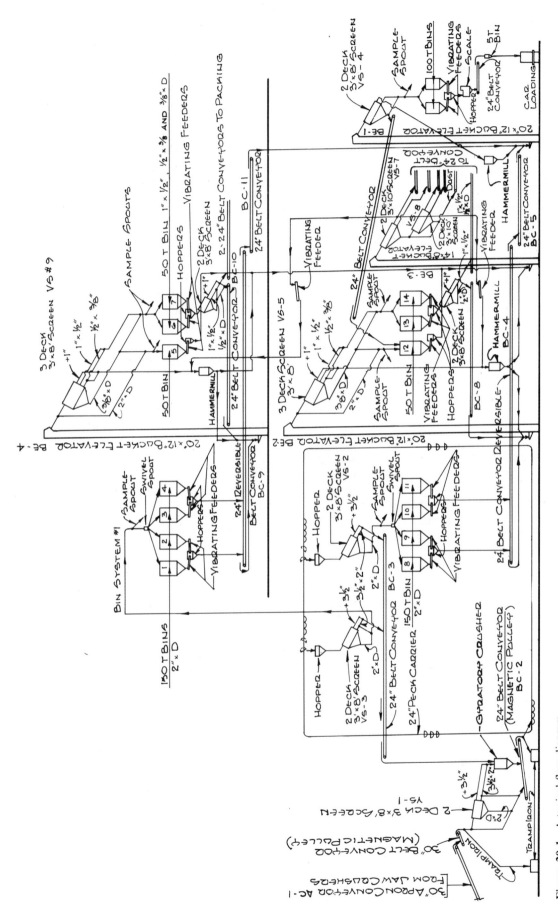

Figure 20.1 A typical flow diagram.

bins provided, for continuity. A failure of the jaw crushers can be overcome temporarily by buying sized material.

20.11   CONTROLS

20.11.1   General

The operation of a material-handling system can be controlled manually (during tests or breakdown of automation) or by automatic controls. Automatic controls can be electrical, electronic, or pneumatic.

20.11.2   Types of Control

Direct digital (electronic) control—proportional, integral, and derivative operation—consists of one or more microprocessors and software application programs to direct the control operation. Controlling elements consist of electronic devices. Controlling devices consist of pneumatic or electronic devices. An operator terminal is necessary as a means of changing programs and limiting functions. Every controlling element can be programmed for alarm. This is cost effective, if accuracy and flexibility is required with annunciation capabilities, time functions, and central multiple monitoring.

Electronic control is predominant in industry today, with direct digital control prevailing over analog control. Sensors and readouts may be analog, but the control is mainly digital. The microprocessors or computers can store information, such as flow rates, temperatures, weights, levels, and pressures, in the form of historical data or set points (control points, upper and lower limits, rate of change, timing, sequencing, interlocking, and annunciation). The digital system operation is readily displayed by cathode-ray tubes.

Pneumatic control—proportional operation—gives the same results as electronic control. In hazardous or high-temperature areas with electromagnetic radiation, pneumatic controls and devices must be used. The pneumatic system is generally more expensive to install than the electronic. It has a limited transmission distance, say, 300 ft, and is slightly slower in response, but is easier to maintain than the electric or electronic system.

High- and low-level indicators are mechanically or electrically actuated.

20.11.3   Emergency Stops

It is wise to provide pushbutton emergency stops at some of the equipment. Pull-cord emergency stops usually are provided along the entire length of long conveyors. The startup buttons should be placed only on the main panel.

20.11.4   Availability of Controls

Control systems can be tailored to suit the plant operators. Change of sequence, interlocking, timing, and recording can be made to suit changes in the process. Troubleshooting can be done by isolating an item or part of the operation. Most bulk-material handling projects involve use of timers and relays to stop the system sequentially. Starting of the system should always be done in the control room, with the operator cutting in each operation manually.

With the two-position control, the start button starts the last item in the system first, and relays start the preceding items (based on timers) until the complete line is in operation. The stop button reverses the process. Where it is desired to leave some material in hoppers upon shutdown, the timers are adjusted to stop the feeders from the hoppers before all the material is removed.

With direct digital control systems, change of sequence, interlocking, and timing can be made with a combination of changing field devices and software programming. There is greater flexibility to the direct digital control system, because the operator can modify numerous programs that are software-based without changing hardware. Recording of activity can be performed via a printer and can be generated by time, function, change of state, or other software-based programs. These systems have built-in diagnostics, allowing troubleshooting to be easier and more accurate. With a remote operator's terminal, an operator can plug into the transmission line and read system activities.

20.12   LAYOUT OF A PLANT

20.12.1   Explanation of Diagrams

To illustrate a flow diagram of a material-handling plant, figures 20.1 and 20.3 are shown. Both diagrams are alike but, for clarity, no electrical information is shown on Figure 20.1. Paragraph 20.12.2 describes the flow. Figures in parentheses refer to sections or paragraphs in the book dealing with the items involved. The electrical drawing (figure 20.2) is explained in paragraph 20.13.

The material handling described here was incorporated in a plant that operated well for about 20 years, but has now been dismantled.

20.12.2   Material Flow (Figure 20.3)

The ore (12 in. lumps) is delivered by rail or truck to a dump hopper (paragraph 13.6) where a dust intake is often required (figure 19.3). Some material should be left in the hopper after a system shutdown. The material travels to a 36-in. jaw crusher where it is reduced to

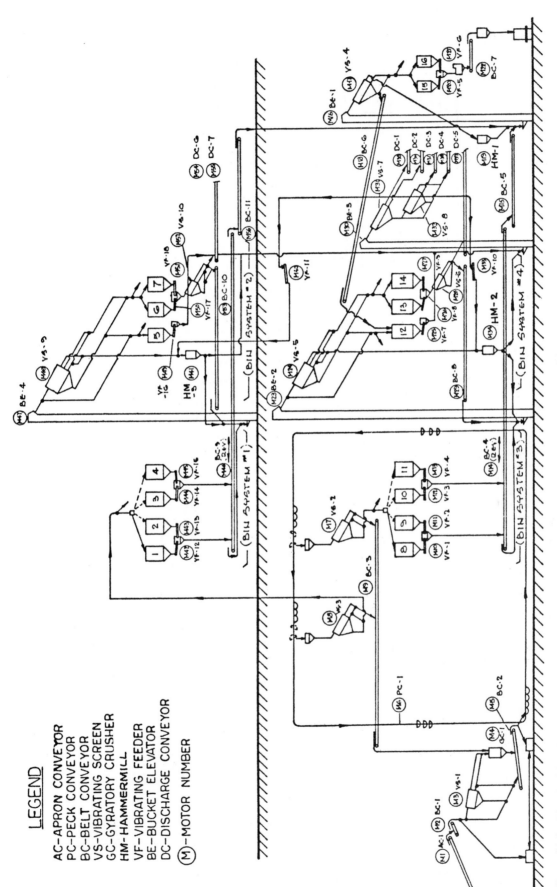

Figure 20.2 Electrical identification diagram.

LEGEND

AC—APRON CONVEYOR
PC—PECK CONVEYOR
BC—BELT CONVEYOR
VS—VIBRATING SCREEN
GC—GYRATORY CRUSHER
HM—HAMMERMILL
VF—VIBRATING FEEDER
BE—BUCKET ELEVATOR
DC—DISCHARGE CONVEYOR

Ⓜ—MOTOR NUMBER

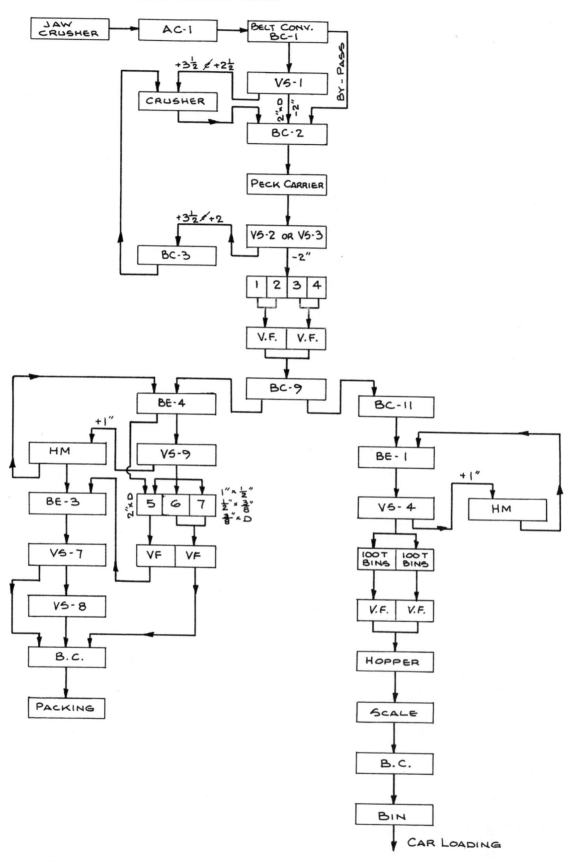

Figure 20.3   Diagram of plant material flow for paragraph 20.12.2. System 1 is shown. System 2 is not shown.

−3½ in. From the jaw crusher, a 30-in. apron conveyor AC-1 (section 5) takes the material to a 30-in. belt conveyor (section 2), where a magnetic pulley (figure 2.8) removes tramp iron and conveys the material through chutes to a two-deck screen VS-1 (section 10). The oversize (+3½ in. and +2 in.) material goes to the gyratory crusher (section 10). The 2 in. × D and −2 in. sizes go to the 24-in. belt conveyor BC-2 (section 2), where a magnetic pulley takes out any tramp iron. In case the vibrating screen VS-1 is out of service, a bypass for the screen and the gyratory crusher is provided. This will take the material via belt conveyor BC-2 to the Peck carrier, where the +2-in. material will get back to the gyratory crusher, as shown on the diagram.

From the 24-in. belt conveyor BC-2, the material is fed to a 24-in. Peck carrier (figure 3.29) where two identical systems are provided. The Peck carrier feeds 3 ft 0 in. × 8 ft 0 in. two-deck screens VS-2 and VS-3 (section 10). The +3½-in. and +2-in. material is brought back to the gyratory crusher via BC-3. Since both systems are alike, only the one with bins no. 1 to 4 will be described. The −2-in material is taken to the four 150-ton storage bins (bin system no. 1) group through a swivel spout (figure 13.3). From the storage bins (section 13), the material is fed by vibrating feeders (paragraph 7.12) to reversible belt conveyor BC-9. This belt conveyor feeds either into bucket elevator BE-4 (section 3) or through belt conveyor BC-11 to bucket elevator BE-1. BE-4 feeds three-deck vibrating screen VS-9 (section 10). The +1-in. material goes to the hammermill, the 1 in. × ½ in., ½ in. × ⅜ in., and ⅜ in. × D travel to separate bins 5, 6, and 7. A bypass can take 2 in. × D material to bin no. 5 and then to elevator BE-3, to screen VS-7 or VS-7 and VS-8, and 24-in. belt conveyor to packing department. The material that went to the hammermill goes back to bucket elevator BE-4 to retrace its steps, or to bucket elevator BE-3 that, in turn, feeds the material to a two-deck 3 ft × 10 ft screen VS-7. VS-7 feeds belt conveyors that lead directly to packing, or to another two-deck 3 ft × 10 ft screen VS-8, which leads directly to packing via 24-in. belt conveyors. This gives an opportunity to obtain a variety of sizes. A bypass for the dust from screen VS-7 is provided. Elevator BE-1 feeds a two-deck 3 ft × 8 ft screen VS-4, where the +1-in. material can be fed to a hammermill that feeds back to BE-1, and the dust can be brought back to bin no. 12. The required sizes are fed to large storage bins, hoppers, vibrating feeders, weighing scales, and 5-ton bins ready for feeding cars.

Belt conveyor BC-11 feeds directly to bucket elevator BE-1, which feeds through screen VS-4 to 100-ton bins. Any size can be fed to this group of bins. The materials go to packing and car loading via scale hopper.

Note that the flow of material from each hammermill can be diverted to the other.

20.13 ELECTRICAL CONTROL FOR A MATERIAL-HANDLING SYSTEM

20.13.1 General

The parameters for design are:

1. Reliability
2. Ease of maintenance
3. Competitive low cost

20.13.2 Reliability

Reliability starts with a durable mechanical system; however, the controls also must be of high quality and intelligent design. In industry, heavy-duty NEMA XII dusttight enclosures are frequently used to ensure protection. Electric starters are best grouped into control centers in separate control rooms in the industrial area. Sometimes explosive atmospheres exist and NEMA VII equipment is required. 480-volt and 120-volt wires may be grouped where 600-volt insulation is used. Communication cable and low-voltage cables must be run in separate wire raceways (not with higher voltage cables). In the heavy industrial areas, metal conduit is used to carry power and control cables.

Control panels are usually installed in air-conditioned control rooms and, in the case of electronic equipment, this is a requirement. Small components of solid-state apparatus can be used in the field by the use of NEMA IV or NEMA VII enclosures; that is, waterproof or explosionproof enclosures.

Interlocking of conveyor controls is critical to prevent pileup or jamming of materials. Sequence starting of motors from downstream to upstream is essential to assure a clear area to receive the materials. Level controls and indicators prevent overfilling and keep the bins filled. Motion detectors will indicate the breakage of a belt or a mechanical device, and subsequently will stop the flow of material to prevent pileup. The detectors may be rotating devices, or noncontact solid state. Level controls must be judiciously chosen to match the materials in the bin or pile. Here the choice is from a wide variety of mechanical and electronic designs. Mechanical rotating paddles have been very successful on a large variety of sizes of materials. Sonar-type sonic level controls work well on ore and coal piles. Resistance probes and capacitance-effect probes are reliable when properly applied. A mechanical electronic plumb-bob device works well on some light granular material.

The main control panel and instrument panel can have a multitude of devices covering a wide range of functions, as follows: pushbutton starting and stopping, running lights, graphic displays, recording charts, statistical printouts, computer programming, and alarms.

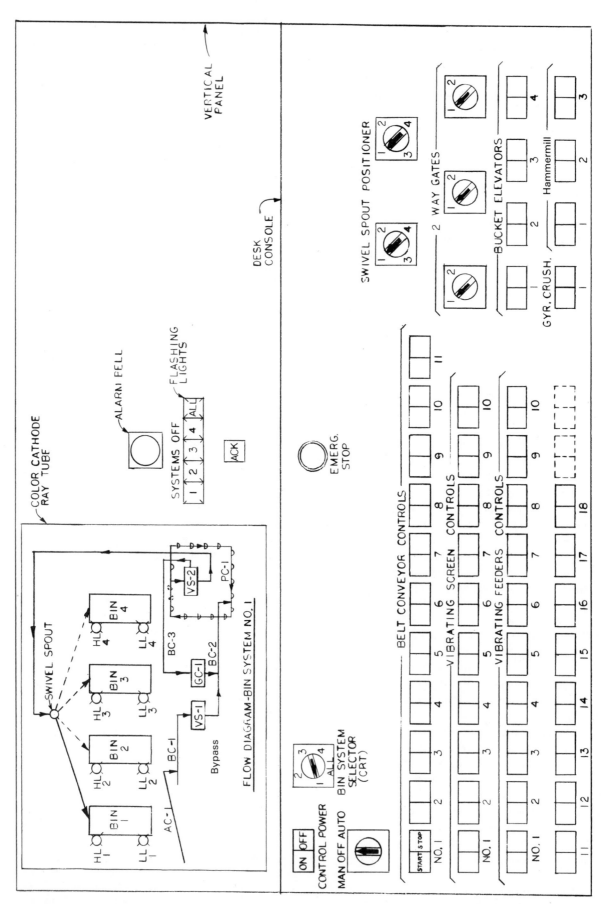

Figure 20.4 Flow diagram of bin system no. 1, main control panel.

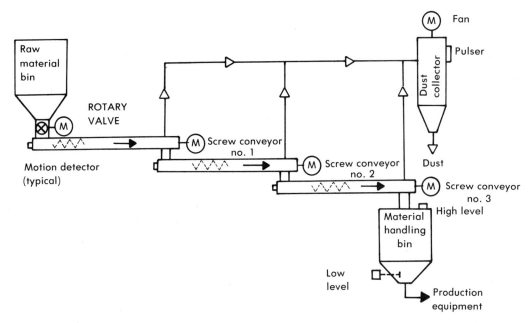

Figure 20.5 Electrical block diagram.
Sequence of operation: 1. Start dust collector; 2. Start screw conveyor no.3; 3. Screw conveyor no. 2, screw conveyor no. 1, and rotary feeder start in sequential order; 4(a). When material reaches the predetermined high level, the system will shut down; 4(b). Dust collector will stop after a timed interval; 5. Motors are sealed in through-motion detector switches; 6. Motors are electrically interlocked so that failure of any will stop all equipment, back to raw material bin; 7. An alarm (audible and visual) will indicate when material reaches high or low level; the audible signal may be silenced, but the visual indication must be retained.

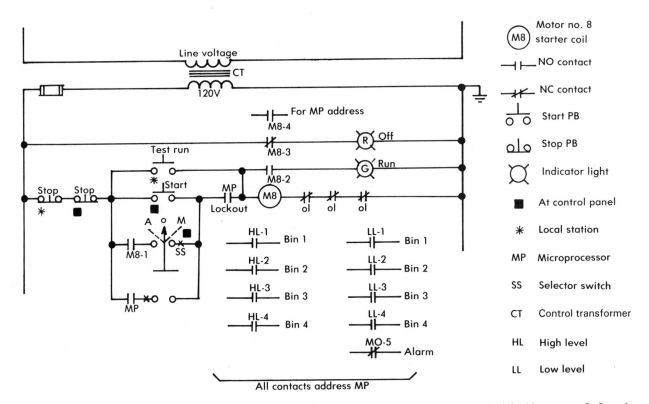

Figure 20.6 Automatic filling sequence for bin system 1 only. However, this is also a typical control for bin systems 2, 3, and 4, as is shown on figure 20.2.

The panels are usually metal housings with flush-mounted instrumentation, accessible from front and rear. Starting and stopping of machinery may be from a master button or from a group of sequential buttons. With modern programming, the sequencing can be monitored and automatically started and stopped according to demand. Miniaturized panels and controls are becoming popular, with microprocessor controls, with computer memories and programming. The latest complex control panels are frequently desk type with vertical CRT (cathode-ray tube) graphics.

Reliability starts with the use of quality equipment from an established supplier. Some redundancy is required in the case of noninterruptible processes. Failure of one control is backed up by a second or third similar device. Computerized checking of controls may be automatic, and the failure localized and identified.

20.13.3 Ease of Maintenance

Ease of maintenance is important. Adequate spare parts must be readily available and adequate working space around electrical equipment must be allowed. General illumination of operating areas should be such that routine inspection is possible during nighttime dark periods. Equipment used should not be so sophisticated as to require specialists at every breakdown. This also ties in with adequately trained maintenance personnel.

20.13.4 Competitive Low Cost

It does not prove an engineer to be a good consultant if he specifies only the most expensive and best equipment that money can buy. A well engineered job is one that does the required operation with reliable and cost-competitive machines.

The preliminary electrical design sheets included herewith exemplify the starting of an electrical control system. This type of design could be presented to the operational management for approval before proceeding with the final design. The planning of the system is as follows:

1. List the electrical components needed, as required by the flow sheet
2. Identify the equipment on the flow diagram
3. Write up the operational requirements and sequences
4. Draw up a control panel, which goes with the operational writeup

20.13.5 Operating Instructions for Conveyors, Crushers, and Storage-Bin Systems

1. Turn on control power (figure 20.4). Study instrumentation on panel.

2. Check condition of all storage-bin levels on CRT. Set swivel spouts and two-way gates to desired positions.
3. If all systems are ready to run, place Auto switch in AUTO position. (Alert siren will sound for one minute prior to any machine start.)
4. All system equipment must be started in sequence from output end to input (figure 20.5). Pushbuttons marked START will start automatic operation of each machine, provided it is operated in sequence. Sequence starting is provided by programmable controller, and a unit will stay in Auto sequence until STOP button is pushed.
5. EMERGENCY STOP turns off and locks the controls. The pushbutton must be pulled out to reset, and all machines must be manually started even though set in Auto sequence.
6. In Auto positions, all bins will be kept full automatically (controlled by level controls) (see figure 20.6). A malfunction will trigger the alarm bell and flashing light. The CRT display will show the condition of the system and also the illuminated buttons. The alarm is silenced with the ACK button (figure 20.4).
7. In the case of failure of the CRT, the machines may be operated by illuminated pushbuttons, in conjunction with the flow diagram.
8. Breakage of a belt or drive train is detected by speed switches, and sequential tripping of the upstream drives will prevent material pileup.

20.13.6 Equipment List

Device Numbers		Description
M1	AC1	30″ Apron conveyor no. 1
M2	BC1	30″ belt conveyor w/magnetic pulley
M3	VS1	2-deck 3′ × 8′ screen
M4	GC1	Gyratory crusher
M5	BC2	24″ belt conveyor w/magnetic pulley
M6	PC1	24″ Peck carrier
M7	VS2	2-deck 3′ × 8′ screen
M8	VS3	2-deck 3′ × 8′ screen
M9	BC3	24″ belt conveyor
M10	VF1	Vibrating feeder
M11	VF2	Vibrating feeder
M12	VF3	Vibrating feeder
M13	VF4	Vibrating feeder
M14	BC4	24″ belt conveyor (reversible)
M15	BC5	24″ belt conveyor
M16	BE1	20″ × 12″ bucket elevator
M17	VS4	2-deck 3′ × 8′ screen
M18	BC6	24″ belt conveyor
M19	HM1	Hammermill
M20	VF5	Vibrating feeder

M21	VF6	Vibrating feeder	M39	DC5	$1'' \times \frac{1}{2}'' \times 1''$ *D* discharge conveyor
M22	BC7	24″ belt conveyor	M40	VF11	Vibrating feeder
M23	BE2	20″ × 12″ bucket elevator	M41	HM3	Hammermill
M24	VS5	3-deck 3′ × 8′ screen	M42	VF12	Vibrating feeder
M25	VF7	Vibrating feeder	M43	VF13	Vibrating feeder
M26	VF8	Vibrating feeder	M44	VF14	Vibrating feeder
M27	VF9	Vibrating feeder	M45	VF15	Vibrating feeder
M28	VS6	2-deck 3′ × 8′ screen	M46	BC9	24″ belt conveyor (reversible)
M29	BC8	24″ belt conveyor	M47	BE4	20″ × 12″ bucket elevator
M30	BE3	14″ × 8″ bucket elevator	M48	VS9	3-deck 3′ × 8′ screen
M31	VS7	2-deck 3′ × 10′ screen	M49	VF16	Vibrating feeder
M32	VS8	2-deck 3′ × 10′ screen	M50	VF17	Vibrating feeder
M33	VF10	Vibrating feeder	M51	VF18	Vibrating feeder
M34	HM2	Hammermill	M52	VS10	2-deck 3′ × 8′ screen
M35	DC1	M.L. discharge conveyor	M53	BC10	24″ belt conveyor
M36	DC2	$\frac{1}{4}'' \times \frac{1}{2}''$ discharge conveyor	M54	DC6	24″ belt conveyor to packing
M37	DC3	14 N.D. discharge conveyor	M55	DC7	24″ belt conveyor to packing
M38	DC4	Dust discharge conveyor	M56	BC11	24″ belt conveyor

Index